AF476853

The Search for Extraterrestrial Intelligence

Theory and practice

Online at: https://doi.org/10.1088/2514-3433/adf5ae

About the program:

AAS-IOP Astronomy ebooks is the official book program of the American Astronomical Society (AAS) and aims to share in depth the most fascinating areas of astronomy, astrophysics, solar physics and planetary science. The program includes publications in the following topics:

Books in the program range in level from short introductory texts on fast-moving areas, graduate and upper-level undergraduate textbooks, research monographs, and practical handbooks.

For a complete list of published and forthcoming titles, please visit iopscience.org/books/aas.

About the American Astronomical Society

The American Astronomical Society (aas.org), established 1899, is the major organization of professional astronomers in North America. The membership (~7,000) also includes physicists, mathematicians, geologists, engineers, and others whose research interests lie within the broad spectrum of subjects now comprising the contemporary astronomical sciences. The mission of the Society is to enhance and share humanity's scientific understanding of the universe.

Editorial Advisory Board

The Search for Extraterrestrial Intelligence

Theory and practice

Jason T Wright

Penn State Extraterrestrial Intelligence Center, Penn State, University Park, PA, USA

IOP Publishing, Bristol, UK

ISBN 978-0-7503-4796-9 (ebook)
ISBN 978-0-7503-4794-5 (print)
ISBN 978-0-7503-4797-6 (myPrint)
ISBN 978-0-7503-4795-2 (mobi)

DOI 10.1088/2514-3433/adf5ae

Version: 20260101

AAS–IOP Astronomy
ISSN 2514-3433 (online)
ISSN 2515-141X (print)

British Library Cataloguing-in-Publication Data: A catalogue record for this book is available from the British Library.

Published by IOP Publishing, wholly owned by The Institute of Physics, London

IOP Publishing, No.2 The Distillery, Glassfields, Avon Street, Bristol, BS2 0GR, UK

US Office: IOP Publishing, Inc., 190 North Independence Mall West, Suite 601, Philadelphia, PA 19106, USA

For Michael

Once my mentor,

Always my muse

How I wish you could read it.

Contents

…we are falling into the future, falling.
M. H. E. *Cape Cod Light*

Author Biography

Jason T Wright

Jason T Wright is a Professor of Astronomy and Astrophysics at Penn State, and director of the Penn State Extraterrestrial Intelligence Center. He is the recipient of the 2019 Drake Award, given by the SETI Institute for his work on exoplanets and SETI.
Photo by Audrey Chambers.

Part I

Background

Chapter 1

Introduction

> The study of extraterrestrial intelligent organisms appears to be the *pièce de résistance* of exobiology, both figuratively and literally. Contact with another sapient species will have the most profound impact on terrestrial science and philosophy. To biologists it will offer the simple practical virtue of providing access, in all likelihood, to an already developed scientific analysis of another highly evolved ecology. It offers a difficult, yet probably the easiest, method of detecting the existence of life, and possibly planetary systems accompanying other stars.
>
> Drake (1965)

1.1 SETI, Technosignatures, and the Search for Extraterrestrial Technology as a Part of Astrobiology

1.1.1 Origins and Jargon

This text is concerned with the entire subject of the search for technological life beyond Earth, and aims to help establish that subject as a scientific discipline within the broader project of astrobiology. The approach here is ecumenical, inclusive of: all forms of technosignature search; the topic of interspecies communication, including message composition and decoding; the theoretical study of the origins, nature, and spread of technological life in the Universe; and the social, historical, and ethical aspects of the search. As the Drake Equation illustrates, both the goal and many of the underlying assumptions of the field are closely connected with the search for microbial life in the Universe, and here we will emphasize those connections.

The modern Search for Extraterrestrial Intelligence (SETI) began in 1959, with the publication of Cocconi & Morrison (1959), and the independent initiation of Project Ozma by Frank Drake. Both efforts illustrated that humanity was capable of

doi:10.1088/2514-3433/adf5aech1

producing signals that humanity was capable of detecting at interstellar distances. Within a few years, four of the branches of the field had been founded: radio SETI, optical/laser SETI by Schwartz & Townes (1961), solar system SETI by Bracewell (1960), and Infrared SETI by Dyson (1960) (a fifth major branch, exoplanetary SETI, would enter the literature decades later with Owen (1980) and Campbell (2006).) On the theory side, the Drake Equation (Drake 1965) would be written down in 1961, the Kardashev scale in 1964 (Kardashev 1964), and 11 years later the Fermi Paradox would be formally articulated (Hart 1975) and named (see Gray 2015 for a discussion).

In parallel to these efforts, NASA and other agencies supported the search for life in the Universe by other means, under names such as bioastronomy, exobiology, and astrobiology. Despite the common aims of SETI and these projects, they had very different trajectories within and beyond NASA. While the search for life in the solar system became a major focus of NASA's planetary science missions, including the Viking lander biological experiments to look for evidence of microbes in the Martian soil, Congressional opprobrium to looking for "Little Green Men" led NASA to avoid association with the search for technological life after the early 1990s. The fields thus diverged culturally and scientifically.

The term "SETI" was coined at the Science Workshops for Interstellar Communication in 1976–1977 (Morrison et al. 1977), to distinguish a passive radio SETI program from earlier efforts that included sending signals (Seeger & Wolfe 1985). It later became the name of a specific program at NASA, in the spirit of Project Ozma, and eventually the name of the SETI Institute.

Other aspects of the search for extraterrestrial technology for years went by other names, such as SETA (the search for extraterrestrial artifacts, i.e., objects in the solar system, Freitas 1983), SETT (Dyson 1966, the search for extraterrestrial technology), or CETI (the name for the approach of communication with ETIs taken in the Soviet Union).[1]

Tarter (2007) coined the term "technosignatures" to broadly describe any sign of alien technology, by analogy with the search for biosignatures, and this term quickly caught on as a way to rebrand the field away from negative connotations associated with SETI's funding difficulties in the US Congress, and to emphasize the contrast to how searches for bio- and technosignatures were treated by NASA. This is how we will use the term in this text.

The rebranding seems to have worked, and today many have argued the field of SETI should simply call itself something like "technosignature search." Others have favored the new coinage as a way to distinguish more "acceptable" forms of SETI from "traditional" radio searches (although Tarter clearly intended the term to be inclusive of radio searches).

[1] As Dan Werheimer likes to quip when favoring terms without the word "intelligence": "I don't know how to detect alien brain waves!" Another tongue-in-cheek suggestion by radio astronomers arguing for more precision in the name is SETEE (the search for extraterrestrial electrical engineers). Amusingly, when spoken aloud, SETI, CETI, SET-T, and SETEE can all be homophones, somewhat obviating the whole discussion! (It has even been pointed out that Penn State's PSETI center could be pronounced with a silent "P" as in *psychology* to join the club.)

Box 1.1: Defining "technosignature"

Defining "technology," and the related problem of distinguishing technology from natural artifacts, is an old and well-studied problem in archaeology. One challenge is that defining a technosignature in terms of technology does not really tell us how to define technology, and defining an artifact in terms of an intelligent designer who built it for a purpose just shifts the definition to what it means to be an "intelligent designer," (which in SETI is an even more open question that it is in, say, studies of animal tool use.) Many attempts, including the definition we use in this text, can rightly be criticized as circular.

In her PhD thesis, Sheikh (2021) explored these issues in an interdisciplinary way, drawing lessons from definitions in evolutionary biology, philosophy, and other fields, and offers a more detailed definition that aspires to be (1) not circular, (2) generally applicable, (3) testable, given a test case, and (4) empirically correct, or at least sensible. Here it is:

> A technosignature is a physical feature whose presence or abundance is a result of the presence of a being that has engaged in goal-directed, intentional manufacture. Here, manufacture is the process of using a tool (something itself manufactured, or gathered from the environment) to manipulate a material in order to create a new form, demonstrating its makers' formal or informal physical knowledge.

One nice thing about such a precise definition is that it draws a line in the sand that gives plausible answers to whether many edge cases of tool use debated by the SETI community in this context constitute technology. Sheikh concludes that tailorbird nests, nitrates in rivers, written language, radio transmissions, and solar system probes all constitute technosignatures, but that beaver dams, spider webs, and spoken language do not.

Recently, the term SETI has often been used in different and inconsistent ways. The existence of terms like SETA and the association with the SETI Institute has caused some to associate the term strictly with radio SETI (although the SETI Institute supports other modes of search as well). In popular culture the term is often associated with a specific organization or set of people (presumably the SETI Institute) rather than a broad field of study or activity. That said, the term has also been used in a manner inclusive of all forms of searching for technological life (for instance by Swift 1993, whose interviewees clearly understand the term to include, for instance, laser SETI and solar system SETI).

Box 1.2: Jill Tarter (1944–)

Image credit: SETI Institute.

Dr. Tarter is an American astronomer who has dedicated her career to the practice and promotion of SETI. She was instrumental in keeping the field moving after the termination of NASA's SETI program.

Tarter's father, who died when she was 12, encouraged her dedication to engineering despite social pressure discouraging this career path for women.

Tarter studied engineering physics at Cornell and did her graduate work in astrophysics at UC Berkeley under Joe Silk. Her thesis work in 1975 was to calculate the emission spectra of substellar objects. Such objects had no widely used name at the time, and since the results of her calculations regarding their colors were inconclusive, she dubbed them "brown dwarfs" in her thesis title.

Stu Boyer, learning she could program a computer he had received to perform a SETI experiment at Hat Creek, gave Tarter a copy of the Project Cyclops report, sparking her career-long interest in the field. She worked on the SERENDIP project, and was project scientist for NASA's short lived HRMS program.

Following the termination of HRMS, Tarter moved her efforts to the newly founded SETI Institute as director of Project Phoenix. Tarter was instrumental in securing philanthropic and government funds to build the Allen Telescope Array at Hat Creek. In 2006, she coined the term "technosignatures" to emphasize the parallels of SETI with the search for biosignatures and to establish a broad vision of how SETI might succeed.

Dr. Tarter retired in 2012 to become, as she put it, "chief cheerleader" for the field. She has been the face of the field for decades, being named one of the 100 most influential people in the world by Time Magazine in 2004 and winning the TED Prize in 2009.

Her biography and career served as one of the inspirations for Ellie Arroway in the film *Contact*, and Tarter worked with actor Jodie Foster on her portrayal of the character. Her personalized California license plate, which she inherited from Barney Oliver, reads `SETI`.

For more on Jill Tarter, see the biography "Making Contact" (Scoles 2017).

Box 1.3: Jargon Review: SETI

- **The Search for Extraterrestrial Intelligence (SETI)**: A subfield of astrobiology focused on searching for signs of non-human technology or technological life beyond Earth; the theory and practice of searching for extraterrestrial technology or technosignatures.
- **Intelligence**: In this book, used poetically and historically in the term SETI, roughly corresponding to the quality of being able to deliberately engineer technology which might be detectable using astronomical observation techniques.
- **SETI Institute**: A research center in Mountain View, California, with an emphasis on SETI, astrobiology, and planetary science.
- **ETI**: The entities that created the technosignatures SETI seeks to find.
- **Technosignature**: Any sign of technology that is not also a biosignature (i.e., it is distinguished from purely natural biological activity.)

1.1.2 SETI: The Name of the Field

This field needs a name, and I positively adopt "SETI" as that name, which in this text we will use broadly to include all aspects of the search for technological life elsewhere, including the question of transmissions from Earth. I make this apologia although many have suggested the field adopt a different term such as "technosignature search," so as to avoid the negative political and social connotations "SETI" developed (Section 3.3.4).

One reason is that the "brand" of SETI is strong enough to be worth reclaiming from any negative connotations it has. This name also acknowledges the role of the SETI Institute in maintaining work in the field through the funding drought of the 90s, 00s, and 2010s, and that the radio program that grew out of the original NASA SETI program has become the most mature and productive aspect of the field.

To be sure, SETI certainly includes technosignature search, but is broader than that, much in the way astrobiology is about much more than just searches for biosignatures. "SETI" also properly refers to the goal of the search (finding another "intelligent" species in the cosmos) rather than the means to that goal (the technosignatures that point to that species).

The primary drawback of the name is that it contains the term "intelligence," and so brings along a whole host of baggage involving the nature of intelligence, its connection to technology, and cultural expectations about the nature of the beings the field is searching for. We will discuss some of these difficulties in Chapter 18. For now, I'll simply state that we will use the term historically, acknowledging the field's foundations, but also metonymically, inviting the listener to consider the nature of beings that could create technology we can detect across the vastness of space.

Being the name of the field, it therefore makes little sense to use it in phrases such as "what is SETI up to?" or "this is what SETI thinks," any more than it would to ask what history is up to or what geochemistry thinks. One should also not use it as shorthand for the SETI Institute, to avoid confusion.

1.1.3 SETI, Biosignatures, and Astrobiology

> *Moore*: Then what should we be doing about this research [the search for extraterrestrial life]?
> *Sagan*: I would say that there are several lines of approach, all of which look promising and should be pursued further.
> One is laboratory work on the question of the origin of life, making the molecules which here were able to reproduce themselves and lead eventually to us.
> Second, a study of the organic chemistry in places like Jupiter, comets, and the interstellar medium. It's remarkable that the molecules of life are littering the cosmos.
> Third, the space vehicle exploration of the neighboring planets, in particular for life or life-related compounds...
> And then there is certainly the serious work—long-term, patient, cautious investigation of other stars for possible signals being sent our way.
> — Carl Sagan, interviewed by Sir Patrick Moore, *The Sky at Night*, broadcast 1974 May 15.

The relationship among SETI, technosignatures, biosignatures, and astrobiology is shaped by the history of their development and the cultures of the communities that have nurtured them. Certainly many of the founders of the field of astrobiology—most notably Carl Sagan, as illustrated in the quote above—considered SETI to be just one component of a comprehensive approach to the study of life in the Universe.

In this text, we will use the term *astrobiology* to describe the entire effort of understanding life in the Universe, i.e., as an umbrella term that includes efforts to find signs of life elsewhere, by whatever means. Because it is often useful to compare and contrast them, we will consider technosignatures and biosignatures to be largely distinct and complementary. SETI is thus the subset of astrobiology that deals with technological life, in theory and practice.

Not all practitioners endorse this framing: some prefer to consider technosignatures a subset of biosignatures, and this taxonomy has its merits, in particular that it emphasizes how technosignatures (presumably) point to life.[2] Others prefer to maintain the historical sense of astrobiology as being exclusive of searches for technology.

One challenge of either taxonomy is that it leaves searches for non-technological life without a name—there is no purely biological counterpart to "SETI" under the astrobiology umbrella. These sorts of issues often arise in interdisciplinary work as fields merge, and will likely resolve themselves organically as SETI matures. For now, it is best to be clear about one's meaning, and lean toward an inclusive sense of terms except when one is deliberately drawing distinctions.

Figure 1.1 shows roughly how this text will organize these ideas; Figure 1.2 shows some alternative arrangements.

[2] It also has the advantage that it allows SETI work to be proposed in funding programs focused on biosignatures.

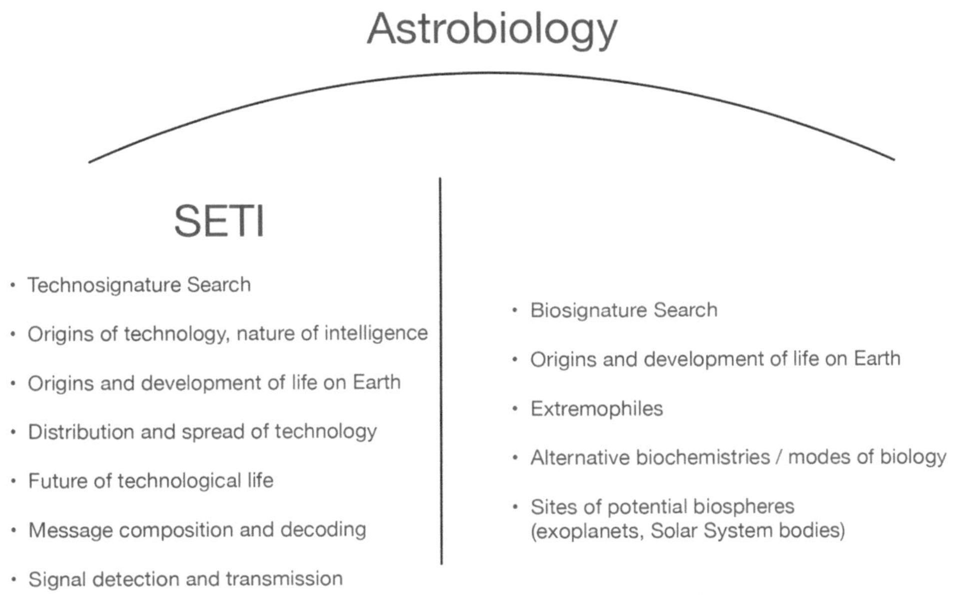

Astrobiology

SETI

- Technosignature Search
- Origins of technology, nature of intelligence
- Origins and development of life on Earth
- Distribution and spread of technology
- Future of technological life
- Message composition and decoding
- Signal detection and transmission

- Biosignature Search
- Origins and development of life on Earth
- Extremophiles
- Alternative biochemistries / modes of biology
- Sites of potential biospheres (exoplanets, Solar System bodies)

Figure 1.1. The organization of astrobiology, SETI, technosignatures, and biosignatures used in this text. The bulleted lists are not intended to be exhaustive.

1.2 SETI as an Academic and Scientific Discipline

1.2.1 The Assumptions and Foundations of the Field

Because SETI has been a small field with a big cultural impact, the views and philosophies of a few of its pioneers (many of whom are profiled in boxes throughout this text) have had a large impact on its shape. These pioneers' personal reasons for engaging in the search, and expectations of the nature of alien life, the nature of intelligence, and how contact might proceed, are baked into the underlying justification for the field.

For instance, many of them felt sure that a Galactic Club existed that humanity was ready to join, that signals would be intelligible and intended for us, and that contact would lead to large benefits for human society. Many modeled their expectations for alien life on science fiction, or their expectations for the future of humanity based on the stories they believed about humanity's history.

Regardless of the validity of these views and beliefs, it is important to establish a firm foundation for SETI that makes as few assumptions as possible. It should not be possible to discredit the whole field by critiquing, say, Frank Drake's understanding of the nature of animal intelligence, credence in the benevolence of alien life, or faith that the signal would be a radio signal. Part of the purpose of this text is to establish that solid foundation.

That said, we will also consider what we are expecting to find and many peoples' personal motivations for the search. We will need to address common ideas of alien contact that are based on antiquated or dubious understandings of human and

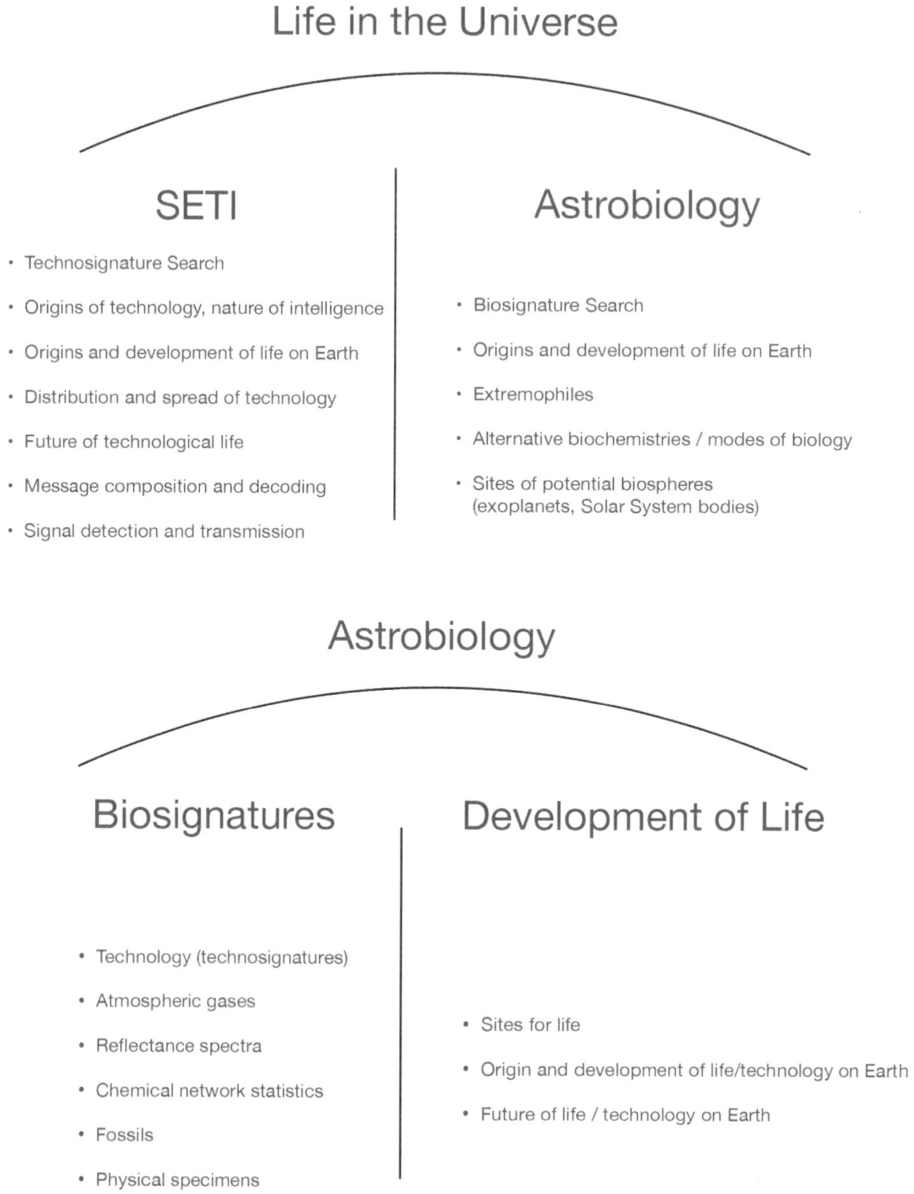

Life in the Universe

SETI

- Technosignature Search
- Origins of technology, nature of intelligence
- Origins and development of life on Earth
- Distribution and spread of technology
- Future of technological life
- Message composition and decoding
- Signal detection and transmission

Astrobiology

- Biosignature Search
- Origins and development of life on Earth
- Extremophiles
- Alternative biochemistries / modes of biology
- Sites of potential biospheres (exoplanets, Solar System bodies)

Astrobiology

Biosignatures

- Technology (technosignatures)
- Atmospheric gases
- Reflectance spectra
- Chemical network statistics
- Fossils
- Physical specimens

Development of Life

- Sites for life
- Origin and development of life/technology on Earth
- Future of life / technology on Earth

Figure 1.2. Two alternative taxonomies sometimes used. The bulleted lists are not intended to be exhaustive.

animal life on Earth, and we will need to anticipate the nature of contact so that we can do post-detection work. But we should be careful not to confuse these discussions for the rationale for SETI, which we discuss in Chapters 8 and 9.

1.2.2 Academics and Curriculum

Despite being over 60 years old, the field is rather young, academically speaking. In terms of curriculum, the only two regular upper level undergraduate or graduate offerings in SETI I am aware of are the courses this text was written for at Penn State (ASTRO 476 and 576), and a course in radio SETI offered by Jean-Luc Margot at UCLA.

As of this writing in 2025, I am aware of only 15 people who have received a PhD in a physical science with a thesis with a heavy focus on SETI. They are listed in Table 1.1.

The primary reason for this is the lack of federal funding, which starved the field of much of its academic basis, forcing it into privately funded centers without a mandate to train students, develop a curriculum, or develop the field into a discipline[3] (Section 3.3.4). As a result, much of the SETI literature remains uncited, with many ideas being re-introduced and reinvented multiple times, multiple jargon terms for the same concepts, and even papers establishing concepts as foundational as the Drake Equation becoming forgotten.[4]

Table 1.1. All PhD Theses in the Physical Sciences Focused on SETI Up to September 2025 (Known to the Author)

Name	Date	Adviser	Link to thesis
Darren Leigh	1988	Paul Horowitz	Thesis
Alexey Arkipov	1998	Leonid Litvinenko	
Stephen Brown	2000	Bob Dixon and John Kraus	Thesis
Chip Coldwell	2000	Paul Horowitz	Thesis
Andrew Howard	2006	Paul Horowitz	Thesis
Andrew Siemion	2012	Geoff Bower and Dan Werthimer	Thesis
Laura Spitler	2013	Jim Cordes	Thesis
Curtis Mead	2013	Paul Horowitz	Thesis
Ian Morrison	2017	Chris Tinney	Thesis
Emilio Enriquez	2019	Heino Falcke	
Sofia Sheikh	2021	Jason Wright	Thesis
Paul Pinchuk	2021	Jean-Luc Margot	Thesis
Macy Huston	2023	Jason Wright	Thesis
Matías Suazo	2024	Erik Zackrisson	
Nick Tusay	2025	Jason Wright	

Leigh's thesis predated Arkipov's by a few months.

[3] Notable exceptions are the MA and PhD students of Paul Horowitz at Harvard, who developed pulsed laser SETI (Box 13.1), and Bob Dixon and John Kraus at Ohio State University, who developed radio SETI at the Big Ear.

[4] The first citeable work by Drake containing the equation is Drake (1965), but this work had no entry in the NASA Astrophysics Data System and apparently no citations until the 2020s.

1.3 Exploring the SETI Literature

1.3.1 ADS and `seti.news`

To remedy the difficulty in organizing and searching the SETI Literature, Reyes & Wright (2019), LaFond et al. (2021), and others have worked to compile and curate the corpus of literature in the physical sciences at the NASA Astrophysics Data Services (ADS) under a "bibgroup." ADS indexes nearly all of the literature relevant to astrophysics, and its website and API offer powerful search, organizational, and analysis tools of the text and metadata of all entries. Users of ADS can access the curated SETI literature by using the search element "`bibgroup:SETI`." This bibliography is maintained by members of the Penn State Extraterrestrial Intelligence Center, and one can subscribe to updates to it in the form of a monthly newsletter from seti.news, offered by Jim Davenport.

Users of ADS should also note the utility of making an account, which allows one to compile libraries which can be searched, linked, and even operated on with set arithmetic. One can also specify one's library proxy server, allowing one to access the full text of nearly all articles behind paywalls with a single click. ADS also provides custom bibliography output for all entries, greatly simplifying the generation of citations in BibTeX and other citation managers.

Some journals, such as the Journal of the British Interplanetary Society or old conference proceedings, are simply unavailable electronically even with a library proxy. In these cases, a good first place to look is archive.org. Asking a librarian to track down a paper can also be quite fruitful, because Interlibrary Loan can find and deliver almost any printed media in the world in a few days. Finally, there is an informal network of SETI researchers with personal caches of PDFs of many particularly hard-to-find SETI papers—if you ask, you can usually get a copy quickly. These researchers are often happy to receive PDFs to add to their files, as well.

1.3.2 The Heterogeneity of the Literature

Because it has been developed mostly outside of any formal academic structure, the existing literature is of highly uneven quality. Many important papers (and many more unimportant ones) were written by hobbyists and dilettantes, engineers, economists, and other nonexperts in astronomy and physics. Many of them dabble in analogies of human history or sociology, even when the authors have no training in these fields.

So some of the citations and further readings I provide in this text will not be of high overall quality, or may contain huge leaps of logic, unjustified assumptions, clear errors of physics, and other howlers. In such cases, I encourage the reader to go "panning for gold," and to find the nugget in the paper that makes it worth reading and citing. Part of the task of making SETI a discipline is collecting these nuggets and refining and working them together into a fine trophy, and along the way acknowledging the priority of ideas by citing those who put them forth first and best.

1.4 Using This Text in a Course

The aim of this text is to establish the jargon, canon, and the scope of the field, and to provide a framework for building an interdisciplinary community of researchers within it. This text is intended for an upper-level undergraduate and graduate course, and was developed specifically for the ASTRO 476 and 576 courses at Penn State, the latter of which is offered as part of a broader astrobiology dual-title PhD program.

Because the field is broad and interdisciplinary, the text is intended to be useful not just for astronomy majors and graduate students, but also for graduate students from related fields in astrobiology, and even historians and philosophers of science. I will therefore provide some background astronomy and attempt to offer heuristics for some mathematical concepts like Fourier transforms, but readers without astronomy, physics, or upper-level mathematics training will likely need to refer to other texts for background.

I both provide problem sets and reading questions for primary sources. The former are intended for undergraduate astronomy majors; the latter for graduate students in astronomy and astrophysics.

The problem sets are largely intended for junior or senior astronomy majors. When I teach astrobiology graduate students I assign reading, especially from the primary literature.

I begin with a history of the field for context, and then move into the theory and overall strategy of the field. We then move into specific search strategies, following the five main branches discussed above.

SETI, like any science, is a human endeavor. More than many physical sciences, though, understanding SETI requires us to dive into the humanities and social sciences to understand and interrogate our own biases as humans from a particular culture. Jill Tarter calls SETI a "Cosmic Mirror" in the many ways that it forces practitioners to understand themselves. Some ways in which this reflection occurs are:

- Benchmarking alien technology against human technology (Section 4.2.1).
- Estimating the chances of success based on our one example of humans and Earth (Chapters 8, 7, and 10).
- How aliens might assume we will act (Chapter 5).
- Our expectations for alien technology and contact (Chapter 18)
- How we should prepare for contact (Chapter 20).
- What our transmissions should say and who and whether we should send them (Chapter 19).

To that end, Chapter 3 and the final part of this book will focus on the social aspects of SETI (which will also crop up throughout the other sections). I encourage readers and instructors to take special note of these sections.

A field this young is ripe with low-hanging fruit. This provides wonderful opportunities for those just entering it: in addition to the excitement of working in a field still on the lower part of the logistic curve of progress, it means that even modest contributions can be publishable. A significant fraction of the individual and class final projects in ASTRO 576 have ended up as peer-reviewed publications.

A drawback of this for an instructor, however, is that this text's content and even structure will quickly become outdated. Instructors are encouraged to explore the recent SETI literature prior to assigning readings to ensure the course is up-to-date.

If you are an instructor using this book in a course, please let me know by emailing astrowright@gmail.com.

1.5 Further Reading

Definitions of SETI terms: Wright et al. (2018).
Report from the NASA Technosignatures workshop: NASA Technosignatures Workshop Participants (2018).

Exercises

1. Set up your account on ADS:
   (a) Go to the ADS website at https://ui.adsabs.harvard.edu/. See Figure 1.3.
   (b) Click on "Sign up" and choose a password, and email address. Note that your email address is your username, so choose one that you're sure you'll have access to for a while. See Figure 1.4.
   (c) Once you've registered, you'll be able to access account options. See Figure 1.5.
   (d) Choose "Settings" and then choose "Library Link Server" and "update" and choose your institution. See Figure 1.6.
   (e) Now navigate to a closed-access paper, like the one linked here. Look for the "My institution" link and click on it. See Figure 1.7.
   (f) You may need to authenticate with your institutional credentials. After you do, you should get a version of the paper you can read. See Figure 1.8.
2. Searching ADS with the SETI bibgroup:

   ADS has a powerful search engine and an extensive library of papers related to astronomy. Use the front page examples for the Modern Form to try some searches and get a feel for it.

   Go to About → ADS Help Pages → Making a Query and explore its capabilities.

   Try your own searches: for instance, find all of the first author papers by an astronomer you know of. Once you have the list, sort it by the number of citations to see their most-cited paper. Limit the list to entries that are refereed using the filters on the left-hand side of the results. Use the explore tab in the upper right to look at the author network and citation metrics for these papers. See Figure 1.9.

   ADS also has a "bibgroup" feature which lets users sort papers into groups. One such group is the SETI bibgroup, described in Reyes & Wright (2019) and LaFond et al. (2021). You can use it to filter your results on just papers relevant to SETI. For instance, to find papers by Frank Drake, use the search `author:"Drake, Frank" bibgroup:SETI`.
   (a) Find the abstract entries for the citations in the bibliography of this chapter. Build a library out of them. Include, if you like, the theses in Table 1.1.

   (b) Explore the citation history of the entire SETI corpus in ADS. What is the *h*-index of the field as a whole? What does that mean? Define the *h*-index and interpret the output of the field in the past 60+ years in terms of the output of a single researcher. How much SETI has been done?
3. Find a classic paper, and explore its impact.
   (a) Find a classic paper on a part of SETI that interests you, download it, and skim it. ADS has lots of metrics you can look at to decide if a paper is a "classic," but it's not an exact term, so use your subjective judgment on which paper you want to pick.
   (b) Write up a 1–2 page report describing:
      i. Your experience with ADS, including (a) things you found tricky to work with and (b) at least one feature you found useful not discussed above.
      ii. Why you chose the paper you did and at least one thing you found new or interesting in or about the paper itself.
      iii. What you hope to learn in this class.
4. Consider the critiques offered in the first three blog posts here or Billings (2012) by Linda Billings, a respected observer of astrobiology and an occasional critic of SETI.
   (a) Critique the definitions and organization of astrobiology, SETI, technosignatures, and biosignatures used in this textbook. You may agree or disagree with this textbook's positions, but back up your critique.
   (b) Which organization do you prefer? Does SETI belong under the astrobiology umbrella?

Figure 1.3. Problem 2. Image Credit: NASA Astrophysics Data System.

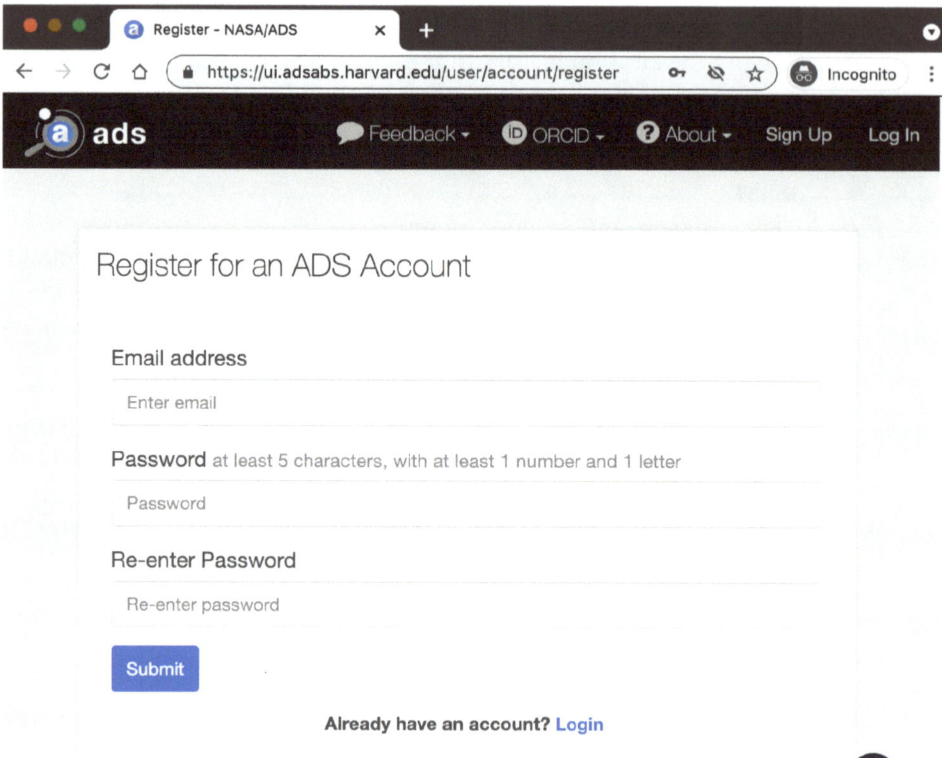

Figure 1.4. Step 1a. Image Credit: NASA Astrophysics Data System.

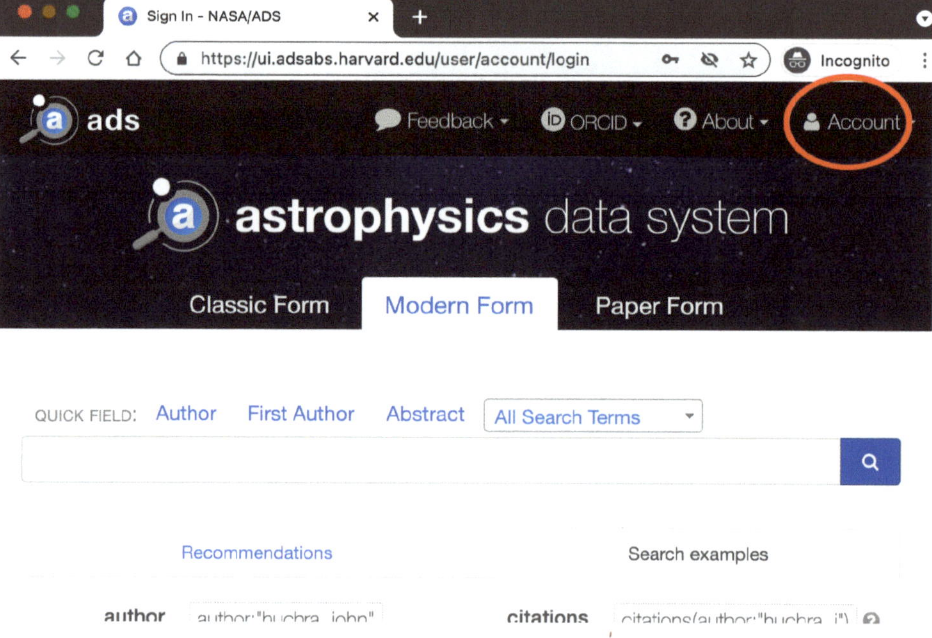

Figure 1.5. Step 1b. Image Credit: NASA Astrophysics Data System.

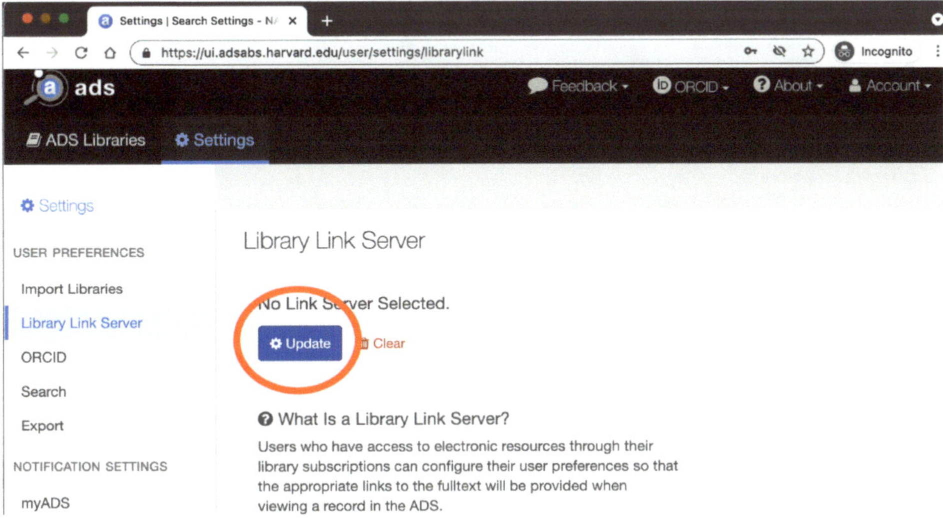

Figure 1.6. Step 1c. Image Credit: NASA Astrophysics Data System.

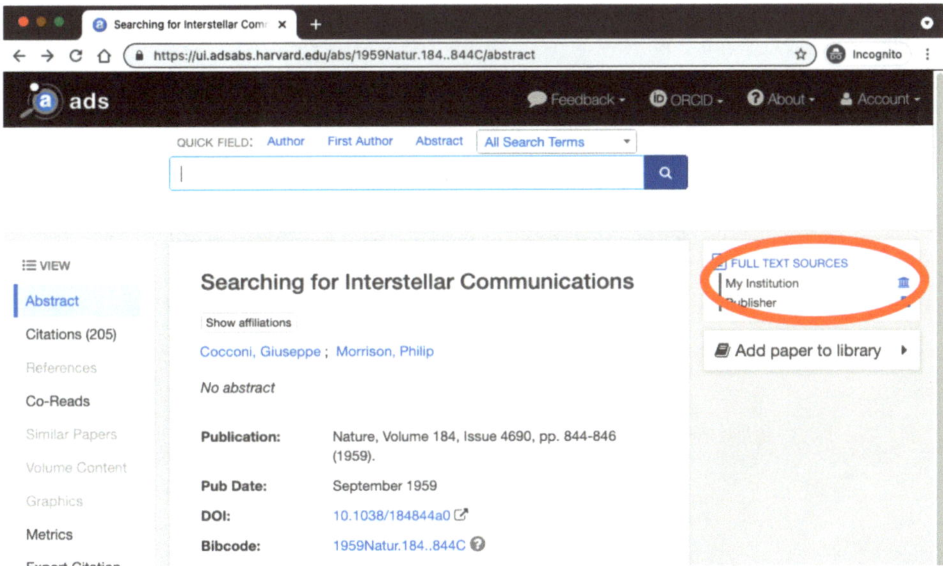

Figure 1.7. Step 1d. Image Credit: NASA Astrophysics Data System.

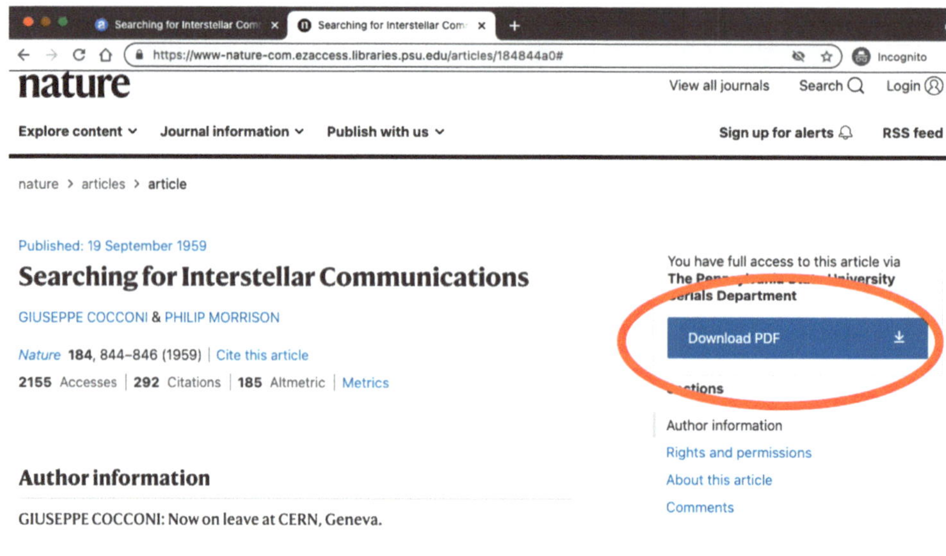

Figure 1.8. Step 1e. Image Credit: NASA Astrophysics Data System.

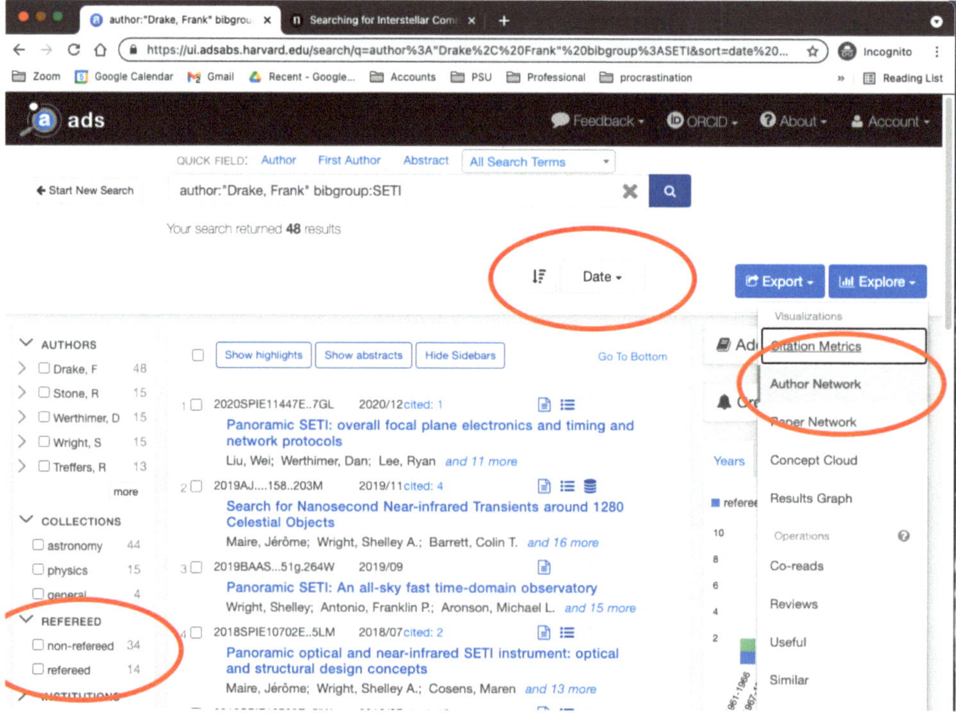

Figure 1.9. Step 1f.

References

Billings, L. 2012, AsBio, 12, 966
Bracewell, R. N. 1960, Natur, 186, 670
Campbell, J. B. 2006, Proc. IAU Colloq, 200, 247
Cocconi, G., & Morrison, P. 1959, Natur, 184, 844
Drake, F. D. 1965, Current Aspects of Exobiology, ed. G. Mamikunian, & M. H. Briggs (Oxford: Pergamon) 323
Dyson, F. J. 1960, Sci., 131, 1667
Dyson, F. J. 1966, Perspectives in Modern Physics: Essays in Honor of Hans A. Bethe on the Occasion of His 60th Birthday, July, 1966, ed. R. E. Marshak, J. W. Blaker, & H. A. Bethe (New York: Wiley) 641
Freitas, R. A. 1983, JBIS, 36, 501
Gray, R. H. 2015, AsBio, 15, 195
Hart, M. H. 1975, QJRAS, 16, 128
Kardashev, N. S. 1964, SvA, 8, 217
LaFond, J., Wright, J. T., & Huston, M. J. 2021, JBIS, 74, 252
Morrison, P., Billingham, J., & Wolfe, J. 1977, The Search for Extraterrestrial Intelligence, SETI, Vol. 419 (Washington, DC: NASA)
NASA Technosignatures Workshop Participants 2018, arXiv:1812.08681
Owen, T. 1980, Strategies for the Search for Life in the Universe, Vol. 83: Astrophysics and Space Science Library, ed. M. D. Papagiannis (Dordrecht: Springer) 177
Reyes, A., & Wright, J. T. 2019, JBIS, 72, 186
Schwartz, R. N., & Townes, C. H. 1961, Natur, 190, 190
Scoles, S. 2017, Making Contact: Jill Tarter and the Search for Extraterrestrial Intelligence (Riverside, NJ: Pegasus Books)
Seeger, C. L., & Wolfe, J. H. 1985, The Search for Extraterrestrial Life: Recent Developments; Proc. of the Symp. (Dordrecht: Reidel Publishing) 391
Sheikh, S. Z. 2021, PhD thesis, Pennsylvania State Univ.
Swift, D. W. 1993, SETI Pioneers: Scientists Talk about Their Search for Extraterrestrial Intelligence (Tucson, AZ: Univ. Arizona Press)
Tarter, J. C. 2007, HiA, 14, 14
Wright, J. T., Sheikh, S., Almár, I., et al. 2018, arXiv:1809.06857

Chapter 2

Some Astronomy Concepts

> … The stars
>
> That glint in the corners of your eyes
> Evaporate looked at straight. Still,
> You can chart a true course from them.
>
> — Michael Elkins Hattersley, *Cape Cod Light*

The aim of this text is to be useful for students with little to no astronomy background, while still providing rigor to those in astrophysics programs. Students may need to pair this text with an introductory astronomy text (at the graduate or advanced undergraduate level) so they can look up unfamiliar concepts.

In this chapter, I provide a whirlwind tour of some astronomy concepts used throughout the book for the benefit of those unfamiliar with astronomical techniques, concepts, and jargon. Readers may wish to work through these ideas before the rest of the book, or simply follow the links in later chapters back to this one as the topics come up.

2.1 The Electromagnetic (EM) Spectrum

Virtually all of the interactions you experience—from things you see to things you feel—are examples of the electromagnetic force of nature. The solid nature of objects, the way things radiate heat and appear to your eyes, the way things collide or rest on one another, and the nerve impulses in your body that control your senses are all examples of how the electrons and protons of atoms interact with one another via their charge and electromagnetic forces.

Electric charges exert forces on one another from a distance: like charges repel each other strongly when close to each other and less as they separate, and opposite charges attract. This attraction or repulsion is described by an electromagnetic field.

doi:10.1088/2514-3433/adf5aech2

A *field* in this sense is a description at each point in space of the way a hypothetical charge at that point would accelerate due to all of the other charges around it.

When a charge in one place moves, this field adjusts everywhere, but this reaction is not instantaneous: the adjustment to the new configuration propagates outward from the charge at the speed of light, and these adjustments *themselves* cause further changes to the EM field.

These changes in time of the field from these propagations can be described as a series of oscillations (Section 12.1.1). The amount of power in the wave at each frequency of oscillation (described via Fourier theory) is known as the *spectrum* of the radiation. When these waves have frequencies a bit above 10^{15} Hz, we call it *light* and can perceive it with our eyes. Astronomers refer to this region of the electromagnetic spectrum as the *visible* or *optical* region.[1]

Light of the lowest perceptible frequency looks red, while light of the highest perceptible frequency looks violet. Intermediate frequencies appear as the familiar colors of the rainbow.[2] Astronomers refer to these limits as *red* and *blue*. Astronomers often use these terms as "directions" on the electromagnetic spectrum, so might say that one frequency of light is *redder* than another if its frequency is lower.

By extension, astronomers often refer to any EM wave as an example of *light*, even if it is invisible to the human eye. At frequencies just above or below where the eye perceives light are ultraviolet and infrared radiation, respectively. See Figure 2.1 for a chart illustrating the entire EM spectrum.

There is a relationship between the frequency of an electromagnetic wave and its wavelength, given by

$$\nu\lambda = c \tag{2.1}$$

where ν is the light's frequency, λ is its wavelength, and c is the speed of light.[3] Figure 2.1 gives the frequencies corresponding to various wavelengths of light.

Astronomers often switch between frequency and wavelength freely, keeping in mind they run in "opposite" directions. In general, radio astronomers specify things in frequencies (with occasional labels such as *meter wave* or *submillimeter* wave for broad ranges of frequencies), while infrared and optical astronomers specify things in wavelength, usually in microns (i.e., micrometers, abbreviated μ or μm) at infrared

---

[1] Some astronomers distinguish these terms, reserving "visible" for light the human eye responds to and "optical" to the region of the EM spectrum wherever detection techniques useful in the visible apply, either because of atmospheric transparency or in terms of detection equipment, so including the infrared and ultraviolet.

[2] Color perception is far more complex than this, and is governed by the response of cones in the eye, which generally come in three varieties, each with a different response to different wavelengths of light. As a result, most humans' perception of color is described as a cycle (the "color wheel") not a linear continuum, and the eye can discern many more colors than are present in the rainbow.

[3] The use of c here is usually that of the vacuum, but technically one should use the local speed of light, which can in principle be slightly lower than c in the presence of matter, as in a transparent medium like air or glass. The reason is that the wave will cause local electric charges to react and oscillate, and this in turn alters the property of the wave. This nuance is generally not important for our purposes, except that when a charged particle exceeds the *local* speed of light (Einstein's relativity only prohibits them exceeding the *vacuum* speed of light) it creates a sort of electromagnetic shock wave akin to a supersonic plane's sonic boom. This is called Cherenkov radiation; it is used to detect cosmic rays in the atmosphere and particle detectors, the byproducts of neutrino interactions in neutrino telescopes, and is a source of background for laser SETI.

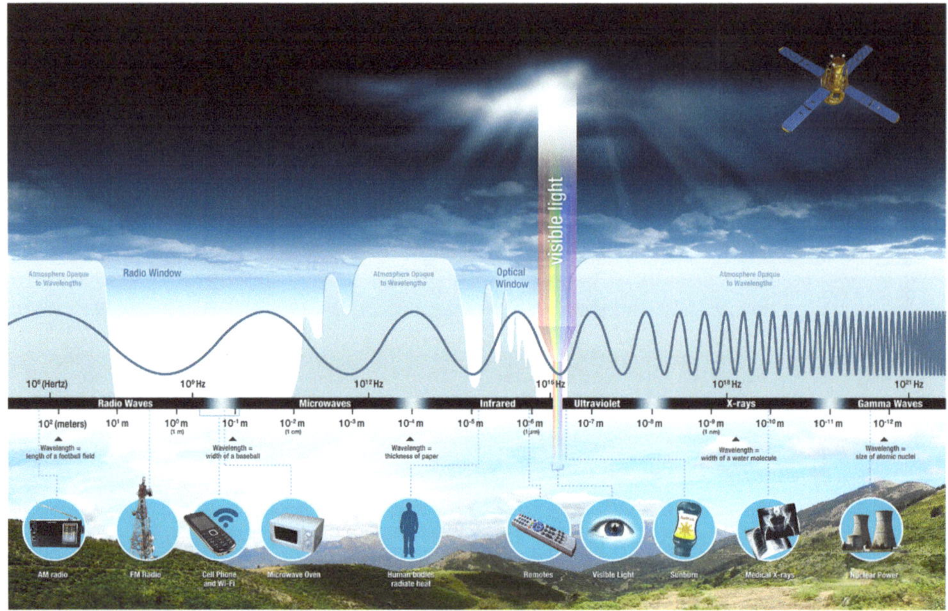

Figure 2.1. The electromagnetic spectrum, and the atmospheric windows. Figure credit: NASA.

wavelengths, and nanometers or Ångstroms ($1\text{Å} = 10^{-10}$ m) at visible and ultraviolet wavelengths. X-ray and gamma-ray astronomers usually specify things in terms of energy of a single photon in electron volts (eV), using the equation

$$E = h\nu \tag{2.2}$$

where h is Planck's constant.[4]

An important practical issue for astronomers is that the atmosphere is only transparent to radiation at radio and optical wavelengths, meaning we can only easily do ground-based astronomy at those wavelengths. This is illustrated in Figure 2.1, which shows the *radio window* and the *optical window*.[5] These windows

---

[4] The fractured and sometimes contradictory nature of astronomy and astrophysics jargon comes from the different disciplines that underlie it. Radio astronomy was long the province of radio and electrical engineers; *high energy* astronomy that of particle physicists; and observational astronomy predates both. Astronomy and astrophysics are a collage of these and many other disciplines.
But this hodgepodge of units has a practical reason behind it: at high frequencies, the particle nature of EM radiation dominates while its wave nature is largely irrelevant, and so light is detected as a discrete series of events, each with distinct, measurable energy. At radio wavelengths the particle nature is hard to discern, and the equipment used to detect light is concerned with measuring oscillatory frequencies. Optical and infrared astronomers generally detect individual photons of indeterminate energy, and determine wavelength using grating spectrographs whose groove spacing determines how light of a given wavelength interacts with the equipment.

[5] At higher frequencies, the sky as seen from the ground is basically black because it is opaque and does not emit, but at low frequencies it is "bright" where it is opaque because it is warm and emits at these wavelengths. Work in the mid-infrared is challenging because the atmosphere is not *completely* opaque, but is opaque enough that it tends to outshine astronomical sources.

are the reason that the two dominant modes of SETI are radio SETI and optical SETI: those are the only wavelengths that reach the ground.

The spectrum is further subdivided into regions not shown in that figure: radio astronomers refer to the *millimeter* and *submillimeter* regions of the microwave region. The infrared is divided into the near-infrared (roughly 1–5 μm, where the atmosphere has many transparent windows), the mid-infrared (5–40 μ, where objects near room temperature shine brightly), and the far-infrared (>40 μ, blending into the submillimeter). The atmosphere starts to become bright at wavelengths longer than around 2.5 μ.

Astronomers use high altitude balloons or satellites that operate above most of the Earth's atmosphere to observe outside the atmospheric windows. Examples are NASA's *JWST*, which operates from the visible into the mid-infrared, NASA's *Swift* Observatory, which has telescopes from the optical through gamma-rays, and NASA's *Chandra* Observatory, which works at X-ray wavelengths.

Box 2.1: Jargon Review: EM Radiation

- **Light**: Any form of electromagnetic (EM) radiation, which is the propagating oscillation of electric and magnetic fields. Sometimes, used to only refer to visible light.
- **Micron**: A micrometer, or 10^{-6} m, symbolized as μm or μ.
- **Electromagnetic Spectrum**: The full range of frequencies (or energies, or wavelengths) of electromagnetic radiation, or the amount of light across that range. Subdivided into regions based in part on wavelength/frequency, and partly on the typical techniques used to detect light:
  - **Radio**: Broadly defined, wavelengths longer than about 100 μ. When distinguished from microwave work, frequencies lower than 1 GHz (wavelengths longer than about 30 cm).
  - **Microwave**: Frequencies between 1 GHz and about 300 GHz (corresponding to 1 mm).
  - **Submillimeter**: Wavelengths from 1 mm to around 300 μ or so.
  - **(Near, Mid, Far) Infrared**: Wavelengths between the limit of human vision, around 0.75 μ, up to 300 μ or so. Subdivided into the near-infrared (up to about 5 μ), mid-infrared (roughly 5 μ to 10 μ), and the far-infrared (the rest). The region from 0.75 to 1 μ is sometimes half-jokingly called the "near optical" by infrared astronomers.
  - **Visible/Optical**: The region of the EM spectrum visible to the human eye, between roughly 0.4–0.75 μ. Optical is often used to refers to a wider region where similar detection methods are used, including some near-infrared and ultraviolet radiation.
  - **Ultraviolet, X-ray, gamma-ray**: Shorter wavelength radiation, with $\lambda < 400$ nm, 10 nm, and 0.01 nm, respectively. X-rays and gamma-rays are usually referred to by their energy in eV, in which case the boundaries are at roughly 100 eV and 1 MeV.
- **Radio/Optical Window**: The frequencies (regions of the EM spectrum) to which the the Earth's atmosphere is fully or mostly transparent, and where ground-based astronomy is effective.

2.2 Radiation

2.2.1 Flux, and the Inverse Square Law

Light carries energy through space. An object that shines has some *luminosity*, or total power output, given by the amount of energy it emits in some amount of time. The cgs units of luminosity are erg s^{-1}.

Consider an imaginary square with some cross-sectional area A floating in space far from an object of luminosity L. The energy that passes through this area from the object in some amount of time is proportional to A, and we call this constant of proportionality the energy *flux* F (or usually, just the "flux"). So, the energy passing through the square per unit time is FA, where F has dimensions of energy per time per area, or erg cm^{-2} s^{-1} in cgs units.

The fraction of the luminosity of the object that intersects this square depends on the square's distance from the object. Imagine a large sphere around the object with radius d, so having area $4\pi d^2$. If our square is at distance d then the fraction of the object's luminosity it encounters is just the fraction of the sphere's surface area it covers, so $A/(4\pi d^2)$. We can then write that the energy per time passing through the square as

$$FA = L\frac{A}{4\pi d^2}$$

or

$$F = \frac{L}{4\pi d^2}. \tag{2.3}$$

This is the *inverse square law of radiation*. It says that the flux drops as the square of the distance from an object, and it is the reason that very distant stars are very faint. Figure 2.2 illustrates the inverse square law.

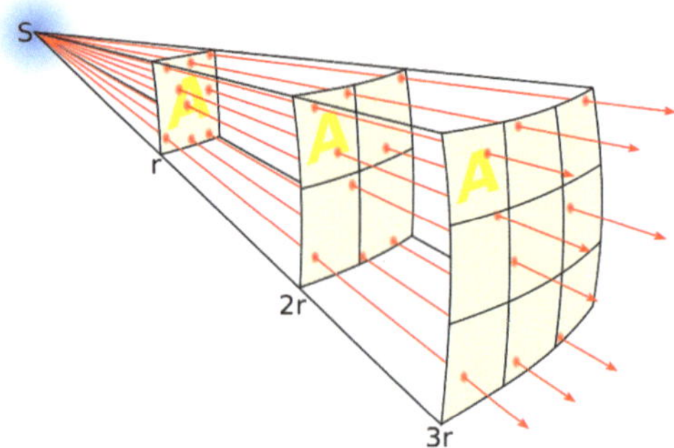

Figure 2.2. The inverse square law. Here a source S having luminosity L radiates in all directions, some of which are shown by the randomly placed red arrows. The three rectangles labeled A have the same area, and are at different distances $d = r$, $2r$, $3r$. If we measure flux as "rays per unit area," we see the flux is $F = 9/A$ at $d = r$, $F = 9/(4A)$ at $d = 2r$, and $F = 9/(9A)$ at $d = 3r$. This means we can write $F = 9/(Ad^2)$, which is the inverse square law. The author obtained this image from the Wikimedia website, where it is attributed to Borb and was made available by Jarvisa under a CC BY-SA 3.0 license.

2.2.2 Angular Size, Solid Angle, and Intensity, Surface Brightness, or Radiance

Things seem smaller when they are farther away. We measure the apparent size of an object by the angle it subtends in our field of view. Rigorously, a sphere of diameter D at distance d subtends an angle $2\sin^{-1}(D/2d)$, but at large distances this reduces via the small angle approximation to

$$\theta = \frac{D}{d}$$

with θ in radians (rad). Because these angles are small, astronomers often instead use smaller units: degrees (such that $360° = 2\pi$ rad), arcminutes ($60' = 1°$), and arcseconds ($60'' = 1'$). There are approximately 206,265 arcseconds in a radian.

Note that the *apparent size* of an object shrinks with distance from an observer as well as $1/d$.

The apparent area of the object also shrinks, and we measure this as a *solid angle*, with units of steradians (sr) or square degrees (sometimes abbreviated □°). Solid angle is to angle as area is to length. There are 4π sr in a sphere, which works out to about 41,000 square degrees.

Consider the light arriving from a well resolved object like the Sun. Each part of the surface contributes some of the flux we receive. The amount of flux coming from a unit solid angle on the solar surface is called its *radiance*, but astronomers generally refer to it by the older name *intensity*. Intensity can also be thought of as a *surface brightness*, and has cgs units erg $\mathrm{cm}^{-2}\ \mathrm{s}^{-1}\ \mathrm{sr}^{-1}$.

As one moves away from the Sun, it appears smaller, and also dimmer. Both the total solid angle of the Sun and the flux drop according to the $1/d^2$ law, so this means the intensity or surface brightness is not changing. This is a law of radiative transfer: the intensity along a beam does not change with distance, but the flux drops as $1/d^2$. This is another way to understand why things get fainter as they get farther away: it is simply that they get smaller, even as their surface brightness stays the same.

2.2.3 Specific Flux or Flux Density

Astronomers often want to know more than just *how much* energy an object emits, and look instead at its emission as a function of frequency ν or wavelength λ. In this case, we can imagine our square only allows light of a narrow range of frequencies $\Delta\nu$ to pass through. The amount of flux we measure is then proportional to the width of this range, and we call this constant of proportionality the *flux density* (or sometimes the *specific flux*), denoted F_ν. If our bins are of equal width in frequency, then flux density has units of erg $\mathrm{cm}^{-2}\ \mathrm{s}^{-1}\ \mathrm{Hz}^{-1}$. We can also measure surface brightness this way, in which case we have a *specific intensity* (or *spectral radiance*) denoted I_ν, having units erg $\mathrm{cm}^{-2}\ \mathrm{s}^{-1}\ \mathrm{sr}^{-1}\ \mathrm{Hz}^{-1}$.

Note that in this convention we are not literally binning the function into discrete bins of, say, 1 Hz each, but imagining the limit of infinitely narrow bins, with the specific flux in each being the flux divided by the bin width. Mathematically, we have that $F_\nu = dF/d\nu$.

Astronomers have several different conventions for flux density. They will often instead use bins of equal wavelength, in which case flux density is denoted F_λ. It is often useful instead to use logarithmically-space bins. This case is a bit tricky because a logarithm is dimensionless, so the flux density in this case has the same units as flux. We use the natural logarithm, so our bins have width $d\ \ln(\lambda) = d\lambda/\lambda$, meaning that this flux density is denoted either νF_ν or λF_λ. In fact, the units and form of the function are the same in either case, and from the equalities $\nu F_\nu = \lambda F_\lambda$ and $\lambda\nu = c$ one can derive the relationship between F_ν and F_λ.

2.2.4 Line and Blackbody Radiation

Light is generated by the motions of electric charges, usually electrons. Two important limits of this are *continuous* emission and *line* emission. Line emission generally occurs when electrons are restricted to discrete energy levels, as in an ion, atom, or molecule, and electrons can transition between two of these levels to produce light of a single frequency. This can occur when electrons recombine with ions and cascade to the ground state (*recombination lines*), when electrons are excited via collisions and de-excite, or when electrons are excited via absorption of photons and de-excite via different energy levels (*fluorescence*).

Continuous emission is most commonly generated when a opaque substance radiates because it is hot. An important fundamental result in physics is that any opaque material of uniform temperature T will radiate energy as light, with a specific intensity that is proportional to the Planck Law:

$$B_\nu(\nu, T) = \frac{2h\nu^3/c^2}{\exp(h\nu/kT) - 1} \tag{2.4}$$

where k is Boltzmann's constant, c is the speed of light, and h is Planck's constant. In terms of wavelength the formula is

$$B_\lambda(\lambda, T) = \frac{2hc^2/\lambda^5}{\exp(hc/\lambda kT) - 1}. \tag{2.5}$$

This kind of radiation (Figure 2.3) is called *blackbody* radiation, referring to a derivation of the Planck Law that invokes a perfectly black substance, i.e., one that absorbs all photons that strike it, independent of frequency. The name is merely a label, however: any substance can radiate as a blackbody if it is of uniform temperature, and indeed an old name for this is *cavity radiation* because it fills any cavity with opaque walls of uniform temperature, regardless of the walls' composition or properties of absorption or reflection.

Astrophysicists will often approximate the spectrum of any warm object as a blackbody, even though the object might have a variety of temperatures. This can be

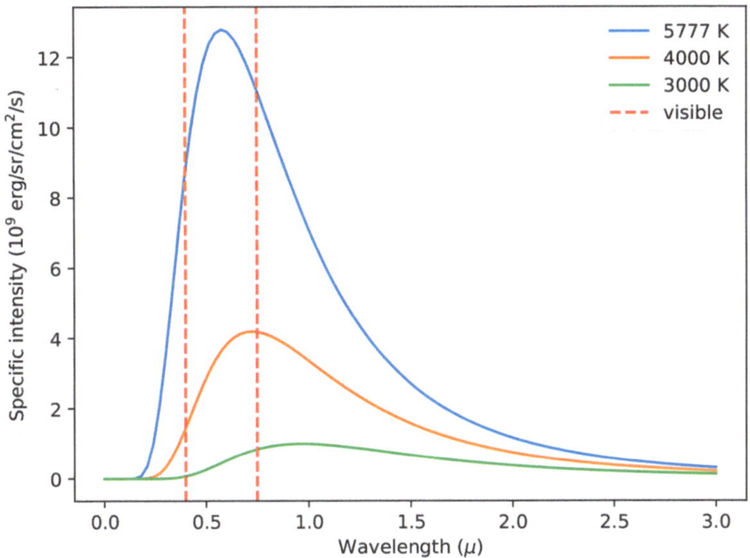

Figure 2.3. The specific intensity of a blackbody at various temperatures. The visible region of the spectrum is demarcated by the red vertical lines; a blackbody at the effective temperature of the Sun (5777 K) is shown to peak there. According to the Wien law, the peak of these curves moves redward as the blackbody gets colder. Blackbody curves are often used as rough approximations of the spectra of any opaque object shining because it is hot.

appropriate when one is interested in only the approximate nature of the light from a source. For instance, one often wishes to approximately determine the predominant kind of light emitted by a source. For a perfect blackbody, the peak of the blackbody curve is given by Wien's law:

$$\lambda_{\max} = \frac{2.9 \text{ mm K}}{T}. \tag{2.6}$$

This can be used to determine which kind of telescope one can use to study a particular phenomenon. For instance, for $T = 5777$ K, (characteristic of the solar atmosphere) one gets 500 nm, which is the wavelength of green light. This illustrates that most of the Sun's luminosity occurs in the visible region of the spectrum. A similar calculation shows that cold interstellar gas is best studied in the microwave or radio regions, and very hot plasmas can be studied in the X-rays.

An important consequence of blackbody radiation is the relationship between the total flux from the surface of an object and its temperature, given by the Stefan–Boltzmann law:

$$F = \sigma T^4 \tag{2.7}$$

where σ is the Stefan–Boltzmann constant, and F is the flux in energy per unit time per unit surface area. Astronomers will often define the *effective temperature* T_{eff} of an object as the temperature of a blackbody with the same flux as the object in question. For example, we can model the Sun as a sphere with radius R and specify its total power output or luminosity as

$$L = 4\pi R^2 \sigma T_{\text{eff}}^4. \tag{2.8}$$

Note that this means that an effective temperature is defined for any object as a relationship between its luminosity and surface area—it need not necessarily relate to the actual temperature anywhere on the object. Nonetheless, for opaque thermal emitters (meaning they are glowing because they are hot) it will approximately correspond to the true temperature.

2.2.5 Light's Interactions with Matter

Light interacts with matter in many ways.

The simplest reaction is *reflection*, where photons simply bounce off of a surface. Light that passes across the surface of a medium like air or glass can also undergo *refraction*, in which it changes direction at the boundary. In most substances, the amount of refraction is a function of the light's frequency, which causes the rays of different colors of light to diverge. Refraction is responsible for the separation of white light into a rainbow of colors in a prism or in drops of water to create a rainbow.

Refraction is an example of the effects of *dispersion*, in which light of different frequencies travels through the medium at different speeds. The speed of light in a medium is given by

$$v_{\text{light}} = \frac{c}{n} \tag{2.9}$$

where c is the speed of light in vacuum. In most media, n is extremely close to 1, but it can be much higher—many glasses have $n \approx 1.5$ in the optical region, and some substances even have $n \approx 4$.

Dispersion by the interstellar medium (ISM) has very different effects on radio and optical radiation. The ISM is a plasma, and the index of refraction of a plasma goes as

$$n = \sqrt{1 - \frac{\nu_{\text{p}}^2}{\nu^2}} \approx 1 - \frac{1}{2}\frac{\nu_{\text{p}}^2}{\nu^2} \tag{2.10}$$

where ν_{p} is the characteristic *plasma frequency*, essentially the frequency below which a plasma is opaque to light. A very short broadband pulse of radio emission passing through interstellar plasma will arrive over a range of times, with the higher frequencies arriving at Earth earlier than the lower frequencies, resulting in a "chirp." The difference in arrival times for photons from the same pulse at two frequencies ν_1 and ν_2 is

$$\Delta t = \frac{d}{c}(n(\nu_1) - n(\nu_2)) = \frac{d\nu_{\text{p}}}{c}\left(\nu_2^{-2} - \nu_1^{-2}\right). \tag{2.11}$$

For scale: for densities typical of the ISM, $\nu_p \approx 1$ kHz. At the radio frequencies where most SETI work is done, between 1 and 10 GHz, this corresponds to $n \approx (1 - 10^{-12})$, which is very close to one but surprisingly important. A signal spanning from 1 to 2 GHz passing through $d = 1000$ pc (roughly 300 ly) of plasma at this density will find itself smeared out, with photons at the two ends of the band arriving at times that differ by around 40 ms. This is easily measured for rapidly spinning pulsars, which can have intrinsic pulse widths much shorter than this.

Light passing through a medium can also suffer *extinction*,[6] meaning some or all of it is removed from the beam. Extinction comes in two flavors: *absorption*, in which photons are destroyed and deposit their energy into the medium, and *scattering*, when the photons change direction, leaving the beam. Scattering is generally more efficient at bluer wavelengths/higher frequencies, which is why a clear daytime sky appears blue, and why sunlight at sunset appears red. Absorption can be highly frequency-dependent, and this dependence lends objects their color, and is the basis of spectroscopy (Section 2.5).

Box 2.2: Jargon Review: Light and Matter

- **Reflection**: Light bouncing from a surface.
- **Refraction**: Light changing direction upon crossing the surface of a medium.
- **Dispersion**: Light of different frequencies interacting differently with matter, including traveling at different speeds through a medium and refracting at different angles.
- **Blackbody**: A theoretically perfect absorber at uniform temperature that emits according to the Planck Law, often used as an approximation of real materials.
- **Continuous emission**: Emission across a large range of wavelengths.
- **Line emission**: Emission at a very narrow range of wavelengths.
- **Extinction, absorption, or scattering**: Extinction (verb: extinguish) is the removal of light from a beam by intervening material, either by absorption (destruction of photons) or scattering (redirection of photons).
- **Linear/circular polarization**: The degree and pattern of oscillation of the direction of the electric field vector in light.
- **Luminosity**: Total power (energy per unit time) emitted by an object, denoted L.
- **Flux**: The luminosity crossing a unit area, denoted F.
- **Intensity** or **radiance**: The flux into a unit solid angle, denoted I.
- **Specific (flux, luminosity, intensity)**: Indicates a quantity is per unit frequency or wavelength, denoted with a subscript ν or λ, as appropriate. If the quantity is per logarithmic flux or wavelength, it is denoted, for instance, λF_λ.
- **Flux density**: A more common term for specific flux.
- **Spectral radiance**: Another term for specific intensity, more common outside of astronomy.
- **Effective temperature**: The temperature of a blackbody with the same surface flux as an object being measured.

[6] Light that has suffered extinction is said to have been *extinguished*, although you will occasionally hear astronomers utter the abomination "extincted." Please don't do this.

2.2.6 Polarization

Light is also characterized by *polarization*, which describes the pattern of directions the electric field makes as it propagates. In linear polarization, the direction charges are accelerated due to the wave is restricted to moving back and forth in a single plane (for instance "up and down" or "left to right"), and in circular polarization it sweeps out a circle as it propagates; elliptical polarization is an intermediate case. Many astronomical detectors are sensitive to only one component of polarization: for linearly polarized light one is sensitive to the *X*- and *Y*-components (i.e., the projection of the direction of polarization onto the axes of the detector); for circular polarization one is sensitive to the left- or right-handed components.[7] Most light is unpolarized, meaning it is composed of equal amounts of light of all polarizations. Figure 2.4 illustrates polarization.

Light becomes polarized when reflected from a surface (linear polarization) or when it is generated by material in a magnetic field (circular polarization). Human radio communication in space is generally purely circularly polarized, which aids with distinguishing the signal from background noise.

2.3 Telescopes

A telescope collects light and interprets it, generally bringing it to a focus and forming an image of a distant source of radiation. Telescopes working at different frequencies can take very different forms.

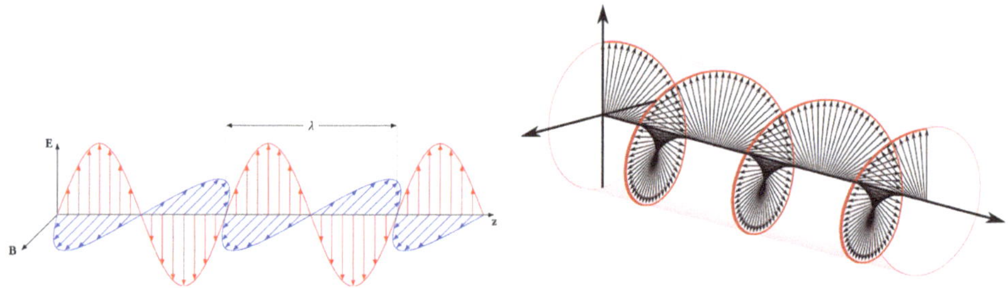

Figure 2.4. Examples of purely polarized light. *Left:* Pure linear polarization. The strength and direction of the electric field at each point is illustrated by a red arrow (or vector). The wave propagates along the *z*-axis, and has wavelength λ. Also shown is the magnetic field, in blue, which is necessarily orthogonal to the electric field. If one measured the electric field at a point along the *z*-axis, the electric field would oscillate sinusoidally, and always point along the *E*-axis. *Right*: Pure circular polarization. In this case, the electric field oscillates in two directions with equal amplitude, and out of phase by 90°, resulting in the vector sweeping out a circle. Here the magnetic field is not shown. An intermediate case, in which the two directions of polarization have different (but nonzero) amplitudes, is elliptical polarization. Unpolarized light is an even mixture of light of all polarizations. The author obtained the left figure from the Wikimedia website, where it is attributed to Francois~frwiki and made available under a CC BY-SA 3.0 license. The author obtained the right figure from the Wikimedia website, where it is attributed to Dave3457 and where it is stated to have been released into the public domain.

[7] Confusingly, astronomers and physicists have different conventions for circular polarization. Astronomers imagine they are looking up into the source of radiation, and describe right-handed circularly polarized light as following the right-hand rule with one's thumb pointed toward one's eyes: the direction of the field rotates in the direction of one's curled fingers. Physicists imagine, like Einstein, that they ride the beam of light, so the thumb is pointed away, in the direction of travel.

The most common image of a telescope is essentially a design of the first telescopes in the optical: a primary lens collects light and brings it to a focus, followed by additional lenses that are used to produce an image of the scale and sort one needs. Today, mirrors are generally preferred over lenses because they reflect all wavelengths of light in the same manner (lenses are dispersive). Mirrors are typically made of glass with a metalic coating, such as aluminum, silver, or gold. At infrared through X-ray frequencies, cameras generally have many pixels and so images can be formed as with one's camera phone.

At radio frequencies, primary mirrors are called *dishes* and can be made of lighter materials with looser figure requirements.[8] At the longest wavelengths, wire mesh can be sufficient to serve as a reflector, and in some cases involving interferometry bare antennae are used with no reflector at all.

Radio telescopes generally have a single "pixel" receiver, and images are formed by building up a map, point by point, although some detectors may have several pixels to speed this up. Unlike photon-counting detectors used at higher frequencies, radio receivers are more properly thought of as antennae that directly measure the electric field variations, as changes in voltage. At far-infrared frequencies, a mixture of radio and higher-frequency techniques can be used, and photon detection is sometimes done via *bolometers* that measure the total amount of energy collected in a wide range of frequencies.

2.3.1 Angular Resolution

In general, larger primary mirrors (or lenses) are preferred for two reasons: they collect more light allowing one to detect fainter sources in less time, and because of the effects of diffraction, which limits the *angular resolution* of a telescope.

Light's wave nature becomes apparent when it interferes with other light, as happens when a mirror brings light from widely spaced parallel beams together in a focus. The result in this case is to create a spot of finite size, which is inversely proportional to the size of the mirror used. Larger spot sizes at focus correspond to "blurrier" pictures, so one's ability to resolve detail in distant objects depends on the size of one's telescope primary mirror. We measure the sharpness of images by the smallest angular size θ that can be resolved, with the usual form being

$$\theta = 1.22\frac{\lambda}{D} \text{ radians} \tag{2.12}$$

$$= 2.5 \times 10^5\frac{\lambda}{D} \text{ arcseconds} \tag{2.13}$$

where D is the telescope diameter and θ is called the *diffraction limit* of the telescope —it measures the blurriness inherent to the system. Radio telescopes typically have

[8] In general, the quality of an optical (i.e., light reflecting) surface must be such that the shape is accurate to a quarter of the wavelength of light being observed.

resolutions measured in arcminutes, and optical telescopes typically have diffraction limits below one arcsecond.

Another form of this equation is

$$A\Omega = \lambda^2 \tag{2.14}$$

where here A is the area of a circular mirror and Ω is the solid angle subtended by a focus spot. This form is useful because $\Omega/4\pi$ represents the fraction of the sky subtended by a single resolution element.

In order to achieve this ultimate limit, however, there must be no other source of "blurrieness" in the system. In the optical window, this limit is rarely achieved because the Earth's atmosphere, which distorts images on rapid timescales, blurring images out to a few arcseconds. Astronomers call this effect (and any measure of its magnitude) *seeing*. To optimize seeing, telescopes are typically put on mountaintops above most of the atmosphere, at locations where the prevailing wind is smooth and steady.

Infrared and optical astronomers employ *adaptive optics* to overcome this atmospheric blurring. Using a bright reference source, astronomers measure the amount of blurring and rapidly manipulate small, flexible mirrors within the telescope to compensate. This source might be a bright star (a *natural guide star*) or, if no such star is available, an *artificial guide star*. The latter is actually a patch of the upper atmosphere illuminated by a laser tuned to the resonant frequency of sodium, which exists as a gas in the upper atmosphere as a result of micrometeorites. Either technique allows a telescope to operate near its diffraction limit.

2.3.2 Transmission Strengths and Gains

Note that all of these principles of angular resolution apply regardless of the direction photons are traveling, that is whether a telescope or radio dish is being used to receive (make an image of the sky) or transmit (as with radar or communication). Radio astronomers in particular refer to the *beam* of a telescope, meaning the pattern of radiation that contributes to a single resolution element when receiving or, equivalently, the pattern of outgoing radiation focused by the dish.

This means that a single transmitter cannot send a signal out in a tighter beam than its imaging diffraction limit. When transmitting, the purpose of a primary mirror is to take light that might otherwise go in every direction and collimate it into a tight beam; when receiving the effect is to take light from that same beam shape and collect in into a single spot at the telescope focus.

The factor by which a mirror or dish improves upon an isotropic transmitter or receiver is called the *geometric* (or *power* or *antenna*) *gain*[9] of the telescope. At the diffraction limit, it is given by

[9] The somewhat odd term "gain" in this context comes via electrical engineering, which considers the strength a signal "gains" or loses through various components of an electrical system, such as via a linear amplifier. Viewed this way, a telescope dish is just another way to amplify the signal, and so has its own "gain" term. The tradition in astronomy at other wavelengths, such as optical or X-ray, is to measure this in terms of the collecting area of the telescope, not a dimensionless "gain."

$$G = \frac{4\pi}{\Omega} = \frac{4\pi A_e}{\lambda^2} = \epsilon \left(\frac{\pi D}{\lambda} \right)^2 \tag{2.15}$$

where Ω is the solid angle of the beam on the sky, A_e is the *effective area* of of the dish (i.e., the area of a perfect dish with the same collecting power), and $\epsilon \leqslant 1$ is a factor that captures the degree to which the mirror approaches the performance of a perfect dish (which is limited by things like surface imperfections, obstructions like secondary mirrors, and non-circularity).

In radio transmissions, one can express P, the power in the beam (i.e., the power needed to generate a signal of a given strength), in terms of the wattage needed to produce that same power level isotropically. This is called the equivalent (or effective) isotropic radiated power (EIRP) and is given by

$$\text{EIRP} = GP. \tag{2.16}$$

This is valuable because it is a property only of the intensity one measures when receiving a transmission and the distance to the transmitter and not none of the other (potentially unknown) properties of the transmitter.

Box 2.3: Jargon Review: Telescopes

- **Primary mirror/lens**: The main collecting surface of a telescope, which determines its collecting area and theoretical maximum resolution.
- **Dish**: The primary mirror of a radio telescope.
- **Angular resolution**: The "sharpness" of the image produced by a telescope; roughly the closest two identical sources can be to one another and still be discerned as distinct. Measured in units of angle, such as arcsecords.
- **Arcseconds, arcminutes**: 1/3600 and 1/60 of a degree, respectively, symbolized by $''$ and $'$.
- **Diffraction limit**: The theoretical best angular resolution of a telescope, determined by the wavelength of light being considered and the size of the primary mirror.
- **Seeing**: The blurring due to the Earth's atmosphere. Also a quantitative estimate of that blurring, measured as an angular resolution.
- **Adaptive optics**: A technique for undoing the blurring of Earth's atmosphere. Can be accomplished with an artificial guide star generated by a laser.
- (**Antenna** or **power** or **geometric**) **gain**: The factor by which a dish concentrates radiation compared to an isotropic transmitter/receiver. Not to be confused with the electrical engineering concept of gain, which is also used ubiquitously in radio astronomy.
- **EIRP**: Equivalent (or effective) isotropic radiated power. The power it would take to transmit a signal of a given strength isotropically.
- **Interferometry**: A technique to combine light from many telescopes to measure features on angular scales below each individual telescope's diffraction limit. When employed with many elements, can simulate the power of a single large telescope.

2.3.3 Interferometry

The diffraction limit can also be overcome by combining the light from many widely spaced telescopes, effectively using them as small pieces of a much larger primary mirror. This is called *interferometry*. To do this, one must allow the light from the many telescopes to interfere. In the optical, this is done with tubes or optical fibers through which light is sent to a common location. A similar technique can be used for radio, for instance with coaxial cables, or voltages can be recorded to disk and "interfered" in software to achieve the same result. In this way, an array of many "single pixel" radio telescopes can work together to create an image (Figure 12.4).

Interferometry comes with many tradeoffs, especially when the number of telescopes is not large, and the gaps between them are large. The sparser the "filling factor" of the telescope array the less information about the image of the sky one obtains, and the more assumptions one must make to reconstruct an image. At high frequencies and at large numbers of telescopes, it can also be extremely computationally intensive.

Optical interferometers are typically composed of just a few elements, and excel at measuring things like the angular diameters of stars. Radio interferometers can have many elements, such as the ALMA array with 66 elements, or the MeerKAT array with 64 elements. The ALMA array, which works at submillimeter wavelengths, can produce outstanding images at resolutions measured in milliarcsecords.

The purpose of dishes or primary mirrors in interferometers is then not to create an image, but to focus light from an area on the sky, which determines the field of view of the interferometer. Some radio interferometers dispense with dishes entirely and simply use a large array of antennae, in which case they can have an extremely large field of view, essentially the entire sky.

These techniques also work in reverse to create a phased array transmitter: if an array of telescopes is *transmitting* in a coherent (coordinated) way, then the outgoing beams will interfere in a way that can be focused to a patch of sky given by the diffraction limit of the entire array.

2.4 Measuring Brightness

2.4.1 Photometric Magnitudes

Astronomers measure the brightness of celestial objects in a way that approximates the method of Hipparchus and Ptolemy, who described the "magnitude" of the stars on a scale from 1 to 6. Since the eye is very roughly a logarithmic detector, the scale is logarithmic.

This *magnitude* system is much maligned—perhaps more so by astronomers than others—because it has many properties that at first seem counterintuitive. It employs a base-10 logarithm, but to match the numbers in the old catalogs the values are scaled by -2.5:

$$m = -2.5 \log_{10}(F/F_0). \tag{2.17}$$

Here, m is the magnitude of an object, and F is the flux we receive, and F_0 is the fiducial flux that sets the *zero point* of the scale. F_0 depends on the specific standardized system being used and the range of wavelengths being considered, usually specified by the *filter* or *bandpass* of the observations.

This zero point is usually chosen to correspond to some convenient, presumably constant source of flux on the sky (the traditional—but obsolete—standard is the star Vega). Some magnitude systems are pegged to particular flux distributions defined in absolute terms (e.g., the AB system defines its standard as having mean flux density of 3631 Jy across a given bandpass) but since such conversions can be quite difficult to achieve accurately in practice, most systems are defined relative to some easily observed photometric standard star.

Equation (2.17) has many advantages. The "backwards" nature of the scale means that as telescopes get more powerful and we see fainter and more numerous objects, the number used to describe them increases—the Hubble Space Telescope can detect stars fainter than $m = 30$. Only a handful of the very brightest stars and planets in the sky have negative magnitudes; the Sun has $m \sim -26.7$.

The 2.5 is also convenient because it makes the scale approximately a natural logarithm, while keeping things locked into base-10 for ease of mental computation. The rule is that five magnitudes are *exactly* a factor of 100, and one magnitude is *roughly* one e-folding. This latter property means that small differences in magnitudes are very close to their true relative differences: a source changing in brightness by $\Delta m = 0.01$ has changed by very nearly 1% (1.08%, to be more precise).[10]

Astronomers also specify *absolute* magnitudes, which refer to the magnitude an object *would* have if viewed from a distance of 10 pc; and *bolometric* magnitudes, in which the bandpass includes all radiation from the object. In planetary science, absolute magnutides refer to the hypothetical brightness of an object viewed from 1 au at full phase when it is at 1 au from the Sun.

Extinction (Section 2.2.4) is exponential: for a uniform material, light loses intensity by a constant factor per unit distance. Since the magnitude system is logarithmic, this means that the magnitude of a star goes up (in the fainter direction) by an amount proportional to the amount of intervening material extinguishing its light. Astronomers will sometimes say things like "the star suffers five magnitudes of extinction" which means it is 100 times fainter than it would appear if there were no intervening material. In that example, if there were twice as much material, there would be 10 magnitudes of extinction, and it would be 10,000 times fainter.

2.4.2 Photometry, Filters/Bands, and Colors

Light comes in many wavelengths, and an object that is bright at one wavelength might be quite faint at another. Roughly speaking, red objects emit or reflect more light at red wavelengths than blue ones.

[10] This compromise is somewhat akin to even-tempering on a piano, in which fifths and other consonant intervals differ very slightly from their perfect ratios, in exchange for the convenience of playing music equally well in any key.

The measurement of the brightness of astronomical sources is called *photometry*. When reporting a measurement it is essential to specify at what wavelengths one made the measurement. The standard way to control this is by using well-calibrated colored filters that restrict the measurement to some standard range of wavelengths. In the optical some standard filter names are UBVRIZYJHKLM, each letter corresponding to a *broadband filter* that spans about 10% of relative bandwidth ($\Delta\lambda/\lambda$) from the near-ultraviolet (U and B, just past the range of human vision) to the optical (VR) to the near-infrared (IYZJHK) to the mid-infrared (LM).[11] *Narrowband filters* might have relative widths of only 0.1%, often selecting out specific spectral features.

Absolute photometry converts measurements to physical units of flux across the band, which is important for comparing with photometry in other bands. *Relative photometry* does not perform this conversion, for instance when measuring just the change in brightness with time of a single source.

The relative brightness of an object through different filters is referred to as a *color* because in the optical comparing, say, the absolute V to the R brightness roughly corresponds to the color perceived by the human eye.

Astronomers usually refer to colors as differences in magnitudes (so, the logarithm of the ratios of fluxes in two filters/bands), with the longer wavelength band subtracted from the shorter. Stars' colors might be expressed in $(B - V)$ color, where small or even negative numbers refer to *blue* objects and large numbers refer to *red* objects. Magnitude zero point standards like Vega have colors of zero. Interstellar dust generally *reddens* colors because it absorbs or scatters shorter wavelengths better, similar to the reddening of sunlight during a sunset. Hotter objects tend to emit more blue light than cooler ones, so color is often a rough proxy for temperature.[12]

An absolutely calibrated photometric compilation of brightnesses across a wide range of wavelengths is called a *spectral energy distribution* or SED. Measuring brightness simultaneously at many wavelengths with resolutions better than about $\lambda/\Delta\lambda \sim 10$ is usually done with dispersive equipment like gratings or prisms and is called *spectroscopy*, which does not use the magnitude system.

[11] Do not confuse these bands with the unrelated band-naming convention in the radio.

[12] n.b. the jargon in the décor lighting industry is the opposite of this, where lights appearing redder to the human eye are said to be "warmer" than light with a more blue hue, which is said to be "cooler." That industry also employs the term *color temperature* (in the same sense astronomers use the term) in which a higher color temperature corresponds to a bluer light.

Box 2.4: Jargon Review: Photometry

- **Photometry**: The practice of measuring the brightnesses of celestial sources of radiation. Also, the brightnesses so measured.
- **Magnitude**: A dimensionless, logarithmic measure of the brightness of an object measured across some range of wavelengths. A difference of 5 magnitudes is exactly a factor of 100, and a difference of 0.01 is approximately 1%. A star with $m = 6.5$ is said to be "sixth magnitude."
- **Filter/band/bandpass**: The region of the electromagnetic spectrum at which a photometric measurement is made. Filters or bands are mathematically described by a filter response function, which quantifies the response of an instrument as a function of wavelength. Filter names usually have short names like "B" (or "B band," for "blue").
- **Broadband/narrowband**: Refers to the relative bandwidth of a filter. This can be expressed as the range of wavelengths over the central wavelength, $\Delta\lambda/\lambda$. In the UV/optical/IR (but not in the radio) broadband filters typically have relative bandwidths of several to tens of percent, intermediate band of around 1%, and narrowband of around 0.1%.
- **Red, blue**: Relative terms describing light of longer or shorter wavelengths, or larger and smaller color indices, respectively. These terms do not necessarily refer to appearance of an object to the human eye.
- **Color or color index**: A comparison of the brightness of an object in two different bands, expressed as a difference in magnitudes (which is a ratio of fluxes). "Red" objects have large (numerical) colors, and "blue" objects have small or even negative colors. In general for objects emitting continuous spectra, hot objects are bluer, and cool objects are redder.
- **Reddening**: The change in apparent color of an object because intervening material has scattered or absorbed more of its blue light than red light.
- **Spectral energy distribution (SED)**: A set of photometric measurements made across a wide range of filters, converted to a common absolute scale. Given enough bands, an SED can describe the wavelength distribution of the bulk of an object's emission.
- **Spectrum**: A measurement of brightnesses across a continuous range of wavelengths, generally measured simultaneously. Unlike an SED, a spectrum might span only a narrow range of wavelengths. That said, a sufficiently high resolution SED could be called a spectrum, and a sufficiently well calibrated and broadband spectrum could be called an SED.

2.5 Spectroscopy

One of astronomers' most powerful tools is spectroscopy, which allows them to determine the chemical makeup, temperature, and velocities of things in space. Astronomers build tools called *spectrometers* or *spectrographs*, which disperse light into its component colors at some *spectral resolution* given by $R = \lambda/\Delta\lambda$. The data so collected compose the *spectrum* of the object.

Spectroscopy works because the electrons in atoms, ions, and molecules tend to interact with photons very strongly when those photons have energies corresponding to transitions between energy levels in the atoms. Light passing through a gas of cool material will thus be absorbed more strongly at a discrete set of wavelengths, which astronomers call *absorption lines.*

When gas is hot, it radiates energy away. It does so best at wavelengths where it is opaque (since those are the wavelengths where, by definition, it interacts most strongly with light). A gas might be virtually transparent at most wavelengths, and so shine most brightly only at the wavelengths corresponding to electronic transitions. These are called *emission lines.*

A mass of material that is mostly transparent is said to be *optically thin*, and material which blocks most of the light at a given wavelength is said to be *optically thick.*[13] The calculation of how light interacts with material is called *radiative transfer*. Whether a line appears in absorption or emission depends on the details of the optical thickness, temperatures, and other physical properties of the materials involved (see Figure 2.5).

The patterns of emission and absorption lines are the same, and are characteristic of the chemical species that compose the gas. Astronomers use this fact to determine the chemical composition of material in space. In detail, the relative strengths of these lines can help determine the *abundances* of the elements in that material. Since the level populations of the electrons in the material depend on the physical conditions in the material, the detailed line strengths can also reveal information about the material's temperature and pressure.[14]

The narrowness of spectral lines depends on many physical factors, but typical intrinsic widths are around $\Delta\lambda/\lambda \approx 10^{-5}$. This means that changes in the locations of the absorption lines can be measured extremely precisely.

This is useful because the wavelength of a spectral line depends on the relative velocity between the source and the observer, given at non-relativistic speeds by the classical *Doppler shift* formula[15]

$$\frac{\Delta\lambda}{\lambda} = -\frac{\Delta\nu}{\nu} = \frac{v_r}{c} \tag{2.18}$$

where v_r is the mutual velocity along the line of sight (the *radial velocity*) and c is the speed of light. The Doppler shift arises in any source that emits waves, and can be

[13] The technical definition is measured with respect to the *optical depth* or *optical thickness* of material, τ, defined as the number of e-foldings of light that is absorbed passing through the material.

[14] The field of astrophysics was arguably born with the PhD dissertation of British–American astronomer Cecilia Payne-Gaposhkin, in the sense that it was the first detailed application of physics beyond gravity to an astronomical object. Payne applied the principles of radiative transfer, ionization balance (the Saha equation), and excitation equilibrium (the Boltzmann equation) to determine the abundances of elements in the Sun from its spectrum. She established that the Sun was made mostly of hydrogen, a result so surprising it was rejected even by her PhD committee. Once it was independently confirmed, Payne-Gaposhkin's result was transformative. Payne-Gaposhkin would go on to be the first woman to be appointed as a professor and the first female department chair at Harvard. She was also the thesis adviser of Frank Drake (Box 8.1).

[15] This formula is correct for small Doppler shifts, where $v/c \ll 1$. For larger shifts, one must make relativistic corrections and use different formulae for λ and ν (Section 12.4.1.)

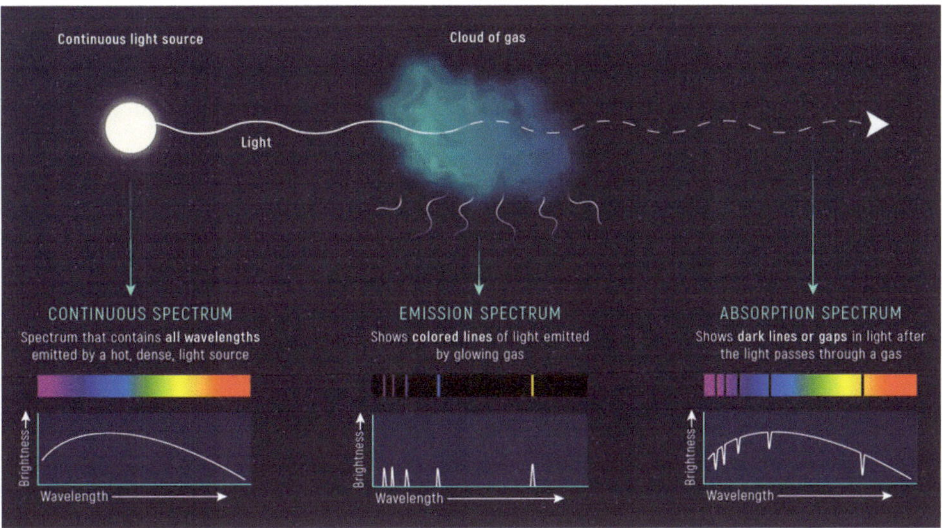

Figure 2.5. Continuous, absorption, and emission spectra. Here, a hot star emits a blackbody spectrum (in truth stellar spectra are more complex than this) near a cooler cloud of gas. Light passing through the cloud is absorbed at discrete wavelengths, resulting in an absorption spectrum that reveals the cloud's composition. Observing the cloud directly (i.e., in directions without a continuous source behind the cloud) results in an emission spectrum, with lines appearing at the same wavelengths as the absorption features in the absorption spectrum. This occurs when light from the star is either *scattered* by the cloud (usually when an atom absorbs and then immediately re-emits a photon), or is absorbed and re-processed (when a photon is absorbed by an atom or molecule and the energy is eventually re-emitted at a different wavelength). Credit: NASA, ESA, CSA, Leah Hustak (STScI).

illustrated well with sound waves: an object emitting a sound with a clear pitch and moving toward or away from you by around 50 miles per hour will appear to have its pitch lowered (if it is moving away) or raised (if it is moving toward you) by around one half step (i.e., from A to A$^{\#}$). The effect is especially obvious in the sound of racecar engines or sirens as they pass by you.

Since we know the wavelength at which light was emitted or absorbed in the rest frame of the relevant atom, we can compute the velocity of that atom with respect to Earth. From this, astronomers can measure the motions of galaxies and stars through space, and the thermal and other motions of the atoms in a gas.[16]

[16] For instance, by measuring very small changes in wavelength, corresponding to just meters per second, astronomers can infer the existence of planets from the "wobble" they induce on their host stars. Note that cosmological redshift has a more subtle interpretation because it originates in the expansion of spacetime itself, and cosmologists argue about the pedagogical value of describing it simply as being due to a recession velocity.

Box 2.5: Jargon Review: Spectroscopy

- **Spectroscopy**: The detailed measurement of light as a function of its wavelength. Used the measure the chemical makeup, physical conditions, and velocity of material.
- **Spectrograph** or **spectrometer**: An instrument for measuring a spectrum.
- **Spectral resolution**: The fineness with which light of different wavelengths can be distinguished, measured as $R = \lambda/\Delta\lambda$.
- **Absorption/emission line**: a narrow range of wavelengths where we receive less/more light due to absorption from intervening material. The wavelength of a line is determined by the electronic structure of the atom/ion/molecule that generated it.
- **Radiative transfer**: The calculation of how light interacts with matter, especially one that describes the spectrum we observe.
- **Optically thick/thin**: Whether material is mostly opaque or transparent at a given wavelength.
- **Abundances**: The detailed amounts of different chemical species present in material.
- **Doppler shift**: The change in wavelength of light between what is emitted and what is measured based on the mutual velocity between the source (or absorber) and the observer. Given in the non-relativistic limit by $\Delta\lambda/\lambda = v/c$.

2.6 Beyond the EM Spectrum

There are four forces of nature, but so far we have only discussed receiving information from space via electromagnetism. There are four other ways however: we can exploit the other three fundamental forces, and we can collect physical things. We will discuss each as a potential source of SETI information later, so here we just go over the basics.

2.6.1 Neutrinos

The weak force of nature was only discovered in the course of studying nuclear decay and radiation; it has no discernible effect on the world beyond its role in shaping the periodic table of the elements. It does, however, produce a long-range particle called the neutrino that astronomers have recently been able to observe.

The neutrino is often said to be a "ghostly" particle because it has no charge and no electric field, so it does not interact with electrons via electromagnetism. This means that for the most part it passes through ordinary matter as if it were not there. The Universe is actually filled with neutrinos, with trillions passing through everyday objects every second.

Neutrinos can, however, occasionally interact with electrons and nuclei via the weak force. These events are extremely rare, however, and are generally counted one-by-one via the charged byproducts of those interactions as they move through a medium, giving off Cherenkov radiation.

There are two kinds of neutrinos that have been detected from beyond the Earth: low-energy neutrinos from ordinary radioactive decay (either from material on Earth or the products of nuclear fusion on the Sun), and high-energy neutrinos from extremely energetic events, presumably like the accretion disks around supermassive black holes and the collisions of neutron stars. Low energy neutrinos are common and can be detected using giant vats of water the size of buildings, for example the Super Kamiokande experiment. High energy neutrinos are extremely rare and so require huge collecting volumes. The king here is the IceCube experiment at the South Pole, which appropriates approximately a cubic kilometer of ice of a glacier for the purpose by filling it with chains of photodetectors to observe the Cherenkov radiation. It detects several high-energy neutrinos from beyond the Earth per year.

Because these detectors have no optics, they stretch the definition of “telescope.” The direction of the source of a neutrino event must be inferred from the direction of the particles detected via Cherenkov radiation, which can only be done roughly. Often, the localization of a neutrino source is only good to a few degrees (unless an optical counterpart can be found).

2.6.2 Gravitational Waves

Just as accelerating a charge generates EM waves, accelerating a mass can generate gravitational waves. Although the detectability of such waves even in principle was unclear even to Einstein (who famously submitted a paper claiming they were just a mathematical artifact, and was saved from this blunder by a referee), today their generation and detection is well understood.

The medium of a gravitational wave is spacetime itself, and its effect is to cause distances between objects to oscillate at the frequency of the wave. The reason this effect is not noticible is that it is *extremley* subtle; we write that matter does not couple to these waves very strongly.

All accelerating masses produce gravitational waves, but only the most massive objects undergoing the strongest accelerations produce anything even remotely detectable. Gravitational waves were first indirectly detected via the slow orbital decay of a binary neutron star, as their orbital energy escaped as gravitational waves. The strength of the escaping waves is strongest when two orbiting objects are closest together.

Gravitational waves can be described by their effect on matter via their *strain*. Strain is dimensionless, and measures the amount an object or length is distorted in one direction relative to an orthogonal direction. When a gravitational wave passes by, spacetime itself (and the objects in it) distort in this way, and the resulting time-variable strain can be expressed in nine numbers expressed as a 3×3 tensor. A gravitational wave antenna consists of a pair of objects whose distances are measured extremely precisely. When a gravitational wave passes by, the resulting strain along the antenna will change the distance between the elements by a very small amount.

The first gravitational waves to be directly detected were by the LIGO experiment, which measures changes in the lengths of a pair of 4 km long vacuum tubes at two sites via laser interferometry at the level of a 10^{-17} cm, which is less than 1/1000 of the size of a proton. The only sources of gravitational waves powerful enough to generate such waves are the mergers of black holes or neutron stars. For instance,

LIGO's first detection was of two black holes, each about 30 times the mass of the Sun, merging in a distant galaxy. The radiation generated from their final in-spiral was equivalent to about three times the mass–energy of the Sun, released in a fraction of a second, for a luminosity of around 10^{55} erg s^{-1}. The mass of the final, merged black hole was less than the combined mass of the progenitor holes by about this amount.

As gravitational wave detectors get more sensitive, less dramatic events will be needed for detection. A space-based version of LIGO, called LISA, will be able to detect gravitational waves from binary black holes and neutron stars long before they merge, when they have orbital periods of 1–1000 s.

More nearby events will produce stronger signals, so in principle these detectors could detect the merger of ordinary stars or even a supernova if it occurred within the Milky Way.

Gravitational wave antennae use interferometry to determine the direction of a source of GWs on the sky, and so improve dramatically with the number of antennae. Typical source localization precisions for LIGO's two detectors alone are around tens of degrees or more; with the addition of new antennae the precision should improve to a few degrees, rivaling neutrino detectors.

2.6.3 The Strong Force: Cosmic Rays

The final force of nature is the strong nuclear force. Unlike the other forces of nature, there is no long-range particle associated with this force, so it cannot be detected at great distances.

However, we do detect the products of strong force interactions in the form of atomic nuclei moving at relativistic speeds through space.[17]

Unlike photons, neutrinos, and gravitational waves (presumably quantized as gravitons), cosmic rays are charged particles.[18] Cosmic rays are accelerated to relativistic speeds in strongly energetic magnetic regions, such as the aftermath of a supernova or in the jet a supermassive black hole. Because they are charged, they can be deflected by magnetic fields, such as that in the Milky Way meaning it can be difficult to localize their origin unless they are so energetic that these fields do not deflect them much.

Cosmic rays are detected when they enter Earth's atmosphere or our cosmic ray "telescopes" because, being charged particles in a physical medium, they emit Cherenkov radiation as they pass through. Telescope arrays like VERITAS can see these flashes of light in the sky and deduce much about the particle that generated them, effectively using the atmosphere as its detector.

[17] This is a bit of a cheat. The "messenger" here is in some sense physical objects, but most cosmic rays are bare atomic nuclei composed of quarks and gluons bound together by the strong force, so it is convenient to think of them as representing our exploration of the Universe via this force of nature. Some cosmic rays are electrons, though.

[18] Except neutrons, but because they decay in 11 minutes unless bound to protons we do not receive them from sources outside the solar system.

2.6.4 Physical Objects

The final way we gather information from space is via physical objects. Meteorites tell us much about the origins of the solar system, and sample return missions from the Moon, comets, and soon other bodies allow us to do "geology" beyond the Earth.

Some of these "messengers" come from outside the solar system. Some of the dust in the solar system is actually interstellar, and with the discovery of 'Oumuama, we now also have the ability to detect larger bodies like interstellar comets and asteroids as they pass through the solar system.

2.7 Lookback Time

Light travels at finite speed, meaning that we technically see things as they were in the past. Across a room this is such a small effect as to be negligible; when communicating in real time electronically around the world it can have a slightly noticeable effect. We see the Sun as it was 8 minutes ago, some of the most distant stars visible to the eye as they were tens of thousands of years ago, and distant galaxies as they were tens of millions or (using powerful telescopes) many billions of years ago.

Cosmologists can use this effect to study the ancient Universe, studying galaxies that formed early in the Universe's history. The cosmic microwave background, the most distant light we can measure, shows us the Universe as it was only 400,000 years after the Big Bang.

Sometimes, too much is made of this effect, especially since, outside of cosmology and studies of the first galaxies, most objects are not at such great distances that the lookback time matters. The signals are, for most practical purposes, happening in our present.

For instance, virtually all of the stars you see at night will continue to shine for millions or billions of years—despite what you may have heard, only a very few "might not exist any more" because they have some chance of having exploded in the past few thousand years the light has been traveling our way.

It matters not at all to astronomers that the variable star they measure pulsating every 13 days was "actually" doing so 2000 years ago any more than solar astronomers worry about the fact that the Sun was "actually" only like we see it 8 minutes ago. If astronauts witness a huge solar flare on the Sun they do not have 8 minutes to avoid the X-rays that accompany it, because they arrive at the same time as the visible light they used to see the flare.

Across the galaxy we might discover alien life, say, 100,000 light years away, and so observe it as it was 100,000 years ago. But since we cannot visit them the fact that it is "old news" will have little practical importance to the discovery. It is a bit, perhaps, like watching a television show that was first released months ago, or a late night show taped earlier in the day, or sporting event broadcast on a few seconds' tape delay. The *lookback time* one experiences in these cases is rarely of any practical interest when interpreting the show.

We will discuss the effects of lookback time in the context of the Drake Equation in Section 8.3.2.

2.8 Distances

2.8.1 Distance Measurement Techniques

One of the most challenging measurements in astronomy is distance. Astronomers use a "ladder" of techniques to determine distances to astronomical sources. Each rung of the ladder is calibrated from the rung below it, so uncertainies accumulate as you climb the ladder.

The first rung of the ladder is radar and radio communications in the solar system. By timing the "ping time" of radio signals bounced off of planetary surfaces from Earth, or from communications with our interplanetary probes, planetary scientists can measure the positions of components of the solar system today with exquisite precision, sometimes to within meters.

The next rung of the ladder is *geometric parallax*, caused by the changing perspective on nearby objects due to Earth's orbital motion. Nearby stars appear to execute an annual reflex motion against background galaxies, and the amplitude of this motion is inversely proportional to their distance. The formula is

$$d = \frac{1}{\varpi} \tag{2.19}$$

where d is the distance to the star and ϖ is the measured parallax. If the equation seems curiously devoid of units, it is because astronomers have chosen their distance units carefully make the constant of proportinality 1. You have two choices: if ϖ is expressed in radians, then d is in *astronomical units* (au or AU, the mean distance between the Earth and the Sun). The more common convention is to express ϖ in units of arcseconds, in which case d is measude in *parsecs* (pc), equal to about 3.26 light years. Astronomers rarely express things in light years, the distance light can travel in a year, except in the context of lookback time and in outreach efforts (because the public is more familiar with it). The lookback time to an object in years is roughly three times its distance in pc.

The nearest stars have parallaxes of about one arcsecond, and today ESA's *Gaia* mission has measured the parallaxes (i.e., the distances) of over a billion stars out to distances of about 10,000 pc, which is a significant fraction of the entire Milky Way Galaxy. Farther away or for stars too faint for *Gaia* to measure, distances must be inferred from how bright a star appears and how bright we expect it to be based on what we can infer about it from its spectrum.

The distances to other galaxies can be inferred from a variety of techniques based on *standard candles*, i.e., stars or other objects within those galaxies whose intrinsic luminosities are known. Distances are then inferred from the object's observed brightness using the inverse square law (Equation (2.3)). Certain extremely bright stars, like Cepheid variables, have well-calibrated luminosities which can be determined from their photometric pulsation periods.

The most distant galaxies appear to be receding from us because of the expansion of the Universe. To first order, this expansion is uniform, so the speed with which a galaxy is moving away from us is proportional to its distance. This relation is called

Hubble's Law, and allows astronomers to infer a galaxy's distance from its recession velocity, which can be measured precisely from the Doppler shift (Section 2.5).

The nearest galaxies (in the Local Group, extending out a few Mpc (= millions of parsecs)) are bound together by gravity, and this group is not expanding—the measured radial velocities of these galaxies are entirely due to their motion within the group. Beyond around 100 Mpc, however, most of a galaxy's measured radial velocity is due to universal expansion, and so can be used as a distance measurement.

2.8.2 Distance Scales: A Quick Tour of the Universe

Space is big. The scale of the inner solar system is just at the edge of where everyday analogies are useful: if one could drive a car to the Sun it would take over 100 years to cover that distance (which is defined as 1 au). The four giant planets in the solar system orbit roughly 5, 10, 20, and 30 au from the Sun.

The distance to the nearest star is almost 300,000 au, or roughly[19] 1 pc.

Stars near the Sun are very roughly about 1 pc apart and uniformly randomly distributed. We have a nearly complete census of stars and brown dwarfs out to about 25 pc or so, beyond which we lack good distance measurements for many fainter objects.

These stars form part of the disk of the Milky Way galaxy, which has a thickness of around 300 pc in the vertical direction.[20]

The Sun sits about 8.2 kpc (kiloparsec = 1000 pc) from the Galactic Center, and the disk has a radius of 10s of kpc. It is orbited by small satellite galaxies, the largest of which, the Large Magellenic Cloud, is 50 kpc away.

The Milky Way is part of the Local Group of Galaxies, the most prominent of which is the Great Galaxy in Andromeda, about 750 kpc distant. The diameter of the Local Group is a few Mpc. Beyond this are many other groups of various sizes, one of the most prominent being the Virgo Cluster, about 16 Mpc away.

Beyond this distance, virtually all galaxies appear to be receding from Earth because the effects of universal expansion, and distances are often expressed in terms of *redshift*, expressed as $z = \Delta\lambda/\lambda$ where $\Delta\lambda$ is the Doppler shift of a spectral feature whose rest wavelength is λ. On scales larger than about 100 Mpc, the Universe is roughly homogeneous; the largest structures of galaxies are around 300 Mpc across.

At very large redshifts ($z > 1$, $\sim$3 Gpc) one is looking back approximately 10 billion years, at a time when the Universe was only a few billion years old. At this point, the details of the dynamics of universal expansion become important in interpreting distance. Only very bright galaxies are generally observable at these distances.

At redshifts of $\sim$6–15 we observe the first galaxies forming; beyond this there are very few things to observe until we get to $z \sim 1100$, the light from which originated when the Universe was around 400,000 years old and filled with a nearly uniform

[19] More precisely, the nearest star, Proxima, is 1.3 pc from the Sun, or 4.2 light years. 1 pc is roughly 206, 265 au; the conversion factor is $360 \times 60 \times 60/(2\pi)$, the number of arcseconds in a radian.

[20] The vertical distribution of stars is well described by an exponential with scale height of a few hundred pc, plus additional components of older and younger stars with different scale heights.

plasma. This is the cosmic microwave background (CMB). This plasma was opaque, and so we cannot observe anything beyond this limit.

We can ask what has become of this material that scattered the light from the CMB as we see it in the 14 billion years since the light we see left it. Because of universal expansion, it is now very far from us. It has cooled and undergone gravitational collapse and formed clusters and groups of galaxies, composed of stars and dust and gas, just as the plasma in our part of the Universe did, but because of universal expansion and lookback time we cannot observe that.

Box 2.6: Jargon Review: Distances

- **(Geometric) Parallax**: The apparent motion of nearby stars from the perspective change due to Earth's orbital motion. Also, the magnitude of this effect, which is inversely proportional to distance.
- **Astronomical unit**: Average distance from Earth to the Sun, abbreviated au or AU.
- **Parsec**: Standard unit of distance in astronomy, abbreviated pc, equal to about 206,625 au or 3.26 lt-yr. Often expressed as kiloparsecs (kpc) or megaparsecs (Mpc).
- **Standard candle**: A star or other object whose luminosity (total power output) is known. Cepheid variable stars are the canonical example of a standard candle, being extremely bright giant stars with a well-calibrated period–luminosity relation.
- **Local Group**: The nearest galaxies, bound together by gravity. The Local Group is a few Mpc across, and contains 3–5 large galaxies like the Milky Way.
- **Redshift**: Used as a proxy for distance for distant galaxies, expressed in units of $z = \Delta\lambda/\lambda$.
- **Cosmic microwave background (CMB)**: light from the plasma that pervaded the Universe when it was around 400,000 years old. Redshifted by a factor of 1100, it is detected in the microwave region of the EM spectrum.

2.9 What the Universe Is Made Of

Studies of the Universe on the largest scales and the dynamics of galaxies reveal that most of the energy and mass in the Universe is of as-yet undiscovered forms, called *dark matter* and *dark energy*. ("Dark" in this sense means "not giving off light." It does not imply something is black or that it absorbs light; both dark energy and dark matter are actually transparent.)

The pace of expansion of the Universe appears to be accelerating, which is naturally explained if a free parameter in Einstein's field equations of general relativity, Λ, has a nonzero value. The value needed to explain this acceleration corresponds to a tiny amount of energy in every volume of space. The nature of the dark energy is unclear; it may represent a whole chapter of new physics and be related to how the Universe began a rapid period of inflation; or it may simply be a fundamental property of spacetime with no deeper explanation than that.

The mass budget of the material in the Universe can be estimated via the gravity in galaxies via several methods, including galaxy rotation rates, the dynamics of galaxies and gas in clusters, and the deflection of light via galaxy clusters' gravity.

All three methods give consistent answers: about 90% of the matter in the Universe has no measureable interaction with light, so it is called "dark" matter. This means it is not made of ordinary matter (because ordinary matter is made of charged particles like protons and electrons, which interact very strongly with light).

The most popular explanation of the dark matter is that it is in the form of an as-yet undiscovered particle not accounted for in the Standard Model of physics. It might be a hypothesized particle such as an axion or supersymmetric particle; particle physicists hope it has some small tendency to interact with ordinary matter or photons so that it might be discovered with sufficently sensitive or precise experiments, but in principle it could have no such interactions, and so not be measureable in any way except via gravity.

The measured amount of dark energy plus dark matter adds up to the proper mass–energy density to explain measurements of the Universe's large scale curvature. Numerical simulations of galaxy and structure formation from the earliest days of the Big Bang to today do a remarkably good job of describing the Universe as we observe it with these two components. The predictions of nucleosynthesis in the Big Bang (i.e., the origins of the elements) are also consistent with this model.

This alignment of multiple independent lines of evidence for the amounts of dark energy and dark matter is called the *Concordance Model* of cosmology, or the ΛCDM model.[21]

The dark energy is, as far as we know, uniform and has no effect on anything except to subtly alter the expansion rate of the Universe. The dark matter, being most of the matter in the Universe, dominates the dynamics of galaxies, although on small scales like the solar system it is too weak to detect. Dark matter attracts dark matter via gravity, and so the smooth uniform distribution of dark matter in the early Universe quickly gave way to clumps, sheets, and filaments of dark matter, forming the structure of galaxies we see today.

The rest of the matter in the Universe is almost entirely in the form of ordinary matter: electrons and atomic nuclei. This is often loosely called *baryonic matter* because the electrons don't contribute significantly to the mass budget.[22] This material is gravitationally attracted to dark matter, so tends to pool up in the centers of dark matter *halos*, where it forms the visible parts of galaxies.

Baryonic matter comes in three major forms. The most pervasive is an extremely rare gas of mostly hydrogen and some helium and other elements; most of it actually occurs in *intracluster gas* between galaxies. When the gas within galaxies collapses gravitationally into clouds, stars form.

Within galaxies, most of the baryonic matter is in the form of stars, which we describe in the next section. Stars are the sites where the primordial hydrogen and helium in the Universe get fused into heavier elements that populate the periodic table.

---

[21] ΛCDM stands for "dark energy plus Cold Dark Matter." There were at one point many competing models of dark matter, including "hot," "warm," and "cold" varieties. We now know the cold dark matter model fits the data best, so rarely bother to qualify it as "cold," but the C remains in the name.

[22] Neutrons and protons are baryons, electrons and neutrinos are examples of leptons.

Stars generate energy by fusing hydrogen into heavier elements in their cores. The lowest mass stars produce only helium in their cores, but stars heavier than about half the mass of the Sun through a variety of mechanisms generate the pressure and temperature needed to fuse heavier elements.

When most stars, especially massive ones, near the ends of their lives, they shed their outer layers and return this material enriched in heavier elements into the *interstellar medium*, where they mix with fresh hydrogen and helium falling into the galaxy. When the rarest, most massive stars reach the end of their lives, they can explode as a *supernova,*[23] which offers an additional mechanism by which elements can fuse into heavier elements, and enrich the ISM.

The elements heavier than hydrogen and helium are called *metals* by astronomers.[24] Metals in the interstellar medium form abundant gases such as CO, but also larger molecules and grains, which astronomers call *dust.*[25] Dust forms an important part of dark clouds of gas, and solid objects like planets are formed out of the dust that collects in disks around stars as they form.

In galaxies like the Milky Way, about 90% of the mass in is the form of dark matter, about 9% is in the form of stars, and about 1% is in the form of gas and dust. Planets and other objects orbiting stars are a trivial fraction of the total mass budget.

Box 2.7: Jargon Review: What the Universe is Made of

- **Interstellar medium**: The gas between the stars, from which stars form, made mostly of hydrogen. This gas is enriched with heavier elements via stars.
- **Intracluster medium**: The gas between galaxies in clusters. Most of the baryonic mass in the Universe is in this phase.
- **Baryonic mass**: Molecules, atoms, and ions, or "ordinary matter." Technically only refers to the nuclei but implicitly includes their electrons as well.
- **(Cold) Dark matter**: A form of matter so far detected only via its gravity, known to not be baryonic and having little or no interaction with light or other matter. Composes 90% of all matter in the Universe.
- **Dark energy**: A hypothesized component of the Universe that explains the accelerating rate of universal expansion. Abbreviated Λ for the term in Einstein's field equations it corresponds to.
- **Metals**: Astronomy jargon for elements heavier than hydrogen and helium.
- **Dust**: Large molecules and tiny solid specs of material formed by metals in the interstellar medium.

---

[23] There is another phenomenon resulting from surface hydrogen fusion on a white dwarf in a mass-exchange tight binary system called a nova. These temporary and occasionally periodic brightenings are intrinsically much fainter than supernova explosions, but are more commonly observable with small telescopes or even the naked eye, so were discovered first.

[24] This annoying bit of jargon is because astronomers measure the amounts of these other elements using the most easily detected of them as proxies, usually iron. The "metal abundance" thus tracks the abundance of all elements, including non-metals, which led to all elements being loosely referred to as "metals." Eventually, this jargon was formalized. Sorry.

[25] In cosmology, *all* baryonic matter, including stars and galaxies, is called "dust," in the sense that it has mass but exerts no cosmologically important pressure on the Universe.

2.10 Gravity and Pressure in the Universe

Mass in the Universe is attracted to other mass in the Universe, causing it to collect into structures such as galaxies, stars, and planets, which are supported against gravity by various forces. As material falls inwards toward other material it accelerates, and this velocity will, absent any dissipative forces or direct collisions, be enough for it to escape again. Material must thus shed this kinetic energy as it falls in order to accrete or merge with other material, and this shedding of energy generally generates outgoing light or material.

The gas and dust that formed the first galaxies have internal viscosity that converts their bulk motion into thermal energy, which they can then shed by radiation. This causes them to collapse, and eventually form stars.

Because close interactions among stars are rare and they suffer very little friction with gas and dust, stars generally cannot shed this energy and so once they form in a galaxy they retain most of their kinetic energy for their entire lives. This kinetic energy of stars provides their support against gravity and prevents galaxies from collapsing.

Internally, stars are supported against gravity largely by gas pressure, in which the thermal energy of the atoms causes them to move quickly and collide with each other, collectively holding each other up. This thermal energy is ultimately radiated away at the surface of the star as luminosity, and so without some internal power source the star would lose this support and slowly shrink under the force of gravity.

Stars power themselves by nuclear fusion reactions in their cores to sustain their gas pressure for billions of years, but eventually this power source is exhausted and they must find some other source of support.

When a star has sufficient mass that its gravity is strong enough to overcome ordinary gas pressure enough that matter achieves very high densities, new forces come into play due to the quantum nature of electrons.

Electrons have spin 1/2 which means, by the Fermi exclusion principle, that no two electrons may share the same quantum state. Practically, this means they cannot have a number density higher than the *quantum concentration* dictated by their energy—in order to pack them together more tightly, one must give them more kinetic energy. This means that electrons can resist further compression through what is called *electron degeneracy pressure*. This is the source of pressure support found in white dwarfs (see below), but also in the cores of planets like Jupiter, and even in ordinary materials like metals. It is closely related to the forces that dictate the nature and strengths of the chemical bonds that dictate the densities and other properties of ordinary matter.

At extremely high pressures, the density of electron-degenerate matter becomes so high that the electrons achieve very high energies, and their velocities approach the speed of light. This causes them to become less stiff, and ultimately unable to provide enough support to prevent gravitational collapse. For the kinds of materials found in white dwarfs, this occurs at the *Chandrasekhar mass* of around 1.4 $M_{\odot}$.

An object that collapses in this way will achieve very high densities, and the high energy of the electrons will allow them to undergo nuclear interactions with the protons to form neutrons. Without these electrons around, electron degeneracy pressure descreases, and so the object will then collapse much farther to a density at which the components of the atomic nuclei—almost entirely neutrons at this point—effectively come into contact.

Neutrons are also spin 1/2 fermions, so they can provide a source of pressure akin to electron degeneracy pressure, called *neutron degeneracy pressure*. This is the source of support for neutron stars.

An interesting and important feature of material held up against gravity largely by degeneracy pressure (from electrons or neutrons) is that adding mass to such an object compresses it so much that it actually shrinks instead of growing as one would expect for ordinary matter.[26] A neutron star that accretes matter thus becomes smaller and smaller.

Neutron stars are so small that the escape velocity (the ballistic velocity needed to completely escape from their surface) can be comparable to the speed of light. In particular, a neutron star with about $3M_\odot$ of mass would be small enough that its surface gravity would be greater than the speed of light, and nothing could escape its surface.

At this point, one needs to invoke the machinery of Einstein's theory of general relativity to describe the object. This theory famously predicts that at such extreme gravitational conditions no material, energy, or information can escape an object and it becomes a *black hole*, which we will discuss in Section 2.12.

2.11 Stars

Within galaxies, almost all of the baryonic mass is in the form of stars. Stars are composed predominantly of hydrogen, helium, and a few percent by mass of the rest of the periodic table.

When stars form, they are very bright as they release gravitational potential energy as they collapse, but they are shrouded in dark clouds of dust and gas so are only visible in infrared wavelengths that dust does not absorb well.

Eventually, the temperatures in their cores are high enough to ignite hydrogen fusion, and they settle into hydrostatic equilibrium. They quickly shed the cloud of material around them, except for a *protoplanetary nebula* that forms in a disk of gas and dust around them, from which planets will eventually form.

For most stars, the hydrogen burning phase[27] is the longest of their lives. The highest mass stars fuse their hydrogen so quickly that they run out even before they have fully contracted, in just a few million years. The lowest mass stars have such low pressures and temperatures in their cores that they will not run out of fuel for hundreds of billions of years, far longer than the current age of the Universe.

---

[26] Jupiter has about the same radius as brown dwarfs and the lowest mass stars for this reason, despite those objects spanning a factor of 100 in mass. Only the cores of these objects are degenerate, and the shrinking of their cores as mass increases is balanced by the expansion of their outer layers (because there is more mass, which occupies volume) so that their total volume stays about the same.

[27] Nothing is oxidizing in stars. Astronomers use the term "burning" to describe how the fusion of hydrogen and other "fuel" generates energy, and often refer to the products of that fusion as "ash."

Stars fusing hydrogen in their centers are called *Main Sequence* stars because[28] of a rather tight relationship between their surface temperature and total luminosity: high mass stars have high surface temperatures (10,000 K or higher) and high luminosities, and low mass stars have low surface temperatures (perhaps 3,000 K) and very low luminosities. *Very* roughly, the mass–luminosity relation is $L \propto M^3$, and the mass–surface temperature relation is $T \propto M^{1/4}$. Main Sequence stars are also called *dwarf* stars to distinguish them from *giant* stars.

High mass stars are rare. Roughly speaking, the number of stars formed as a function of mass follows the "Salpeter" *initial mass function* as $dN/dM \propto M^{-2.35}$. They are also rare because they do not last long: most high mast stars that ever formed no longer exist; virtually every low-mass star (roughly less massive than the Sun) that ever formed is still around. But despite being rare, high mass stars contribute most of the luminosity of stars on the Main sequence, because they are so bright. The brightest stars in a stellar population are usually giant stars, which we describe next.

When stars run out of hyrdogen in their centers they begin to lose the thermal energy that supports them against gravity and they begin to contract. This causes their core pressures and temperatures to rise until they begin fusing hydrogen in a shell around a core of helium. This is a more powerful power source than core hydrogen fusion, so the stars grow in brightness, grow in size, and drop in surface temperature.[29] This is the giant phase of stellar evolution, which can last millions of years. Most giant stars have very cool surfaces (red giants). Giant star evolution can be quite complex, especially for high mass stars, involving many changes in size and brightness before final collapse.

The end state of stellar evolution depends almost entirely on the mass of a star.[30] The highest mass stars (greater than around $8M_{\odot}$) have sufficient central pressures and temperatures to fuse a sequence of elements exothermally until they produce an iron core. At that point, they have no further sources of energy to support against gravity and they collapse catastrophically. This produces a supernova explosion,[31] returning much of their mass to the interstellar medium, and enriching it with the nuclear products of central fusion and the explosion itself.

---

[28] This relationship forms a clean locus on a color(temperature)–magnitude(luminosity) diagram. Before stellar evolution was understood, it was thought this was an *evolutionary* sequence, with massive, luminous stars being younger (*early type*) and the other end of the sequence being older (*late type*). We still use this language, although now we understand stars do not move along the sequence at all.

[29] Objects like stars supported by thermal pressure against gravity have negative "gravitothermal" heat capacity. This means that when you add energy to them they grow and cool. An analogy is a satellite: if you add energy to it, it will go into a higher orbit (by analogy with a larger star), and its average velocity will drop (by analogy with lower thermal/kinetic energy of the atoms in a star). By the virial theorem, the average kinetic energy supporting a system bound by gravity is always half of the (negative) gravitational potential energy binding the system. Thus, when you add energy to a gravitational system, it becomes less bound by an amount equal to all of the energy input plus an equal amount converted from its kinetic energy.

[30] Except in tight binary systems. Most stars are bound by gravity to at least one other star, and in some cases stars are so close that when one enters the giant phase its outer envelope can fall within the gravitational influence of the other. In these cases mass exchange can occur, greatly complicating stellar evolution for both stars. Especially if one member of a tight binary is a stellar remnant (see below), such systems can have very high luminosities.

[31] It is possible that some stars collapse completely into a black hole without an explosion.

Not all of the material from such a *core-collapse supernova* is ejected in the explosion, and so the remaining mass collapses into a *compact object*. The most massive stars leave behind compact objects of such high mass that they become black holes. Less massive stars (but more than around $3M_\odot$) have enough leftover material to leave behind a neutron star, having a few solar masses but a radius of around 10 km. Such objects are born rapidly rotating and highly magnetic, and when the geometry is right we can detect them as pulsars (Box 3.1) from the radio emission they send out in opposing beams that sweep the sky as they rotate.[32]

Less massive stars do not explode as a supernova. As giants, they fuse helium in their cores and hydrogen in a shell, and as these energy sources run out they become unstable and very bright, ejecting their outer envelope as a *planetary nebula*.[33] The remnant in this case is a *white dwarf* star, an extremely compact object about the mass of the Sun and the size of the Earth. These are composed almost entirely of either helium (for lower mass stars) or carbon and oxygen (for higher mass stars), often with some hydrogen on their surfaces. Such objects are born very hot and slowly cool, so are visible as faint but very blue objects.

White dwarfs, neutron stars, and black holes formed this way are collectively called *stellar remnants* and *compact objects*.

Figure 2.6 schematically shows some more details of the various paths of stellar evolution for stars of various masses.

Box 2.8: Jargon Review: Stars

- **Main Sequence**: As an adjective, describes a star in the main phase of its evolution, fusing hydrogen it its core. Such stars are also called **dwarf** stars. As a noun, the locus of such stars on a color–magnitude diagram.
- **(Red) giant stars**: A late phase of evolution in which a star brightens and grows as a result of switching from central hydrogen fusion to other modes, such as hydrogen shell fusion or central helium fusion.
- **White dwarf, neutron star, black hole**: the final product of stellar evolution, with masses of one to a few times that of the Sun, and supported by electron degeneracy pressure, neutron degeneracy pressure, and nothing, respectively.
- **Stellar remnant** or **compact object**: A black hole, white dwarf, or neutron star.
- **Planetary nebula**: The final phase of a low mass star's evolution, not to be confused with a **protoplanetary nebula**, in which planets form in a disk around young stars.
- **Supernova**: A cataclysmic explosion that marks the end of a high mass star's evolution, leaving behind a compact object.
- **Initial mass function**: Quantitatively describes the distribution of stellar masses in a star-forming region, with low-mass stars being more common.

[32] The usual analogy for this effect is that of a lighthouse.

[33] So-called because they somewhat resembled planets when viewed by eye through early telescopes. This name is unfortunate: they have little to do with planets and nothing to do with a protoplanetary nebula from which planets form, which in turn should not be confused with a pre-planetary nebula, which is a precursor to a planetary nebula. I'm sorry.

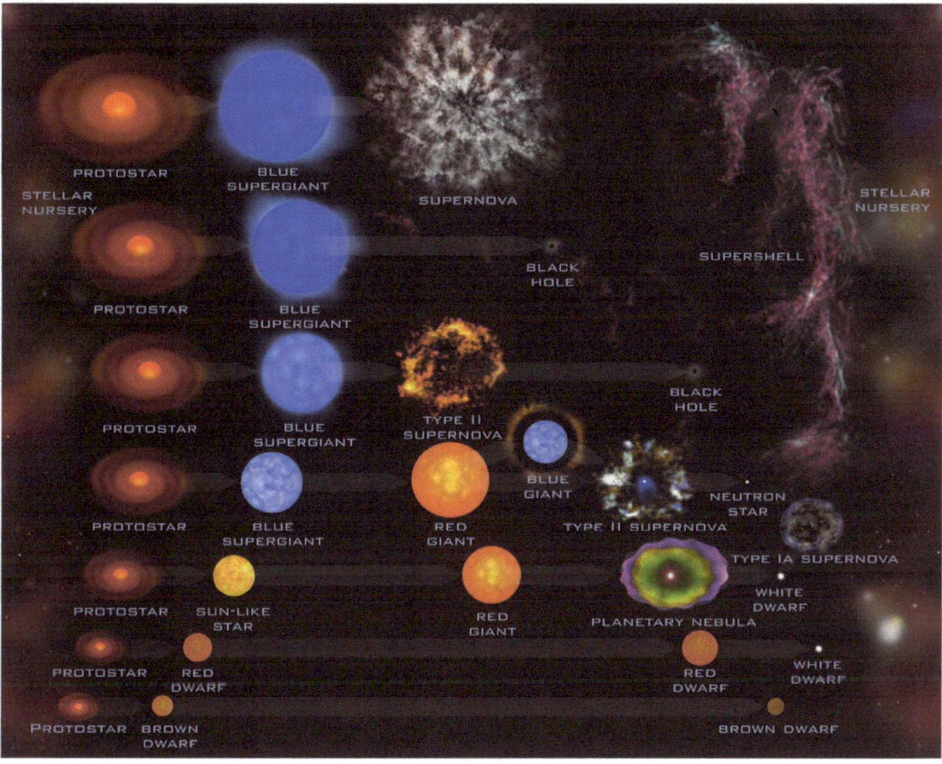

Figure 2.6. Some outcomes of stellar evolution. Image credit: NASA, JPL.

2.12 Black Holes and Active Galactic Nuclei

2.12.1 Black Hole Basics

Despite their portrayal in much science fiction and popular science descriptions, black holes do not possess any special gravitational forces far away from them—they are not "cosmic vacuum cleaners." Rather, it is best to think of them as being extremely small for their mass, such that it is possible to get so close that one cannot escape.

For instance, if one could somehow compress the Earth to half its current radius, its surface gravity would be larger because things on its surface would be closer to the other parts of the Earth. But the Moon's orbit would not change because that depends only on the mass of the Earth, not its size. If one continued the thought experiment and compressed the Earth to its Schwarzschild radius given by:

$$R_{\mathrm{s}} = \frac{2GM}{c^2} \tag{2.20}$$

then the orbit of the Moon and our satellites would not change, but the Earth would seem to vanish. It would have no well defined size, except that there would be a surface called an *event horizon* (with radius R_{s}) within which any matter or energy could not escape. It would be a black hole.

2.12.2 Black Holes in the Universe

Mass falling toward a black hole is accelerated to relativistic speeds, approaching the speed of light near the event horizon. This will generally cause it to be sheared and stretched into a hot plasma that forms an accretion disk around the black hole. The viscosity of this material will heat it to very high temperatures and cause much of it to accrete into the black hole. Before being accreted and trapped within the event horizon forever, this material radiates brightly across the electromagnetic spectrum from X-rays to radio.

Stellar-mass black holes that form after a supernova are sometimes in binary systems, in which case their stellar binary companion might, when it evolves into a giant star, begin to donate mass to the black hole, which then accretes. This results in a bright X-ray source, and indeed this is how the first well-documented black hole, Cygnus X-1, was discovered.

In the early Universe, galaxies formed via mergers and accretion of gas from the intergalactic medium, and along the way some black holes accreted very large amounts of mass. Through a hierarchical process of merging with other black holes, accreting gas, and interacting gravitationally with the stars in their galaxy, these black holes largely ended up in the centers of their galaxies growing to be millions or even billions of solar masses. The Milky Way, for instance, is known to host a supermassive black hole at its center, usually referred to by the name of the associated radio source Sgr A^*.[34]

Supermassive black holes are capable of accreting very large amounts of material, and such material is often abundant in galactic centers, especially when galaxies are growing due to mergers and accretion of gas from the intergalactic medium, which was a common phenomenon in earlier times in the Universe. The intense temperature, radiation and magnetic fields associated with accreting supermassive black holes make them some of the most luminous objects in the Universe: in addition to the light from the accretion disk, there are often jets of material that emerge from the poles of the system.

This radiation is responsible for the phenomenon of *active galactic nuclei*,[35] in which the central region of a galaxy is bright, usually. The details of the spectrum and emission of these objects depend on the angle we view the system, in particular whether we are looking near pole-on into the jet ("down the dragon's throat"), edge-on such that the accretion disk blocks our view of the black hole, or at an intermediate angle.

[34] Pronounced "'Sadge-A-star." The name comes from it being in the constellation Sagittarius, and the star refers to the bright central radio source within a larger region of emission. Demonstrating that Sgr A^* must be a black hole and determining its mass won German astronomer Reinhard Genzel and American astronomer Andrea Ghez the Nobel Prize in Physics in 2020.

[35] There is a whole menagerie of active galactic nuclei, with designations that reflect the nomenclature given to various objects before they were appreciated to be different manifestations of the same phenomenon. These include Seyfert galaxies, quasars, radio galaxies, and blazars.

2.13 Planets

2.13.1 The Classes of Planets and Planet Formation

When stars are forming, not all of the gas manages to be accreted onto the star. Some is left over in a disk of material when the star finally ignites fusion and its radiation begins to blow away this material. Over millions of years, the dust in this disk, called a *protoplanetary nebula* begins to agglomerate until many planetary embryos are in the disk. These become large enough that they can interact with the disk and each other gravitationally.[36]

Some become large enough that they can begin to accrete hydrogen from the disk in a thin envelope, which rapidly increases their mass, leading to runaway accretion until they have exhausted their local supply of gas. This is the formation channel for *gas giants* or *Jovian* planets, with total masses from roughly 100 times the mass of Earth ($M_{\oplus}$) to upwards of 1000 $M_{\oplus}$.

Not all embryos will become large enough to trigger runaway accretion. If they accrete large amounts of volatile gases with some hydrogen, we call them *ice giants*, and Neptune and Uranus are the best examples of this, having around a dozen times the mass of Earth, and a few times its radius. Around other stars, we have discovered a class of planets not present in the solar system: sub-Neptunes, which seem to be scaled down versions of our ice giants.

Closer to the star, where it is too hot for volatiles to form, the refractory elements form ever larger rocky bodies. These are the *terrestrial planets*, composed mostly of metals and silicates, with small amounts of volatiles that form atmospheres. Mercury, Venus, Earth, and Mars are examples of this class, and orbiting other stars we have found examples of even larger terrestrial planets, dubbed *super-Earths*.

Planets less massive than Earth can have lower surface gravities and so can have a hard time retaining atmospheres against stellar wind and radiation.

Planetary systems have a large number of smaller objects that never became part of planets. The largest of these have sufficient gravity to be round and differentiate with their metals falling to the center to form a core, and we call these *dwarf planets*, but most are smaller and irregular. Their compositions reflect what part of the protoplanetary nebula they were formed in, with dry, rocky bodies close to the star and icy bodies farther out.

The rocky bodies we call *asteroids*, and those with sufficient volatile composition that they begin to lose material when warmed as they near the Sun are called *comets*. Far from the Sun, there are a large number of icy bodies that never come near the Sun called *Kuiper Belt Objects*, which include Pluto, Charon, Haumeamea, Eris, and others.

Planets do not necessarily remain at the orbital distance that they were formed at. Interactions with the protoplanetary disk and other material can cause them to

[36] This section describes the *core accretion* model of planet formation. Some large planets may for via other mechanisms, like *disk instability* where they collapse directly out of the disk.

migrate toward or away from their star, and close encounters resulting in planet–planet scattering can even eject planets from the system entirely, resulting in *rogue planets*. In rare cases (around 1% of all stars) a gas giant will migrate in to within a few stellar radii and park close to the star with an orbital period of just a few days. These are called *hot Jupiters*.

2.14 Basics of Exoplanetary Astronomy

2.14.1 Radial Velocities and Transits

Prior to 1995, virtually nothing was known about the prevalence or nature of potentially habitable planets orbiting other stars like the Sun. From 1995 to 2000, astronomers found around 40 planets orbiting other stars (*exoplanets*) via the radial velocity method, measuring the subtle Doppler shift in a star's spectrum due to the star's motion under the influence of their planets' gravity. This method continues to be used to discover and measure the masses of exoplanets, especially giant planets and those close to their host stars.

In 2000, astronomers discovered a planet *transiting* its host star, meaning that its orbit's orientation happened to make it pass between its star and Earth once per orbit. This effect is quite subtle: for a planet the size of Jupiter orbiting a star the size of the Sun, the star's light is dimmed by about 1% during the transit. For terrestrial planets like Earth, the effect would be around 1% of 1% (one part in 10^4), which is an extremely challenging effect to detect, even from space.

This ushered in a new era of exoplanet discovery and analysis, and in 2006 NASA launched the *Kepler* observatory to monitor 100,000 stars for any exoplanets orbiting them in fortuitous configurations amenable to transit detection. *Kepler* discovered thousands of transiting planets, almost all of them orbiting very close to their host stars and so extremely hot. *Kepler* found that most stars host planets, and that smaller planets are generally more abundant than giant planets like Jupiter. *Kepler* was most sensitive to larger planets, and most of its discoveries were of planets intermediate in size between Neptune and Jupiter. Because the transit signals of Earth-sized planets are very weak, *Kepler*'s statistics for terrestrial planets like Earth are poor, and mostly confined to close-in rocky planets orbiting small, M dwarf stars.

Exoplanets that transit their host star as seen from Earth offer an opportunity to study their atmospheres via *transmission spectroscopy*. By comparing the spectrum of starlight during and outside of the transit event, which typically lasts a few hours per orbital period, one can in principle detect the spectroscopic absorption features associated with the light filtered through the small annulus formed by the planet's atmosphere. Figure 2.7 shows how an artist's impression of how exoplanetary transit might look up close, illustrating how subtle the effect is.

2.14.2 Direct Imaging

The gold standard of exoplanet detection is *direct imaging*, meaning that a telescope images the planet as a point of light distinct from its host star. This allows for the

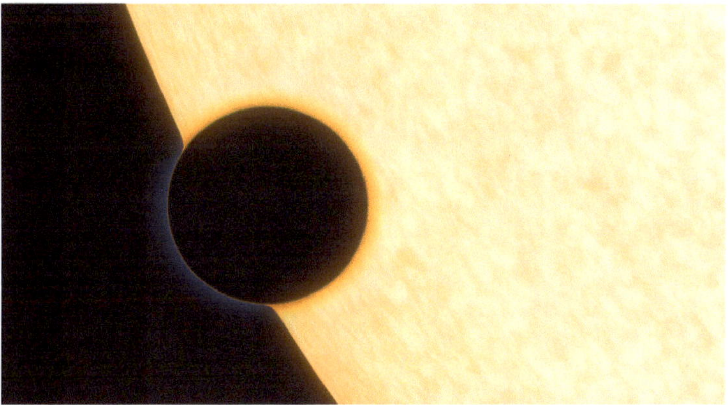

Figure 2.7. Artist's impression of an exoplanet transiting its host star, as seen from Earth. We do not image the transit like this, we merely witness the brightness of the star dim slightly while the planet passes in front of it, completely blocking the light behind it from reaching Earth. There is an even more subtle effect due to the exoplanet's atmosphere, which filters specific wavelengths of light in a thin annulus according to its composition, shown here as an orange haze. Comparison of the spectrum of the star's light in and out of transit can reveal the composition of this atmosphere. Credit NASA/JPL-Caltech.

planetary orbit to be fully characterized, and the reflected light from the planet's atmosphere and surface to be studied spectroscopically.

The maximum apparent angular distance between a planet and its star is given by $\theta = a/d$ where a is its orbital radius and d is the distance to the star. For stars in the Sun's neighborhood—say, within 100 pc—exoplanets orbiting at 1 au are about $0\overset{''}{.}01$ (recall from Section 2.8 that when a is in au and d is in pc, θ is in arcseconds). Since the angular resolution of a 10 m telescope at 1 μ is about $0\overset{'}{.}02$ (Section 2.3.1), only planets orbiting the Sun's very nearest neighbors—at about 10 pc—can be imaged.

But even this understates the magnitude of the problem. While *most* of a star's light appears within the diffraction limit, a significant amount is diffracted to larger angular distances. The extreme brightness contrast between stars and the reflected light of their terrestrial planets—which can approach 10^9—means that this diffracted light must be known and controlled to that precision for the planet to be distinguished. This puts stringent requirements on the figure of the mirror and all optical components, which push the limits of engineering, even for space observatories without the atmosphere to contend with. The instruments which perform this careful management of stray light are called *coronagraphs* because the principles were first used to image and study the corona of the Sun.

The first exoplanets to be imaged in this way were young, very massive, giant planets still glowing brightly at infrared wavelengths from the heat of their formation, and so which had more manageable contrast ratios. Figure 2.8 shows an example of a coronagraphic image of HD 8799, a face-on system of four giant planets orbiting a very nearby star. The scale in the lower right indicates that these orbit quite far from their host star, akin to the outer giant planets in the solar system. The central blotch is the imperfectly-subtracted image of the central star.

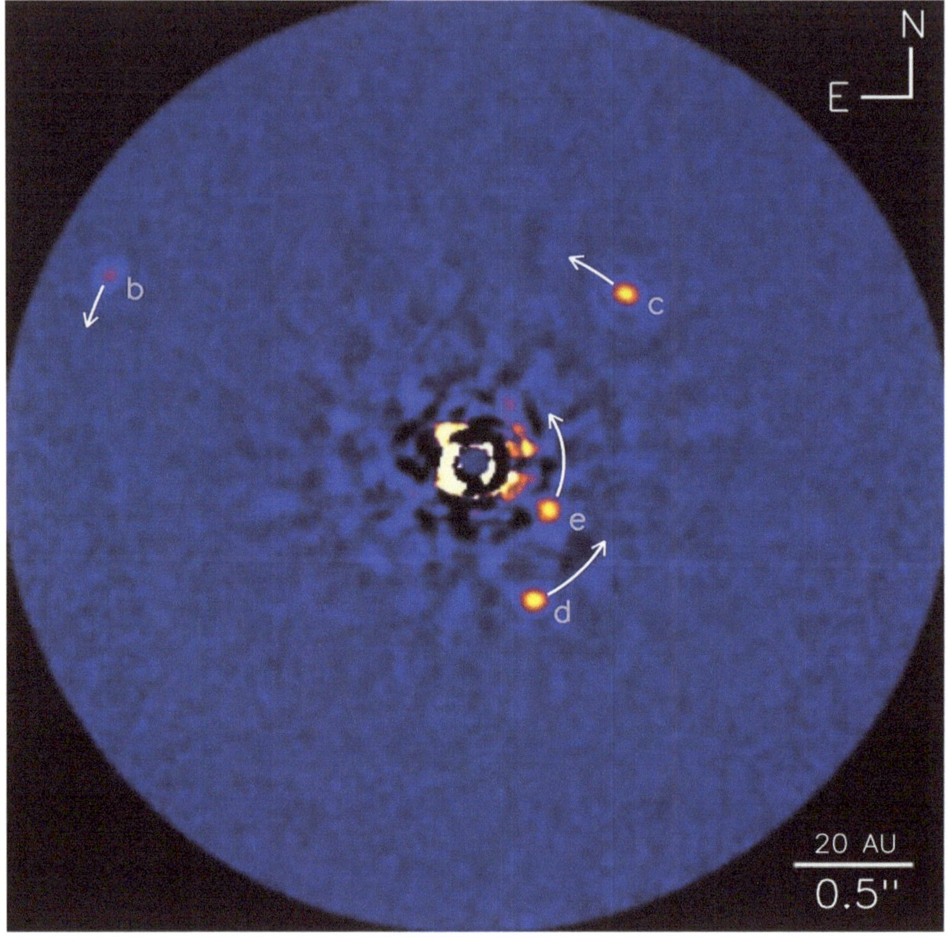

Figure 2.8. Image of the exoplanetary system HR 8799 from the Keck Telescope. The central splotchiness is an imperfectly-subtracted central star. The four labeled points represent four planets and their orbital motion around the star, which have been tracked for over 10 yr. Terrestrial planets in reflected light at 1 au from their stars will have contrasts about 10^3 higher than these planets, requiring more sophisticated coronagraphs and space telescopes. Credit: NRC-HAA, C. Marois, and W. M. Keck Observatory.

2.14.3 Planetary Conditions and Habitability

The temperature of planets and other orbiting material is set by their distance from the Sun, determined by the balance of incoming and outgoing radiation. The simplest expression involves the mean distance from the star, r, and the object's reflectivity or *albedo* a. The amount of incoming radiation depends on the luminosity of the star L and the size of the object R:

$$E_{\text{in}} = \frac{L}{4\pi r^2}\pi R^2(1 - a) \tag{2.21}$$

where we have corrected for the fraction of light that is reflected away.

The outgoing radiation depends on its effective temperature T_{eff} and total surface area:

$$E_{\text{out}} = 4\pi R^2 \sigma T_{\text{eff}}{}^4 \tag{2.22}$$

where σ is the Stefan–Boltzmann constant. Equating these, we calculate:

$$T_{\text{eff}} = \left(\frac{L(1 - a)}{16\pi\sigma r^2}\right)^{\frac{1}{4}} \tag{2.23}$$

The actual temperature on the surface of a terrestrial planet is modified by its atmosphere, which can insulate the surface from radiative cooling effects via opacity in carbon dioxide, water, and other gases. This is the *greenhouse effect*, and warms the surface of the Earth by about 20° compared to its equilibrium temperature.

There is a region around a star where terrestrial planets can support liquid water on their surfaces. The details depend on the planet, but we can define the *Habitable Zone* as the range of distances where the Earth wold have at least some liquid water on its surface.

This is not to say that all planets in the Habitable Zone will have liquid water: the Moon is squarely in the Sun's Habitable Zone but has no water. Nor is it to say that all planets with liquid water will be in the Habitable Zone: planets with different atmospheres might support liquid water elsewhere, and subsurface water might be maintained via internal heat sources, as with the moons of Jupiter and Saturn.

Nonetheless, the Habitable Zone remains a useful target for life-detection experiments because it is approximately the region around a star where we might hope to someday find a planet with liquid surface water that can be analyzed spectroscopically for signs of life.

2.14.4 Newtonian Gravity and Planetary Orbits

For centuries, astronomy was largely the art of tracking the planets'[37] motion. In Europe, Aristotle and others taught that the motion of the planets was perfect and circular. To reconcile this with the decidedly nonuniform motion of the planets, astronomers invoked epicycles (circles upon circles) moving at different speeds. As planetary orbits were tracked longer and more precisely, more and more epicycles were needed, sometimes off center, to describe and predict planetary positions into the future.

Copernicus's model of the solar system was revolutionary because it showed how placing the Sun at the center of the solar system reduced the need for a large epicycle in each of the superior planets' orbits to explain their retrograde motion—when during opposition (when they are opposite the Sun in the sky) they seem to reverse direction. Copernicus's model still relied on epicycles, however, and had no explanatory power.

[37] The original Greek concept of a planet (πλανήτης) was any heavenly object that moved against the fixed background of stars, so included the Sun and the Moon. For a detailed account of how the definition of planet has changed, especially in the 20th century and into the IAU redefinition, see Metzger et al. (2022).

German astronomer Johannes Kepler found that he could explain and predict the orbits of all of the solar system's planets extremely precisely with three mathematical principles, now called Kepler's Laws of Motion:

I Their orbits have the shape of ellipses with the Sun at one focus (a circle is a special case of an ellipse).

II The speeds of the planets along their orbits are higher when closer to the Sun. Specifically: their *angular* speed around the Sun is inversely proportional to their distance.

III The period of the orbit P is related to the semimajor axis a of the ellipse as $P^2 \propto a^3$.

This model was celebrated because of its simplicity and correctness, having no epicycles and requiring only seven parameters to perfectly describe the orbit of each planet (to within the measurement errors of the time).

The English physicist Isaac Newton was the first to properly describe the force of gravity, which he demonstrated by *deriving* Kepler's laws from first principles. Newton argued that all masses attract each other according to the *inverse square law* for gravitation:

$$F = \frac{GM_1 M_2}{r^2} \tag{2.24}$$

where r is the distance between them and F here is the force. From this and Netwon's laws of motion, one can generalize Kepler's first and third laws. The general shape of the motion of *any* two objects, including those that are just passing by each other, is always given by a *conic section*. If the objects are bound, then the shape will be an ellipse or circle. If the objects are not bound (they are just passing by) the "orbit" (really just a path) is that of a hyperbola, and the critical case between these two situations is a parabola.

The third law is generalized for *any* two objects orbiting each other as

$$P^2(M_1 + M_2) = a^3 \tag{2.25}$$

when the units are expressed in years, solar masses, and au. In the limit of planets orbiting the Sun, we have $M_2 \ll M_1$, so $M_1 + M_2 \approx M_1$ and we recover Kepler's version of the law (that is, it appears to be independent of planet mass).

Pace Copernicus, Newton's third law of motion says that the planets also pull on the Sun, and in so doing cause it to move about the center of the solar system. By Newton's second law of motion, $F = ma$, the effect of this on the Sun, that is, the Sun's acceleration, is inversely proportional to its mass. Because the Sun is very massive, its "orbit" about the planets is thus very small, and amounts to a "wobble" about the scale of its own radius. The true center of the solar system is called the *solar system barycenter*, and is calculated as the center of mass of the Sun and all of the objects bound to it.[38]

[38] From βαρύς meaning "heavy," the *barycenter* is the center of mass of a system. Strictly speaking, the a in this version of the equation is the sum of the semimajor axes of the orbit of the planet about the solar system barycenter and that of the Sun about the same.

Finally, Newton's formulation also reveals that the planets pull on *each other*, so Kepler's first law technically only applies for a single planet. Over long timescales, these much weaker forces can alter the orbits of the planets, making them slowly change shape.

Box 2.9: Jargon Review: Planets

- **Gas giants** or **Jovian planets**: The largest planets, composed mostly of thick hydrogen envelopes.
- **Hot Jupiters**: Gas giants in very short period orbits, just a few days long.
- **Ice giants**: The next largest planets, composed of significant amounts of hydrogen must mostly volatile elements like oxygen, nitrogen, and carbon.
- **Terrestrial planets**: The smallest class of planets, being almost entirely rocky, often with metallic cores, and at most a veneer of atmosphere.
- **Super Earths**: Terrestrial planets larger than Earth. We have no examples of these in the solar system but they are common elsewhere.
- **Dwarf planets**: Small icy or rocky bodies too small to be considered planets by the IAU but still round.
- **Migration**: The slow change in the size of a planet's orbit, usually early in its life due to interactions with its protoplanetary disk.
- **Asteroids**: Small rocky bodies in the inner part of the solar system.
- **Comets**: Small icy bodies that sublimate and grow comae and tails when coming close to the Sun.
- **Kuiper Belt Objects**: Icy bodies beyond the orbit of Neptune.
- **Albedo**: The fraction of light that strikes a body that is reflected.
- **Greenhouse effect**: The warming of a planetary surface because of the radiative insulating effect of gases in its atmosphere.
- **Habitable zone**: The region around a star where a planet like Earth would have at least some liquid water on its surface.
- **Kelper's laws**: Describe the motions of planets in orbit around the Sun. The third law relates orbital period and size, $P^2 \propto a^3$.
- **Inverse square law of gravitation**: The gravitational force between two objects drops as the square of the distance between them.
- **Solar System barycenter**: The center of mass of all things in the solar system. The Sun moves slightly about this barycenter due to the gravitational pulls of the planets.

2.15 Further Reading

General astronomy: Carroll & Ostlie (2007), Maoz (2016)
Photometry and magnitudes: Hogg (2022)

Exercises

1. Parallax
   Derive, using a diagram, Equation (2.19), for both sets of units. Some hints:
   - You may assume that the star is much farther away from the Sun than the Earth is. This means you'll do an approximation at some point, which you should justify with real numbers.

- You are measuring the change in the apparent position of the star as measured from Earth compared to objects at infinity (i.e., background objects with unmeasurably small parallax).
- For this derivation, you should assume the Earth is in a circular orbit with radius 1 au.
- You'll need to look up the value of the au in meters. This problem is equivalent to *deriving* the value of the parsec in meters, so you can use the known value to check your answer.

2. Kepler's third law and detecting exoplanets.
   (a) Starting with Newton's second law, $F = ma$, Newton's law of gravitation, and the acceleration of an object in uniform circular motion, $a = v^2/r$, derive Kepler's third law for the special case of a circular orbit. Just solve for the motion of the planet; treat the star as fixed for now.
   (b) Now that we've solved for the planet's motion, solve for the Sun's motion due to the gravity of the orbiting planet. What is the velocity of the Sun's motion, in terms of the masses of the star and planet, and the orbital distance of the planet? *Hint: Unlike in the last part, the radius of the Sun's orbit is different from the distance to the source of gravity, i.e., the planet—in fact you may assume it is much, much smaller. You may also assume the Sun will move in a small "counter-orbit" such that the distance to the planet is constant.*
   (c) What is the orbital velocity of the Sun due to the gravitational influence of Jupiter? Of Earth? Of a hot Jupiter orbiting a Sun-like star? (You'll need to approximate this last one).
   (d) What is the magnitude of the Doppler shift for each of the above three cases? How does this compare to the resolution of a typical spectrograph?
   (e) State of the art spectrographs can be calibrated well enough that shifts in stellar spectral lines can be measured at a precision 1000 to 10,000 times better than a resolution element. Which sorts of planets can be detected around other stars with this method?

References

Carroll, B. W., & Ostlie, D. A. 2007, An Introduction to Modern Astrophysics (2nd ed.; San Francisco: Addison-Wesley)

Hogg, D. W. 2022, arXiv:2206.00989

Maoz, D. 2016, Astrophysics in a Nutshell: Second Edition, In a Nutshell (Princeton, NJ: Princeton Univ. Press)

Metzger, P. T., Grundy, W. M., Sykes, M. V., et al. 2022, Icar, 374, 114768

AAS | IOP Astronomy

The Search for Extraterrestrial Intelligence
Theory and practice
Jason T. Wright

Chapter 3

History

> Bernard of Chartres used to compare us to [puny] dwarfs perched on the shoulders of giants. He pointed out that we see more and farther than our predecessors, not because we have keener vision or greater height, but because we are lifted up and borne aloft on their gigantic stature.
>
> — John of Salisbury, *The Metalogican* (1159), as translated in McGarry (1955)

3.1 Pre-history of SETI: Before 1959

3.1.1 Plurality of Worlds Prior to Copernicus and Galileo

The idea that there might be other conscious creatures throughout the cosmos on other planets is quite old, and is called *cosmic pluralism* or the *plurality of worlds*.

Western thought on the possibility of ordinary life elsewhere in the cosmos goes back at least as far as the ancient Greek atomists such as Leucippus, Democritus, and Epicurus. The last of these wrote in his *Letter to Herodotus*

> …there are infinite worlds both like and unlike this world of ours. For the atoms being infinite in number…are borne far out into space. For those atoms…have not been used up either on one world or on a limited number of worlds, nor on all the worlds which are alike, or on those which are different from these.
>
> …We must believe that in all worlds there are living creatures and plants and other things we see in this world…
>
> —quoted in Crowe (1986), 3

doi:10.1088/2514-3433/adf5aech3

Their influence was extensive. For instance, the Roman poet Lucretius described the Doctrine of Infinite Worlds in *De Rerum Natura* (On the Nature of Things), and this position is famously described by Plutarch in the *Moralia*:

> It is reported that King Alexander the Great, hearing Anaxarchus the philosopher discoursing and maintaining this position: That there were worlds innumerable: fell a-weeping: and when his friends and familiars about him asked what he ailed. Have I not (quoth he) good cause to weep, that being as there are an infinite number of worlds, I am not yet the lord of one?
>
> —*On Tranquillity of Mind*, translation by Philemon Holland, 1603.

Interpreting such writings prior to Copernicus can be tricky, because his work inspired people to take the possibility of other planets as worlds like Earth quite seriously and literally. Before this, the other "worlds" in question were not necessarily other planets like Earth, but often entirely other cosmoses (κόσμοι) beyond human ken. In other contexts, there was not necessarily an important distinction between aliens in outer space, and a more spiritual or metaphorical sense of divine beings in the heavens.

That said, Crowe (1986) (Chapter 1 footnote 4) quotes Grant McColley quoting Eusebius:

> Heraclitus and the Pythagoreans consider that every star is a world in an infinite aether, and encompasseth air, earth, and aether. This opinion is current among the disciples or Orpheus, for they suppose that each of the stars does make a world.

Regardless of whether these "worlds" were planets as we think of them or abstract systems beyond our perception, the atomists did not dominate Western thought after the rise of Christianity; that honor went to Aristotle, who taught that the heavenly bodies were of different stuff than the Earth, which uniquely sat at the center of the Universe. This was compatible with Christian belief that Earth and the Universe were God's special creation for the benefit of humans, specifically. In the Aristotelian framework of Christianity, heavenly beings were necessarily divine.

Unsurprisingly, the influence of the atomists led some medieval Islamic scholars to discuss the plurality of worlds, although there as in Europe the Aristotelian view largely dominated. Weintraub (2014) writes:

> The Qur'an an does not leave the faithful guessing as to whether extraterrestrial life exists. As early as the first century after the birth of Islam, one of the most revered of the early leaders of Islam, Imam Muhammad al-Baqir (676–733 CE), wrote, "Maybe you see that God

> created only this single world and that God did not create Homo sapiens besides you. Well, I swear by God that God created thousands and thousands of worlds and thousands and thousands of humankind.
>
> —citing Golshani (2002) for the quoted translation.

Eastern traditions, such as Daoism and Mahayana and Vajrayana Buddhism, similarly placed sentient beings in the heavens, and have stories of them visiting Earth and communing with humans. In some cases, these beings are even associated with specific places among the stars. From the Shiyi Ji[1] (拾遺記, “Record of Retrieved Traces”) a Chinese text containing apocryphal stories dating from the first centuries CE during the Han era and beyond, compiled by the Daoist scholar, Wang Jia (王嘉, died 390 CE):

> After [legendary, ancient] Emperor Yao had ascended the throne for 30 years, a giant raft began floating in the Western Sea. Aboard the raft was a light which shone at night and was extinguished during the daylight hours. The people of the sea would turn toward the light, which was flickering at once brightly then dimly, as though the stars and moons were entering and exiting [the craft]. The raft revolved regularly around the Four Seas, every 12 years constituting one heavenly revolution, and starting again after completing its revolution. Its name was the “Moon-piercing Raft,” but it was also called the “Star-suspended Raft.” Feathered people perched and rested on it. Flocks of xian-immortals gargled with its morning dew collected in their jowls, and the gleam of its Sun and moon was akin to the Deep Dark.[2] After the time of Emperor Shun and the Xia, there were no more records of its appearance. It was probably the people who swam in the seas who transmitted [knowledge of] these divine prodigies.

But again, since these can be interpreted as divine beings, it is perhaps not really what we mean when contemplating aliens from other planets.

At any rate, the plurality of worlds was also an explicit topic of Chinese philosophy later as well. Needham (1959) translates the 13th century philosopher Deng Mu (鄧牧) writing in the Bo yo quin (伯牙琴):

> Heaven and Earth are large, yet in the whole of empty space they are but as a small grain of rice…It is as if the whole of empty space were a tree and heaven and earth were one of its fruits. Empty space is like a kingdom and heaven and earth no more than a single individual person

[1] Retrieved from the Chinese Text Project at https://ctext.org/ in 2023. Translations and annotation here by Erica Brindley.

[2] A reference to ming (冥), a description of the Daoist notion of the Dao or the greatest cosmic mystery.

> in that kingdom. Upon one tree there are many fruits, and in one kingdom many people. How unreasonable it would be to suppose that besides the heaven and earth which we can see there are no other heavens and no other earths!

3.1.2 Plurality of Worlds After Copernicus and Galileo

When wide appreciation of Copernicus's work and Galileo's telescopic observations of the Moon and Venus moved Earth from the center of the Universe to being just one of many planets in many people's minds, the possibility of ordinary beings on other planets moved back into Western thought, and the plurality of worlds came to be a heresy under the doctrines of the Roman Catholic Church.

The most famous heretic to proclaim that aliens existed was the influential philosopher and gadfly Giordano Bruno, a heterodox priest who frequently defied church authority by promoting his interpretations of Copernicus and Galileo's works. He wrote:

> There are countless suns and countless earths all rotating round their suns in exactly the same way as the seven planets of our system. We see only the suns because they are the largest bodies and are luminous, but their planets remain invisible to us because they are smaller and non-luminous. The countless worlds in the universe are no worse and no less inhabited than our earth. For it is utterly unreasonable to suppose that those teeming worlds which are as magnificent as our own, perhaps more so, and which enjoy the fructifying rays of a Sun just as we do, should be uninhabited and should not bear similar or even more perfect inhabitants than our earth.
>
> —As quoted in Bettex (1965)

Bruno, of course, had no empirical reason to believe this, so was hardly a paragon of the scientific method, but his execution by burning at the stake in 1600 CE for his heretical philosophies, including the plurality of worlds, has made him a martyr to the principles of freedom of thought (Martinez 2016).

Note that this is not to say that the idea of alien minds in the heavens is anathema to Roman Catholic theology. No less an authority than Albertus Magnus, one of the Doctors of the Church, mused in the 13th century:

> Do there exist many worlds, or is there but a single world? This is one of the most noble and exalted questions in the study of Nature.
>
> —Attributed in Bless (1996)

although he was likely thinking more abstractly on questions of angels than undertaking a literal consideration of other terrestrial planets in space. Indeed, on the topic his own conclusion was that God did not, in fact, create other worlds (Dick 1982).

And despite the Bruno affair, and contrary to popular perception, today Roman Catholic theology is largely receptive to the possibility that other intelligent species exist in space (Vakoch 2000).

Soon after Bruno's execution, speculation about life on the Moon became more common among astronomers. English Anglican bishop John Wilkins of England wrote popular accounts of the new telescopic observations of the heavens, including *The Discovery of a World in the Moone* in 1638, in which he argued the Moon was likely inhabited. Going farther than this, he considered the parameters required for a vessel that could actually make the journey to the Moon (based on the pre-Newtonian physics known at the time, and a belief in human technological progress).

An influential contribution to the topic was Bernard le Bovier de Fontenelle's *Entretiens sur la pluralité des mondes* (Conversations on the Plurality of Worlds). Published in the 1680s, it was an accessible and widely read and translated discussion on the topic.

The Dutch astronomer Christiaan Huygens wrote *Kosmotheoros* in 1697, in which he deduced the nature of intelligent beings in the solar system. He discounted the Moon as a good place for life, but imagined that technological civilizations would thrive on the other planets. He reasoned that they would be physically similar to humans, develop astronomy and other sciences, have telescopes that can observe the Earth, and that they would have trade and other social structures similar to 17th century Holland.

3.1.3 Communicating with Martians, Venusians, and Lunarians

3.1.3.1 Early Proposals

Some of the first proposals to communicate with extraterrestrial technological societies were on Mars, Venus, and the Moon, which were thought by many people (if not many scientists) to be not just habitable but inhabited, sometimes motivated by claims of telescopic observations of vegetation or structures on these worlds.

The German polymath Carl Gauss invented the heliotrope, which used mirrors to redirect sunlight as a signaling mechanism over great distances. In 1822, he suggested to the German astronomer Heinrich Olbers that a sufficient number of large mirrors would be sufficient to signal the surface of the Moon, including any inhabitants there. He was apparently quite excited about this idea, writing "This would be a discovery even greater than that of America, if we could get in touch with our neighbors on the Moon." Occasional reports of flashes seen on the Martian surface would thereafter be considered by some to be evidence of Martians' attempts to communicate with Earth.

Gauss is also widely credited with a popular idea at the time to use patterns of trees across a huge expanse to signal neighboring planets' astronomers that technological life was here. The specific proposal mentioned is usually to plant trees in the outline of three squares touching at their corners to form a right triangle, as a common figure in a visual proof of the Pythagorean theorem. This shape would be huge, spanning a large part of the Siberian tundra, so as to be visible from the Moon or Mars. There is a similar proposal attributed to his contemporary, the Austrian astronomer Joseph von Littrow, to fill a circular canal 30 km across with kerosene and light it ablaze at night.

The French inventor and poet Charles Cros, perhaps best known for having a claim to the priority for both color photography and sound recording and playback, was convinced of the reality of Martian technological life. In his 1869 book *Études sur les moyens de communication avec les planètes* (Studies on Means of Communication with the Planets) he imagined getting Martians' or Venusians' attention via a giant Earthbound mirror that would focus sunlight on the surface of the other planet. The intense light would burn the surface of the planet in intelligible patterns that would inform our neighbors of our presence (if not our good intentions).

The possibility of interplanetary communication became popularized as the 19th century went on. In 1891, Camille Flammarion, a proponent of the plurality of worlds hypothesis, announced a prize of 100,000 francs to anyone who "will find the means within the next ten years of communicating with a star (planet or otherwise) and receiving a response" (quoted in Vakoch & Dowd 2015).

3.1.3.2 Lowell and the Martian "Canals"

The reality of Martian civilization became even more popular in the early 20th century thanks to the efforts of American astronomer Percival Lowell.

Lowell, a largely self-taught "gentleman astronomer" and scion of the wealthy Boston Lowell family, became interested in Mars in his late thirties and dedicated much of his life to studying it and documenting its civilization. He was inspired in part by the maps of Mars made by the Italian astronomer Giovanni Schiaparelli.

Schiaparelli's maps (see Figure 3.1) included many linear features he referred to, in Italian, as *canali*, meaning "channels" or "gullies," but also plausibly "canals." Schiaparelli suggested these were waterways that moved water about the Martian surface, and was a proponent of the idea that Mars had significant biota. In English, the term was translated as "canals," implying that they were artificial waterways created by Martians.

Lowell eventually founded his eponymous observatory in Flagstaff, Arizona and began extensively mapping the canals and what he believed was vegetation on the Martian surface (see Figure 3.2). He became convinced that he was documenting alien technological life. His popular books, especially *Mars and its Canals*, had a large influence on public perceptions of Mars and convinced many it was inhabited. These ideas did not find much purchase in the scientific community, but there were exceptions, for instance Soviet astronomer G. A. Tikhov, founder of "astrobotany" (e.g., Tikhov 1949; Gindilis & Gurvits 2019).

As late as the early 1960's, linear features were prominent in the best maps of Mars being made, by American planetary scientist Earl Slipher at Lowell Observatory for the US government (see Figure 3.3). Today we understand that

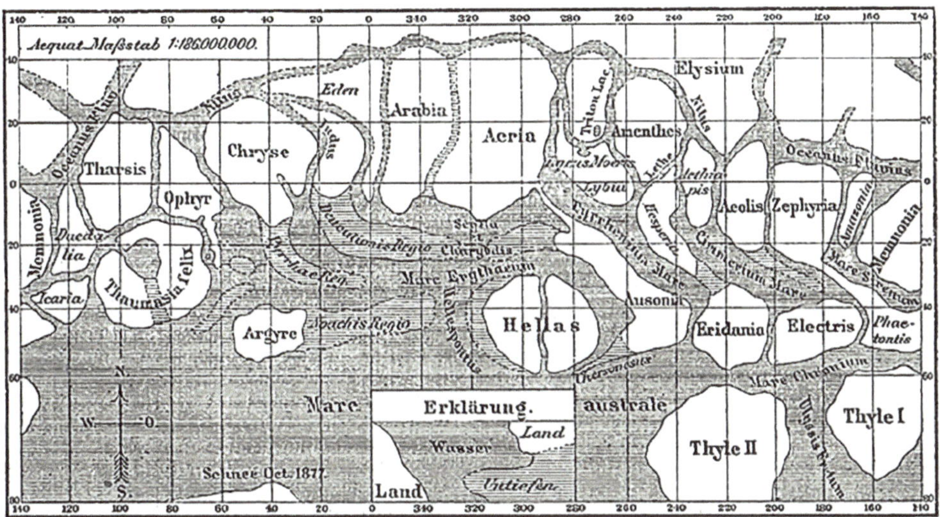

Figure 3.1. One of Schiaparelli's maps of the Martian surface, showing canals. The author obtained this image from the Wikimedia website, where it is stated to have been released into the public domain.

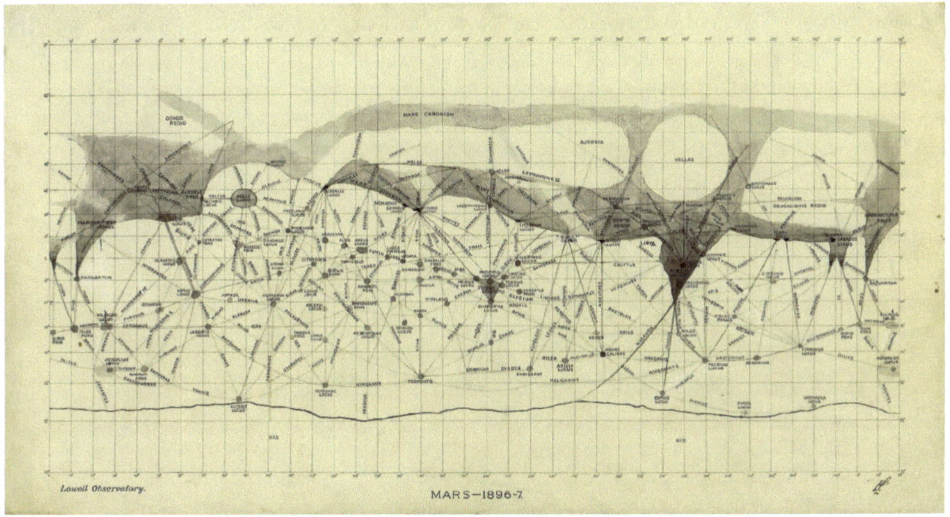

Figure 3.2. Lowell's 1896 map of Mars and its canals. Image credit: Lowell Observatory Archives.

no such linear features actually exist on Mars, and Lowell's skeptics were correct that his maps amounted mostly to wishful thinking.

3.1.3.3 Communicating with the Martians

The development of wireless communication via radio waves by Serbian-American inventor Nikola Tesla, Italian inventor Guglielmo Marconi, and others revealed the possibility that technological contact might be possible with beings from space, and Mars was the natural target of such work.

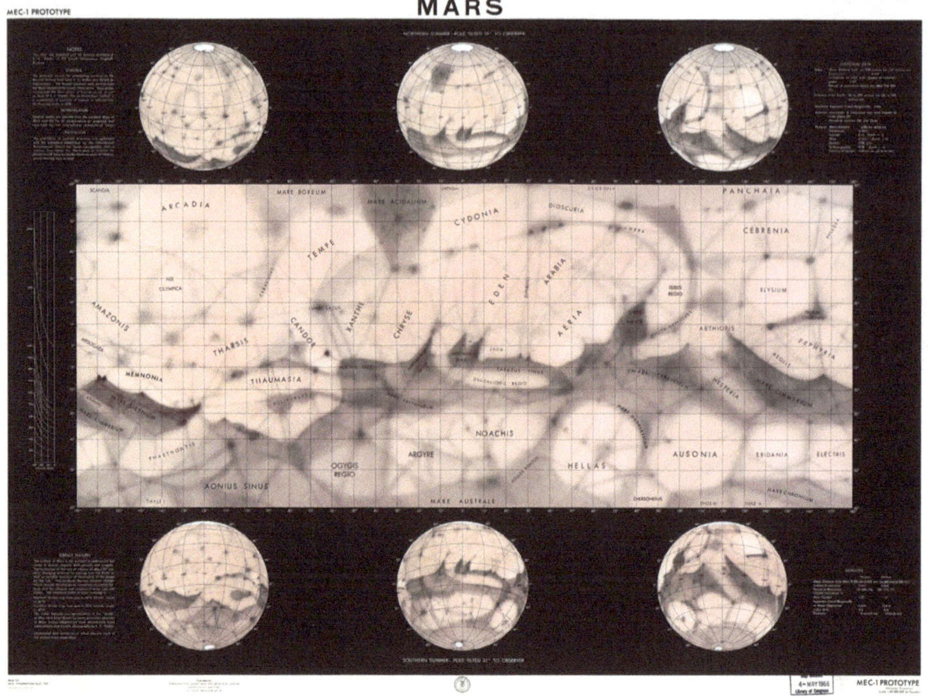

Figure 3.3. 1965 map by Earl Slipher of Mars, based on his own observations and older maps. It notes: "the linear features are representative of the 'canals' as they have been drawn by many prominent observers of Mars." Image Credit: Library of Congress, Geography and Map Division.

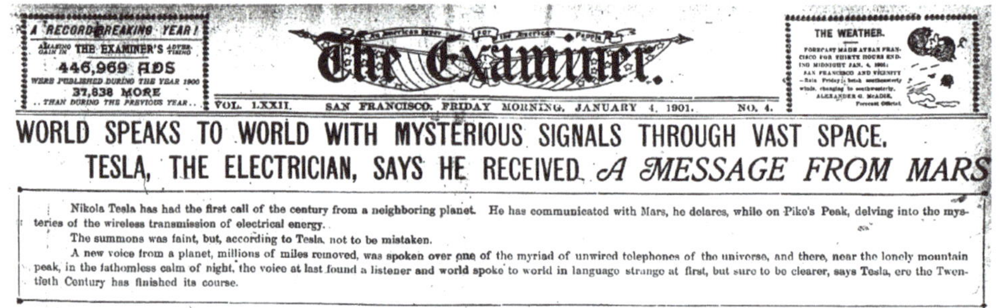

A RECORD BREAKING YEAR!
THE EXAMINER'S
446,969 ADS
WERE PUBLISHED DURING THE YEAR 1900
37,838 MORE
THAN DURING THE PREVIOUS YEAR

The Examiner.

THE WEATHER.

VOL. LXXII. SAN FRANCISCO, FRIDAY MORNING, JANUARY 4, 1901. NO. 4.

WORLD SPEAKS TO WORLD WITH MYSTERIOUS SIGNALS THROUGH VAST SPACE.
TESLA, THE ELECTRICIAN, SAYS HE RECEIVED A MESSAGE FROM MARS

Nikola Tesla has had the first call of the century from a neighboring planet. He has communicated with Mars, he dclares, while on Pike's Peak, delving into the mysteries of the wireless transmission of electrical energy.

The summons was faint, but, according to Tesla, not to be mistaken.

A new voice from a planet, millions of miles removed, was spoken over one of the myriad of unwired telephones of the universe, and there, near the lonely mountain peak, in the fathomless calm of night, the voice at last found a listener and world spoke to world in language strange at first, but sure to be clearer, says Tesla, ere the Twentieth Century has finished its course.

Figure 3.4. Headline in the 1901 January 4 *San Francisco Examiner* announcing Nikoli Tesla's "discovery" of radio waves from Mars.

Both Tesla and Marconi believed that they had detected radio waves from Mars, having heard signals they could not explain when Mars was at opposition (i.e., when it was closest to the Earth in their respective orbits). Both had very high standing in the scientific community and their discoveries were widely reported (Figures 3.4 and 3.5).

These efforts caught the attention of American astronomer David Peck Todd, who devised a plan to record these signals in a permanent way during the next close approach of Mars in 1924. He worked with American inventor Charles Francis

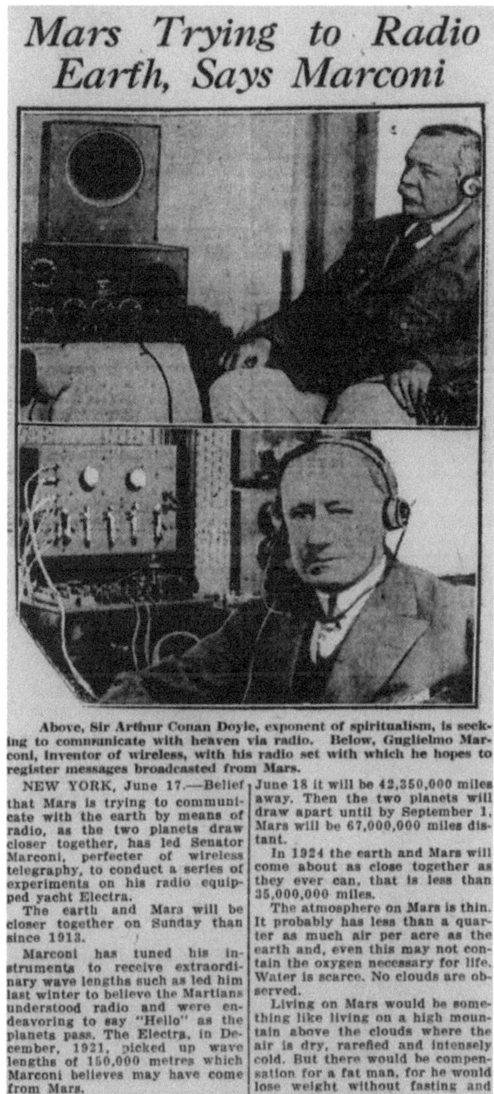

Mars Trying to Radio Earth, Says Marconi

Above, Sir Arthur Conan Doyle, exponent of spiritualism, is seeking to communicate with heaven via radio. Below, Guglielmo Marconi, inventor of wireless, with his radio set with which he hopes to register messages broadcasted from Mars.

NEW YORK, June 17.—Belief that Mars is trying to communicate with the earth by means of radio, as the two planets draw closer together, has led Senator Marconi, perfecter of wireless telegraphy, to conduct a series of experiments on his radio equipped yacht Electra.

The earth and Mars will be closer together on Sunday than since 1913.

Marconi has tuned his instruments to receive extraordinary wave lengths such as led him last winter to believe the Martians understood radio and were endeavoring to say "Hello" as the planets pass. The Electra, in December, 1921, picked up wave lengths of 150,000 metres which Marconi believes may have come from Mars.

June 18 it will be 42,350,000 miles away. Then the two planets will draw apart until by September 1, Mars will be 67,000,000 miles distant.

In 1924 the earth and Mars will come about as close together as they ever can, that is less than 35,000,000 miles.

The atmosphere on Mars is thin. It probably has less than a quarter as much air per acre as the earth and, even this may not contain the oxygen necessary for life. Water is scarce. No clouds are observed.

Living on Mars would be something like living on a high mountain above the clouds where the air is dry, rarefied and intensely cold. But there would be compensation for a fat man, for he would lose weight without fasting and

Figure 3.5. Article in the 1922 June 17 Norfolk Post about Marconi's attempts to confirm what he believed were radio signals from Mars. Note the juxtaposition with Doyle (of *Sherlock Holmes* fame) using similar equipment to listen for spirits. Courtesy of the Library of Virginia, 800 East Broad St. Richmond, VA.

Jenkins, who had devised a way to produce a photographic image from signals received on a radio, and convinced the US Navy to help. The Navy launched a dirigible with an antenna pointed at Mars that relayed signals to Jenkins' radio, and also issued a telegram instructing shore transmitters to cease transmissions for three days and receivers to note "any electrical phenomenon [of] unusual character" during that time. Figure 3.6 shows the result of Jenkins' radio work, which was at the time widely promoted as tantalizing evidence of a radio signal from Mars (although not by Jenkins, who dismissed such suggestions).

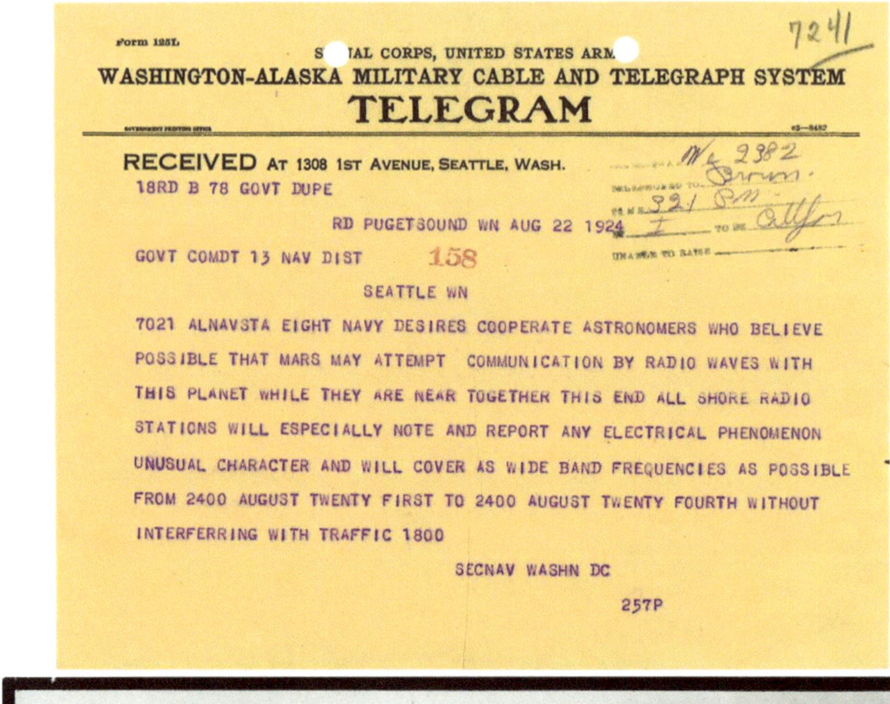

Form 125L

S IAL CORPS, UNITED STATES ARM

WASHINGTON-ALASKA MILITARY CABLE AND TELEGRAPH SYSTEM

TELEGRAM

RECEIVED AT 1308 1ST AVENUE, SEATTLE, WASH.

18RD B 78 GOVT DUPE

RD PUGETSOUND WN AUG 22 1924

GOVT COMDT 13 NAV DIST 158

SEATTLE WN

7021 ALNAVSTA EIGHT NAVY DESIRES COOPERATE ASTRONOMERS WHO BELIEVE POSSIBLE THAT MARS MAY ATTEMPT COMMUNICATION BY RADIO WAVES WITH THIS PLANET WHILE THEY ARE NEAR TOGETHER THIS END ALL SHORE RADIO STATIONS WILL ESPECIALLY NOTE AND REPORT ANY ELECTRICAL PHENOMENON UNUSUAL CHARACTER AND WILL COVER AS WIDE BAND FREQUENCIES AS POSSIBLE FROM 2400 AUGUST TWENTY FIRST TO 2400 AUGUST TWENTY FOURTH WITHOUT INTERFERRING WITH TRAFFIC 1800

SECNAV WASHN DC

257P

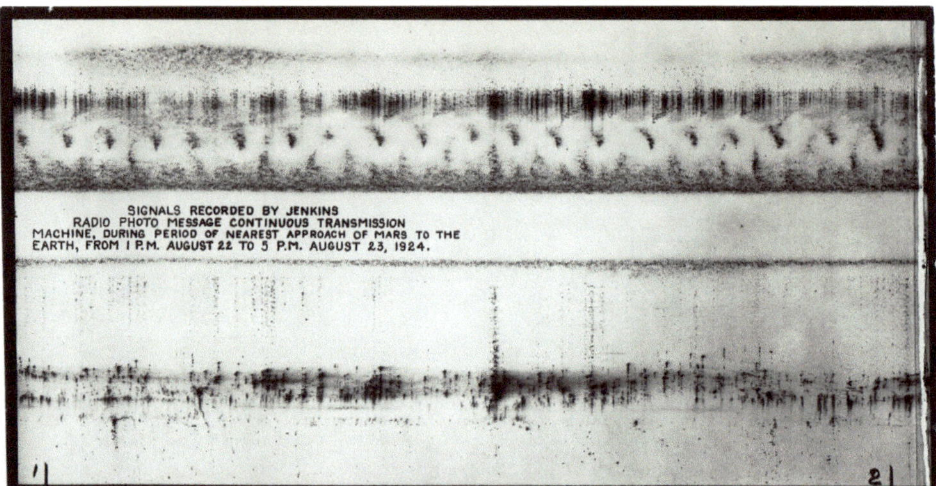

Figure 3.6. *Top:* Telegram signed by the Secretary of the US Navy instructing the cessation of transmissions for three days and to monitor for any potential transmissions from Mars during its closest approach to Earth in 1924. *Bottom:* Recording of radio signals during the 1924 experiments by Charles Jenkins. Top image credit: The U.S. National Archives and Records Administration. Bottom figure image credit: David Peck Todd Papers (MS 496B). Manuscripts and Archives, Yale University Library.

Today we understand that the frequencies they were working at were far too low to penetrate the Earth's ionosphere and so these signals must have had some sort of terrestrial origin.

3.2 1959–1969: The Birth of Modern SETI

3.2.1 Radio SETI

3.2.1.1 Cocconi and Morrison

In 1959, Italian particle physicist Giuseppe Cocconi was working on problems of gamma-ray astrophysics with MIT physicist Phillip Morrison[3] when their conversation turned to the problem of interstellar communication with alien species. Cocconi proposed gamma-rays as a likely medium for communication, but Morrison pointed out that radio probably worked better. The two of them did the math, and wrote a short paper for *Nature* entitled "Searching for Interstellar Communications" (Cocconi & Morrison 1959).

The paper laid out the logic of interstellar communication via radio, and pointed out that existing and planned telescopes and transmitters would be capable of two-way communication at a distance of a few light years. The 21 cm line of hydrogen had just been discovered in 1951 and was still the only radio emission line known, and they suggested that since it lay in a region with especially low background continuum radio emission that this would be a good channel to check. They proposed focusing first on the nearest Sun-like stars, including τ Ceti and ϵ Eridani, among others.

Meanwhile, at the new National Radio Astronomical Observatory (NRAO) in Green Bank, West Virginia, Frank Drake (Box 8.1) had already begun Project Ozma, essentially and coincidentally the very project Cocconi & Morrison (1959) had described. Drake had concluded that the new 85 foot Tatel telescope equipment he was using could detect artificial radio signals from nearby stars, and so he convinced his superiors, including Llody Berkner, the acting observatory director, to give him a small budget and let him repurpose some equipment used for magnetic field measurements to observe τ Ceti and ϵ Eridani.

As Drake and observatory engineers worked on the hardware for the project, the observatory gained a permanent director, the Russian-American astronomer Otto Struve. Struve was extremely hesitant to advertise Drake's work on Project Ozma,[4] lest it delegitimize the work going on at the new facility. But the large amount of positive press and publicity showered on Cocconi and Morrison for their paper changed this reluctance to annoyance that NRAO had been "scooped" on the idea. Struve began to advertise the project, and Drake received many hours of telescope time for it.

The publicity paid off, and soon Microwave Associates had provided Drake with new state-of-the art amplifiers for his work.

[3] Morrison had a distinguished career, including playing key roles in the Manhattan Project, and later becoming a vocal champion of nuclear non-proliferation. He was a prolific science popularizer, and his distinctive voice narrates the classic 1977 short film *Powers of Ten*, https://www.youtube.com/watch?v=0fKBhvDjuy0.

[4] Struve apparently misunderstood the term to be an acronym, not catching Drake's reference to the Oz books, see Section 18.2.

Box 3.1: LGM-1

Jocelyn Bell (Burnell) in 1967. The author obtained this image from the Wikimedia website, where it is attributed to Roger W. Haworth and made available under CC BY SA 2.0.

In 1967, Jocelyn Bell (Burnell) was a graduate student working on a radio astronomy project near Cambridge, England under her adviser Tony Hewish. When analyzing charts from the new array, which she had helped to construct, she noticed some "scruff" appearing periodically. She had discovered the first pulsar, a then-only-theorized object as massive as a star but the size of a city. Pulsars emit beams of light and radio emission from their poles, which sweep across the sky like a lighthouse as they spin.

At first the team did not know what to make of this exciting discovery, and in the wake of Project Ozma the radio astronomers naturally wondered if they might have discovered an alien radio transmitter.

Internally, and only half-jokingly, they referred to the source as "LGM-1," meaning the first signal coming from "Little Green Men." They kept the discovery quiet for weeks pending measurements to establish whether the signals originated from a radio transmitter on the surface of a planet via its drift rate (Section 12.4.3). Accepting the Nobel Prize, Hewish said: "Without doubt those weeks in December 1967 were the most exciting in my life." In their discovery paper, Hewish et al. (1968) wrote:

> It is also interesting to note the possibility of detecting a variable Doppler shift caused by the motion of the source itself. Such an effect might arise if the source formed one component of a binary system, or if the signals were associated with a planet in orbit about some parent star. For the present, the systematic increase of N is regular to about 1 part in 2×10^7 so that there is no evidence for an additional orbital motion comparable with that of the Earth.

As oblique and understated as it is, this appears to be the first published SETI false positive in the peer-reviewed literature.

Bell Burnell notoriously did not share in the Nobel Prize awarded to Hewish for this discovery. Regardless of where the blame for this oversight lies, the incident has become emblematic of the sexism that Bell Burnell and others faced in the physical sciences. She would go on to have a distinguished career, including as the president of the Royal Astronomical Society, the winner of its Copely Medal, and recipient of the $3 million Special Breakthrough Prize in Fundamental Physics.

3.2.2 SETI Beyond Radio

Almost immediately after the publicity of Cocconi and Morrison's paper and Project Ozma, many scientists began thinking about the problem. Otto Struve commented

"This project has been given an unreasonable amount of publicity, often incorrect or distorted and always with the wrong emphasis. It has aroused more vitriolic criticisms and more laudatory comments than any other recent astronomical venture, and it has divided the astronomers into two camps: those who are all for it and those who regard it as the worst evil of our generation" (Struve 1960).

Among those in the first camp, the first out of the gate to publish was American engineer Ron Bracewell, a professor of electrical engineering at Stanford who felt the problem had a different natural solution in a May 1960 paper in *Nature*, "Communications from Superior Galactic Communities." Bracewell (1960) argued that if communication was the goal, then sending a radio signal across the vastness of space from one's home planet was an inefficient way to go about things: it required the recipient to properly guess at which stars might be sending signals, and would result only in weak signals with very high latency (two-way communication would take many years between messages).

Instead, Bracewell imagined it would make more sense to send a fleet of probes capable of autonomous communication to all of the nearby stellar systems. Recipients would then merely have to search their *own* systems for probes, and could have faster communication using much stronger signals. Thus, modern "solar system SETI" was born.

The very next month, Freeman Dyson (Box 15.1) proposed his own solution in *Science*. Dyson disliked the assumptions that an alien species would be deliberately trying to communicate with strong signals, and sought a way to detect a species via its passive emissions. Thinking "like a physicist" Dyson (1960) contemplated that an extremely old technological species might expand its technology and biosphere far beyond its home planet, and calculated the ultimate limit: a swarm of biospheres orbiting a star that collect all of the star's luminosity. Today, we call such structures Dyson Spheres. Infrared SETI was born.

The problem inspired the Dutch mathematician Hans Freudenthal (then visiting faculty at Yale) to tackle the problem of how to communicate with aliens via an assumed shared language of mathematics. In 1960 he published the book: *Lincos: Design of a Language for Cosmic Intercourse, Part 1* detailing a system for building up a language via radio transmissions, starting with simple mathematics and working up to complex communication on any topic. The book would inspire multiple efforts at transmitting intelligible signals to alien species. METI was born.

Less than a year later, American physicist Charlie Townes (Schwartz & Townes 1961) argued that optical transmissions were a better bet. Townes had already invented the maser in 1953, and in mid-1960 another group had announced the creation of an "optical maser" (today, we would call it a "laser"). Such a light source was coherent and narrowband, like radio communication signals, but having a shorter wavelength could be beamed with much higher gain. The essence of their argument was that a sufficiently narrowband and powerful continuous laser could outshine the Sun (which spreads its light over a huge range of frequencies), and that the power requirements were reasonable (of order kW). Laser SETI was born.

3.2.3 The 1961 Green Bank Workshop

Amid the press surrounding Project Ozma, Drake gave a lecture at the Philosophical Society of Washington in March 1961, which attracted the attention of J. P. T. ("Pete") Pearman, a ballistics expert and a staff officer for the National Academy of Sciences' Space Science Board. Pearman immediately suggested to Drake that the Academies sponsor a workshop on the topic, and Drake quickly arranged permission with Struve.

The original invitation list included 18 scientists that had published on the topic, including Ron Bracewell, Freeman Dyson (Section 15.1), American molecular biologist Joshua Lederberg, American physical chemist Harold Urey, and Charlie Townes, none of whom could attend.[5]

The final meeting was organized largely by Pearman, who handled logistics, and Drake, who handled the scientific content. It was held 1961 November 1–2. Sagan (1963) and Pearman (1963) list the attendees as:

Attendees of the 1961 Green Bank Workshop

1. Dana Atchley, president of Microwave Associates, Inc., who had supplied the amplifiers for Project Ozma.
2. Melvin Calvin, an American biochemist who won the Nobel Prize in Chemistry for his work on photosynthesis during the meeting.
3. Giuseppe Cocconi, Italian particle physicist and co-author of Cocconi & Morrison (1959).
4. Frank Drake.
5. Su-Shu Huang, a Chinese-American astrophysicist responsible for foundational work on astrobiology and who had formally defined the Habitable Zone.
6. John Lilly, an American polymath and expert in dolphin communication.
7. Phillip Morrison, an American physicist and co-author of Cocconi & Morrison (1959).
8. Barney Oliver, an American scientist and head of R&D at Hewlett Packard who would later lead the Cyclops report and endow the SETI Institute.
9. Pete Pearman.
10. Carl Sagan, at the time a Miller Postdoctoral Fellow at UC Berkeley.
11. Otto Struve.

To organize the agenda for the Workshop, held in what today is called the Drake Lounge of the NRAO Residence Hall, Drake outlined the whole problem, expressed as an equation (Chapter 8) that he wrote on the chalkboard

[5] Some sources claim Lederberg was in attendance, and he indeed appears on the final list of expected attendees in the NRAO archives (Rebecca Charbonneau, private communication), but Dick (2020) notes that according to Lederberg's own papers, he was in fact a late scratch. Lederberg and Urey were Nobel Laureates, and Townes would win the prize in 1964; if they had all attended, the meeting would have included four eventual laureates, including Calvin.

$$N = R_* f_p n_e f_l f_i f_c L$$

Each term of the equation represented a requirement for the development of a communicative species. The first represented the rate of star formation in the Milky Way, which was reasonably well known at the time. The next five terms successively narrowed the fraction of these stars that form by requirements for life: that the star host planets, that the planets be habitable, that life arises on them, that the life develops intelligence, and that it communicates in a way we could detect. The final term, L, captured the duration of the communications. Each attendee had some expertise in at least one of the terms, and the workshop participants went about estimating conservative and optimistic values for each term.

Asked about how he came to write the equation down, Drake responded:

> The Drake Equation was a way of organizing the meeting. I thought we should organize the meeting and categorize topics and establish themes for various sessions at the meeting, and that caused me simply to think about what we needed to discuss and how these things were related. And it was easy to see that they were interrelated in the way that is the described in the equation…There was really not a great deal of profound thought or anything involved in putting it together (quoted in Swift 1993).

Even before the workshop, there had been calculations in the literature along the lines of the Drake Equation. For instance Huang had just two years earlier discussed many of the underlying issues and defined the Habitable Zone (Huang 1959). In 1960, Iosif Shklovskiĭ (Box 3.2) published a paper outlining a similar argument (translated and reprinted in Shklovski, 1963), as did Ron Bracewell (Bracewell 1960). Perhaps the earliest example, identified by Lorenz (2020), is by Maunder (1913), who used the term "Habitable Zone" and performs a calculation that roughly tracks the logic of the Drake Equation.

Drake's version of the exercise invigorated the attendees, who afterwords began describing a version of the calculation in various fora. After a conversation with Drake, Sebastian van Hoerner, an NRAO astronomer, published a calculation inspired by the equation in *Science* just a month after the workshop (von Hoerner 1961), and Sagan (1963) published it nearly verbatim shortly thereafter. Drake himself published what we might call the definitive version in Drake (1965).

The equation has achieved celebrity and influence that Drake could hardly have imagined when he conceived it. Indeed, it has been called one of the most famous equations in all of science, perhaps second only to $E = mc^2$.

3.2.4 The Orders of the Dolphin and Octopus

At the workshop, Lilly's description of dolphin intelligence captured the imaginations of the attendees, and gave them confidence that intelligence was a convergent feature of evolution that had a good chance of being repeated on other planets

(Section 7.5). Lilly's engaging stories and the camaraderie engendered from Calvin's win of the Nobel Prize during the meeting led the attendees to form a whimsical society dedicated to the topic, which they called "The Order of the Dolphin." Lilly even made commemorative pins for the group.

Over the years, other members were added by majority votes of "rump sessions" of this largely apocryphal society, including the Harvard biophysicist Arthur Solomon, University of Pennsylvania biophysicist Britton Chance, John's wife Elisabeth Lilly, Iosif Shklovskiĭ, and Nikolai Kardashev (Box 4.1). Carl Sagan also wrote to British-Indian biologist J. B. S. Haldane inviting him to join the Order; Sagan recalled his response: "… he (Haldane) wrote me that membership in an organization that had no dues, no meetings, no responsibilities was the sort of organization he appreciated; he promised to try hard to live up to the duties of membership." (Kellermann et al. 2020, citing Sagan 1973)

Later, Sagan and others would fall out of Lilly's spell as the results of his experiments on dolphins failed to find traction in the broader community, and as his methods were revealed to be inhumane (among other things, he was dosing the dolphins with LSD).

In May 2021 a group of early career scientists in SETI reconstituted the idea as the Order of the Octopus, which held its first "assembly" in July 2021 at the first Penn State SETI Symposium. The group chose to rename itself after the octopus both to distance itself from Lilly's methods, and to acknowledge the extraordinarily convergent nature of octopus intelligence, which descends from a completely different phylum than mammalian intelligence (Section 7.5.3).

Today, the Order of the Octopus continues as a way for early career scientists in SETI to find support and community as the field grows.

3.2.5 CETI in the USSR

Interest in SETI began in the Soviet Union almost immediately after Cocconi & Morrison's paper. In 1960, Iosef Shklovskiĭ (Box 3.2) wrote "Is Communication Possible with Intelligent Beings on Other Planets?" for *Priroda* (Croatian and Russian for "nature"), a popular science journal of the Soviet Academy of Sciences (translated in Shklovskiĭ 1963). The excitement over the prospect led him to prepare a book *Universe, Life, Mind* (Вселенная Жизнь Разум) in 1962 commemorating the fifth anniversary of the launch of *Sputnik*, illustrating how the Space Race between the USSR and US influenced the development of SETI.

The book was extremely popular and influential in the Soviet Union, but was locked behind the Iron Curtain.[6] Rumors of it, however, made it to the West where Carl Sagan championed its publication.

Sagan, who had not met Shklovskiĭ, worked with an English translation and made extensive annotations and additions, roughly doubling its length. In a

[6] The term comes from Winston Churchill, who warned of the growing sharp divide between countries allied with the Soviet Union and the United States. This divide was often physical and literal, as with the Berlin Wall, but included the sharp curtailment of cultural and scientific exchanges between the two sides.

remarkable and unusual arrangement, he had the book published in English with Shklovskiĭ's text virtually unchanged, and with Sagan's text clearly identified with bracketing triangles. The book, *Intelligent Life in the Universe*, was widely successful and has had multiple editions.

Box 3.2: Iosif Shklovskiĭ (1916–1985)

Image credit: NRAO/ AUI/NSF (CC BY-SA 3.0).

(Ио́сиф Самуилович, Шкло́ский, also sometimes rendered as Josif, Shklovsky, Shklovskii, or Shklovskij)

Prof. Shklovskiĭ was a Soviet astrophysicist born in present-day Ukraine. As a graduate student in Moscow at the Shternberg Astronomical Institute he avoided the purge of "disloyal" astronomers in the USSR in 1935.

Upon hearing of van de Hult's work predicting the 21 cm line of HI in 1947 (but being unable to access the paper across the Iron Curtain) he performed and published the calculation himself, launching a new era of Soviet radio astronomy. He later founded the radio astronomy department at the Shternberg Institute. He established the temperature of the solar corona and explained the radio emission from the Crab Nebula.

When the Soviet Union realized it needed powerful radar to track satellites after the launch of *Sputnik*, it turned to Shklovskiĭ, whose work on rocket telemetry for the *Luna* program earned him the Lenin Prize, the highest scientific honor in the USSR.

His work with Sagan to turn his Russian-language book *Universe, Light, Mind* into the English-language collaboration *Intelligent Life in the Universe* was unusual and challenging. He and Sagan did not meet until after publication, and government censorship interfered with their communication, so when he finally received a copy he was surprised at the volume of additional material Sagan provided.

Shklovskiĭ played a key role in promoting space science, radio astronomy, and CETI in the USSR, and his students included Nikolai Kardashev (Box 4.1). His suggestion that Phobos might be artificial as a way to explain its apparently low density is an early example of modern day solar system SETI. He was a charming and well-loved figure in the Soviet astronomy community.

Despite his stature, Shklovskiĭ was often denied the freedoms of travel and publications that his peers enjoyed, was never elected to the Soviet Academy of Sciences, and at many points may have been in mortal danger from the authorities. This is likely because of his Jewish heritage and strident public stances on human rights and defenses of persecuted scientists. Towards the end of his life he became despondent at the state of global affairs and the prospect for nuclear war, and something of a SETI pessimist.

For more on Shklovskiĭ, see his autobiography, "Five Billion Vodka Bottles to the Moon" Shklovskiĭ et al. (1991).

In 1964, Shklovskiĭ's student Nikolai Kardashev (Box 4.1) laid out the requirements for isotropic radio signal transmission in an effort to establish interstellar communication (Kardashev 1964). This work established the system now referred to as the Kardashev scale for classifying the energy consumption of civilizations at different levels of technological advancement. It establishes types I, II, and III, as civilizations with energy supplies on the order of a planet, a star (here coining the term "Dyson sphere"), and a galaxy, respectively.

In that same work, Kardashev identified two sources of radio emission, CTA-21 and CTA-102, that roughly fit his predictions for the transmission spectra of galaxy-spanning civilizations.[7] This and the discovery by Gennady Sholomitskiĭ that it had very fast flux variations that seemed impossible to explain by natural means were picked up by Soviet state media, catapulting Kardashev (and the new field of SETI) into the popular consciousness not just in the Soviet Union but around the world. Not long after, the American folk band The Byrds would include a song "CTA-102" on their album *Younger than Yesterday* in an homage to world (indeed, intergalactic) peace.[8]

Also in 1964, Kardashev organized the First All-Union Conference on Extraterrestrial Civilizations at the Byurakan Astrophysical Observatory in Armenia. There, Soviet radio astronomers including Shklovskiĭ discussed the problem of communicating with ETIs.

The USSR Academy of Sciences then formed a section named "Search for Extraterrestrial Civilizations" to develop the problem of "Communications with Extraterrestrial Intelligence" (CETI), which helped launch a significant Soviet radio SETI program.

3.3 1969–1993: The Era of Government Support

3.3.1 The First Joint Soviet–American Symposium on Communication with Extraterrestrial Intelligence (CETI) and Its Successors

The Cold War made collaboration between the US and Soviet SETI efforts challenging, but some interaction among scientists on purely civilian science projects was possible and even diplomatically desirable as a way to ease geopolitical tensions. In this spirit, US and Soviet astronomers met 1971 September 5–11 at Byurakan for the First Soviet–American Conference on Communication with Extraterrestrial Intelligence (CETI; this is often referred to as the "Second Byurakan conference," following the earlier all-union conference in 1964. See Figure 3.7.)

Sagan, Drake, and Morrison led the US organizing committee, and Kardashev and Shklovskiĭ were two of the Soviet organizers; other notable attendees included English molecular biologist Francis Crick (of double helix fame), Barney Oliver, and

[7] The objects are now understood to be active galactic nuclei, see Section 2.12.

[8] The song, sung to the inhabitants of CTA-102, includes the lines "Signals tell us that you're there/We can hear them loud and clear" and "Science tells us that there's hope/Life on other planets might exist." The song ends with a psychedelic journey along with the radio waves carrying the song on American radio out to CTA-102, where we hear a discussion in an alien language, presumably between recipients of the message.

Figure 3.7. Attendees at the Second Byurakan Conference in 1971. Reprinted with permission from the Byurakan Astrophysical Observatory

Freeman Dyson. The discussion and resulting conference proceedings led to an invigoration of interest in conducting a serious search for extraterrestrial signals in the US and the USSR.

The Soviet and post-Soviet CETI community would meet again, often with their American counterparts, in a series of workshops over the next two decades in Zelenchukskay, Russia (1975); Tallinn, Estonia (1981); Vilnius, Lithuania (1987); Santa Cruz, California (1991); Nizhny Arkhyz, Russia (2005); and St. Petersburg, Russia (2011). Up until the early 70's, most radio SETI programs were happening in the USSR.

3.3.2 NASA Gets in the Game: Project Cyclops

British physician John Billingham, working at NASA Ames in Mountain View, California, became interested in SETI from Sagan and Shklovskiĭ's book and a reprinting of several SETI papers in the Alastair Cameron anthology *Interstellar Communication*. In 1969, he made it the topic of that year's annual "summer study" he organized at Stanford University. The feasibility of the project determined at that workshop convinced him and his superiors that it was worth pursuing the problem more vigorously.

So in 1970, he approached Barney Oliver, a vice president at Hewlett Packard, to help lead a second summer study. Oliver had attended the 1961 Green Bank Workshop and had remained interested in the topic. The project bloomed under Oliver and Billingham's leadership into a large feasibility study over the next two years, which Oliver named Project Cyclops.

The Cyclops Report set the standard for a major national effort to detect radio signals from ETIs, and became in many ways the symbol of the field's ambitions. It was a "maximal" concept, outlining what a major governmental effort could accomplish in principle to have a realistic chance of detecting a signal.

The Cyclops Report laid out Drake Equation-style arguments to determine the number of target stars and sensitivity needed to have a reasonable chance of success.

Figure 3.8. Artist's conception of the full Cyclops array. Each dish was to be about as large as the largest fully steerable radio telescope on Earth today. Image Credit: NASA.

It did not give hard numbers about the size, number, or distribution of dishes—the authors concluded the program would grow as resources and technology allowed.

Roughly speaking the report described a telescope array composed of hundreds of 100-m fully steerable radio dishes to be operated as a phased array across several kilometers. (See Figure 3.8. For scale, the single 100 m Robert C. Byrd Green Bank Radio Telescope is the largest articulating object in the world.) The telescope would observe in 100,000 channels simultaneously (an extremely large number number at the time) in the "Water Hole" in L band near 1 GHz.

The Cyclops report proved enormously influential, though not entirely in a good way. It provided a concrete path forward for SETI, and developed the rationale for large telescope arrays that would eventually evolve into today's Allen Telescope Array and MeerKAT.

But, being a "maximal" program, it also came with significant sticker shock. The pricetag was around $10 billion, the equivalent of around $80 billion today and about half of the eventual cost of the entire *Apollo* program (ongoing at the time). The immediate perception at NASA and in Congress was that the project was infeasibly expensive, and that a significantly smaller program would have little chance of success.

Box 3.3: The Flag of Earth

The author obtained this image from the Wikimedia website, where it is stated to have been released into the public domain.

Many flags have been proposed to represent the entirety of the planet, as opposed to any one nation or political entity, and although none have widespread acceptance, this flag has been informally adopted by the SETI community.

James Cadle created the simple design in 1970 in response to the American flag being planted in lunar soil by the Apollo astronauts. He felt a more inclusive, universal flag would be a more appropriate symbol of human expansion into space.

The design is a schematic and stylized representation of the Earth at the center, the Moon near it, and the Sun and deep space in the background. The relative size of the Earth and Moon are approximately correct; the Sun is far too small to be to scale, although it works if one imagines the image is taken from a vantage in which the Sun is about 30 times farther away than the Earth-Moon system. It is, in some sense, a family portrait of the solar system objects most important to humanity and life on Earth.

Cadle hoped that the flag would find purchase in popular culture and be widely used in space exploration. For many years he sold copies of the flag, but it remains obscure. It did, however, find resonance among SETI practitioners, who generally recognize the global nature of the search and how the significance of contact will transcend national boundaries. For decades, the flag has been a common feature at observatories and offices around the world where SETI is performed.

Cadle released his design for the flag into the public domain in 2003, and today the flag can be purchased from the North American AstroPhysical Observatory at http://flagofearth.com.

3.3.3 American Searches

3.3.3.1 NRAO Continues the Search

Several searches continued at Green Bank using the 140 ft and 300 ft antennae, including a search in 1971–1972 by Verschuur (1973), a search of over 674 nearby stars called Ozma II by Palmer & Zuckerman (Gray 2021), and the SERENDIP II experiment in the late 1980s (Werthimer et al. 1988).

Figure 3.9. Aerial view of the Big Ear. The flat reflector is on the right, which could tilt to provide y-axis pointing. The focusing, parabolic dish is on the left, which sent radio waves to the pair of feed horns at the base of the flat reflector. These were moved along tracks, which provided x-axis pointing. Two horns were used as a way to isolate sources on the sky from radio frequency interference. Photo credit: NAAPO and The Ohio State University Radio Observatory. From the papers of John Daniel Kraus, "Big Ear aerial," *NRAO/AUI Archives, Papers of John D. Kraus*, accessed 2025 February 15 and licensed under CC BY 4.0.

3.3.3.2 The Big Ear and the Wow! Signal

The Cyclops Report inspired American radio astronomers Bob Dixon and John Kraus to conduct a radio SETI program on the idle Ohio State University Radio Observatory, nicknamed the "Big Ear," designed by John Kraus. Unlike most radio telescopes, the Big Ear did not employ a pointable dish, but instead had a long rectangular flat reflector roughly 100 m × 33 m, which tipped to provide elevation control, feeding a fixed, large parabolic reflector roughly 110 m × 21 m. A pair of feed horns moved along tracks to provide an orthogonal direction of pointing (see Figure 3.9). Signals from the two horns were differenced, with the resulting power recorded by alphanumeric characters on chart paper. Radio frequency interference would generally be present in both horns, and so be removed by this differencing. As a second check, the feed horns were kept stationary while the Earth rotated, so any persistent source would appear first in one horn, then the other, resulting in a pair of detections, each lasting about 1 minute, or as long as the source took to drift across the field of view, and separated by about 5 minutes.

In 1973, Dixon and Kraus began what would become the longest-running SETI program in the world, running for 22 years. The project produced some of the first (and only) advanced degrees in engineering or the physical sciences in SETI, including Master's Degrees in engineering in 1976 to Dennis Cole and in 1988 to Jim Bolinger, and a PhD to Stephen Brown in 2000 (just the third ever). Much of this work was on the "Argus array," a proposed (but never built) interferometer for SETI.[9]

In 1977, the Big Ear produced the Wow! Signal, which has since become the quintessential potential SETI signal (or, if you like, the quintessential false positive, see Box 3.4).

[9] Not to be confused with the SETI League's *Project Argus*, an distributed amateur effort.

Box 3.4: The Wow! Signal

The author obtained this image from the Wikimedia website, where it is stated to have been released into the public domain.

Jerry Ehman was reviewing charts from the Big Ear taken a few nights earlier on 1977 August 15, when he came across an extraordinary signal. The chart recorded the statistical significance in standard deviations of signals in each of its 10 kHz channels (recorded in columns) in 12 s intervals (one per row). The recorder used alphabetic characters to record values greater than 9.

Ehman was stunned to see the sequence "6EQUJ5," indicating a signal that had grown and peaked at 26 standard deviations above the noise, then fallen back to zero, exactly as a sidereal source would. Ehman circled the signal and scrawled "Wow!" in the margin, and among the team, and later the world, it became known as the "Wow! Signal."

The signal was detected at 1420.4556 MHz, near the neutral hydrogen line. Frustratingly, the signal did not repeat 5 minutes later, meaning it was present only during the passage of one of the two feed horns. Because it did not repeat, there is no way to know which horn it was in.

The Wow! Signal has since become famous as a tantalizing hint of a successful radio SETI detection. Several studies have sought to detect it repeating, including many hours-long stares with the Very Large Array by Gray & Ellingsen (2002). The nature of the signal has been much debated, with hypotheses ranging from idiosyncratic radio frequency interference, a glitch in the electronics, to a genuine SETI signal.

Today, the Wow! Signal is somewhat notorious in the SETI community, serving as a reminder of the importance of good source localization, saving the rawest possible data for candidate signals, excellent RFI rejection, and prompt follow-up/confirmation capabilities in SETI.

For more on nature, history, and follow-up efforts of the Wow! Signal, see *The Elusive Wow!: Searching for Extraterrestrial Intelligence* by Bob Gray (2012), a self-taught radio astronomer who dedicated years to studying and searching for the Wow! Signal.

3.3.3.3 Demonstrations

Scientists were also thinking about what a message might contain, and what sort of message we might send. In the early 1970s Carl Sagan and Frank Drake helped design the Pioneer Plaques (Section 19.2.2) and Golden Records (Section 19.2.4) as physical, interstellar messages to potential alien species. In 1974 Drake, now head of the organization running Arecibo Observatory, co-designed, sent, and advertised the Arecibo message (Section 19.2.3).

These demonstrations served to illustrate and popularize the concept of SETI, but also served to raise concerns that the field of SETI was not completely passive and harmless, but might lead to actually attracting the attention of an alien species, which might harm Earth or humans. We will return to this in Chapter 19.

3.3.3.4 SERENDIP and Berkeley

In the early 1970s, American astrophysicist Stu Bowyer at UC Berkeley read the Cyclops report and began a project called SERENDIP to "piggyback" on radio observations made for other purposes. He compiled a team to install hardware at Hat Creek Radio Observatory and Goldstone, employing a 100-channel backend. Having heard of her facility with the necessary computer hardware from an earlier project, Bowyer recruited Jill Tarter (Box 1.2) onto the project, who created the backronym "Search for Extraterrestrial Emission from Nearby Developed Intelligent Populations."

This began a heritage of commensal SETI projects led by Bowyer, American radio astronomer Dan Werthimer, and others including SERENDIP II at Green Bank with 65,000 channels from 1986 to 1988, SERENDIP III at Arecibo with 4.2 million channels from 1992 to 1996, SERENDIP IV, also at Arecibo, with 168 million channels from 1997 to 2009, SERENDIP V at Arecibo with 120 million channels from 2009 to 2016, and SERENDIP VI installed at Green Bank and Arecibo in 2016. These later SERENDIP projects did real-time analysis to flag potentially significant signals, which were then saved and shipped to Berkeley for further analysis. Hardware and other developments for these various "piggyback" programs were developed largely at the Berkeley Radio Lab, and the Berkeley SETI Research Center.

Berkeley researchers would later contribute significantly to the development of the Allen Telescope Array and methods for radio interferometric SETI, in particular with the PhD work of Andrew Siemion in 2012 (the sixth such PhD ever, who would later go on to lead the Breakthrough Listen Initiative (Section 3.5.1) and win the 2024 Drake Award.)

3.3.3.5 NASA Moves Along

By 1975, a Congressionally requested report on the topic discussing US and Soviet SETI programs by Marcia Smith and the Library of Congress served to legitimize the topic in the eyes of legislators (Smith 1975). In addition, the International Astronomical Union set up a "Bioastronomy" division, dedicated to search for extraterrestrial life, including technological life, and the effort was endorsed in the decadal surveys of astronomy produced by the National Academy of Sciences in 1980, 1991, and 2001.

Meanwhile, John Billingham continued to push SETI at Ames. He commissioned Phillip Morrison to lead a design study on a NASA radio SETI mission, and held a series of workshops (Morrison et al. 1977) that included Tarter. It was at one of the workshops organized by Morrison that the term "SETI" was coined and began to be used widely.[10] These workshops concluded with a strong recommendation to

[10] One reason was to distinguish the work from the Soviet CETI program and efforts like the Arecibo Message, in that it would be a purely passive search, and not attempt to make contact via broadcasting.

immediately begin a SETI program at NASA. JPL had also begin its own analysis, and by the end of the 1970s Ames and JPL had joined forces to recommend a coherent two-part search strategy, with deep stares at some stars likely to have life, paired with a broad survey strategy (Billingham 1999, see also Figure 11.7).

3.3.3.6 Horowitz and Harvard

In 1981, Harvard physicist Paul Horowitz (Box 13.1), inspired by a lecture by American physicist Ed Purcell on SETI, asked Frank Drake if he could help with SETI at Arecibo. There, he developed what he called "Suitcase SETI," a real-time spectrum analyzer that could search over 100,000 narrow channels simultaneously. Horowitz and his colleagues searched 250 stars at Arecibo with this equipment and then, with some modifications, in 1983–1985 performed a search of the entire northern sky with the Harvard-Smithsonian Radio Telescope with Project Sentinel.

Horowitz would continue to innovate radio backends, greatly increasing the number of channels. He and his colleagues, including Sagan, operated META ("Megachannel Extraterrestrial Assay"), which had over 8 million channels and operated from 1983 into the 1990s at sites in the Northern and Southern Hemispheres. Project BETA, which began in 1995, searched the sky in over 240 million channels simultaneously. This project also produced the first PhD in the field, by Darren Leigh in 1988.

3.3.4 Political Trouble

Funding for the NASA SETI programs hit a roadblock in 1978, when the program caught the eye of Senator William Proxmire, who was prone to grandstanding on matters of the budget deficit as a way to bolster his fiscal bona fides. One of Proxmire's regular stunts was to bestow some silly-sounding government program his monthly "Golden Fleece Award," for "fleecing" the American taxpayer with frivolous spending.

The NASA SETI program "won" this dubious honor in 1978, and Proxmire chided NASA for taking science fiction fantasies of the time like *Star Wars* too seriously. Although the actual text of the award made debatable but reasonable arguments for why SETI might not succeed, the press reported that Proxmire infuriatingly "suggested that the project be postponed until the federal budget is balanced and income and Social Security taxes are reduced to zero, or a few million light-years—whichever comes first."[11]

Despite this negative publicity, Proxmire did not succeed in zeroing out NASA's budget for SETI, and NASA was able to keep the project lightly funded and below congressional radar for years as the more technically-named Microwave Observing Project (MOP).

[11] This last flourish is frustrating because light-years are a unit of distance, not time, which makes the criticism ignorant in addition to specious. The actual press release from Proxmire's office does not include this final jab, which might be an invention of the Miami News.

Carl Sagan, who had worked with Proxmire on matters related to nuclear non-proliferation and the prospects of "nuclear winter," was able to persuade Proxmire that SETI was worthwhile. Over time, Proxmire's opposition to the funding relented, and it was again included in federal funding in 1983.

NASA's plan was to conduct two surveys, one broad, using the JPL Deep Space Network of radio telescopes, and a targeted survey with telescopes that included the NRAO 140 foot telescope and Arecibo.

It was clear at this point, however, that SETI would have a hard time depending on large sums of money from Congress. Billingham also fretted over the very large overheads associated with money flowing through JPL and NASA Ames, which greatly increased the price tag of any governmental funds to do SETI. In 1984, he therefore worked with Oliver and others to found the nonprofit SETI Institute nearby in Mountain View, to be a low-overhead clearinghouse for federal money dedicated to SETI and astrobiology. Its board members included Frank Drake, and Jill Tarter served as its SETI Scientist.

Finally, in 1992, NASA was ready to launch its SETI program, now renamed the High Resolution Microwave Survey (HRMS). There was a sense on Capitol Hill that this was an attempt to avoid too much negative attention, and the joke ran that HRMS stood for "He Really Means SETI." The formal launch of the program was October 12, Columbus Day, chosen (and criticized, see Section 18.4) for its symbolism of discovery and contact.

The program was to cost around $10 million per year, and proponents including Tarter and Sagan lobbied Congress heavily to ensure it would not be cut. The effort for such a small program may have backfired, however, and Senator Richard Bryan, an opponent of the program, saw an opportunity to repeat Proxmire's stunt. He introduced a late amendment in the 1994 budget process, frustratingly mocking the program as "The Great Martian Chase," and claiming "as of today millions have been spent and we have yet to bag a single little green fellow. Not a single Martian has said take me to your leader, and not a single flying saucer has applied for FAA approval." The amendment ultimately passed, and the program was ended.

NASA had learned its lesson: it would not provide more than token support for SETI programs for another 25 years, and developed an institutional hostility to the field, excluding such studies even from external grant programs designed to find life in the Universe.[12] As Tarter put it "we instantly became the four-letter S-word that you couldn't say" at NASA.[13]

3.4 1993–2015: A Turn to Philanthropy

With the NSF having ceded the field to NASA, and with NASA all but refusing to fund the program, SETI's supporters sought to fund the program privately.

[12] Neither the NSF nor NASA completely stopped funding SETI projects. NASA granted funds for a few external projects over the next few decades, almost entirely for novel radio SETI hardware development that had broader astronomical application.

[13] "Congress Is Quietly Nudging NASA to Look for Aliens," Marina Koren, *The Atlantic* online, 2018 May 9.

Sagan, Tarter, and others became prolific fundraisers, especially in California seeking Silicon Valley and Hollywood backers, and with philanthropic foundations. Sagan began working on a screenplay that would popularize the importance of the project. This would eventually evolve into the book *Contact* (Box 18.1), and in 1997 into a major motion picture.

The book and film follow a fictionalized history of the field, incorporating many real elements, including its loss of government support and turn to philanthropy. The protagonist, Ellie Arroway, is partly based on Tarter (but with elements from other real scientists, e.g., Arroway, like Horowitz, was a ham radio operator at age 9). Eventually, Arroway convinces an eccentric and reclusive Howard Hughes-like billionaire to fund her search, which is ultimately successful. The second half of the film follows the fallout of success, including the decoding of an alien signal and construction of a machine to contact them directly.

In 1995, Oliver died and left a substantial fraction of his estate to endow the SETI Institute. With these funds, along with those of other major donors including the founders of Hewlett Packard, the SETI program was relaunched as Project Phoenix, to conduct the targeted portion of the NASA SETI plan.

Project Phoenix operated at three observatories: first at the Parkes 210 ft telescope in Australia for 16 weeks on about 200 stars. In 1996, the project then moved to the 140 foot telescope at Green Bank, where it operated for nearly 2 years in a targeted search of hundreds of nearby stars. It finished at Arecibo where from 1998 to 2004 it received a small fraction of the observing time over the course of the year. Overall, Project Phoenix observed roughly 800 nearby stars between 1.2 and 3.0 GHz at 1 Hz resolution.

Ambitions were still high, however, for a Cyclops-style instrument. In 1999, the SETI Institute convened a workshop of scientists and Silicon Valley entrepreneurs to envision what the future of SETI would look like. The team came up with a twist on the Cyclops concept: instead of vary large dishes with their small fields of view, a SETI survey telescope could instead be composed of a large number (509) of relatively small (6 m) dishes, backed by powerful computers. This concept, which they dubbed the One Hectare Array, pioneered the "large N/small D," strategy, which makes use of a very large field of view, and large amounts of computing power to observe multiple stars simultaneously over a very wide bandwidth. The resulting report, *SETI 2020*, mapped the strategy for radio SETI for the next 20 yr, one that anticipated and depended on Moore's law[14] eventually making the "large N/small D" concept feasible.

The estimated cost at the time was $25 million. UC Berkeley and the SETI Institute partnered to share the proposed array of 350 telescopes 50/50 on SETI and other science, and to use the Hat Creek Radio Observatory as a site after the existing telescopes there moved to a new home. In 2000, the SETI Institute obtained a commitment from Microsoft Founder Paul Allen and others to prototype the array, and later to build a 30 element prototype array at the site. The project was renamed the Allen Telescope Array.

---

[14] Moore's law, named for Intel co-founder Gordon Moore, states that integrated circuit capacity on a microchip doubles every 2 years. Moore made this claim in 1975 after a decade of growth at this rate. The gamble of the *SETI 2020* authors was correct: the "law" indeed held through 2020.

Cost and schedule overruns eventually led to the end Allen's involvement and the divestment of UC Berkeley from the project before the 42 dishes that were eventually built could be fully outfitted and operational. Crowdfunding eventually provided some funds to keep the array going, but the project suffered from a series of hiatuses and operated on a skeleton crew only part time for the next several years.

3.4.1 SETI@Home

Data collected commensally at Arecibo was too voluminous to be analyzed efficiently, and so scientists at the Berkeley SETI Research Center, including Dan Werthimer and Dave Anderson, came up with a novel solution involving "citizen science." The idea was that SETI had broad popular salience, and that users of personal computers around the world might be willing to donate their unused cycles to the cause.

In 1999 BSRC released the distributed computing platform SETI@Home, one of the first scientific distributed computing projects and perhaps the most famous and successful. The program ran when a user's computer was otherwise idle, originally as a screensaver[15] displaying the results of the analysis it was performing (Figure 3.10).

Data were collected from Arecibo and sent to SETI@Home users as a background process, where they were analyzed during idle time, and with results returned to BSRC.

The program was extremely popular, with over 5 million users collectively donating around 1 million years of computer time, and was at one point considered the largest computation ever performed.

The program briefly accepted data from other telescopes used by Breakthrough Listen, but was eventually discontinued in 2020, as the BSRC changed the nature of its analysis focus.

3.4.2 Advances in Laser SETI

Charlie Townes and others, in a series of technical papers in the 1980s and 1990s, laid out the foundational ideas of laser SETI. In work funded by NASA, Townes and Albert Betz at UC Berkeley developed a program to look for CO lasers at 10μ using small telescopes and a CO laser to perform heterodyning spectroscopy.

Townes and others gave a series of workshops at the SETI Institute, which cemented the idea to look for pulsed laser signals with Paul Horowitz, who had earlier worked on the META and BETA projects, among others. Horowitz's group at Harvard included Chip Coldwell, Andrew Howard, and Curtis Mead who finished their PhD's on the projects in 2002, 2006, and 2012 (just the third, fourth, and eighth such degrees in the field). With funding from the Planetary Society, the SETI Institute, and others, they used the 1.5 m telescope at Oak Ridge Observatory

[15] Older computer monitors based on cathode ray tube technology were susceptible to "burn in" if they displayed the same image for too long. Monitors were therefore supposed to be turned off when not in use, but this was hard to accomplish reliably. A common solution was a program that detected periods of idleness and either turned the monitor black or (to avoid the appearance that the computer was turned off) showed some sort of moving graphic that "exercised" all of the pixels more or less uniformly.

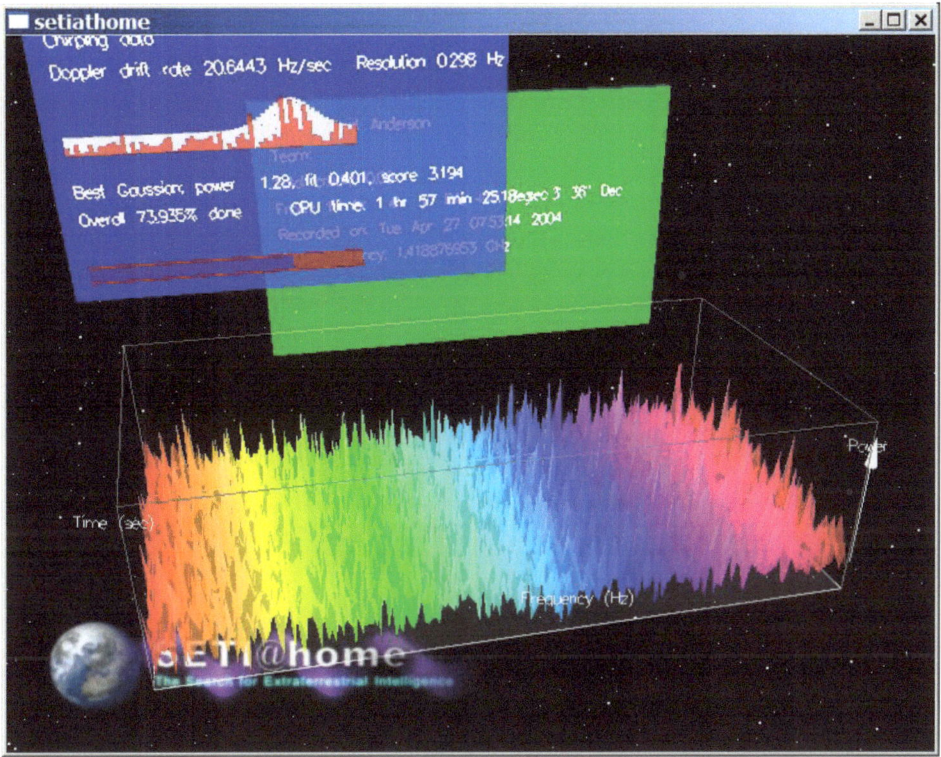

Figure 3.10. The screensaver mode of SETI@Home. Images like these became a fixture of idle computers around the world. Image credit: SETI@home, University of California, Berkeley.

to survey over 13,000 stars over 2400 hr of observations. Horowitz's group would continue to innovate in pulsed optical SETI for many years.

We will return to laser SETI in Chapter 13.

3.4.3 Advances in Dysonian SETI

Dyson's 1960 idea to look for infrared emission from energy use was far ahead of its time: infrared detectors were still in their infancy, and the first sensitive maps of the sky at these wavelengths were still over 25 years away.

So it was with some anticipation that the launch of the *IRAS* satellite by NASA in 1983 allowed the first searches for such technosignatures, conducted by Slysh (1985). Work progressed slowly from there, with the first full interpretation of the data set not published until Carrigan (2009).

In the meantime, astronomers were anticipating the launch of a new space telescope, *Kepler*, which was ostensibly designed to find Earth like planets. Arnold (2005) pointed out that if there existed large (planet-sized) structures orbiting other stars at short orbital periods, that *Kepler* would find them, and could even distinguish them from planets if they were not round.

We will return to these ideas in Chapter 15.

3.4.4 The Cloistering of the Field

The ability of the field of SETI to survive through the drought of federal funding is a testament to the drive and ambition of those, like Jill Tarter, who managed to keep the field going through relentless fundraising and personal effort. This reliance on philanthropic dollars, however, had an unfortunate effect on the field. With the notable exceptions of the efforts at Ohio State and Harvard, much of the work in the field was being done by scientists who were not faculty at universities, and so outside of the publish-or-perish culture of academia, and without significant student training. Many searches were performed without any results published as rigorous upper limits on well-defined target lists, or after undergoing peer review.

The result was that the field contained few young people, and had virtually no presence in graduate curricula. The lack of a publishing culture in astronomy journals meant few papers were published, and those that did appeared not in common astronomy journals where they would be widely read, but as conference proceedings or in journals like *Acta Astronautica*, and the *Journal of the British Interplanetary Society*, both dedicated largely to astronautics.

Also likely contributing to this is an accident of history, that most SETI searches were in the radio and the origins of radio astronomy lie in electrical engineering, where it matured thanks to radar efforts during World War II. Radio astronomers have their own conventions for things like mathematical notation for waves, the sense of circular polarization, and how flux density measurements are recorded and interpreted at telescopes. This cultural separation, already a minor problem in mainstream radio astronomy, and the fact that much SETI money was spent on developing specialized hardware, and was thus focused on engineering and not astronomical applications, likely contributed to the alienation of the field.

As the field faded from salience in astronomical circles, many myths arose about its status. Many astronomers were unaware of any kinds of SETI other than radio SETI, or that SETI was practiced anywhere other than the SETI Institute, or that there was no actual statutory prohibition on federal money being spent on SETI. It was not even clear to many that SETI belonged in mainstream astronomy journals, or that SETI papers sent there would even be accepted. Many had the impression either that SETI was well-funded philanthropically, and so needed no additional funding, or that it had "looked and found nothing" and so was largely done with its task.

These misperceptions fed back into the problem of lack of training of new students, since the lack of legitimacy and funding in the field were reasonable excuses not to allow students to work in the field, lest they not have a future in astronomy.

3.4.5 The Tide Begins to Turn

There were several factors working in the other direction, however, that would reverse the tide. One was the discovery of large numbers of exoplanets, especially by the *Kepler* mission, which showed that temperate terrestrial planets were likely common throughout the Galaxy. This invigorated the astronomical community's interest in finding life beyond the solar system. The widespread use of the parameter $\eta_\oplus$ ("eta sub Earth") as the symbol for the prevalence of such planets connected the work of these astronomers directly to the Drake Equation (Equation (8.1)).

Another was NASA's strong push to develop the field of astrobiology (a term which has now mostly supplanted the earlier terms "exobiology" and "bioastronomy"). NASA's nurturing of this field in its centers and at universities was in line with popular hopes and expectations about astronomical exploration in space, and created a large interdisciplinary community thinking about the problem of finding life beyond Earth (Dick & Strick 2004).

This coupled with NASA's allergy to SETI created a strange demarcation, where, as Tarter liked to put it, NASA would spend large amounts on the search for "stupid" life, but not a dime on the much more interesting intelligent life. In 2006, Tarter found a way to make the contrast stark by coining the term "technosignature" as an analogy to the term "biosignature" used widely in astrobiology. This emphasized the complementary nature of the two strategies, and gave the field a new way to refer to its search, one that lacked the cultural baggage of "SETI." The term caught on quickly, as astronomers discovered the term did not seem to trigger the usual SETI antibodies at NASA.

Also, by the 2010's a whole generation of astronomers who had grown up with Sagan's *Cosmos* and, later, *Contact*, began to fill the ranks. SETI's banishment and cloistering for so long had given this new crop of scientists no strongly held preconceived notions of what SETI was or why it might not belong in NASA's portfolio, or in the halls of Academe. This set the stage for a comeback in the US.

3.5 2015–: Resurgence

3.5.1 Breakthrough Listen

In 2015, fundraising by the SETI Institute and Berkeley SETI Research Center paid off in a major way when, as if following the script of *Contact*, they caught the attention of Yuri Milner, a Silicon Valley billionaire. Milner had dropped out of graduate school in theoretical physics in Moscow to study business at the Wharton School in Philadelphia, and made his fortune as a venture capitalist in Russian and eventually American tech companies. By a strange twist of fate, Milner had recently purchased a large home in Silicon Valley that had previously belonged to Barney Oliver.

Milner had already co-founded the Breakthrough Prizes in fundamental physics, mathematics, and life sciences with other Silicon Valley billionaires, intended to compete with the Nobel Prize for prestige (it is worth three times as much). Milner was persuaded that SETI was a worthwhile endeavor, so he decided to fund a major new project, dubbed "Breakthrough Listen," and described its scope as being of the order of $100 million over 10 years.[16]

The initiative was centered at the Berkeley SETI Research Center, and primarily focused on purchasing time and equipment at major radio observatories. The project was briefly led by American astronomer Geoff Marcy, a pioneer in exoplanet discovery who was widely considered a contender for the Nobel Prize. Marcy would

[16] This nominal pair of numbers turned out to be roughly accurate, but the initiative is not limited by either number. The original announcement also included a "Breakthrough Message" component about composing an outgoing message, but this was later dropped.

resign in disgrace later that year in a sexual harassment scandal, at which point the initiative was taken over by Andrew Siemion, who has run the project ever since.

The injection of millions of dollars per year into SETI at a university transformed the field. Although most of the early funding went to hardware development and telescope time rental, Breakthrough Listen also supported and trained many junior scientists. Among its first employees were graduate student Emilio Enriquez, who earned the tenth PhD in SETI while working on the project, and undergraduate student Sofia Sheikh, who would go on to earn the eleventh at Penn State, and return to the project as a postdoc.

Breakthrough Listen also became a highly visible presence in the spheres of mainstream astronomy. It generated a strong publication record in astronomy journals and *Nature*, demonstrating that SETI was still part of astronomy. It also supported conferences and travel for attendees, and sponsored SETI sessions in mainstream astronomy conferences. This normalization of SETI in the astronomy community has brought it a new credibility and acceptance across the field.

The existence of this well-funded project also provided a plausible destination for students and junior scientists interested in entering the field, removing an obstacle for scientists to train new students in it.

Today, Breakthrough Listen is a large international effort, still primarily focused on radio SETI, but also supporting other projects by its team members to some degree, and employing many of the most powerful telescopes around the world.

3.5.2 NASA's Re-entrance

In 2018, draft language appeared in a authorization bill in the US House instructing NASA to report to Congress on its efforts to find life in the Universe, and to spend $10 million on the search for technosignatures.

The Office of Management and Budget then instructed NASA to be ready to produce such a report, including how it would spend these funds. NASA hastily convened the "NASA Technosignatures Workshop" in September 2018 in Houston, Texas at the Lunar and Planetary Institute, where attendees were to prepare a report on how NASA could be involved in technosignature research.

Although the language supporting technosignatures did not survive the budgeting process, the workshop served in many ways as a turning point for NASA's involvement in SETI.

NASA officials had a vigorous discussion with participants about the degree to which NASA external grant programs implicitly allowed technosignature proposals, or explicitly disallowed them. Shortly thereafter, the language in multiple grant programs was changed to explicitly include technosignatures in their scope, and some Exoplanet Research Program and Exobiology grants were awarded to SETI programs over the next few years.

The report produced by the workshop (NASA Technosignatures Workshop Participants 2018) served as a survey of the field, including a broad scope across and beyond the electromagnetic spectrum. The report has served as an introduction to the field for those in and beyond NASA.

The workshop itself also provided a rare opportunity for scientists in and around SETI to meet and discuss the future of the field, inspiring new research directions.

But perhaps most importantly, it became a concrete symbol of NASA's new involvement in SETI, puncturing the inaccurate sense that NASA was in some way legally disallowed from supporting the effort, and creating an expectation that technosignatures work would be allowed as part of NASA's broader astrobiology and exoplanetary research portfolio.

Box 3.5: Carl Sagan (1934-1996)

Sagan c. 1987. The author obtained and cropped this image from the Wikimedia website, where it is attributed to Kenneth C. Zirkel and made available under a CC BY-SA 4.0 license.

Prof. Sagan was a celebrated American astronomer, and perhaps the most important science popularizer of the 20th century. His PBS documentary series and its companion book *Cosmos* inspired a generation of astronomers and helped shape the world's views on science, astronomy, and life in the Universe.

Sagan was born and raised in New York City, and at 16 attended the University of Chicago where he received four degrees, including a PhD in astrophysics on *Physical Studies of the Planets*. There he worked with Harold Urey and Hermann Muller as an undergraduate, and with Gerard Kuiper, George Gamow, and Melvin Calvin in graduate school, after which he went to Harvard as an assistant professor.

His significant focus on outreach and his self-promotion was not received well in many quarters, and was reportedly the reason he was denied tenure at Harvard and lost a contentious bid to become a member of the the National Academy of Sciences (though he did win its Public Welfare Medal in 1994). He joined the faculty at Cornell in 1968 and never left.

Sagan established many important ideas in astrobiology and planetary science, and is among the most cited planetary scientists. He was among the first to correctly identify Venus as being extremely hot due to greenhouse warming, and that Titan would have standing liquid on its surface.

He was a strong and highly visible advocate for nuclear disarmament, and a prominent voice for preventing climate change due to fossil fuel use and nuclear winter due to nuclear weapons. He was a prolific writer of popular science books, co-founded the Planetary Society, sat on the SETI Institute's Board of Trustees, and was the recipient of numerous prizes and awards.

He played a foundational role in SETI, attending and popularizing the 1961 Green Bank workshop, working with Paul Horowitz on many radio SETI projects, and helping develop the Pioneer Plaques and Voyager Records. He put SETI into the popular consciousness with his novel *Contact*, later made into a major film (Box 18.1).

3.6 Further Reading

History of the Plurality of Worlds: Crowe (1986), Dick (1982).
History of SETI in the USSR: Gindilis & Gurvits (2019).
History of SETI at NRAO and NASA: Dick (1982), Dick (1996), Garber (1999), Dick & Strick (2004), Kellermann et al. (2020), Kellermann (2023).
Oral histories: SETI Pioneers Swift (1993).
SETI, CETI, and the Cold War: Charbonneau (2024).
Background and history of the Drake Equation: Vakoch & Dowd (2015).

References

Arnold, L. F. A. 2005, ApJ, 627, 534
Bettex, A. W. 1965, The Discovery of Nature (Riverside, NY: Simon and Schuster)
Billingham, J. 1999, Social Implications of the Detection of an Extraterrestrial Civilization: A Report of the Workshops on the Cultural Aspects of SETI Held in Three Sessions, October 1991, May 1992, and September 1992 at Santa Cruz, California (Mountain View, CA: SETI Press)
Bless, R. C. 1996, Discovering the Cosmos (Mill Valley, CA: Univ. Science Books)
Bracewell, R. N. 1960, Natur, 186, 670
Carrigan, R. A. 2009, ApJ, 698, 2075
Charbonneau, R. 2024, Mixed Signals: Alien Communication across the Iron Curtain (Cambridge, UK: Polity Press)
Cocconi, G., & Morrison, P. 1959, Natur, 184, 844
Crowe, M. J. 1986, The Extraterrestrial Life Debate, 1750-1900 (New York: Dover)
Dick, S. J. 1982, Plurality of Words: The Extraterrestrial Life Debate from Democritus to Kant (Cambridge: Cambridge Univ. Press)
Dick, S. J. 1996, The Biological Universe: The Twentieth Century Extraterrestrial Life Debate and the Limits of Science (Cambridge: Cambridge Univ. Press)
Dick, S. J., & Strick, J. E. 2004, The Living Universe: NASA and the Development of Astrobiology (New Brunswick, NJ: Rutgers Univ. Press)
Dick, S. J. 2020, Space, Time, and Aliens: Collected Works on Cosmos and Culture (Berlin: Springer) 81
Drake, F. D. 1965, Current Aspects of Exobiology, ed. G. Mamikunian, & M. H. Briggs (Oxford: Pergamon) 323
Dyson, F. J. 1960, Sci., 131, 1667
Garber, S. J. 1999, JBIS, 52, 3
Gindilis, L. M., & Gurvits, L. I. 2019, AcAau, 162, 1
Golshani, M. 2002, God, Life, and the Cosmos, ed. T. Peters, M. Iqbal, & S. N. Haq (London: Taylor and Francis)
Gray, R. H., & Ellingsen, S. 2002, ApJ, 578, 967
Gray, R. H. 2012, The Elusive Wow: Searching for Extraterrestrial Intelligence (Mission Woods, KS: Palmer Square Press)
Gray, R. H. 2021, JAHH, 24, 981
Hewish, A., Bell, S. J., Pilkington, J. D. H., et al. 1968, Natur, 217, 709
Huang, S.-S. 1959, AmSci., 47, 397
Kardashev, N. S. 1964, SvA, 8, 217

Kellermann, K. I. 2023, arXiv:2302.06446
Kellermann, K. I., Bouton, E. N., & Brandt, S. S. 2020, Open Skies; The National Radio Astronomy Observatory and Its Impact on US Radio Astronomy (Berlin: Springer)
Lorenz, R. D. 2020, RNAAS, 4, 79
Martinez, A. A. 2016, AnSci., 73, 345
Maunder, E. W. 1913, Are the planets inhabited? (New York) Harper & Brother)
McGarry, D. D. 1955, The Metalogicon of John Salisbury (Berkeley, CA: Univ. California Press)
Morrison, P., Billingham, J., & Wolfe, J. 1977, The Search for Extraterrestrial Intelligence, SETI. (vol. 419, Washington, DC: NASA)
NASA Technosignatures Workshop Participants 2018, arXiv: 1812.08681
Needham, J. 1959, Science and Civilisation in China: Volume 3, Mathematics and the Sciences of the Heavens and the Earth (Cambridge: Cambridge Univ. Press)
Pearman, J. P. T. 1963, Interstellar Communication: A Collection of Reprints and Original Contributions, ed. A. G. W. Cameron (New York: Benjamin)
Sagan, C. 1973, The Cosmic Connection. An Extraterrestrial Perspective (Garden City, NY: Anchor Press, Doubleday)
Sagan, C. 1963, P&SS, 11, 485
Schwartz, R. N., & Townes, C. H. 1961, Natur, 190, 205
Shklovskiĭ, I. S. 1963, Translated in Interstellar Communication. The Search for Extraterrestrial Life, ed. A. G. W. Cameron (New York: Benjamin)
Shklovskiĭ, I. S., Zirin, M. F., & Zirin, H. 1991, Five Billion Vodka Bottles to The Moon: Tales of a Soviet Scientist (New York: W.W. Norton)
Slysh, V. I. 1985, IAU Symp. 112, The Search for Extraterrestrial Life: Recent Developments, ed. M. D. Papagiannis (Dordrecht: Reidel Publishing Co.) 315
Smith, M. 1975, in The Possibility Of Intelligent Life Elsewhere in the Universe: Report: Prepared for the Committee on Science and Technology, U.S. House of Representatives, Ninety-fourth Congress, First Session (Washington, DC: US Govt. Printing Office)
Struve, O. 1960, PhT, 13, 18
Swift, D. W. 1993, SETI Pioneers: Scientists Talk about Their Search for Extraterrestrial Intelligence (Tucson, AZ: Univ. of Arizona Press)
Tikhov, G. A. 1949, Astrobotony (Almaty: Publishing House of the Acad. Sci. of KazSSR)
Vakoch, D. A., & Dowd, M. F. 2015, The Drake Equation: Estimating the Prevalence of Extraterrestrial Life through the Ages (Cambridge: Cambridge Univ. Press)
Vakoch, D. A. 2000, Citeseer, 145, 165
Verschuur, G. L. 1973, Icar, 19, 329
von Hoerner, S. 1961, Sci., 134, 1839
Weintraub, D. A. 2014, Islam (Cham: Springer) 161
Werthimer, D., Brady, R., Berezin, A., & Bowyer, S. 1988, AcAau, 17, 123

Part II

Theory

Chapter 4

The Big Picture

> …none of us can know what any alien civilization would be like, even if…we try to proceed in a logical way to keep as many variables constant as possible.
>
> But the point is not to "guess right." The point is to make the guesses. Our job before contact is to refine the art of asking and answering all manner of questions…
>
> That is how we will uncover some of our own unconscious assumptions about life, intelligence, and language, and that is what will move us further along the path of preparing for contact with intelligent beings that are truly different from us.
>
> —Sheri Wells-Jenson (2023)

4.1 The Catalytic Insight of Modern SETI

Modern SETI began in 1959–1960 with the publication of Cocconi & Morrison (1959) and the commencement of Frank Drake's Project Ozma at Green Bank Observatory (Drake 1961); these two efforts remain the theoretical and observational foundations of the field.

While stories of and even genuine attempts to detect extraterrestrial life predate these efforts (Section 3.1), Cocconi, Morrison, and Drake's work mark an epoch for two reasons. The first is that they are the first application of modern concepts that continue to shape the field today, using methods that are still considered to have realistic chances of success.

The second is that they represent a watershed moment in science: they demonstrated that humanity was capable of producing signals that humanity was capable of detecting at interstellar distances. This proved to be the *Catalytic Insight of Modern SETI.*

Although some proposals that shortly followed—such as Dyson spheres—imagined species with technology far beyond what humanity is capable of (at least, in the near future), the fact that humanity's current technology level was sufficient

doi:10.1088/2514-3433/adf5aech4

for interstellar communication provided the existence proof that convinced many it was at least worth pursuing.

In response to the catalysis, most of the major strains of SETI were rapidly proposed, and many of its theoretical outlines were quickly drawn. In the span of 5 years, papers establishing laser SETI (Schwartz & Townes 1961), infrared SETI (Dyson 1960b), solar system SETI (Bracewell 1960), the Drake Equation (Drake 1965), and the Kardashev Scale (Kardashev 1964) were published, most of them in major journals such as *Nature* and *Science*. The idea of technological alien species went from science fiction and flights of fancy to a serious scientific endeavor almost overnight.

To offer another metaphor from chemistry, Cocconi, Morrison, and Drake were the seed crystal from which the field grew. There are likely many reasons that scientists were so primed to follow Cocconi, Morrison, and Drake's lead and offer an immediate torrent of activity and papers (why the solution of the culture of science was "supersaturated" with SETI), including the dawn of the Space Race, the popularity of science fiction, the scientific competition of the Cold War, and even the Roswell Incident. And the fact that Drake and Cocconi and Morrison had the same idea independently and simultaneously illustrates that it was "in the zeitgeist."

What Cocconi, Morrison, and Drake offered in their work (and at the Green Bank Workshop in 1961) was a compelling, sound, and easily understood argument for why the search was worthy of scientific study. Knowing that it did not require aliens to be much more "advanced" than humans to be detectable offered a powerful point of reference that has driven the field ever since. One final push for the field is how Cocconi & Morrison also finished their paper with a powerful and oft-quoted counterargument to those who would yet dismiss the effort:

> The probability of success is difficult to estimate; but if we never search, the chance of success is zero.

4.2 Quantifying Technology: The Inchnoscale and the Kardashev Scale

> If it is true, as many chemists and biologists believe, that there are millions of places in the universe where technology might develop, then we are not interested in guessing what an average technological society might look like. We have to think instead of what the most conspicuous out of a million technologies might look like. The technology which we have a chance to detect is by definition one which has grown to the greatest possible extent. So the first rule of my game is: think of the biggest possible artificial activities, within limits set only by the laws of physics and engineering, and look for those. I do not need to discuss questions of motivation, who would want to do these things or why. Why does the human species explode hydrogen bombs or send rockets to the Moon? It is difficult to say exactly why. My rule is, there is nothing so big nor so crazy that one out of a million technological societies may not feel itself driven to do, provided it is physically possible.
>
> —Dyson (1966)

4.2.1 The Ichnoscale

A useful framework for considering this insight, and tracking its relevance across the field as it matures, is the *ichnoscale* of Socas-Navarro et al. (2021), which aims to quantify the "footprint" of technology.[1]

Socas-Navarro et al. (2021) proposed that we consider humanity's current technological footprint on Earth as a unit of measure. For instance, if we wish to consider radio transmissions, then our current total radio emissions would have an ichnoscale factor of $\iota = 1$, while 100 years ago when there were many fewer radio transmitters on Earth it would have been proportionally smaller.

The ichnoscale factors for the sum of different human technologies are all $\iota \sim 1$ today, but they need not move in lockstep. For instance, humanity's energy generation and use has roughly doubled in the past 35 years, meaning that our heat signature on the planet was about $\iota = 0.5$ 35 years ago. Our other techno-signatures, such as our atmospheric pollutants like CFCs or our nighttime city lights, likely have very different growth rates and patterns.

The value of the ichnoscale is that it allows us to determine the magnitude of a technosignature we could detect in units of our own. For instance, although 1960s radio transmission technology was *capable* of generating a signal it was capable of detecting at a few light years, there were very few if any actual transmissions of that strength. Project Ozma would only have succeeded if it had pointed at a system with a planet with a radio transmission ichnoscale many orders of magnitude above our own.

As humanity's technology grows in scope and our remote detection capabilities improve, we are moving toward even more compelling versions of the Catalytic Insight of Modern SETI, where we would be capable of detecting not just deliberate, brief, or contrived transmissions, but even passively emitted evidence of our existence.

There is thus an upcoming watershed moment when our technosignature detection abilities will allow us to detect a twin of Earth, with $\iota = 1$ technology that is not doing any special or ostentatious signaling. This may have a similarly important psychological effect on SETI as the Catalytic Insight, convincing many current skeptics that the time to search for alien technology is ripe.

One difficulty with the ichnoscale is that it was not specified by Socas-Navarro et al. (2021) whether the scale was fixed to be normalized for Earth technosignatures in 2020, or whether "current" would change with the times. For very large scales the difference in the near future will be small, but for the scale to remain useful it will be good practice in the future to specify the approximate year being used to pin $\iota = 1$. I recommend the scale be frozen at the strength of Earth technosignatures in 2020, around when the scale was conceived.

[1] *Ichnos* (ἴχνος) is Greek for "footstep" and we use the Greek lowercase iota (ι) to represent it.

4.2.2 The Kardashev Scale

> [This vast community] began to avail itself of the energies of its stars upon a scale hitherto unimagined. Not only was every solar system now surrounded by a gauze of light traps, which focused the escaping solar energy for intelligent use, so that the whole galaxy was dimmed, but many stars that were not suited to be suns were disintegrated, and rifled of their prodigious stores of subatomic energy...
>
> —Olaf Stapledon, *Starmaker* (1937)

Kardashev (1964) (Box 4.1) proposed another scale to describe the scope of technology, determined simply by the total energy supply[2] available to a species. Kardashev's purpose was to estimate the total energy available for radio transmissions in an attempt to estimate the detectability of an "extraterrestrial civilization."

Kardashev noted that humanity's energy supply had grown exponentially for decades (he estimated a $\sim$20 yr doubling time) and, following Dyson (1960b)'s argument for Dyson spheres (Chapter 15) argued that there was no physical reason such doubling could not increase indefinitely.

He then roughly categorized all hypothetical civilizations into three types, into what is now known as the Kardashev Scale. Quoting him:

I technological level close to the level presently attained on Earth, with energy consumption at $\approx 4 \times 10^{19}$ erg s^{-1}.

II a civilization capable of harnessing the energy radiated by its own star (for example, the stage of successful construction of a "Dyson sphere"[3]; Dyson 1960b); energy consumption at $\approx 4 \times 10^{33}$ erg s^{-1}.

III a civilization in possession of energy on the scale of its own galaxy, with energy consumption at $\approx 4 \times 10^{44}$ erg s^{-1}.

In this book *The Cosmic Connection*, Sagan (1973) describes a reformulation to Kardashev's scale that makes it continuous, rather than discrete, noting that the jumps between the steps are conveniently close to factors of 10^{10}. This implies the formula

$$K = \frac{\log_{10}(P(\mathrm{MW}))}{10} \tag{4.1}$$

where P in MW is the species' total energy supply.[4] $K = 1$ then corresponds to an energy supply of 10^{10} MW $= 10^{23}$ erg s^{-1}, which is much larger than Kardashev's

[2] *Energy supply* is jargon for the total amount of energy produced and used by some entity.

[3] This appears to be the first use of the term "Dyson sphere" in the literature.

[4] Sagan's scale essentially measures power output in the sensible but improper unit of decabels (DB), or hundreds of decibels.

estimate of 1960s energy supply, and indeed puts humanity in 1960 at around $K = 0.66$.[5]

Conveniently, the steps in Sagan's version of Kardashev's scale correspond within a factor of a few to the energy scales of the total sunlight that strikes the Earth, total solar output, and total luminosity of the Milky Way. Kardashev's three types thus often serve as rough markers not just for energy supply, but for the physical extent of a species, describing those that are confined to their planet, have spread throughout their stellar system, and have spread throughout their galaxy (for instance, Zubrin 2000).

Many other authors have extended and modified the Kardashev scale, for instance invoking a fourth level (i.e., an intergalactic species; Gray 2020) or additional dimensions describing other aspects of their energy use (for instance, the scale of matter manipulation, see Vidal 2011).

4.2.3 A Note on "Advanced" Technology

4.2.3.1 Alien Technology Will Likely Be Older Than Ours

There are few things we can say with much certainty about the nature of alien technology, but one is that it will very likely be much *older* than our technology. Consider the entire temporal span τ of a technological species. We might detect this species at any point in its history when it is detectable, which might be random or favor later times in its span if its technosignatures grow with time.

If τ is much more than a few thousand years (i.e., most species have technology much longer than we have already) we are much more likely to detect them when they are older than us than younger. In addition, for many species with many lifetimes, we are more likely to detect the longer lived ones exactly because they are around longer for us to find. Both factors strongly favor older species. This argument has been formalized by Kipping et al. (2020).

4.2.3.2 Human Analogies

Socas-Navarro et al. were deliberate in their choice of terms and phrasing of the ichnoscale, and clear that they were merely describing the "footprint" of a technology (its literal scale in the sense relevant for its detection) and making no claim about the inevitability or level of "advancement" of the technology involved.

This is not true, however, of Kardashev and Dyson, who made arguments based on their understanding that such large energy supplies were a natural consequence of human, and perhaps alien, nature.

Dyson (1960a) wrote:

> I consider that an open society would be likely to expand by proliferation of 'city states' each pursuing an independent orbit in space. Such expansion need not be planned or dictatorially imposed; unless it were forcibly stopped it would result in the gradual emergence of an artificial biosphere of the kind I have suggested.

---

[5] According to the International Energy Agency, humanity's total energy supply in 2019 was around 606 EJ, for an average power of 1.9×10^{20} erg s^{-1}, moving humanity up to $K = 0.73$.

Kardashev (1964), extrapolating humanity's exponential energy growth into the future wrote:

> ...we find that the energy consumption per second will be equal to the output of the Sun per second, 3200 yr from now, i.e., 4×10^{33} erg s^{-1}, and that in 5800 yr the energy consumed will equal the output of 10^{11} stars like the Sun. The figures arrived at seem to be inordinately high when compared to the present level of development, but we see no reasons why the tempo of increase in energy consumption should fall substantially than predicted. Moreover, the availability of a large amount of information forthcoming from other and more highly developed civilizations might contribute to a staggering increase in energy consumption.

For reasons relevant to the dominant political philosophies of their places and eras, Dyson and Kardashev subscribed to different but related sorts of a unilinear, hierarchical view of technological advancement, imagining that the scale and sophistication of human technology has grown and will continue to inevitably grow with time.

The view of the progress of human history, perhaps according to a particular order (stone age, iron age, space age, etc.) has been largely abandoned as inaccurate and harmful by the social sciences that study the matter. It is inaccurate because human technological progress, while it certainly can be cumulative, is nonlinear and multidimensional. It is harmful because claims that a culture is "primitive," "backwards," or "barbaric" illogically connect the technology used by a culture with the moral standing of its members, and have been and still are regularly used to justify political domination, violence, and even genocide.

We should be careful not to bring these views along when applying the ideas they have contributed to SETI for two reasons. The first is that it is important that the rationale for SETI be based on a firm scientific foundation, and not on false or highly debatable beliefs about human history and technology. The second is to avoid perpetuating harmful ideologies here on Earth.

So, we should not assume that any species (or human society) is or is not "advanced" in a general or specific sense based on the kind of technology they routinely apply to a particular task. We should also not assume a species' ichnoscale will necessarily grow with time, or that technologies at a similar ichnoscale will be of similar character or sophistication across species.

We will return to these themes in Section 18.5.

4.2.3.3 A Helpful Analogy from Evolution

Kardashev and Dyson's ideas are still useful, however, even without the thinking that motivated them. To see how we can separate these things, consider an analogy from the evolution of life on Earth.

Evolution of plants and animals on Earth is not directional—there is no generic tendency for creatures to become larger, for instance. Nonetheless, a few species have selected for size for long enough that they are enormous—the blue whale and giant sequoia are ostentatious outliers in the animal and plant kingdoms.

We could consider a chart of the size of creatures on Earth with time. For 3 billion years, all life on Earth was microscopic (or, at least, single-celled) and so the average size of organisms (or, if you like, the size of a typical member of the largest species) was measured in microns.

About 1.5 billion years ago, multicellular life developed and the size of organisms grew. Far from an inevitable step on the march of evolution, the late arrival of macroscopic organisms might even suggest that it was a highly contingent event (Section 7.5). Importantly, *single-celled life did not go away* and today still composes a significant fraction of all biomass on Earth: being large is not so advantageous that macroscopic life could outcompete microscopic life across all of Earth's biosphere's niches (far from it).

Eventually, some organisms became quite large, and this took time. Our chart of the sizes of organisms might show an upward trend until evolution found a natural maximum useful size in the largest creatures, and from there would flutter up and down as various species of the largest megafauna and megaflora come and go. Certainly, the path from microscopic to enormous must include intermediate steps, and passing through those steps requires time. In this sense we do expect that larger species must be old (in the sense that they are all the products of at least 3 billion yr of evolution) but it would not be *generally* true that older species are larger.

To develop the analogy further, imagine that we are on safari and interested in the kinds of life that exist in the wild. Most of the life is microscopic, and we will not be able to see it with our unaided eyes no matter where we go or in what direction we look. Macroscopic life we have a better chance of finding, and the larger it is, the greater our search space because the farther it can be and still be detectable. Although they are rare, we will readily spot distant giant trees, giraffes, and elephants. We would not call these larger species more "evolved" than the other, smaller species (they are not!), and we would be incorrect to think that they are somehow older than other species. We do, however, expect that they took a long time to evolve to such size, and that many of their ancestors in the past must have been smaller.

Similarly, we might expect a few species elsewhere in the Universe to develop very large energy supplies or technologies with large ichnoscales for whatever reasons happen to drive them to exploit the limits of exponential growth in that direction, without invoking analogies based on outdated ideas of human progress. We expect that it takes time to build technologies with large ichnoscales, and to develop the scientific and engineering understandings necessary for such technology. Any species that does so will be old, and will have had technology longer than humanity, but we should not assume they have followed a similar path through technology that we have or that we will in the future. We will be most sensitive to those species with the largest ichnoscales and, depending on how abundant they are, we might find those presumably rare manifestations of ETI first, but this will not imply that they are the oldest, or even the most "advanced" species in the Galaxy.

We will return to this theme in some detail in Section 18.5, and here simply emphasize the utility of being able to quantify the scale of a technology or signal to determine its detectability, without making any implicit judgment on the level of "advancement" or inevitability of that technology. In this text we will thus avoid referring to "advanced" technology in favor of describing its ichnoscale value, or otherwise describing its relevant scale.

4.3 Cultural Expectations of Contact

As we will examine in more detail in Chapter 18, much of the historical expectation for the nature ETIs is based on expectations of humanity's future (as Dyson and Kardashev's ideas make clear). This expectation is implicit in the Catalytic Insight, and explicit in the ichnoscale itself. This is a very useful heuristic: it is straightforward to begin thinking about SETI by extrapolating humanity's current technosignatures to larger scales and searching for those.

Science fiction, especially, has primed us to expect to eventually be in communication with highly intelligent aliens with social structures and technology reminiscent of ours but far more advanced. We expect them to have spacecraft, abundant energy supply, and an appreciation for science, technology, and mathematics. If they are friendly, they will have much to teach us, we are told, if they deign to interact with a lowly species like ours at all. If hostile, they will be formidable and potentially unstoppable foes.

We should be careful, though, of making too much of the "aliens as future humans" analogy and consider that Earth provides many examples of intelligent, social, communicative, and tool-using species that do not fit this mold. Beavers, ant colonies, dolphins, elephants, octopi, and the great apes are just a few examples of beings with what might be described as having an "alien" intelligence or cognition that might provide better templates for interspecies communication with real aliens than any sci-fi movie has contemplated. Despite centuries of study at close range, we still cannot communicate with them, except superficially. How difficult will an alien consciousness be to communicate with?

Box 4.1: Nikolai Kardashev (1932–2019)

Kardashev in Santa Cruz in 1991. Image credit: Seth Shostak, SETI Institute.

(Никола́й Кардашёв, roughly "kerr-duh-SHOAF" in American English).

Prof. Kardashev was a Soviet radio astronomer and prominent member of the Soviet CETI program. Born to Bolshevik parents, he lost his father and sister to Stalin's purges when he was very young; he grew up variously in an orphanage, with extended family, and on his own.

His PhD thesis adviser at Moscow State University was Iosif Shklovskiĭ (Box 3.2), a charismatic pioneer in the field of radio astronomy and the author of an influential text on extraterrestrial intelligence with Carl Sagan. Kardashev's thesis included a theoretical treatment of recombination lines at radio frequencies; these were discovered just a few years later, earning him the USSR State Prize. His thesis also included an early theoretical discussion of neutron stars that effectively predicted the existence of pulsars.

Kardashev was a pioneer of very long baseline interferometry, and spent much of his career championing and overseeing the RadioAstron space interferometer (finally launched in 2011!). He became the deputy director of the Soviet Space Research Institute in 1977, served as vice president of the Committee on Space Research (COSPAR) from 1982–1986, and was chair of the Russian Academy of Sciences Council on Astronomy from 1999 until his death.

In SETI, Kardashev is most famous for his eponymous scale of civilizations described in Section 4.2.2. He briefly achieved worldwide fame for suggesting (as described in that same paper, Kardashev 1964) that the spectrum of the strange, pointlike radio source CTA-102 had a spectrum consistent with his predictions for the combined radio chatter of transmitters in a distant Galaxy (the topic of the eponymous song by the Byrds in 1967; CTA-102 is an active galactic nucleus).

For more on his life and career, see Balega et al. (2020).

Even these terrestrial analogs might be too close to home to be of use: humans share more evolutionary descent with an elm tree than with any alien species we might detect, and even our wildest speculations on what we find might end up reeking of what Sagan called "Earth chauvinism."

This is just one example of the ways in which the nature of the contact SETI seeks might defy our expectations. Keeping an open mind in SETI can be a challenge, both because it is hard to shake popular narratives of aliens, and because one must start *somewhere* when starting the search, so it would be foolish to ignore the only examples of intelligent life we have.

We will return to this theme in Chapter 18, but for now we will simply be alert for unwarranted assumptions based on popularized versions of Earth and human history, especially those that have been baked into the founding ideas of the field.

4.4 Physics As We Don't Know It

> Any sufficiently advanced technology is indistinguishable from magic.
> —Arthur C. Clarke's "Third Law," from Clarke (1962)

4.4.1 What Might Be Possible

A common trope in science fiction is advanced technology that violates the laws of physics as we know them. *Star Trek* helped establish many of these in popular culture: teleportation (via a "transporter"); faster-than-light communication (via "subspace"); faster-than-light spaceflight (via "warp drive" or a "wormhole"); arbitrary energy-to-matter conversion (via "replicators"); arbitrary force generation ("containment fields," "artificial gravity"); and time travel, to name a few.

It is important to remember, though, that these technologies aren't in science fiction because physicists expect them to be developed in the future; many of them are used purely as plot and production conveniences. Artificial gravity is ubiquitous in television and film because simulating zero-g is awkward and expensive; the transporter on *Star Trek* was reportedly there simply because models of shuttlecraft were beyond the budget of the first season. Though they violate special relativity, faster-than-light communication and warp drive allow for familiar narrative conventions to be played out across enormous gaps of space. We will discuss this in more detail in Section 18.3.2, but for now we will simply note that we should not let these contrivances have an outsized impact on our expectations for actual alien life.

To be sure, we know that our understanding of the laws of nature is missing important pieces: the nature of dark matter and dark energy both suggest either our understanding of spacetime (general relativity) or the components of the Universe (the Standard Model) or both is incomplete. The Standard Model also fails to account for the mass of the neutrino, and many physicists believe (or at least hope) that much of the complexity of the Standard Model can be explained with a simpler model (explaining, e.g., why there are three generations of particles, or why various physical constants have the values they do).

Box 4.2: Jargon Review: The Big Picture

- **The Catalytic Insight of Modern SETI**: The realization that humanity is able to produce signals that humanity would be capable of detecting at interstellar distances.
- **Energy supply**: The total amount of energy generated and used by an entity, such as a city, nation, or species. Usually expressed as a power, calculated as the energy generated or used per unit time.
- **Ichnoscale**: A linear quantification of the "footprint" or scale of a technology, normalized such that humanity's total footprint has $\iota = 1$ (in the year 2020).
- **Kardashev Scale**: Originally, a three-step scale of "types" of species based on total energy supply, with type I having roughly humanity's level of energy use, type II having energy equal to a solar luminosity, and type III having the total energy of the Milky Way Galaxy.

  A popular revision by Sagan makes the scale continuous, with $K = \log_{10}(P/\mathrm{MW})/10$, with integers separated by powers of 10^{10}.

  Many extensions have been proposed, and Kardashev's types are now often used to refer to the spatial extent of a species, rather than its energy supply.
- **Advanced technology**: A term to be used with care, since it carries along antiquated and often harmful ideas of human progress. Better to refer to the scale or form of technology being described.

On a grander scale, cosmologists seek a model of the Universe that explains the origins of the Big Bang (such as inflation), and physicists hope that a Grand Unified Theory or Theory of Everything might further unify the disparate theories of the fundamental forces.

We thus have an expectation that there is more physics to discover. We also have an expectation that work within the laws of physics as we know them will enable technology that would seem like magic to us, much the way we imagine our modern conveniences like computers, rockets, televisions, smartphones, antibiotics, plastics, and teleconferencing might seem like magic to humans that lived just 200 years ago.

But we should be careful with such expectations, because if we allow that any of the laws of physics as we know them might be overcome with sufficient understanding and technology, then we are left without any foundation to do a scientific study. If we cannot assume the laws of physics are true, then we cannot really do physics at all.

4.4.2 How Should We Deal with Physics We Don't Know?

> Cursed is he that does not know when to shut his mind. An open mind is all very well in its way, but it ought not to be so open that there is no keeping anything in or out of it. It should be capable of shutting its doors sometimes, or it may be found a little draughty.
>
> —Samuel Butler (1908)

Freeman Dyson was fond of exploring the physical limits of technology according to the laws of physics as we know them (his articulation of Dyson spheres in his 1966 essay (Dyson 1966) is an excellent example of this). He thus distinguished between laws of nature that he was willing to accept as immutable and universal, and problems that were "just engineering."

Put another way: no matter how sophisticated science and technology are, they are bound by the laws of physics, whatever they may be. No amount of "advanced" technology can create a warp drive if a warp drive is physically impossible.

We thus need to have various levels of credence for which laws of physics we believe bind all species and technologies, and which we only accept provisionally. We might accept as absolute, for instance, the fundamental limits imposed by the speed if light (since faster-than-light signaling or travel would violate causality), the laws of thermodynamics (which can be as rigorously derived from first principles as anything in physics), and the well-measured parts of the Standard Model. We might accept as more provisional things like the number of fundamental particles, the maximum strength of physical materials, and our understanding of extreme environments like black holes or the early Universe.

If one wishes to go beyond what appears to be bedrock physics and still be doing science, it is important to be clear about the parameters of the investigation. For instance, if we ask whether warp drives might produce a technosignature in gravitational waves, then we are first hypothesizing that warp drive is physically possible, then that it is practical from an engineering perspective, then that such engineering exists. There is nothing wrong with this line of inquiry in general, but that first step requires us to clearly state and constrain which laws of physics-as-we-know-it we are willing to suspend or change. If we cannot describe how warp drives would operate under general relativity, then we must justify why it makes sense to use general relativity to calculate their gravitational radiation, if only as a matter of self-consistency.

Ultimately, this text will try to follow Dyson's philosophy of striking a balance between keeping an open mind about what technology could accomplish ("just engineering") and what we must assume to be said to be doing science at all (the widely accepted foundations of physics).

4.4.3 Looking For "Advanced" Technology

An occasional complaint about SETI is that it looks only for technology like ours, and thus is blind to "advanced" technology that alien life will undoubtedly use. An analogy might be radio waves, which we use routinely to communicate but which would have been completely undetectable and inexplicable to people prior to the development of electromagnetic theory in the 19th century. Looking for alien transmissions or other technosignatures, this argument goes, would be as fruitless as a 15th century human looking for a WiFi signal.

There are several responses to this argument.

One is to acknowledge this may be true, in which case there is nothing we can do to find such signals, so there is not much point in worrying about it. They may as well not exist.

Another is to deploy part of a rebuttal to the Fermi Paradox (the monocultural fallacy, see Section 6.3.3) and say that unless *all* alien species *exclusively* use such technology, we will still be able to find the other forms. After all, humans do not *exclusively* use radio waves, and a 15th century human would have no trouble discovering and recognizing our letters in the postal service, our conversations, and our clock tower bells for what they are.

Finally, there are good reasons to think that photons are the optimal channel for interstellar communication, as we will discuss in Section 17.3. We should of course have the humility to acknowledge this may not be true, while also recognizing the unlikelihood.

Briefly, if there exists another way to send information across space, it must either be a particle we know of and have studied, or a new one. Of the ones we have studied, photons offer many clear advantages. If there exists a new particle, then it must interact with normal matter in order for aliens to use it, in which case we need to understand why we have not discovered it yet.

By "normal matter" we mean electrons and perhaps nuclei. If this new particle interacts with them then we should see the particle emerge in high energy particle experiments. If it interacts very strongly we would have seen it by now; if it interacts very subtly then it will be not just difficult to detect, but also to employ.

Another possibility is that there are quantum properties of the photon that can be exploited that we haven't figure out yet that let communication photons be distinguished from natural ones, perhaps involving entanglement or other quantum properties like orbital angular momentum. This would provide a new way beyond frequency and time compression to make those photons stand out against noise, and explain why we have missed them. This is perhaps more likely, and provides a potentially fruitful avenue for future research, although it is still subject to the monocultural fallacy rebuttal.

4.4.4 Is SETI Science?

SETI, like much of astrobiology, is sometimes subjected to criticisms that it is not a proper science.

One objection is that astrobiology is a "study without a subject" since we have no examples of non-Earth life to study. One rebuttal to this argument is given nicely by Chyba & Hand (2005):

> Astro-physicists...spent decades studying and searching for black holes before accumulating today's compelling evidence that they exist. The same can be said for the search for room-temperature superconductors, proton decay, violations of special relativity, or for that matter the Higgs boson. Indeed, much of the most important and exciting research in astronomy and physics is concerned exactly with the study of objects or phenomena whose existence has not been demonstrated—and that may, in fact, turn out not to exist. In this sense astrobiology merely confronts what is a familiar, even commonplace situation in many of its sister sciences.

Since that writing, the Higgs boson has been detected, and one might similarly point to black holes and gravitational waves as hypothesized phenomena that were extensively studied theoretically and hunted for systematically for decades before they were eventually and decisively discovered.

Another analogy is to the dark matter particle. We have multiple independent lines of evidence that there exists some form of dark matter (Section 2.9) that does not (strongly) interact with ordinary matter or light. The most popular explanation is that it comprises one or more as-yet-undiscovered subatomic particles.

While there is no *a priori* reason such particles need have *any* cross-section with ordinary matter—in which case we can only ever study them observationally via their gravity—many well-motivated extensions of the Standard Model of physics, such as supersymmetry, actually predict such particles should exist with small cross sections.

As such, many experimentalists have hypothesized a wide range of potential interactions that might be detected between the ambient dark matter and experimental equipment and searched for. Non-detections then rule out cross sections above some threshold, leading to exclusions of large amounts of parameter space where such particles might exist, akin to how upper limits can proceed with SETI (Section 11.6).

SETI is actually somewhat *better* motivated than searches for the dark matter particle(s), or even for searches of black holes and gravitational waves before they were successful. This is because in humanity we at least know with certainty that the phenomenon we seek—technological life—exists. The question for SETI, then, is simply one of abundance, not existence.

Another objection is that SETI and METI are in some sense quasi-religious endeavors, seeking heavenly beings that are in some sense superior or beyond humanity who might save or destroy us (Traphagan 2018). While one can certainly find examples of this thinking in the SETI literature, this is essentially a critique of particular motivations for pursuing SETI or hypothesizing certain technosignatures or kinds of alien life. While we should certainly examine such motivations, this objection does not strike at SETI's scientific foundations.

4.5 Further Reading

Physics of Star Trek: Krauss & Hawking (1997).

Exercises

1. The total energy supply of all of humanity increases both because the population grows and because the amount of energy used per capita increases. Historically, for many centuries the growth has been such that energy supply doubles very roughly every 20 years or so, which is what Kardashev assumed. Let's check Kardashev's math.
   (a) Write down an equation for the average power in MW (megawatts) (energy use per time) used by humanity as a function of time given a

doubling time $\tau = 20$ yr (you may also do your own research to get a better value for this number).

(b) Write down another equation for Sagan's version Kardashev's type as a function of year. Solve it to determine: in what year will we hit $K = 0.8$? 0.9? 1? 2? 3?

(c) Calculate the total power of sunlight that falls on the Earth. In how long (in what year) will half of the Earth need to be covered in solar panels to satisfy humanity's energy needs? How long until even 100% of Earth's surface area is insufficient?

(d) Taking the equation at face value, how long until the entire output of the Sun is insufficient to satisfy humanity's energy needs?

(e) How long until the entire output of all 10^{11} stars in the Milky Way is not enough? (You may do this rigorously if you like, or you can use the rule of thumb that the average luminosity of a star in the Milky way is about 1/10 that of the Sun's, to within a factor of a few.)

(f) How long would it take to actually *reach* all of those stars, if the Milky Way is roughly 30 kpc across?

(g) What are some of the limiting factors that might thwart this progression toward becoming a type II species? When do you think it will stop?

2. *Choose one of Earth's more obvious technosignatures, such as radio transmissions, number of satellites in orbit, number of deep space probes, level of atmospheric pollution from CFCs, or level of infrared emission from energy use.

   (a) Use ADS or a web search to find credible estimates for how the magnitude of this technosignature has changed in Earth's past. Preferably, these will come from credible governmental reports or peer-reviewed papers.

   (b) Make a plot of the ichnoscale value for this technosignature with time.

   (c) Extrapolate this plot into the future. What might ultimately limit the growth of this technosignature in the future? Can you use the ADS SETI bibgroup to find an estimate of how large it would need to be to be detectable at interstellar distances? Keep in mind most literature does not use the ichnoscale formalism; you'll have to do the translation.

References

Balega, Y., Gurvits, L. I., Dagkesamanskii, R. D., et al. 2020, PhyU, 63, 622

Bracewell, R. N. 1960, Natur, 186, 670

Butler, S. 1908, The Note-Books of Samuel Butler, Vol. 1 (London: J. M. Dent & Company)

Chyba, C. F., & Hand, K. P. 2005, ARA&A, 43, 31

Clarke, A. C. 1962, Profiles of the Future: An Enquiry into the Limits of the Possible (London: Gollancz)

Cocconi, G., & Morrison, P. 1959, Natur, 184, 844

Drake, F. D. 1961, PhT, 14, 40

Drake, F. D. 1965, Current Aspects of Exobiology, ed. G. Mamikunian, & M. H. Briggs (Oxford: Pergamon) 323
Dyson, F. J. 1960a, Sci, 132, 250
Dyson, F. J. 1960b, Sci, 131, 1667
Dyson, F. J. 1966, Perspectives in Modern Physics: Essays in Honor of Hans A. Bethe on the Occasion of His 60th Birthday, July, 1966, ed. R. E. Marshak, J. W. Blaker, & H. A. Bethe (New York: Wiley) 641
Gray, R. H. 2020, AJ, 159, 228
Kardashev, N. S. 1964, SvA, 8, 217
Kipping, D., Frank, A., & Scharf, C. 2020, IJAsB, 19, 430
Krauss, L. M., & Hawking, S. 1997, The Physics of Star Trek (New York: Flamingo)
Sagan, C. 1973, The Cosmic Connection. An Extraterrestrial Perspective (Garden City, NY: Anchor Press, Doubleday)
Schwartz, R. N., & Townes, C. H. 1961, Natur, 190, 205
Socas-Navarro, H., Haqq-Misra, J., Wright, J. T., et al. 2021, AcAau, 182, 446
Traphagan, J. W. 2018, SpPol, 46, 18
Vidal, C. 2011, arXiv:1104.4362
Wells-Jenson, S. 2023, Xenolinguistics: Towards a Science of Extraterrestrial Language, ed. D. A. Vakoch, & J. Punske (London: Taylor and Francis)
Zubrin, R. 2000, Entering Space: Creating a Spacefaring Civilization (New York, USA: Jeremy P. Tarcher/Putnam)

Chapter 5

Schelling Points

> [A similar problem to the one under discussion] is meeting on the same radio frequency with whoever may be signaling to us from outer space. "At what frequency shall we look? A long spectrum search for a weak signal of unknown frequency is difficult. But, just in the most favored radio region there lies a unique, objective standard of frequency, which must be known to every observer in the universe: the outstanding radio emission line at 1420 megacycles of neutral hydrogen" (Giuseppe Cocconi and Philip Morrison, *Nature*, 1959, 844–846). The reasoning is amplified by John Lear: "Any astronomer on earth would say, "Why, 1420 megacycles of course! That's the characteristic radio emission line of neutral hydrogen. Hydrogen being the most plentiful element beyond the earth, our neighbors would expect it to be looked for even by tyros in astronomy'" ("The Search for Intelligent Life on Other Planets," *Saturday Review*, 1960, 39–43). What signal to look for? Cocconi and Morrison suggest a sequence of small prime numbers of pulses, or simple arithmetic sums.
>
> —Thomas Schelling (1960), The Strategy of Conflict

5.1 Schelling Points

Driven by the necessities of early equipment with limited sensitivity and computational power, the first SETI searches were motivated largely by the hope that strong signals were being sent directly to Earth, at easily guessed frequencies. Much work in the early literature is thus focused on the problem of *beacons*: strong, "we are here" broadcasts intended to be found by us.

doi:10.1088/2514-3433/adf5aech5

In 1960, American economist Thomas Schelling[1] published *The Strategy of Conflict* (Schelling 1960) in which he described a non-communicative, cooperative game. That is, a game in which players must work together to all win or all lose together, but in which they cannot communicate in any way.

Schelling's example was meeting someone in New York City without having prearranged a place or time to do so, in an age before mobile phones and other methods of easy communication. At first the task seems impossible: with no plan and no communication, there is no way to know where to go, or when, and so no way to find one person out of millions.

But some consideration reveals that there are certainly *bad* strategies: one does not search under a random overpass at 3:26 AM. There are other times and places that are more salient, and which give the searchers a better chance at success. When teaching the concept, Schelling would poll his students on where they would go, and when, to find their hypothetical counterparts in the game. Popular answers were famous New York City landmarks and noon; the fact that so many students guessed the same answer implied that many of them would, indeed, have found each other if actually playing the game.

Indeed, the news program ABC Primetime around 2006 actually had six pairs of strangers play this game, attempting to find each other in Manhattan. All of the players had found at least one other team by the end of the day, having met on the observation deck of the Empire State Building and in Times Square (after a near miss that morning in Grand Central Station).[2]

The reason this game is winnable, Schelling explained, is that it is possible to guess the others' strategy, and for them to guess what you would guess about their strategy, and so on. He cites the (antiquated) practice of "meeting" someone in the personal column of a newspaper as being a good example, and then includes the extraordinary footnote quoted at the beginning of this Chapter.[3]

---

[1] Schelling (1921–2016) won the 2005 Nobel Prize in Economics "for having enhanced our understanding of conflict and cooperation through game-theory analysis." His more well-known contributions in foreign relations include formalizing the "madman" or "credible irrationality" strategies (briefly, if you can convince your opponents that you might irrationally harm both of you if you do not get your way, their optimum counter-strategy may be appeasement) and "mutually assured destruction" (a credible promise to use overwhelming and mutually destructive nuclear force in response to any nuclear aggression). Both strategies are, of course, extremely dangerous (the latter played strongly into fears that nuclear war would extinguish the human species, which in turn led to pessimistic values for Drake's L), but their identification and analysis were important steps toward modern theories of nuclear non-proliferation and rogue state containment.

[2] See https://rowan.illumira.net/show.php?pid=njcore:17377; the episode is variously called "Mission Impossible: Lost in New York" or "Game Theory: Lost in New York."

[3] One wonders if Schelling added the footnote in proofs after reading the *Saturday Review* article, which was published just months before his own book!

The point of the exercise is not that it is *easy* to win such a game, just that one can improve one's odds if one can successfully guess what your partner will do, and vice versa. The reason the players of "Lost in New York" were able to win is that they had a shared cultural knowledge base with each other, for instance knowing that the observation deck of the Empire State Building is a famous meeting spot from the films *An Affair to Remember* and *Sleepless in Seattle*, and that noon is the most salient time of day.

The term has had many names in the literature, and has been apparently re-invented many times. Schelling called these equilibria in his game "focal points," but that term is already well known jargon in astronomy. The concept was also called the "Principle of Anti-Cryptography" by Dixon (1973); a "mutual strategy of search" by Makovetskii (1980) and "a convergent strategy of mutual search" by Filippova et al. (1991); signals coordinated in time were called "synchrosignals" by Makovetskii (1977); and Filippova & Strelnitskij (1988) referred to these points as "attractors" for SETI. Today, all such suggestions for the how, where, and when of finding beacons are best known as "Schelling Points."

5.2 Frequencies

Most work on Schelling points in SETI has been on the appropriate frequencies to tune radio telescopes to, driven by the relatively narrow bandwidth available to early radio backends. By the late 1970s, Tarter et al. (1980) and others were performing million- and billion-channel searches that lessened the need to consider what "magic frequencies" to search in, in favor of choosing broad regions to search.

5.2.1 1.420 GHz (21 cm) and the Water Hole

Cocconi & Morrison (1959) originally suggested 1.420 GHz as a frequency for beacons to transmit at not just because ETIs might expect us to look there for such beacons, but because "it is reasonable to expect that sensitive receivers for that frequency will be made at an early stage of development of radio astronomy. That would be the expectation of the operators of the assumed source, and the present state of terrestrial instruments indeed justifies the expectation." In other words, ETIs might catch the attention of *all* radio astronomers in the Galaxy, not just those looking for them.

There was a second reason to search near this frequency: it was in the region of the spectrum between 1–10 GHz where the noise from the Galactic synchrotron emission and Earth's atmosphere were both low, part of the Terrestrial Microwave Window, shown in Figure 5.1.

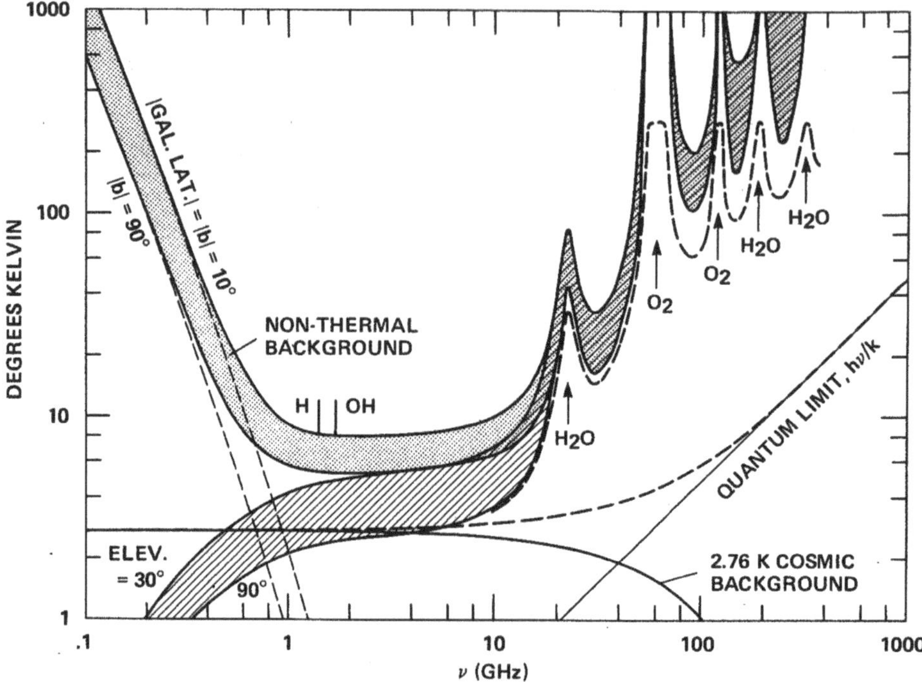

Figure 5.1. From Morrison et al. (1977), an illustration of the terrestrial microwave window. The *y*-axis is the intensity of radio emission detected on Earth expressed as a brightness temperature[4] in Kelvins, a common unit in radio astronomy. Shown are the values of the cosmic microwave background, typical values of Galactic synchrotron emission at the Galactic poles (latitude $|b| = 90°$) and near the Galactic plane ($|b| = 10°$), and typical contributions from emission from water and molecular oxygen in Earth's atmosphere at various elevations (angle above the horizon). The quantum limit refers to the noise introduced from the discrete nature of photons, which provides a floor on the fidelity with which a weak signal can be detected via radio techniques involving linear amplification (see Section 13.2). Also shown are the locations of the H and OH^- lines at 1.420 and 1.665 GHz. The region from 1 to 10 GHz is the *terrestrial microwave window*. The region between the emission lines is the Water Hole. Image Credit: NASA Ames Research Center.

The discovery of maser emission from the hydroxyl (OH^-) radical at 1.663 GHz in 1963 led to another Schelling point.[5] Barney Oliver in the Cyclops report (Oliver 1971, Section 3.3.2) put it best:

> Nature has provided us with a rather narrow quiet band in this best part of the spectrum that seems especially marked for interstellar contact. It lies between the spectral lines of hydrogen (1420 MHz) and the hydroxyl radical (1662 MHz). Standing like the Om and the Um on either side of a

[4] That is, the temperature at which a blackbody would have the measured intensity.

[5] The hydroxyl radical has four transitions, at 1.612, 1.665, 1.667, and 1.720 GHz. The 1.665 and 1.667 GHz lines are the strongest, and there was at first theoretical and observational uncertainty about what the exact frequencies were; they are often listed in the early literature as a single feature at 1.662 GHz.

> gate,[6] these two emissions of the disassociation products of water[7] beckon all water-based life to search for its kind at the age-old meeting place of all species: the water hole.

Later, Oliver (1979) explained the reasoning here, making the need for exploiting a common cultural currency to find Schelling points explicit:

> It is easy to dismiss this as romantic, chauvinistic nonsense, but is it? We suggest that it is chauvinistic and romantic but that it may not be nonsense.
>
> It is certainly chauvinistic to water-based life, but how restrictive is such chauvinism? Water is certain to be outgassed from the crusts of all terrestrial planets that have appreciable vulcanism and, therefore, a primitive atmosphere capable of producing the chemical precursors to life. We can expect seas to be a common feature of habitable planets. Exobiologists are becoming increasingly disenchanted with ammonia and silicon chemistries as bases for life. Water-based life is almost certainly the most common form and well may be the only (naturally occurring) form.
>
> Romantic? Certainly. But is not romance itself a quality peculiar to intelligence? Should we not expect advanced beings elsewhere to show such perceptions? By the dead reckoning of physics we have narrowed all the decades of the electromagnetic spectrum down to a single octave where conditions are best for interstellar contact. There, right in the middle, stand two sign posts that taken together symbolize the medium in which all life we know began. Is it sensible not to heed such sign posts? To say, in effect: I do not trust your message, it is too good to be true!

Drake & Sagan (1973) suggested that there is a natural frequency within the Water Hole that most corresponds to water: the weighted mean of the frequencies (or wavelengths, or log wavelengths) where the weights correspond to the atomic weights of the components (so, the line for H would have weight 1 and that of OH^- would have weight 17; that is, the frequency is 17/18 of the way from H to OH^-).

5.2.2 Physics- and Math-based Schelling Points

Most efforts to find Schelling points with beacons have dispensed with the romance and stuck to arguably less sentimental points of potential commonality: physics and

[6] Oliver is referring to the first and final syllables of the *Oṃ maṇi padme hūṃ* (ॐ मणि पद्मे हूँ) mantra of Tibetan Buddhism, and to the Niō (仁王) and komainu (狛犬) guardian statues at the doorways of Japanese and other East Asian temples, which mouth these syllables, symbolizing the beginning and end of all things.

[7] Liquid water consists mostly of H_2O, but the molecules exist in equilibrium with their dissociation products $H_2O \rightleftharpoons H^+ + OH^-$. The metaphor is not precise, however: the more accurate reaction is $2\ H_2O \rightleftharpoons H_3O^+ + OH^-$, and anyway the 1.420 GHz line is from *neutral* (atomic) hydrogen, and is generated by the electron that the H^+ ion lacks.

mathematics. The reasoning is that any species that has constructed a powerful radio transmitter will necessarily understand engineering, which implies mastery of these fields.

An early articulation of this idea actually dates back to Planck (1900) in his foundational work describing "natural units" derived from Planck's constant, Newton's gravitational constant, and the speed of light (which we today call the Planck length, time, and mass). He wrote:[8]

> It is interesting to note that with the help of [these physical constants] it is possible to introduce units…which…remain meaningful for all times and also for extraterrestrial and non-human cultures, and therefore can be understood as "natural units."

and noted[9]

> …these units keep their values as long as the laws of gravitation, the speed of light in vacuum, and the two laws of thermodynamics hold; therefore they must, when measured by other intelligences with different methods, always yield the same.

Planck's definition of "natural" is that which is free of arbitrary human choices and conventions, and therefore that which we could know in common with extraterrestrial species.

Wright (2020) explored whether, therefore, the Planck units could be used as magic frequencies. Planck's "natural" frequency might correspond to the photons at the Planck energy, yielding:

$$\nu_{\mathrm{P}} = \frac{E_{\mathrm{P}}}{h} = \frac{1}{2\pi t_{\mathrm{P}}} = \sqrt{\frac{c^5}{2\pi\hbar G}} \tag{5.1}$$

where $\hbar = h/2\pi$ is the "reduced" Planck constant, E_P is the Planck energy $\sqrt{\hbar c^2/G}$, and $t_{\mathrm{P}} = \sqrt{\hbar G/c^5}$. This frequency is extremely large, corresponding to energies of $\sim 10^{28}$ eV, and indeed it is unclear if photons of this energy can even exist, since the Planck units represent in some sense the scales at which general relativity and quantum mechanics come into conflict.

Wright (2020) explored whether there might be other combinations of physical constants that generate more useful frequencies, and noted there are at least two dimensionless quantities that seem "natural" and which might multiply this frequency down to the radio region: the natural logarithm e and the fine structure constant $\alpha = e^2/c\hbar$, where e is the electron charge in electrostatic units.

One can thus generate a series of photon energies $E = E_{\mathrm{P}}\alpha^n$ or $E = E_{\mathrm{P}}e^n$ where n is some positive integer. This generates a series of evenly-spaced energies (or

[8] Translation by Sabine Hossenfelder.

[9] Translation by Michael Hippke.

"comb") from the very highest energies allowed by nature down through the EM spectrum down to arbitrarily low energies.

Wright (2020) noted that there are still many anthropocentric choices being made in this comb's construction: one might instead choose to define Planck units in terms of h, not $\hbar$; one might use the Rydberg ($\frac{1}{2}m_e c^2 \alpha^2$) as the basis of energy; one might define the comb to start at $\nu = 1/t_P$; and so on. At best, the exercise can identify a large number of *potential* Schelling points to search.

Makovetskii (1976) suggested using mathematical multiples of the 1.420 GHz line, multiplying it by e, $\sqrt{2}$, ln(2) or π. This last suggestion was particularly influential, being the frequency of alien transmissions detected in *Contact* (Box 18.1). Blair & Zadnik (1993) collected a large number of similar suggestions, combining the frequencies of H, OH^-, and many mathematical constants to generate a large list of potential frequencies.

Zuckerman (1985) declared that given the large number of options, no real "magic frequencies" exist, and indeed the advent of million- and billion-channel receivers somewhat obviates the problem. The exercise might still be useful, however: searches at a small number of potential frequencies can be done to higher sensitivity than searches across billions of channels, so one might lower one's threshold for significance at those frequencies in a broadband search.

Wright (2020) also pointed out that the large number of options might imply that any beacons would broadcast on *multiple* frequencies simultaneously, sacrificing some signal strength in favor of providing many more opportunities for a signal to be discovered. Indeed, much communication on Earth across cultures and species is multimodal, and we might expect alien communications to be the same.

5.2.3 Beyond the Radio

Shortly after radio communications were proposed by Cocconi & Morrison (1959), searches elsewhere in the EM spectrum were suggested. Schwartz & Townes (1961) suggested that "optical masers" would be useful for interplanetary and interstellar communication. They wrote:

> It may be both an encouraging enlargement of possibilities and at the same time an unwelcomed complication, to find that the frequency of the hydrogen line in the micro-wave region is not the only reasonable place at which to search for possible interstellar communications, and that the optical region also seems a logical one. What other methods are we overlooking which might appear natural to some other civilization? Far ultra-violet and infra-red are absorbed by most imaginable atmospheres friendly to life, and hence would be avoided unless reception above the atmosphere is to be expected. Beams of charged particles would be subject to inconvenient bending in interstellar fields. We see no way of producing neutron beams or electromagnetic waves shorter than the ultra-violet with adequate intensity.

5.2.4 Correcting for Motion and Doppler Shift

A transmission at a particular frequency will be Doppler shifted (Equation (2.18)) by the motion of the transmitter. Presumably, if a species hopes it is transmitting at a frequency that will be guessed, it will correct for this motion either by altering the frequency of its transmissions in a way that counteracts the Doppler shift, or by placing its transmitter in either deep space or in orbital space and using some sort of thrust to counteract orbital and other accelerations.

The question then arises: which frame would an alien species expect that we would expect to find their carefully chosen frequency? Fortunately, there seem to be some clear answers.

There is a clearly preferred frame of the Universe: that of the cosmic microwave background (Section 2.8.2).[10] We see a clear pattern in this radiation on the sky with one side of the sky being redshifted and the other blueshifted, indicating that the Sun is moving at 369 ± 0.9 km s^{-1} with respect to this frame, due mostly to its motion around the Galaxy and the Galaxy's motion within the Local Group of galaxies.[11]

If, however, the sender and receiver are both in the Milky Way, one might expect that they would use a frame in which our shared galaxy is at rest. This is not as well defined a frame as the CMB, but one might attempt to determine its barycentric frame[12] (Section 2.14.4), or use the frame of the supermassive black hole at its center.

Alternatively, if the signal is intended for the Sun or Earth, one might choose the Local Standard of Rest, which is defined by the circular velocity of material at our distance from the Galactic center, and known to around 1 km s^{-1}. Another frame relevant to Earth is that barycenter of the solar system, which is known exquisitely well, around or better than 1 mm s^{-1}.

The solar system barycenter, Galactic barycenter, and CMB frames were all commonly used; Horowitz & Sagan (1993) checked for signals in all three frames.

5.3 When to Look Where: The SETI Ellipsoid

Pace & Walker (1975) noted that "The chances that two civilizations establish contact with each other by means of interstellar radio communication are exceedingly small in the absence of time markers which will tell the two civilizations when to search for one another." Indeed, always-on, omnidirectional beacons are extremely energy intensive. A beacon could be stronger at fixed energy cost if it sent a signal in a particular direction only occasionally, either as part of a pattern of sweeping the sky or as a periodic omnidirectional broadcast. Coordination in time

---

[10] It is curious that although the laws of the nature seem to have no preferred frame of reference (this is the Lorenz invariance of general relativity), the *material* in the Universe clearly does.

[11] Some of the power in this "dipole term" in the cosmic microwave background is actually intrinsic to the inhomogeneities in our patch of the Universe, and some of it is due to the gravitational effects of matter between us and the surface of last scattering for the CMB (the "integrated Sachs–Wolfe effect"). These effects are smaller than the uncertainty in the magnitude of the dipole term (Planck Collaboration et al. 2014).

[12] Known, we believe, to around 5 km s^{-1}.

also allows a searching party to properly husband their search resources on a given target to likely times of transmission.

In one of the simplest schemes, one saves one's transmission efforts until a coordinating event occurs, such as a bright supernova. Upon detecting the burst, beacons transmit omnidirectionally for some brief period. If a receiver knows the distance to potential transmitters, it then checks those for which the coordinating event appears to have just been noticed.

Tang (1976), McLaughlin (1977), and Makovetskii (1977) suggested coordinating signals via bright transient events visible across the galaxy, such as novae or supernovae. The receiver then searches only along the surface of what Hilton (1981) dubbed the "SETI ellipsoid," the locus of stars which we see, because of light travel time, to have just witnessed the event, as shown in Figure 5.2. As time progresses, the ellipsoid expands, and the receiver has new stars to search.

Figure 5.2 illustrates the geometry, where we put the coordinating event (the burst) at the origin at time $t = 0$. The target/receiver of the coordinated signal sits on the x-axis at a distance $D_{\oplus}$ from the burst. It detects the burst at time $D_{\oplus}/c$.

The transmitter sits in the x–y plane at distance D_* from the burst and polar angle ϕ from the x-axis. The burst–receiver–transmitter angle is

$$\theta = \alpha + \phi - \frac{\pi}{2}. \tag{5.2}$$

The transmitter receives the burst signal at time D_*/c, and then transmits to all potential receivers, omnidirectionally. The receiver then knows to observe a

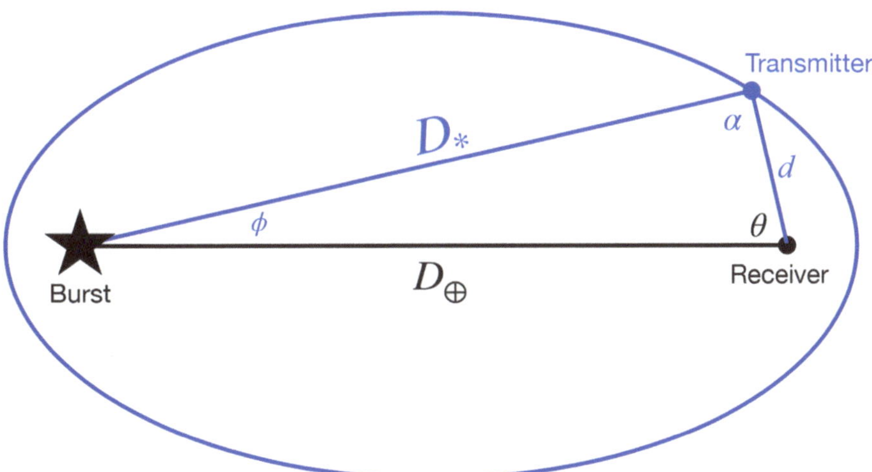

Figure 5.2. Geometry of the SETI Ellipsoid. The Earth/receiver and the burst are on the x-axis, with the burst at the origin. The transmitter transmits at angle α from the burst, and the receiver observes at angle θ from the burst to receive the signal. In the simplest scheme, transmissions are sent out omnidirectionally upon detection of the burst, and so signals arriving at the receiver at any given moment will have originated from transmitters on the surface of the ellipsoid. In the Seto scheme, transmitters wait some time before transmitting in the direction α to coordinate all transmissions in that direction.

transmission sent at angle α from the burst by the transmitter at time $t = (D_* + d)/c > D_\oplus/c$.

Stars outside the SETI ellipsoid have not yet seen the burst from our perspective, and so cannot be expected to be coordinating their signals. The signals from stars within the ellipsiod in this scheme have presumably ceased their coordinated transmissions.

This scheme is simple but has some disadvantages: one must wait for such an event to transmit, and when one does transmit it is done omnidirectionally or to all potential receivers, which is expensive. Receivers also save little effort in this scheme, because stars on the surface of the SETI ellipsoid lie in all directions, necessitating an all-sky survey. Determining which stars lie on the SETI ellipsoid also requires precise knowledge of those stars' distances, which may only be practical for nearby stars (Davenport et al. 2022; Seto & Kashiyama 2020).

5.4 When to Look Where: The Seto Scheme

Seto (2021) described a coordination scheme that eliminates the need for the receiver to know the distance to the transmitters, and provides a narrower range of places to search. In this scheme, one searches only in directions such that the burst–ellipsoid–receiver angle is $\alpha = \frac{\pi}{2}$, and all stars *within* the ellipsoid along that receiver–ellipsoid line transmit at a time appropriate for such reception. Observing at the appropriate angle θ then collects signals from all potential transmitters along the line from the receiver to the ellipsoid in this direction, not just those on the surface of the ellipsoid.

5.4.1 Derivation of the Seto Scheme

Let us derive the search and transmission patterns that allow this to work. Given this burst's properties, which need to be known by astronomers participating in the scheme, we want all transmitters along a given line to transmit along the line in order, each sending their signal when the signal from the previous transmitter arrives, adding their signal to the train. A receiver pointed in the correct direction can thus receive them all at once, without knowledge of their transmitters' distances.

Consider the geometry in Figure 5.3, which shows a line of transmitters in a given direction and its relationship with a coordinating burst. The most salient unique place on this line is its closest approach to the burst, at a point C and at distance L. The most salient unique time at this place is when the signal from the burst reaches C, at time L/c after the burst event.

Each transmitter in this train must thus send a signal in direction α from the burst at the time after the burst goes off

$$t_{\text{sig}} = \frac{L}{c} + \frac{l}{c} = \frac{D_*}{c}\left(\cos\left(\alpha - \frac{\pi}{2}\right) + \sin\left(\alpha - \frac{\pi}{2}\right)\right) = \frac{D_*}{c}(\sin\alpha - \cos\alpha). \quad (5.3)$$

All receivers on the line must receive their signals in direction θ at times given by

$$t_\oplus = \frac{L}{c} + \frac{l + d}{c} = \frac{D_\oplus}{c}(\sin\theta + \cos\theta) \quad (5.4)$$

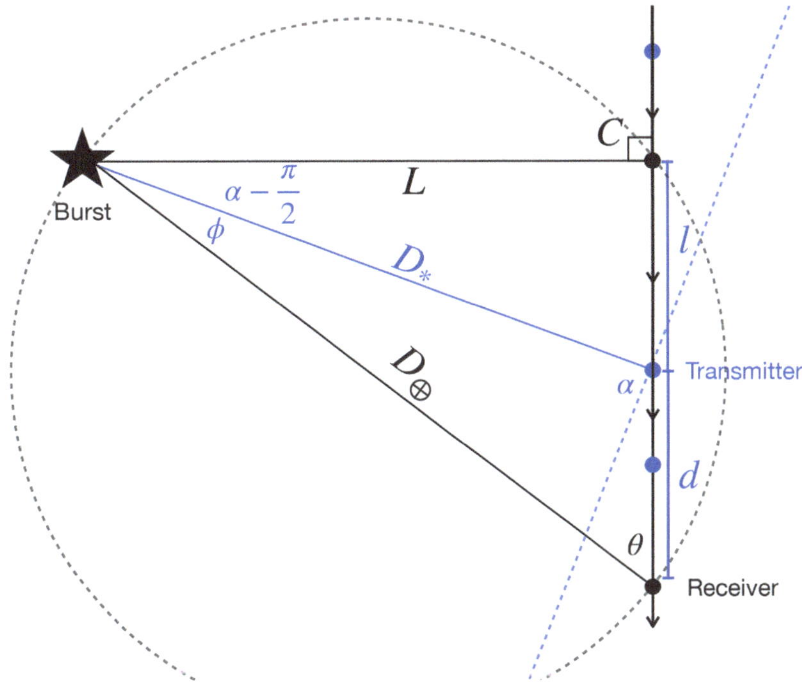

Figure 5.3. Geometry of the Seto transmission scheme. Here, Seto beacons (blue dots) along a given transmission line coordinate their transmissions such that they all arrive at any receiver farther down the line at the same time. A burst occurring at time $t = 0$ allows this coordination. The angles ϕ and α and distances D_* and l are shown in blue for one of the transmitters, using the same geometry and variables as Figure 5.2. The signals are timed such that they are coincident with a theoretical signal passing point C, the closest approach of the line to the burst, at time $t = L/c$ (i.e., when the burst signal arrives there). Transmitters farther from the receiver than C (i.e., those above the horizontal line from the burst to C) cannot participate in the scheme unless they can somehow anticipate the burst. The blue dashed line shows the plane normal to the burst-transmitter line; over the course of its transmissions the transmitter will send signals everywhere to the right/below this plane. The receiver will be able to use this scheme to detect any transmitters within the gray dashed circle that transmit after detecting the burst.

As promised, both the transmission and the search patterns involve only the distance to the burst and the angle from the burst, and so are agnostic about the distance to the communication partner.

Stars within the ellipsoid will, by definition, have already seen the burst, so they will be able to participate in the scheme by transmitting at some angle $\alpha > \frac{\pi}{2}$. Stars outside the ellipse have not seen the signal in time to participate, so they cannot.

5.4.2 Strategies Implied by the Seto Scheme and Its Relationship with the SETI Ellipsoid

Let us examine the solutions to Equations (5.3) and (5.4). Since θ and α are bounded between 0 and π,[13] we have a finite range of times to transmit and receive.

[13] Remember α and θ are angles on the sky away from the burst, and so cannot exceed 180°.

If we require that the transmitter and receiver have already seen the burst when they begin communicating, then we need $t_{\text{sig}} > D_*/c$ and $t_\oplus > D_\oplus/c$, meaning that both points must be below point C in Figure 5.3, which only happens when $0 \leqslant \theta \leqslant \frac{\pi}{2}$ and $\frac{\pi}{2} < \alpha < \pi$. That is, the receiver only searches the hemisphere centered on the burst, and the transmitter only transmits in directions in the hemisphere opposite the burst.

In this range, there are two solutions to both equations, resulting in a pair of rings on the sky centered on the burst for both transmission and reception. The search angles as a function of time after the burst are:

$$\theta_\pm = \frac{\pi}{4} \pm \arccos\left(\frac{ct_\oplus}{\sqrt{2}D_\oplus}\right) \tag{5.5}$$

and the transmission angles are:

$$\alpha_\pm = \frac{3\pi}{4} \pm \arccos\left(\frac{ct_{\text{sig}}}{\sqrt{2}D_*}\right). \tag{5.6}$$

The search angles are illustrated in Figure 5.4. The two rings start with one at the burst and the other at right angles to it. The inner ring grows and the outer ring shrinks until the two rings merge at $\theta_+ = \theta_- = \frac{\pi}{4}$ a time $\sqrt{2}D_\oplus/c$ after it begins, at which point the search pattern is complete. The transmission scheme is the same, but centered on the hemisphere opposite the burst, and occurs over a time $\sqrt{2}D_*/c$.

We can visualize this evolution of the geometry as the SETI ellipsoid grows in Figure 5.5. Because the point C in Figure 5.3 represents the point that, from the receiver's perspective, has just received the burst signal, it must lie on the SETI ellipsoid. One can thus visualize the angle θ in the Seto scheme as the angle toward the point on the surface of the ellipsoid where $\alpha = \frac{\pi}{2}$.

5.4.3 Transmission and Search Volumes

As the ellipse evolves, the transmission lines shown in Figure 5.5 sweep out a volume. Since all triangles containing the diameter of a circle and a point on that

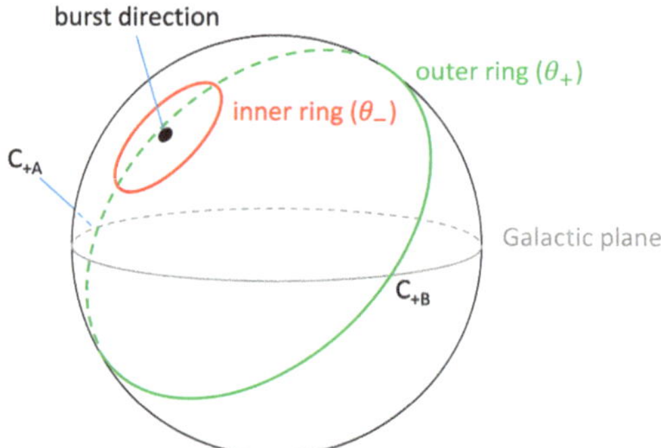

Figure 5.4. The search rings on the sky after a burst. The Galactic plane and its intersection with the outer ring is illustrated here by the symbols C_{+A} and C_{+B}. Reproduced from Seto (2021) © the American Astronomical Society. All rights reserved.

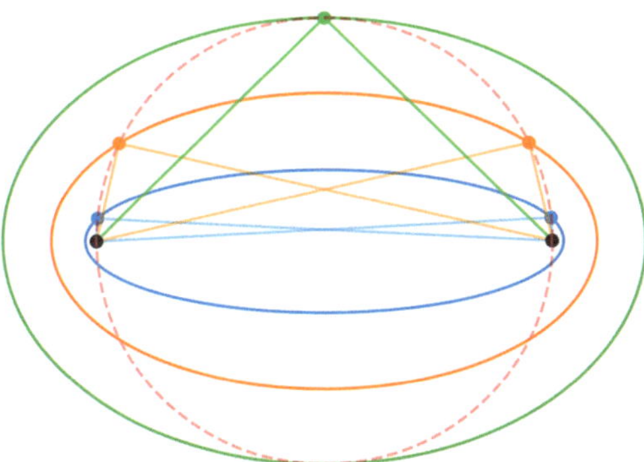

Figure 5.5. An extension of Figure 5.2, showing the geometry of the Seto scheme as it evolves in time. Each color ellipse represents the SETI ellipsoid at a different time as it grows, from blue to orange to green. The colored points represent the special points on each ellipse where the burst–ellipse–receiver angle $\alpha = \frac{\pi}{2}$ (there is a corresponding point on the lower half of the figure not shown; in 3D these points form rings on the sky.) For later times/larger ellipsoids, no such angle exists. The locus of these points from all of the ellipsoids is given by the red dashed line, which is a circle, meaning that the full search volume allowed by such collaboration is a sphere.

circle arc right triangles, the total search volume is a sphere of diameter $D_{\oplus}$, as shown by the red dashed line in Figure 5.5.

The transmissions all are in the range $\alpha > \frac{\pi}{2}$. This means that transmissions will reach all potential receivers in the half of the sky opposite the burst as seen from the transmitter.

5.4.4 Consequences of the Seto Scheme

Galactic supernovae occur at distances of order kpc, and so the transmission and search sweeps coordinated with them are typically of order 10^5 yr. Given that supernovae occur roughly every 100 years in a galaxy like the Milky Way, there should be at any given moment ~1000 transmission rings for any given transmitter, and the same number of receiving rings for us here on Earth.

The width of the rings is a function of the distance to the burst and its uncertainty, the duration one anticipates the coordinated signals will last, the opening angle of the transmissions, and the angular resolution of the receiver. If the width of these search rings is of order 0.001 radian (i.e., a few arcminutes), the many overlapping rings will cover most of the sky.

In other words, even though humanity only knows of a handful of recent Galactic supernovae, and even fewer with good distance measurements, there are likely of order 1000 past supernovae that may have triggered Seto beacons throughout the Galaxy that are still arriving on Earth, from essentially every part of the sky. One advantage of focusing on the recent supernovae we know about as Schelling points for reception is that their SETI ellipsoids are close to Earth, meaning any potential transmitters are nearby and their signals might be more easily detectable.

5.4.5 Extensions of the Seto Scheme

Waiting for a burst is necessarily a strict requirement: if the burst is that of some event that is easily forecasted, one could in principle use it to extend the Seto scheme. For instance if the burst is due to the coalescence of a pair of stars or compact objects, one might detect the final inspiral via gravitational waves and predict its date (indeed, this was the original scheme suggested by Seto 2019).

In this case, both the receiver and transmitter can extend their participation. The receiver can begin its sweep directly opposite the burst at time $t = 0$, i.e., a time $-D_{\oplus}/c$ before the burst signal arrives, and sweep a ring that reaches $\theta = \frac{\pi}{2}$ upon receiving the signal. Similarly, the transmitter can begin transmitting *toward* the burst at $t = 0$, reaching $\alpha = \frac{\pi}{2}$ when the burst signal arrives.

This allows the transmitter to transmit over its entire sky, and the receiver to search for transmitters anywhere in the Universe. In Figure 5.3, this allows the transmitter above point C to send a signal for the receiver to receive at angle θ (even though $\alpha < \frac{\pi}{2}$), and for the receiver to search along transmission paths with $\theta > \frac{\pi}{2}$.

As a practical matter, one might suspect neutron star-black hole mergers, which are visible with LIGO at extragalactic distances, to be good coordinating events that could be detectable prior to their coalescence with a detector like LISA. If such systems could be somehow localized well in space, one could search the ring in the hemisphere opposite them for Seto beacons.

5.5 When to Look Where: Orbital Considerations

Pace & Walker (1975) suggested that beacons in binary star systems be activated during periastron and apastron[14] of the stars. Pace (1979) expanded this scheme to include transmitters orbiting single stars, which might use binary stars that appear near them on the sky from the perspective of the receiver to time their transmissions. This latter scheme presumes one has good distance measurements to stars, so that the light travel time can be corrected for.

The apparently non-constant nature of the Wow! Signal inspired Gray (2020) and others to consider that transmitters might focus their transmissions along their local meridian.[15] Over the course of a day, this long beam would then sweep out the entire Celestial Sphere, being an omnidirectional beacon with a period of one day (24 hours for Earth, but presumably a different value for other planets). This suggests that to perform a thorough search, one should dwell on sources for a few times the length of typical planetary day, which is measured in hours.

Kipping & Teachey (2016) suggested that transmissions, especially optical transmissions, be sent directly opposite the host star, especially along the local ecliptic, so that others observing the planet transit its star would be more likely to be observing during transmission. This has the advantage that the signal might be observed by astronomers studying the planet for reasons other than SETI, but also provides a

[14] The times of closest and farthest approach in their mutual, elliptical orbit.

[15] The meridian is the line from horizon to horizon passing through the Celestial Pole and the zenith—that is, running from due north, through overhead, to due south.

Schelling point for those looking for beacons regarding when to observe. Sheikh et al. (2023) combined these ideas to consider transmitters placed on the dark sides of tidally-locked planets discovered by *Kepler*.

5.6 The Earth Transit Zone: Ecliptic Plane SETI

Filippova & Strelnitskij (1988) suggested an inverted form of the Kipping & Teachey (2016) strategy, that transmissions to the Earth would be timed to arrive here when *Earth* is between the transmitter and the Sun. They referred to the ecliptic thus as an "attractor for SETI" and advocated for observing stars in the ecliptic directly opposite the Sun (this strategy is also thus a "when to look" strategy, although it can be employed all the time). Corbet (2003) refined this idea, noting that an ETI might detect Earth and target it because it is in its star's Habitable Zone. He suggested they might be likely to time such transmissions to the Earth for our detection, and so recommended looking all along the meridian, not just along the ecliptic. Note that this strategy requires the sender to know the precise distance to the solar system to account for light travel time, but our own parallax measurements from *Gaia* can accomplish this to a precision of better than a year out to distances of tens or hundreds of light years.

Heller & Pudritz (2016) defined the "Earth Transit Zone," (ETZ) to be the region of the sky in which stars will have seen Earth transit the Sun, and so might know of our biosphere's existence. Wells et al. (2018) and later authors have computed lists of stars that have seen Earth or other solar system planets transit in the past, mentioning them as potential SETI targets (see Figure 5.6). Sheikh et al. (2020) summarized the reasons for focusing efforts in this area, and performed the first dedicated radio search of nearby stars in it.

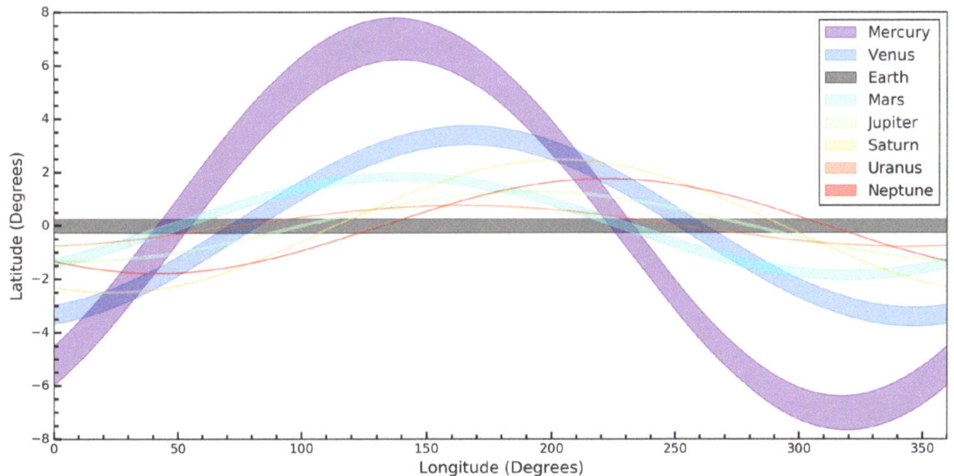

Figure 5.6. From Wells et al. (2018), regions of the sky in ecliptic coordinates that contain stars from which one the solar system planets could be seen to cleanly transit the Sun (that is, such that the disk of the planet is fully within the solar disk, yielding deep transits.) Note the very different scales of the *x*- and *y*-axes.

5.7 Schelling Points in the Solar System

Bracewell (1960) suggested that there would be probes in the solar system waiting for humans to discover and contact them. Such probes might therefore be waiting at easy to guess locations. However, finding transmitting objects in the solar system is even more challenging than finding which stars are transmitting because solar system objects orbit the Sun, and so their positions are constantly changing from our perspective. It is thus arduous to perform an exhaustive search: instead of searching every part of the sky, we must search every set of orbital parameters.

Freitas (1983) suggested that probes would be found in the vicinity the Earth's various Lagrange points, including the Sun–Earth L4 and L5 regions, and all of the Moon–Earth Lagrange points, shown schematically in Figure 5.7. Lagrange points are places where the gravitational forces of two bodies balance out, allowing an object to stay stationary with respect to them both with minimal propulsion. The L1, L2, and L3 points are not stable, but orbits around them can still be maintained with minimal propulsion.

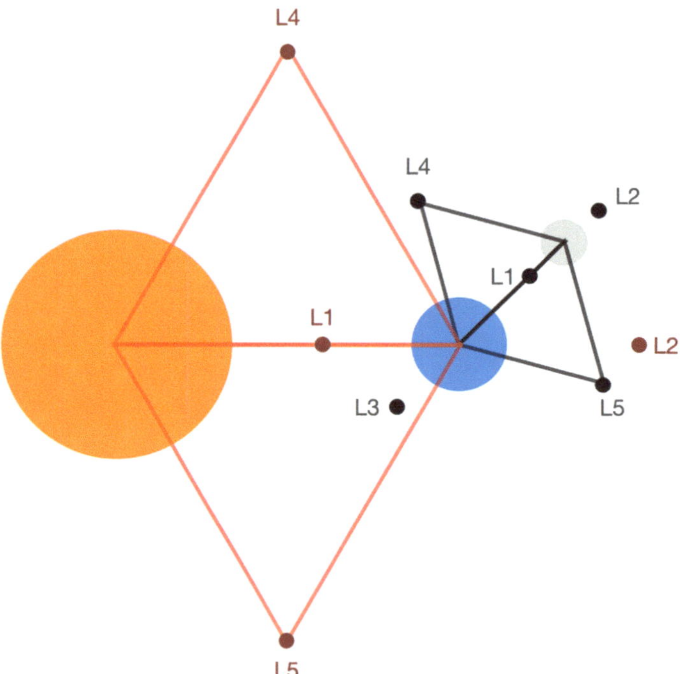

Figure 5.7. Schematic showing the locations of the Earth–Sun (red colors) and Earth–Moon (black) Lagrange points. These points rotate as the bodies orbit, so they remain fixed with respect to the geometry of the Sun–Earth or Sun–Moon system. They thus represent special locations in the solar system with respect to Earth, as well as locations requiring minimal propulsion to maintain. There are actually large regions around each point that are stable, meaning the actual search space for probes in them is quite large. Nothing is to scale here, except that L4 and L5 points do form the points of equilateral triangles. The Sun-Earth L3 point is far to the left, not shown.

Box 5.1: Jargon Review: Schelling Points

- **Schelling points**: Also called *focal points*, an optimal strategy in a non-communicative, cooperative game, as with finding another person looking for you in a city, or another species trying to get your attention in the Galaxy.
- **Beacon**: A strong, "we are here" broadcast announcing a species' existence, intended to be found by us or another species.
- **Terrestrial microwave window**: The region of the electromagnetic spectrum around 1–10 GHz where both background Galactic synchrotron radiation and atmospheric opacity are minimized.
- **The Water Hole**: Frequencies between 1.420 GHz and 1.667 GHz, bookended by the strong natural emission lines by neutral hydrogen and the hydroxyl radical, the dissociation products of water. Oliver suggested this region was a Schelling point in frequency space.
- **SETI ellipsoid**: The region of space from which a signal sent in our direction at the moment of the reception of light from some notable astronomical event would be arriving at Earth now. Stars on the SETI ellipsoid coordinating their transmissions to such events make particularly interesting SETI targets.
- **Earth transit zone**: The region of the sky near the ecliptic in which distant observers would see the Earth transit the Sun, and therefore have relatively easy access to information about Earth's existence and biosphere.

The L4 and L5 points form equilateral triangles with the two objects and are quite stable, and there are a variety of stable "tadpole" and "horseshoe" orbits around and between these points. Any object in such an orbit will remain in a 1:1 orbital resonance with the Earth or Moon (meaning over many orbits it will share the same orbital period) and so be said to occupy that Lagrange point, even if it does not sit at the point exactly. Determining whether an object is occupying the Lagrange point requires an analysis of the dynamics of its orbit, and cannot be deduced solely from its location in space at a given time.

Because these regions can be more stable than other orbits, asteroids that find themselves perturbed into these regions may linger there for a long time. Indeed, Jupiter has entire families of asteroids occupying the L4 and L5 points called the Trojan asteroids (the "Trojan camp" is at L4 and asteroids there are named after Trojan characters (as from *The Iliad*); the "Greek camp" is at L5). In general, though, perturbations from the planets will eventually kick out any asteroids occupying the Earth–Moon or Earth–Sun Lagrange points, meaning anything found in such an orbit will be a rare interloping asteroid or something artificially maintaining that orbit.

In Section 14.4.2 we will explore the possibility of using stars as lenses to facilitate interstellar communication by focusing outgoing or incoming communications across interstellar distances. If a network of communication relays exists in the Galaxy, then stars acting as nodes should have relay probes that sit at the focal line of the star's gravitational lens, directly opposite the star the lens is focused on. These thus represent special locations around a star, especially the focal lines of its nearest neighbors. Transmitters at these locations thus serve dual functions: they can be used to relay communications to a Galactic network, and they can serve as beacons sitting at a Schelling point in the system. One advantage to this location is that it requires very little propulsion to remain stable (Kerby & Wright 2021).

Exercises

1. After completing the in-class Schelling Point exercise, do the post-exercise worksheet provided by your instructor.
2. Think about the geometry of the Earth Transit Zone. What sets the thickness of the lines in Figure 5.6? Estimate the thickness of each line you expect from properties of the solar system planets, and the actual thicknesses, and compare.
3. Could *Kepler* detect the solar system?
   (a) For each of the solar system planets, how much dimmer does the Sun appear to get during transit?
   (b) How often do the transits occur?
   (c) Look up the mission specifications for *Kepler*. Which solar system planets would *Kepler* have reasonably been able to detect from a nearby star?
   (d) Use the Exoplanet Archive to explore the radii and orbital periods of planets discovered by *Kelper*. Does this match your expectations?
4. Temporal Schelling Points

   Imagine we discover a planet transiting a star in the ecliptic. That means that a species on that planet will also witness the Earth transiting the Sun from their perspective. You should draw a picture.

   Let's say we wanted to send a strong signal to them, but only have enough resources to do so occasionally, for a few hours a year. Consider some potential Schelling points on when to transmit:

   I. When we see them transit their star.
   II. When we are between them and the Sun.
   III. Timing things so that our signal arrives when they are between us and their star.
   IV. Timing things so that our signal arrives when they see us transiting the Sun.

   Answer the following five questions.
   (a) Consider light travel time. Are there really four distinct schemes above, or are some of them the same?
   (b) Typical planetary transits last a few hours. How precisely do you need to know the distance to the other star to correct for light travel time?
   (c) *Gaia* measured parallaxes to most bright stars within 100 pc to better than 0.1 mas. Can we correct for light travel time to such stars to a precision of hours?
   (d) Which of the four schemes requires precise distance knowledge? Which schemes require that knowledge of us, and which require it of them? Do any schemes not require this at all?
   (e) Which scheme is best, and why?

5. The compatibility of Equations (5.3) and (5.4) is enforced by the geometry of Figure 5.3, but we can also derive it with algebra and trigonometry.
   (a) We know that the arrival time of a signal at Earth $t_\oplus$ must be independent of the distance to a transmitter in direction θ. Write down how $t_{\rm sig}$ must then relate to $t_\oplus$ as a function of d, the distance to the transmitter.
   (b) Derive the relationship among d, $D_\oplus$, a D_*, ϕ, α, and θ using the law of sines from Figure 5.2.
   (c) Using only the results from (a) and (b) algebra, and trigonometric identities, start with one of Equations (5.3) and (5.4) and derive the other.
6. Derive Equations (5.5) and (5.6).
7. Merging neutron stars and black holes are bursts that are detectable across the Universe. LIGO now regularly detects such events at distances of a significant fraction of the visible Universe, and with upgrades will be able to detect them much farther than this.

   Neutron stars merge with black holes at a volumetric rate of, very roughly, 100 $\rm Gpc^{-3}\ yr^{-1}$.

   Let's do an order of magnitude calculation to determine the fastest, best-case scenario for using one of them as a Seto burst to search the entire Seto search volume.

   (a) How many mergers occur in a box of volume V in a time interval ΔT?
   (b) What is the length d of the side of a box expected to have only one event in that time frame ΔT? Argue that d represents the closest event one can expect in a time frame ΔT.
   (c) How long τ does it take to complete a Seto scheme search across the whole hemisphere, given a coordinating event at distance d? Also express τ in terms of ΔT.
   (d) Suppose you'd like to complete your search in 100 years. How often, ΔT do such events occur?
   (e) Let's say that we don't want to wait any longer to start the search than it would take to conduct the search. Set $\tau = \Delta T$. How long will your search take?
   (f) Comment on the practical issues of such a plan.

References

Blair, D. G., & Zadnik, M. G. 1993, A&A, 278, 669
Bracewell, R. N. 1960, Natur, 186, 670
Cocconi, G., & Morrison, P. 1959, Natur, 184, 844
Cocconi, G., & Morrison, P. 1959, Natur, 184, 844
Corbet, R. H. D. 2003, AsBio, 3, 305
Davenport, J. R. A., Cabrales, B., Sheikh, S., et al. 2022, AJ, 164, 117
Dixon, R. S. 1973, Icar, 20, 187
Drake, F. D., & Sagan, C. 1973, Natur, 245, 257

Filippova, L. N., & Strelnitskij, V. S. 1988, ATsir, 1531, 31
Filippova, L. N., Kardashev, N. S., Likhachev, S. F., & Strelnitskij, V. S. 1991, Bioastronomy: The Search for Extraterrestrial Life—The Exploration Broadens Lecture Notes in Physics, ed. J. Heidmann, Vol. 390 (Berlin: Springer) 254
Freitas, R. A. 1983, JBIS, 36, 501
Gray, R. H. 2020, IJAsB, 19, 299
Heller, R., & Pudritz, R. E. 2016, AsBio, 16, 259
Hilton, W. F. 1981, JBIS, 34, 400
Horowitz, P., & Sagan, C. 1993, ApJ, 415, 218
Kerby, S., & Wright, J. T. 2021, AJ, 162, 252
Kipping, D. M., & Teachey, A. 2016, MNRAS, 459, 1233
Makovetskii, P. V. 1976, SvA, 20, 123
Makovetskii, P. V. 1977, AZh, 54, 449
Makovetskii, P. V. 1980, Icar, 41, 178
McLaughlin, W. I. 1977, Icar, 32, 464
Morrison, P., Billingham, J., & Wolfe, J. 1977, The Search for Extraterrestrial Intelligence, SETI, Vol. 419 (Washington, DC: NASA)
Oliver, B. M. 1979, AcAau, 6, 71
Oliver, B. The 1971 NASA/ASEE Summer Fac. Fellowship Program (NASA-CR-114445)
Pace, G. W. 1979, JBIS, 32, 215
Pace, G. W., & Walker, J. C. G. 1975, Natur, 254, 400
Planck, M. 1900, AnP, 306, 69
Planck CollaborationAghanim, N., Armitage-Caplan, C., et al. 2014, A&A, 571, A27
Schelling, T. C. 1960, The Strategy Of Conflict (Galaxy Books) (Cambridge, MA: Harvard Univ. Press)
Schwartz, R. N., & Townes, C. H. 1961, Natur, 190, 205
Seto, N. 2019, ApL, 875, L10
Seto, N. 2021, ApJ, 917, 96
Seto, N., & Kashiyama, K. 2020, IJAsB, 19, 308
Sheikh, S. Z., Siemion, A., Enriquez, J. E., et al. 2020, AJ, 160, 29
Sheikh, S. Z., Kanodia, S., Lubar, E., et al. 2023, AJ, 165, 61
Tang, T. B. 1976, JBIS, 29, 469
Tarter, J. C., Cuzzi, J., Black, D., & Clark, T. 1980, Icar, 42, 136
Wells, R., Poppenhaeger, K., Watson, C. A., & Heller, R. 2018, MNRAS, 473, 345
Wright, J. T. 2020, arXiv:2008.01817
Zuckerman, B. 1985, AcAau, 12, 127

Chapter 6

The Case Against SETI: The Fermi Paradox

> Where is everybody?
>
> —Attributed to Enrico Fermi, 1950

6.1 The Fermi Paradox

6.1.1 Origins of the Paradox

> The Fermi Paradox is neither Fermi's, nor a paradox.
>
> —Gray (2015)

The *Fermi Paradox* is, along with the Drake Equation, one of the founding ideas of SETI, and one of the most discussed and debated topics in both the literature and popular discussions of the topic.

The name apparently comes from Stephenson (1977), who connected arguments made shortly prior by Hart (1975) (see below) to a 1950s lunchtime discussion at Los Alamos National Laboratory among Enrico Fermi and fellow physicists Emil Konopinski, Edward Teller, and Herbert York (see Gray 2015 and references therein).

The discussion was about UFOs and alien life, likely inspired by the Roswell Incident a few years prior (Section 18.7.4). Fermi, who was famous for his

doi:10.1088/2514-3433/adf5aech6

eponymous order-of-magnitude problems (Box 6.1) and inspired leaps of reasoning, asked the table almost as a non-sequitur "Where is Everybody?"[1]

The table understood that Fermi was wondering why alien spacecraft *weren't* already on Earth. The implication was that the age of the Galaxy was so great and the time it would take even slow ships so short in comparison, that it seemed odd that the Earth was not regularly visited by "them."

The consensus recollection of those present was that Fermi concluded that, given the distances involved and lack of obvious alien activity here, interstellar spaceflight was genuinely prohibitively difficult. This is a notably different conclusion from Hart, Tipler, and other proponents of Fermi Paradox reasoning, who instead concluded that extraterrestrial technological life must be rare or nonexistent in the Galaxy.

6.1.2 Not a Paradox?

Gray (2015) points out that the Fermi Paradox is doubly misnamed: not only can it hardly be said to be Fermi's, but it is not a valid logical paradox.

To the first point, Enrico Fermi never published on the topic, and it is unclear to what degree he was aware that the lunchtime discussion was even remembered by its participants; he died in 1954, just four years later.

To the second point: in logic a paradox is a self-contradictory statement, which Fermi's question (and Hart and Tipler's arguments below) are not. In everyday speech, a paradox is a pair of truths that appear to be in tension or inconsistent, but which may not actually be so, which is a perhaps fair description of the Fermi Paradox.

Gray therefore favors the name "Hart–Tipler argument" for the Fermi Paradox, although in this work we will continue to call in the Fermi Paradox because that is by far its most common appellation, and we will heed King Cnut's lesson about not ordering the tide to ebb when it flows.

6.2 The Calculation behind the Fermi Paradox

6.2.1 A Model for Interstellar Expansion

We cannot be sure what Fermi's reasoning was, but given his penchant for *Fermi Problems* (Box 6.1) and the recollections of the attendees, we can reason out a likely version of it.

Consider that a single alien species develops the technology for interstellar travel and settlement, and uses it. At this point, there will be two locations in the Galaxy with spacefaring technological life. Given enough time, we might expect both to expand to nearby star systems, and so on, perhaps in analogy to humans populating the islands of the Pacific thousands of years ago.

At some point, the number of settlements becomes so great that they become self-perpetuating, robust against the collapse of technology or even extinction of the species on one of the settlements, which could then be "back-filled" by others.

[1] This question by Fermi was apparently first described in print in a footnote in Sagan (1963), who called it "now well known."

At this point, the population can spread through the Galaxy, akin to an invasive species introduced to a new favorable habitat.[2] As long as there are ample new stars to migrate to, we expect, on average, the rate of new settlements to be proportional to the number that currently exist, each capable of settling nearby stars.

6.2.2 First Calculation: Exponential Growth

We can thus take a first stab at solving the Fermi Problem of the number of settlements in the Galaxy. If the number of settlements as a function of time is $N(t)$ then we have

$$\frac{dN}{dt} = N/\tau \tag{6.1}$$

where τ represents the characteristic time between new settlements being launched by a typical settlement. This is the famous exponential equation. If settlement begins at time $t = 0$ when $N = 1$, then the solution is:

$$N = e^{t/\tau} \tag{6.2}$$

If a new settlement ship is launched every, say, 10,000 years, then in 1 million years one expects 100 e-foldings of settlements, or about 10^{43}.

The age of the Galaxy is around 10 billion years, and the age of typical stars like the Sun is around 5 billion years. For any modest ship launch rate—even as long as one every 10,000 years, we conclude that there should by now be many more settlements than there are stars in the Milky Way ("only" 10^{11}!)

6.2.3 Second Calculation: Travel–Time Limited Growth

Of course, this is absurd. All exponential growth is ultimately checked by some limiting reagent—a petri dish with mold is eventually checked by the amount of agar and the edges of the dish; eventually humans run out of islands in the Pacific, etc. Once that limiting factor slows growth, we say growth has become *saturated.*

In the Galaxy, saturation most naturally occurs when all of the nearby stars are settled, and only those stars on the edges of the settled region are available. At this point, new settlements expand at the speed of the ships involved, since those on the interior of the settled region have nowhere nearby to settle.

This leads us to our second calculation: how fast might this settlement front expand? We can bound the problem quite tightly on two ends.

The fastest one could possibly settle the Galaxy is at the speed of light, although the challenges of relativistic spaceflight are extremely steep. This is a hard upper limit on the speed of the settlement front.

What is the slowest a settlement ship might travel? Our chemical rockets can propel a large probe at speeds very similar to the orbital velocity of Earth around the Sun: the highest launch velocity ever recorded by a human-made object is that of the

[2] All of these analogies to Earth life are purely to illustrate the principles at work; they nonetheless come with semantic, rhetorical, and even political baggage that is not essential to the argument.

New Horizons probe, around 16 km s^{-1}, which, roughly added to the 30 km s^{-1} of the Earth's orbital motion, is in principle enough energy to escape the Sun's gravity with around 18 km s^{-1} of velocity to spare for interstellar travel. In fact, New Horizons left at an angle to Earth's motion, and also used a gravitational slingshot maneuver at Jupiter to acquire some of its kinetic energy. When it is far from the solar system it will have a speed of around 11 km s^{-1}.

So, we might put a lower bound on spacecraft speeds based on this, our best effort to date. It is worth noting that this speed comes in part from these probes' gravity assists from Jupiter, and we might imagine that another species could similarly use giant planets in their planetary systems to launch interstellar probes. Doing so requires no special technology beyond interplanetary spaceflight, only knowledge of the physics of gravitation.

These gravitational assists require a planet whose orbital velocity is not much greater than its escape velocity, i.e., where

$$v_{\rm orb} \lesssim v_{\rm esc} \tag{6.3}$$

where the escape velocity is given by

$$v_{\rm esc} = \sqrt{2GM/R} \tag{6.4}$$

where M and R are the planet's mass and radius. For Jupiter, this is 60 km s^{-1}. The orbital velocity of the planet, if it is in a circular orbit, is given by

$$v_{\rm orb} = \sqrt{GM_*/r} \tag{6.5}$$

where M_* is the star's mass and r is the planet's orbital distance. For Jupiter, this is 13 km s^{-1}. Since this is less than its escape velocity, Jupiter can provide gravitational assists that can eject a probe from the solar system.[3]

The result of such an encounter is that the probe velocity changes direction, and increases by an amount of order $v_{\rm orb}$. If it had a similar velocity going into the

[3] Gravitational encounters can be modeled as, in the jargon of physics, "elastic collisions." Nothing is actually colliding and certainly nothing is stretching or compressing, of course. The term applies to collisions in which no energy is lost to deformation of the objects, and the total kinetic energy of both objects is conserved. A superball toy is extremely elastic and so returns from a bounce against a wall with almost the same energy it went in with.

When two objects have a close gravitational encounter, their total kinetic and gravitational potential energies are usually conserved, so we can model them the same way. In this case, the best-case scenario for a gravitational assist is a head-on "collision" in which a slow probe approaches the planet opposite its direction of motion, swings around the back, and leaves the way it came, picking up all of the planet's orbital motion in the process, like a superball thrown directly at an oncoming train.

Its ability to do this depends on how close it can get to the planet on the back side in order to feel enough gravitational acceleration to "turn the corner" as sharply as possible. The probe will need to experience a change in velocity of order the encounter velocity $v_{\rm enc}$. At closest approach to a planet of radius R, it will experience an acceleration of order $a \sim GM/R^2 \sim v_{\rm esc}^2/R$, over a duration of order $\Delta t \sim R/v_{\rm enc}$, for a total change of $\Delta v \sim v_{\rm esc}^2/v_{\rm enc}$. If $v_{\rm esc} \gg v_{\rm orb}$, then in the frame of the planet most of the velocity of the probe will be from falling into the planet's gravity well, so $v_{\rm enc} \sim v_{\rm esc}$, and the change in velocity will be $\Delta v \sim v_{\rm esc} \sim v_{\rm enc}$, as required. In the other extreme, the encounter velocity is of order the orbital velocity, so we have $v_{\rm enc} \sim v_{\rm orb} \gg v_{\rm esc}$ and $\Delta v \sim v_{\rm esc}^2/v_{\rm orb} < v_{\rm enc}$, so it doesn't work.

encounter, then its velocity will be doubled, giving it a kinetic energy of order $KE = \frac{1}{2}m(2v_{orb}{}^2) = 2mv_{orb}{}^2$. To escape a star, the probe needs to have a kinetic energy that exceeds the gravitational potential energy from the star at the distance of the assist, which is $GmM_*/r = mv_{orb}^2$. Since this is satisfied, we see how gravitational assists can provide the necessary acceleration to turn inter*planetary* spacecraft into inter*stellar* spacecraft.

So, a reasonable lower bound for any interstellar probe is probably around 10 km s^{-1}, available to any spacefaring species or technology with a Jupiter-like planet in its system.

Now, the Fermi Problem: the expansion front can move no faster than the speed of the ships. The Galaxy is, very generously, 50 kpc from side to side (and only 8.5 kpc from the Sun to the Galactic Center), so let's choose 20 kpc as the diameter within which most of the stars can be found. Let's first calculate the slowest the front could expand: since 1 km $s^{-1} \approx$ 1 pc Myr^{-1}, we have the time it takes a saturated exponential growth to cross the Galaxy of

$$T = \frac{2 \times 10^4 \text{ pc}}{10 \text{ pc}/10^6 \text{ yr}} = 2 \text{ Gyr}. \tag{6.6}$$

This is less than the age of the Earth, which is around 4.4 Gyr. We could easily imagine, if time were a factor, using ordinary chemical propulsion to launch ships perhaps 10 times faster than our estimate, so a more practical number might be more like 200 Myr.

At the faster end, a relativistic ship moves near the speed of light, which is 300,000 km s^{-1}, or 30,000 times faster than our gravity-assist-limited craft. The crossing time is then of order 150,000 yr!

So, regardless of the methods employed, "they" have had plenty of time to get here. Indeed, if their ships are faster than ours, then unless the first such travelers were relatively recent (the Galaxy is 10 Gyr old) they were here quite a long time ago and have criss-crossed the Galaxy many times.

This is the essence of the Fermi Paradox: even for interstellar flight technology as rudimentary as ours, the time to get to Earth from anywhere in the Galaxy is much less than the age of the Earth, and much less than the age of the Galaxy. So where are they?

6.2.4 Validating the Calculation

The calculation above is a rough-order-of-magnitude one that ignores many details, including the practicalities of how stars move and are distributed in the Galaxy, their typical distance (which is important if ships have finite range), and how often settlement ships are launched.

There is a substantial literature discussing these nuances and doing analytic and numerical simulations to verify them (see Carrol-Nelenback et al. (2019) and the references therein for an extensive review of the literature on the topic of more careful calculations than Fermi's). One such simulation is that of Carroll-Nellenback et al.

(2019) and Wright et al. (2021) who accounted for stellar motion, finite ship range, and variations in stellar density throughout the Galaxy (Figure 6.1).

For instance, in their simulation the finite ship range meant that in regions of low stellar density, such as the outskirts of a Galaxy, τ was long and growth was frustrated because stars had to "wait" for a close stellar encounter to settle the new star system. Closer to the center of the Galaxy, growth proceeded exponentially (Figure 5.1).

Such simulations can model expansion under various assumptions of alien behavior, but ultimately Fermi's original calculation holds: unless spaceflight is hard and exponential growth never occurs, the timescales for Galactic settlement are much shorter than the age of the Galaxy, which is all that is required for the paradox to hold.

6.3 Hart's Formulation of the Fermi Paradox

6.3.1 "Fact A" and Categories of Explanations for it

The most influential formulation of the Fermi Paradox is due to American astronomer Michael H. Hart,[4] who in a 1975 paper (Hart 1975) laid out an early and trenchant case for why SETI would fail.

Hart's argument begins with what he called *Fact A*, that "There are no intelligent beings from outer space on Earth now," which is consistent with the spirit of Fermi's lunchtime question. Hart then asserts "If…there were intelligent beings elsewhere in our Galaxy, then they would eventually have achieved space travel, and would have explored and colonized the Galaxy, as we have explored and colonized the Earth." Hart's argument concludes by saying that since this conditional is contradicted by Fact A, then it follows that "they do not exist."

The argument at first blush seems far too facile: why would they have done that? But on closer examination the most straightforward counterarguments founder. Hart grouped all counterarguments into four categories:

1. **Physical explanations:** Space travel is infeasible because of "physical, astronomical, biological, or engineering difficulty." In other words, they are not here because traveling here is too hard or impossible.
2. **Sociological explanations:** Alien species choose not to colonize Earth because they lack the interest, motivation, political motive, or organization to do so.
3. **Temporal explanations**: There has not been enough time for them to have been expected to reach us yet.
4. **Fact A is false**: They are here or have been here.

[4] It should not go without mention that Hart is an astrophysicist-turned political activist, known for his advocacy for "white separatism" (his term; p. 201 Swain & Nieli 2003). We will here focus on the argument in his paper and its influence, but there is an interesting discussion to be had about how scientists' views of humanity and its future affect the perspectives they bring to our expectations of alien contact, based on analogies to human behavior. See Chapter 18 and Wright & Oman-Reagan (2018). It is also worth considering who gets cited and has their work canonized, and why.

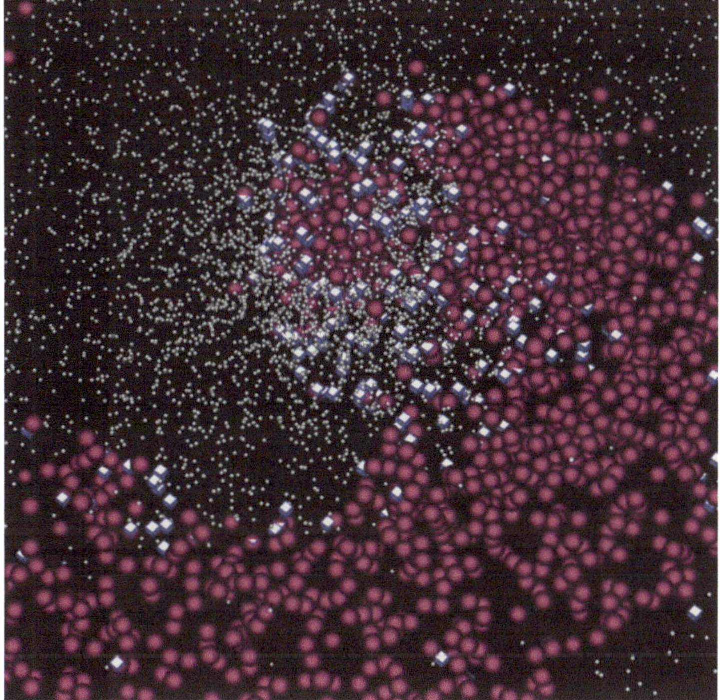

Figure 6.1. From Wright et al. (2021), a snapshot of a simulation showing stars in the central part of a galaxy (small white dots), stars that have been settled (large magenta spheres), and the settlement ships between them (white diamonds). The parameters of this simulation were tuned to show interesting dynamics due to the motions of stars, especially the Galaxy's differential rotation. To keep the simulation manageable, the number of stars was reduced, and the minimum range of ships extended, to reproduce the qualitative behavior in a more realistic Galaxy. CC BY 4.0.

6.3.2 Physical Explanations

Hart dismissed the first of these, physical explanations, with a discussion of the physical possibility of space travel. Here, the basic logic of Fermi's rough calculation suffices to show it is not physically impossible for them to have arrived, even on "slow" ships under chemical propulsion.

The main challenge of interstellar space travel is that of time: how might a ship and its occupants survive in space for the very long travel times. Hart offered a few solutions: suspended animation of occupants, aliens with very long lifespans, extremely fast ships,[5] autonomous ships that grow a crew from zygotes upon arrival, and generation ships.[6]

[5] Hart is referring to ships traveling at relativistic speeds, such that clocks on the ship run slow and the occupants experience very little passage of time. Such travel seems borderline physically impossible given the energy costs involved, and Hart acknowledges "the difficulties in this approach seems extremely great to me."
[6] Generation ships are self-contained biospheres in which the occupants carry out lives for many generations, and the original occupants die before arriving at their destination.

Hart's point is that the challenges of interstellar travel and settlement[7] are "just engineering." While such engineering may be far beyond what humans could accomplish in the near future, the fact that we can contemplate and outline plausible solutions to the problem at all implies they are not insurmountable given enough time, resources, and motivation.

6.3.3 Sociological Explanations and the Monocultural Fallacy

Hart acknowledged many reasons that a species might not undertake the difficult task of space travel, including lack of interest and self-destruction, which we will discuss later. A common and fatal flaw to all such solutions, Hart claims, is that they only succeed if they apply to *all members of all species for all time.* That is, that they represent a sort of ironclad law of "xenosociology." This is what Wright et al. (2014) call the *monocultural fallacy*: presuming that all alien behavior across all species will be characterized by a single set of traits, instead of, like humans, comprising a variety of cultures and behaviors across time and space.

To illustrate this, Hart asks us to consider why any proposed solution to the Fermi Paradox, which might apply to the inhabitants of a hypothetical planet "Vega III" would apply to them not just now, but 1000 years ago, or 2000 years ago, or in fact at any point in that species' history. And if it applies to the Vegans, what rule of xenosociology says it would apply to the inhabitants of "Procyon VI" or "Altair IV" or any other inhabited planet in the Galaxy?

The point is that the logic of the Fermi Paradox hinges on the inevitability of exponential growth, which requires only a single species to start the chain. Even if a rule of alien behavior applies 99% of the time to a species for its entire history, that still means 1 in 100 species will settle the entire Galaxy.

Brin (1983) calls this the *Principle of non-Exclusiveness*: "diversity will tend to prevail unless there is some mechanism to enforce conformity." We might call proffered solutions that violate this principle *exclusive solutions*, because they posit alien behavior that is exclusive of any variation that would cause alien species to spread.

In other words: it only takes one. To put a finer point on the matter, consider that humanity, as a species, holds no particularly strong desire to settle the Moon. Very few people make that a priority of their lives or seek to become lunar astronauts; most nations do not pursue the project. Even the nations that lead space exploration, like the United States, have not devoted the resources necessary to make permanent lunar settlement a reality in the 50+ years since the technology became available.

However, plans to settle the Moon exist and are being pursued apparently seriously not just by NASA but by a few large companies and extremely wealthy individuals. It is not at all hard to imagine that in the future one such private venture

[7] We will generally use cognates of "settle" instead of "colonize" because they have fewer imprecise political connotations, except where such connotations are appropriate. See Chapter 18 and Wright & Oman-Reagan (2018).

might succeed in settling the Moon and mining it for resources, in spite of the indifference, or even objection, of many or even most humans.

The point is, again, "it only takes one" to initiate a pattern of exponential growth, so pointing to any sociological tendency fails as a "solution" to the Fermi Paradox unless it is absolute. A good way to keep this objection of Hart's in mind is to rephrase any proffered solution to the Fermi Paradox of the form "they are not here because they X" as "they are not here because *all members* of *all alien species* in the Galaxy *always* X *for all time*." Phrased this way, many seemingly reasonable solutions appear less compelling.

6.3.4 Temporal Explanations

This is simply the standard Fermi Paradox calculation: there has been more than enough time for them to arrive on Earth. Hart gave an example for ships traveling at $0.1c$, and argued the age of the Galaxy is thus about 5000 crossing times, far too many for them to have overlooked Earth.

6.3.5 Refutation of Fact A

Hart rebutted this class of explanations by noting that arguing "they have been here in the distant past but did not remain" is not a rebuttal to his argument because even if it were true, it would not explain why no species *returned* to Earth to settle it subsequently.

He also discussed and dismissed claims of alien artifacts on Earth (which are thoroughly refuted by anthropologists, see Chapter 18) and UFOs as evidence that they are, in fact, here today.

6.4 Tipler's Formulation of the Fermi Paradox

In 1980 American physicist Frank J. Tipler[8] expanded on Hart's arguments by introducing the idea of autonomous robotic explorers of the Galaxy.

Tipler (1980) begins by arguing that biologists believe the development of intelligence from early life is a fantastically unlikely event, and that humans are likely alone in the Galaxy. He then asserts "if they did exist and possessed the technology for interstellar communication, they would also have developed interstellar travel and thus would already be present in our solar system."

Tipler argues that any species we might detect via radio waves would necessarily[9] have developed rocketry and computers, and that these combined allow for the construction of a "'self-replicating 'universal constructor" with intelligence comparable to the human level."

[8] Tipler is a mathematician and physicist at Tulane University. He holds heterodox views on cosmology and theology, and espouses his pseudoscientific "Omega Point" theory unifying them (Eliis, 1994).

[9] By the Principle of Non-Exclusiveness, Tipler's argument holds even if this assertion is weakened to only some such species developing rocketry and computers.

6.4.1 Von Neumann Machines

The idea of a universal constructor was developed by Johnny von Neumann,[10] to mean a self-replicating machine.

Box 6.1: Fermi Problems

Enrico Fermi, taken sometime around 1945. Image credit: U.S. National Archives and Records Administration.

Enrico Fermi (1901–1954) was an Italian-American physicist and Nobel laureate who made foundational contributions to quantum mechanics and particle physics. He is also famous for an eponymous style of calculation to generate estimates.

Fermi problems are a kind of "(R)OOM" ("(rough) order of magnitude") calculation concerned with bounding or estimating some quantity, especially when one does not have the need or time for a more precise calculation. The goal is to get an answer to a precision of a factor of a few (or at least within a factor of 10, which is one order of magnitude.)

Fermi Problems are intended to performed quickly, often entirely in one's head or in a few lines on a napkin or blackboard, and are often of a surprising or frivolous nature, the classic being, "How many piano tuners are there in Chicago?" The point is not to get the answer precisely, but to bound and/or estimate a seemingly impossible-to-guess quantity based on things you know roughly, such as the population of Chicago, what fraction of people own tuned pianos, how often they need tuning, how long it takes to tune one, and how many hours people work in a year. One only need know each of those things to within a factor of a few to get a final answer to within a factor of 10 or so.

A more specific example involving Fermi himself is how he (correctly) quickly estimated the yield of the Trinity nuclear explosion by dropping strips of paper into the wind from the blast and measuring how far they flew (Katz 2021).[11]

Such exercises can greatly sharpen one's ability to hone in on the relevant aspects of a problem, and so is a favorite intellectual pastime among certain kinds of physicists. Played as a kind of game, answering the question by contacting the Chicago chapter of the Piano Technician's Guild would be considered cheating, or at best missing the point.

Once one gets good at it, being able to generate OOM estimates quickly can be a powerful component of a scientist's toolbox, allowing them to quickly identify the relevant factors in a problem and avoid unnecessary calculations. Insights gained from checking against the "right" answer can also sharpen one's physics intuition more generally.

[10] von Neumann (1903–1957) was a Hungarian-American mathematician and polymath known for his foundational contributions to quantum theory, computer science, and game theory. He also served many roles in US defense and the development of the US nuclear arsenal, from his work on the Manhattan project to planning the development and deployment of ICBMs. He was a strong proponent of the concept of "mutually assured destruction" as a deterrent against the Soviet Union, first developed by Schelling (Chapter 5).

[11] https://fermatslibrary.com/s/my-observations-during-the-explosion-at-trinity.

The term today usually refers to von Neumann's abstract concept of a "machine" in the context of cellular automata theory,[12] but he originally intended it to be a physical object capable of self-reproduction by purely mechanical means. Von Neumann imagined the machines would have many subsystems, including a separate means for reproducing themselves from a store of parts and materials, and for transmitting the instructions for their behavior and construction to their "offspring."

Tipler extended this idea to a self-replicating set of specialized space probes. He imagined that they could be launched to a source of resources, such as an asteroid, and use the materials there to make many copies of themselves. From there, they would expand to additional asteroids and repeat the process.

Tipler imagined that these probes would be able to go to nearby star systems and there set up any kind of machine or settlement the designers of the probes liked, including machines for raising biological creatures from "scratch," and for making more machines to continue the spread.

Today, such spacefaring machines are known as *von Neumann probes*, and they greatly extend the logic of Hart by overcoming arguments about the difficulty of acquiring the resources necessary for large scale interstellar flight or sustaining biology on long trips. In fact, Tipler saw human settlement of the Galaxy as simply a special case of von Neumann probes, in the sense that humans themselves are a sort of self-replicating machine.

Other variations on von Neumann machines also exist, including Bracewell Probes, which we will discuss in Chapter 14, and the "astrochicken," a specific suggestion for a very small cyborg-like probe for exploring the solar system suggested by Freeman Dyson.

6.4.2 Objections to Tipler's Argument

Von Neumann imagined that universal constructors would be subject to evolution because of inevitable transcription errors in their "genetic code." Tipler acknowledged that this evolution might cause the probes might at some point change their behavior. He suggested that the creators of the machines would have various fail-safes in place to prevent this, or else that such autonomy might be allowed or even favored ethically, as the machines become a species of their own.

In a series of papers Sagan and Newman argued with Tipler about the validity of many of his conclusions, referring in Sagan & Newman (1983) to the Fermi Paradox argument as the "solipsist"[13] approach to SETI. By this they mean it would be extremely anti-Copernican[14] if "our intelligence and technology are unparalleled in the history of the cosmos." They pointed out that "absence of evidence is not evidence of absence"[15] and we should draw few if any conclusions from it.

[12] Most famously exhibited in Conway's Game of Life.

[13] Using the figurative sense of an extreme egoist, not in the philosophical sense of denying true knowledge of anything except oneself.

[14] See Chapter 10.

[15] A favorite aphorism of Sagan's, although its origins go back at least a century earlier.

On the merits, Sagan and Newman suggested that von Neumann probes that "escaped" their programming would become extremely dangerous, akin to locusts consuming all of the available resources everywhere they went. As such, they would be far too dangerous to be built or to be allowed to exist, and therefore would not exist. They also quibbled with Tipler's estimates of growth rates for a species, based on estimates of the rates of human migrations on Earth, and concluded that Tipler's calculations were too crude to draw any conclusions from.

6.5 The Mystery of the Great Silence

6.5.1 A Variation of the Fermi Paradox, and Fact B

Despite the rather clear history and definition of the Fermi Paradox, the term has become more loosely used to refer to the puzzle of finding *any* answer to Fermi's question "Where is Everybody?" in light of the existence of SETI programs looking for them. That is, instead of starting from Fact A, one begins from the fact that we have done some looking for signs of ETI with telescopes and have not found them yet. To distinguish this argument from the proper Fermi Paradox, we might say it relies instead on *Fact B*: "We have not yet discovered proof of extraterrestrial technology."

The logic of this argument is that if alien life existed in the Galaxy they might *not* be here on Earth today, but in that case we would have noticed them remotely by some means by now. In particular, it is often phrased in terms of the many decades of searching by various SETI programs, and why they have "failed" to find anything to date.

In this book, we will not call this argument the "Fermi Paradox" in part because it is based on a very different premise from Fact A, but also because this could not have been what Fermi meant: he asked his question in 1950, ten years before Project Ozma even began! Brin (1983) refers to this argument as "The Mystery of the Great Silence," a term we adopt here.[16]

Drawing any conclusions from Fact B is dubious, largely relying on a popular misperception that (1) a great deal of searching has been done and (2) the lack of detections to date proves that alien technology does not exist around the targets that have been searched.

The latter of these is easy to dismiss: most searches for alien life to date would only have detected extremely powerful transmissions directed at Earth, or assume some specific pattern of behavior detectable at interstellar distances that would be obvious to us. The lack of a strong 21 cm radio beacon detectable from Earth from somewhere in the Proxima system hardly proves that there is no intelligent life there; after all, we are technological and send no such beacons to any system.

[16] Ćirković (2018) refers to conclusions based on Fact A as the "Proto-Fermi Paradox" (ProtoFP), conclusions based the lack of evidence for them or any of their artifacts in the solar system as the "Weak Fermi Paradox" (WeakFP), and conclusions based on our Fact B as the "Strong Fermi Paradox" (StrongFP), *Silentium Universi*, or "The Great Silence."

6.5.2 The Cosmic Haystack

The former argument, that a lot of searching has been done, is quantifiable in terms of the Cosmic Haystack which, as we will see in Section 11.3, has hardly been searched, equivalent to the volume of a hot tub out of the world's oceans.

The lesson here is not that the search is hopelessly big—one does not need to drain the oceans to find a fish, one only needs search where fish are abundant! The lesson is that any inferences from Fact B are premature, because one can hardly be said to have put constraints on the likelihood of the future success of SETI with so much space left to search.

The Mystery of the Great Silence is thus not even a problem that necessarily demands a solution. Referring to it as a "Mystery" (as the Fermi "Paradox") gives the false impression that Fact B is difficult to explain.

That said, some have argued that some remotely detectable manifestations of alien technology are just as inevitable as alien technology on Earth today, such as the construction of Dyson Spheres. Ćirković (2018), for instance, argues that the climb of species up Kardashev's ladder is so inevitable that if they exist they should be "no less obvious than the (counterfactual) brightness of the night sky" in Olbers' Paradox.[17] Under such an assumption, the Fact B is properly a premise for the Fermi Paradox.

6.6 "Solutions" to the Fermi Paradox and the Mystery of the Great Silence

There is something of a cottage industry of generating "solutions" to the Fermi Paradox and the Mystery of the Great Silence. Brin (1983) cataloged many of them, and by 2000, American author Stephen Webb in his book "If the Universe is Teeming with Aliens…Where is Everybody?" (Webb 2002) was able to count 50 distinct proffered solutions (a number which had grown to 75 in Webb 2015). Ćirković (2016) performed a deep dive into many of the underlying philosophical assumptions behind the Fermi Paradox and its solutions.

Here, we will not attempt to categorize all the solutions, but we will briefly discuss some of the more commonly discussed ones.

[17] Olbers' Paradox asks why every line of sight in the Universe does not end on the surface of a star, leading the entire sky to be as bright as the Sun. The fact that stars get fainter as the inverse square law is exactly canceled by the fact that their numbers per unit solid angle increase with distance squared, so the darkness of the night sky must mean stars are not uniformly and infinitely distributed. The paradox is sometimes "resolved" by invoking lookback time and the finite age of the Universe, because at large distances we see a time before stars formed. But this does not resolve the paradox at all: all lines of sight that do not encounter a star actually end near the edge of the observable Universe where, because of lookback time and the conditions of the early Universe, we see the "surface of last scattering," a bright plasma at thousands of degrees, which is brighter than most stars! The most important reason the sky is dark is that the Universe is *expanding* and this surface is redshifted by a factor of over 1000, making it effectively invisible at optical wavelengths (but detectable as the cosmic microwave background).

6.6.1 The Parable of the Anthill

A useful thought experiment is to apply the logic of the Fermi Paradox to Earth.

Imagine an intelligent, conscious ant colony in the wilderness somewhere—perhaps in the Serengeti, on a llano in South America, or in the Siberian forests. Through clever observation of the skies and soil from its vicinity, this unusually inquisitive colony has developed an impressive store of scientific knowledge about the Earth, and wonders if there might be any other intelligent species upon it, perhaps any that use tools. Such a species, it speculates, might even be able to construct machines that could travel great distances around the Earth, rearrange the land itself as they like on scales orders of magnitude larger than an anthill, and—who knows?—even fly through the air or soar around the planet beyond the atmosphere.

"Definitely not!" the colony concludes. If such species exist, then they could certainly travel at least as fast as an ant, around 1 km h^{-1}. The circumference of the Earth (our hyperintelligent ants have deduced) is "only" 40,000 km, meaning it would take of order 40,000 hours or 4.5 years to cross the Earth. But this is a *hundred million* times older than land life on Earth. Surely they would have settled this area of the Earth by now if they existed! But as far as the colony can see there is no hint of such technology.

A doubt crosses the colony's consciousness—perhaps such a species does exist, but only recently became technologically developed? But even if a species were, say, 100,000 years old, they would have still had 20,000 crossing times to settle this area.

And with technology, they could travel quickly. With land machines that could reach 100 km h^{-1} for the past 100 years, there have been over 200 Earth-crossing times—more than enough time to have settled nearby. With air machines traveling at 1,000 km h^{-1} for the past 50 years there have been 10,000 crossing times. And spacecraft could reach the anthill in a matter of minutes from anywhere in the world—enough for 300,000 crossing times in the past 50 years.

Still, some parts of the colony are not sure. Perhaps they have been here but chose not to stay? Perhaps the other species has collectively agreed not to visit the colony's part of the planet, perhaps even to ensure the colony is left alone. Perhaps they simply do not travel much. Perhaps their population is too small and grows too slowly to have spread this far.

Unsure, the colony continues to search for such technology in the soil and the skies. Perhaps some of the lights in the sky are actually aircraft or spacecraft. Perhaps some of the particles or chemicals in the soil are actually of technological origin. Perhaps even some of the clouds in the sky are products of propulsion?

And so the colony continues to speculate on the reasons there are no technological species visiting their colony today. Perhaps some reasons are right, perhaps all are wrong. One can't help but wonder, though.

6.6.2 They Do Not Exist (Drake's $N < 1$)

The conclusion Hart and Tipler come to, and the simplest resolution to the Fermi Paradox and Mystery of the Great Silence, is that we are alone in the Galaxy. We will discuss many of these solutions in Chapter 7.

This category of solution comprises many possibilities: abiogenesis is rare, multicellular life is rare, analogies to neurons and brains are rare, tool use is rare, and the sort of technology we use that might generate technosignatures or spacecraft is either rare or extremely short lived or has a very low "duty cycle."

6.6.3 They Avoid Contact With Us

6.6.3.1 The Zoo Hypothesis

First proposed and named by Ball (1973), *the Zoo Hypothesis* states that we are like an anthill in a protected part of the Serengeti where humans are not or only rarely allowed to travel—we have no alien technology here because they have collectively decided to leave Earth unvisited, although they can observe us from afar like animals in a zoo. Specifically, the hypothesis states that "we shall never find them because they do not want to be found and they have the technological ability to insure [sic] this."

Ball's suggestion was rather brief, and he did not opine with much detail about why ETIs would not visit us. He suggested they might be simply so uninterested in us that they would never bother to visit, or that we might be in a "laboratory simulation." Ball also lamented that this hypothesis and its cousins were "falsifiable, but not, in principle, confirmable by future observations," and pointed out that guessing at the motivations for us being in a "Zoo" "is outside the purview of science" and that "because it leads nowhere, it immediately calls into question the premises on which it is based, and it makes no predictions" (beyond, presumably, that we will never find evidence of ETIs).

Many authors have nonetheless suggested many reasons we might be protected from contact: they respect life on Earth and do not want to damage or interfere with it; they wish to study us; they are waiting for us to reach a certain level of technological and/or moral development,[18] and so on. Brin (1983) refers to this as the "quarantine" class of solutions.

This hypothesis generally assumes that ETIs are well aware of life on Earth, or perhaps humans specifically, and this is the reason they avoid us.

The Zoo Hypothesis goes beyond the Fermi Paradox, however, and also addresses the Mystery of the Great Silence, since it hypothesizes not just that they do not visit us, but that we are kept deliberately ignorant of their presence. This might mean that they simply avoid doing anything flamboyant that we might detect, like sending strong radio beacons in our direction, but some would argue it also means they might avoid building things like Dyson Spheres which we might detect.

6.6.3.2 The Dark Forest Hypothesis and Deadly Probes

A popular variant of the Zoo Hypothesis focuses on just the idea that all ETIs attempt to conceal their presence from us out of fear.

Perhaps the most commonly known variant today goes by the name *The Dark Forest* Hypothesis, coined in the Liu Cixin science-fiction novel of that same name.

[18] Akin to the Prime Directive in *Star Trek*.

This posits that resource scarcity in the Galaxy means that any species that makes itself known will inevitably become the target of neighboring species seeking to remove competition for those resources. Most species either realize this and hide, or else become the targets of their neighbors and are destroyed.

A variation of this hypothesis is what Brin (1983) refers to as the "Deadly Probes" class of solutions: essentially, some subset of von Neumann probes are hunters that seek and destroy technology, or life, everywhere in the Galaxy, echoing the concerns of Sagan & Newman (1983) about Tipler's proposal. Such probes that specifically seek out technological life foolish enough to make itself known are sometimes known as *berserker probes* after the Saberhagen series of novels.

We will revisit this concern in Chapter 19, but point out that the strategy of waiting for a species to become technological and announce itself before destroying them is somewhat contrived. Would not a better strategy be to go "weeding," and destroy intelligent species before they develop technology that might thwart them? And why would such a malevolent species be quiet, as opposed to baiting other species to announce themselves by sending false welcome messages? Why would they hide and *avoid* contact? Worse, why would *all* such species avoid such contact *forever*? These questions have plausible answers of course (and have been explored in science fiction), but do have a whiff of the monocultural fallacy about them.

The idea that the deadly probes are purely reactive and go into action only when a technological species reaches a certain level of technological sophistication or become sufficiently "loud" leads into fears that if we send strong, deliberate, target signals in the wrong direction we will trigger humanity's extinction (Section 19.4.1).

6.6.3.3 The Planetarium Hypothesis

The most extreme version of this class of "solution" is the Planetarium Hypothesis of Baxter (2001) which posits that all of humanity's observations of the Universe are actually the product of a detailed computer simulation, within which we are the only technological species. This hypothesis is a special case of the "Simulation Hypothesis" that all human experience is merely an elaborate virtual reality, as popularized in, for instance, *The Matrix* film series. The Planetarium Hypothesis is distinguished from these broader ideas in that it is explicitly framed as a solution to the Fermi Paradox and the Mystery of the Great Silence, and in that the creators of the simulation have explicitly chosen to have us live in a universe otherwise empty of intelligent life.

This idea that "reality is all an elaborate illusion" is much older than this, of course, and goes back at least as far as Irish philosopher George Berkeley's "subjective idealism" which argued that only experiences are real and material objects essentially do not really exist,[19] and even has much in common with Plato's

[19] English writer Samuel Johnson, upon being told Berkeley's philosophy was impossible to empirically refute, reportedly kicked a heavy stone with great force, saying "I refute it *thus*." Since it is not a logically valid refutation, this colorful demonstration became the source of the name of the logical fallacy "Appeal to the Stone," where one simply asserts that a hypothesis is obviously ridiculous, and so must be false.

Allegory of the Cave. The Planetarium Hypothesis, then, is perhaps the more appropriate target for Sagan & Newman (1983)'s accusation of being "the solipsist approach to SETI," and this is in fact how Brin (1983) uses that term.

6.6.4 They All Go Elsewhere or Become Something Else

A broad category of solution is that we will fail to find ETIs because they have "left" or taken some form or use some technology that we cannot detect.

For instance, some have suggested that alien species most naturally will live near black holes, in intergalactic space, or might be hibernating until the Universe is more hospitable to their forms, or might have completely retreated into computer simulations of their own design.

These solutions all suffer from the monocultural fallacy, and do not even apply to humans. For instance, the Hawaiian island of Maui is a paradise, and a far easier and more pleasant place to live than most places on Earth, most humans would agree.[20] So one might forgive our ants from Section 6.6.1 for concluding the reason they see no people is that they all live in Maui! But they would be very wrong—only a tiny fraction of humans live there, or in tropical paradises in general. This is for many sociological reasons our ants would have a hard time guessing or even comprehending (and indeed most humans would have a hard time accurately saying exactly why).

Similarly, one might conclude that all ETIs use technology other than those we can detect, perhaps because they employ physics we have not yet discovered. While this is certainly possible (although we will see in Section 17.3 why photons are actually probably still a good bet), it again would require *all* species to be using such technology now for them all to be invisible. So while this explanation might reduce the number of species we can communicate with, it does not explain why we see none of them.

6.6.5 Exponential Growth is Inherently Unsustainable

The idea of life spreading through the Galaxy and inhabiting every star strikes many as having parallels to human exploitation of the Earth. Environmental concerns then make one wonder if every species that behaves in such a manner might simply go extinct for various reasons.

This is perhaps most succinctly expressed in the *sustainability solution* to the Fermi Paradox of Haqq-Misra & Baum (2009), who suggest that any exponential growth in the Galaxy fast enough to justify the Fermi Paradox might be coupled with a predilection for civilization collapse and extinction.

As pointed out by Wright et al. (2014), to avoid being tautological ("exponential growth is never sustained because exponential growth is not sustainable") this solution requires some mechanism by which all members of fast-spreading species *permanently* stop spreading fast, and to avoid the monocultural fallacy it must explain why this mechanism is universal. After all, it is possible for a species to be locally sustainable in an ecological sense, and still send out settlement ships that create exponentially more sustainable settlements on a regular basis. Indeed, if one

[20] Mauians like to boast *Mauinōka'oi* ("Maui is the best") and they are probably not wrong!

settlement proves unsustainable and goes completely extinct, there is no reason that other, more sustainable settlements could not subsequently "backfill" the failed settlements later. In other words, this solution conflates, or at least couples, the issue of *local* (i.e., planetary) sustainability with *global* (i.e., interstellar) expansion.

For instance, a common assumption is that the primary driver of settlement and expansion would be overpopulation or resource depletion, and this is sometimes used to create mathematical models of expansion, or to argue that all fast-spreading species are necessarily resource-hungry or unsustainable. There are, however, many other justifications for expansion that do not rely on these motivators (which, after all, probably do not describe humanity's initial peopling of the globe, which was driven by many factors such as following herds of animals for food).

In addition, unsustainability does not necessarily imply extinction. It would take a truly apocalyptic ecological collapse for all of humanity to die out, probably including the extinction of most animal life on land in the process. The worst-case, long-term effects of climate change or nuclear war are catastrophic, but it is far from clear they would cause human *extinction* (and even farther from clear that such a mechanism would be necessarily universal across the Galaxy.)

6.6.6 All Technology Is Recent

One assumption buried in the Fermi Paradox is that the relevant timescales of the problem are something like the age of the Galaxy or the typical age of Sun-like stars, around 5–10 Gyr. The logic of the Fermi Paradox assumes that life has had a very long time to arise, evolve, and spread.

One astrophysical way out pointed out by Annis (1999) is to posit that the Galaxy only recently became hospitable for life, and so while we may not be unique in the Galaxy, we are in the first generation of species that could become technological. Ćirković & Vukotić (2008) call this an *astrobiological phase transition* of the Galaxy, and it may have happened quite recently.

This idea is closely related to the concept of the "Galactic Habitable Zone" (Chapter 7), which claims that there is a special region in space and time in the Galaxy where life can arise for a variety of reasons. The general idea is that the early Galaxy was a violent place inhospitable for life, and so any planets on which life managed to arise would be regularly sterilized by events such as gamma-ray bursts and supernovae. Only within the past few billion years have the rates of these events slowed enough that complex life can arise anywhere.

The idea of a phase transition comes from the hypothesis that, absent these destructive events, life is common and develops technology quickly, meaning that there could be a rapid transition from a Galaxy devoid of technological life to one filled with it once any part of the Galaxy becomes habitable for a sustained duration.

If this tipping point time in the Galaxy is around now, then the relevant timescale in the Fermi Paradox is the time since that tipping point, which could be quite recent. If the fastest ships have speeds of, say, 300 km s^{-1} and the tipping point occurred 100 Myr ago, then the first wave of settlement ships would be crossing the Galaxy right about now.

Box 6.2: Jargon Review: The Fermi Paradox

- **Fact A**: The premise of the Fermi Paradox: There are no aliens or alien technologies on Earth today.
- **The Fermi Paradox**: *Formally*: The supposed inconsistency between Fact A and the fact that the even "slow" ships have had plenty of time to get here. *Loosely*: Often used to refer to the "Mystery" of the Great Silence.
- **The Great Silence, Fact B**: Defined here as the fact that humans have not yet found proof of any ETIs, i.e., the absence of extraterrestrial technosignatures so strong that we would have detected them by now.
- **The Mystery of the Great Silence**: The supposed inconsistency between optimistic estimates from the Drake Equation of the number of species to be found, and the "failure" of SETI to date to find them.
- **Monocultural fallacy**: Ascribing a single set of behaviors to all members of alien species for all time, especially in an attempt to explain the Fermi Paradox.
- **Exclusive solutions/Principle of non-exclusiveness**: Solutions that commit the monocultural fallacy, which asserts that all aliens will forever exclusively follow a single set of behaviors, and the principle that such solutions fail to resolve the Fermi Paradox.
- **von Neumann probe**: A self-replicating, spacefaring machine or set of machines used to explore and settle the solar system and the Galaxy.
- **Fermi Problem**: A quantitative question, especially one of a surprising or seemingly impossible nature, that admits a quick rough order of magnitude solution.

Box 6.3: Jargon Review: A Few "Solutions" to the Fermi Paradox and Mystery of the Great Silence

- **The Zoo Hypothesis**: We will never find ETI because they do not want to be found by young species such as us and they have the technological ability to hide.
- **Sustainability solutions**: Exponential growth throughout the Galaxy is inherently unsustainable because the drivers of that growth inevitably lead to collapse and extinction.
- **The Dark Forest Hypothesis**: All other technological species stay "silent" and do not spread to avoid conflict with hostile species.
- **Deadly probes/Berserker probes**: Von Neumann probes whose purpose is to destroy technological life.
- **Astrobiological phase transition**: The transition from a sterile (or non-technological) Galaxy to one endemic with technological species. If this transition were frustrated by cataclysmic events which recently abated, this transition could be happening now.

6.7 Further Reading

David Brin's analysis of the Fermi Paradox: Brin (1983).
Solving Fermi's Paradox: (Forgan et al. 2019).
The Great Silence: Science and Philosophy of Fermi's Paradox (Ćirković 2018).

If the Universe Is Teeming with Aliens … WHERE IS EVERYBODY?: Fifty Seventy-Five Solutions to the Fermi Paradox and the Problem of Extraterrestrial Life (Webb 2015).

Exercises

1. Practice some Fermi Problems
   (a) How many grains of sand are there on the world's beaches? Start without looking anything up.
      i. Estimate the total length of coastlines on Earth.
      ii. Guess at the fraction that are sandy.
      iii. Use your experience at beaches to estimate how far from the waterline it is sandy.
      iv. Estimate, or at least bound, how deep the sand goes.
      v. Compute the volume of sandy beaches.
      vi. Estimate the size of a grain of sand.
      vii. Compute the number of grains on Earth's beaches.
      viii. Go back and get better estimates of each step in your calculation using some Googling. How well did you do? How much does your answer change based on the updated numbers?
   (b) How long would it take to meet everybody on Earth?
   (c) How long would it take to read all of the books in your local library?
   (d) How many times have people laughed?
   (e) How large of a vessel would it take to hold all of the tears humans cried in 2022?
   (f) How much energy would it take to clear New York City of snow after a blizzard by melting it? How much would that much energy cost?
2. How much harder or easier would interstellar travel be for humans if Jupiter did not exist? What if it were a hot Jupiter, at an orbital distance of 0.1 au? What if it were at the orbital distance of Pluto?
3. Use the ADS SETI bibliography to find and catalog a few papers that offer solutions to the Fermi Paradox. Do they fall into one of Hart's four categories? Do they survive his critique?
4. What are some *valid* solutions to the Fermi Paradox?
5. A More Detailed Fermi Paradox Calculation

   Apply the method of Fermi Problem calculations to a more detailed calculation of the Fermi Paradox, breaking the problem up into parts and making "conservative" estimates.

   Assume Earth-level technology today as a lower bound on how fast interstellar ships could go, and that they can only reach the very nearest stars. You may need to look some numbers up for this problem. Stars are roughly 1 pc apart in this part of the Galaxy, and the Galaxy is roughly 17 kpc across. The galaxy contains something like 100 billion stars.

(a) How long will it take such ships to reach the nearest stars?
(b) Assume it takes some amount of time on arrival at each of these stars to settle on a new planet and start building ships. What numbers seem safely conservative to you here?
(c) Which time matters more: the time to travel to a nearby star or the time to settle in and start building ships?
(d) How long will it take for ships to reach the other side of the Galaxy? Remember for each hop they must travel to the nearest star and stop to start building new ships, both of which take time, but one of which probably matters more (if you need it, go ahead and assume the distance between stars is about the same everywhere, since the local value is a rough average).
(e) But we might have a different bottleneck: how long until we have enough ships to visit every star? Every ship makes a settlement, which makes ships. Once a settlement starts making ships, it does so at some rate, r. Make what seems like a conservative estimate of that rate to you (that is, how long between ships being built?)
(f) Call the number of settlements s. The number of new settlements created per unit time depends on the number of existing settlements and how often new ones are made. Write down the differential equation this implies, in terms of s, t (time), and r. (For this part, ignore the travel time between stars. Assume that a settlement is made once a new ship is built and the new settlement starts building ships immediately.)
(g) Solve the equation. If there is 1 settlement at $t = 0$, how long until there are as many settlements as stars?
(h) Which takes longer: making enough settlements, or getting to the other side of the Galaxy?

References

Annis, J. 1999, JBIS, 52, 19
Ball, J. A. 1973, Icar, 19, 347
Baxter, S. 2001, JBIS, 54, 210
Brin, G. D. 1983, QJRAS, 24, 283
Carroll-Nellenback, J., Frank, A., Wright, J., et al. 2019, AJ, 158, 117
Ćirković, M. M. 2016, AsBio, 16, 737
Ćirković, M. M. 2018, The Great Silence: Science and Philosophy of Fermi's Paradox (Oxford: Oxford Univ. Press)
Ćirković, M. M., & Vukotić, B. 2008, Origins Life, 38, 535
Eliis, G. 1994, Natur, 371, 115
Forgan, D., Wright, J., Tarter, J., et al. 2019, IJAsB, 18, 336
Gray, R. H. 2015, AsBio, 15, 195
Haqq-Misra, J. D., & Baum, S. D. 2009, JBIS, 62, 47
Hart, M. H. 1975, QJRAS, 16, 128
Katz, J. I. 2021, NucTe, 207, S326

Sagan, C., & Newman, W. I. 1983, QJRAS, 24, 113
Sagan, C. 1963, P&SS, 11, 485
Stephenson, D. G. 1977, JBIS, 30, 105
Swain, C. M., & Nieli, R. 2003, Contemporary Voices of White Nationalism in America (Cambridge: Cambridge Univ. Press)
Tipler, F. J. 1980, QJRAS, 21, 267
Webb, S. 2002, If the Universe Is Teeming with Aliens … WHERE IS EVERYBODY?: Fifty Solutions to the Fermi Paradox and the Problem of Extraterrestrial Life Copernicus Series (Berlin: Springer)
Webb, S. 2015, If the Universe Is Teeming with Aliens …WHERE IS EVERYBODY?: Seventy-Five Solutions to the Fermi Paradox and the Problem of Extraterrestrial Life, Science and Fiction (Berlin: Springer)
Wright, J. T., Mullan, B., Sigurdsson, S., & Povich, M. S. 2014, ApJ, 792, 26
Wright, J. T., & Oman-Reagan, M. P. 2018, IJAsB, 17, 177
Wright, J. T., Carroll-Nellenback, J., Frank, A., & Scharf, C. 2021, RNAAS, 5, 141

AAS | IOP Astronomy

The Search for Extraterrestrial Intelligence
Theory and practice
Jason T Wright

Chapter 7

The Case Against SETI: Rare Earth

> Curiously enough, the dolphins had long known of the impending demolition of Earth and had made many attempts to alert mankind to the danger. But most of their communications were misinterpreted as amusing attempts to punch footballs, or whistle for titbits, so they eventually gave up and left the Earth by their own means—shortly before the Vogons arrived. The last ever dolphin message was misinterpreted as a surprisingly sophisticated attempt to do a double backwards somersault through a hoop, while whistling the "Star-Spangled Banner." But, in fact, the message was this: "So long and thanks for all the fish."
> —Douglas Adams, *The Hitchhiker's Guide to the Galaxy* BBC radio program, Episode 3.)

The more specific we make our definition of life, the rarer we expect it to be. The logic of the Copernican Principle and the mathematics of infinity suggest that somewhere in an unbounded homogeneous universe there exists a near-perfect copy of you reading this very text. But since such copies must be so very rare, we do not expect there to be one in any of the nearby stars—or really, anywhere within the observable Universe.[1]

One question for SETI is whether the defining characteristics of technological life that could be detected at interstellar distances—as we can be—are so narrow that the nearest example to too far away to matter, or sufficiently broad that we might make contact.

[1] In the Concordance Model of cosmology, the Universe is infinite in extent, but our ability to observe it is bounded by the speed of light—the more distant areas are too far away for light to have had time to reach us in the age of the Universe. The limiting distance out to which we can see is called a *horizon* and the space within that horizon is called the *observable Universe*. Cosmology does not really teach there is a near copy of you somewhere—there are many ways out of this logic. For instance, the Universe may be finite in size (either "closed" or simply bounded), the laws of physics may be different at large distances, or because of some other effect.

doi:10.1088/2514-3433/adf5aech7

7.1 Ward and Brownlee's Formulation

The *Rare Earth* hypothesis—named by American paleontologist Peter Ward and American cosmologist Donald Brownlee in their book on the topic (Ward & Brownlee 2000)—states that there are so many unlikely contingencies that led to animal life on Earth, that it is unlikely it would have similarly arisen anywhere else in the Galaxy, and so SETI is doomed to fail.[2]

In Ward & Brownlee's formulation, some of these rare contingencies include:

- The location of the Sun within the Galaxy, providing an optimal balance between an environment free from sterilizing events such as supernovae, and chemical enrichment sufficient to produce planets like Earth.
- The organization of the solar system, with Jupiter limiting the number of large impactors that could have sterilized Earth, the Moon stabilizing Earth's spin and climate, and the Earth's orbit being just right for the sustained existence of liquid surface water.
- The size and composition of the Earth, which seems finely tuned[3] to produce plate tectonics (which are an essential part of Earth's carbonate–silicate cycle and life's evolution) and magnetic field (which protects Earth's atmosphere and life from damage by the Sun.)
- The development of complex life, such as eukaryotes and, later, animals. Ward & Brownlee argue the complex nature of such development and its "late" arrival in Earth's history indicates it is an extremely improbable event.
- The unlikely survival of Earth's complex life across many cataclysms, including asteroid impacts and climactic shifts such as "Snowball Earth" during which the entire planet was glaciated.

Multiplying the chances of each of these contingencies together, Ward and Browlee argue, yields a number smaller than the number of stars in the Galaxy, meaning we should expect no other planets with animals, much less technological creatures, nearby.

7.2 The Galactic Habitable Zone

The region within the Galaxy suitable for the long-term development of complex life was named the "Galactic Habitable Zone" by American astronomer Guillermo Gonzalez[4] in collaboration with Ward and Brownlee (Gonzalez et al. 2001). This

[2] This is an oversimplification of the conclusion: we could in principle discover life in other galaxies. Also, if SETI can show that technological life does *not* exist then it could hardly be said to have failed, since it will have answered the question it was designed to resolve.

[3] "Fine tuning" in physics describes a parameter that has a value which, if only slightly different, would lead to qualitatively different physics. Although the term is usually used neutrally, it is sometimes meant as part of teleological or anthropic arguments (Sections 7.2 and 10.2).

[4] Some advocates of the Rare Earth hypothesis may be victims of motivated reasoning. For instance, Gonzalez was involved in significant controversy over his support of "Intelligent Design," i.e., the necessity of God to explain the existence and evolution of humans. Gonzalez served as a fellow at the Discovery Institute and at the International Society for Complexity, Information and Design, which are or were think tanks that promote Creationism, i.e., the belief that the cosmos was created and designed specifically for humans by God.

concept has been further developed by others, including Lineweaver et al. (2004) who bounded it as an annulus in the disk 7–9 kpc in radius.

The Galactic Habitable Zone arises from the fact that the different parts of the Galaxy sports very different kinds of stars and environments.

The oldest parts of the Galaxy are in its halo, whose stars formed before the Galaxy had fully formed. These stars generally are almost pure hydrogen and helium, hosting very few "metals"[5] like iron, silicon, carbon, oxygen, and nitrogen. Since planets like Earth are made of these elements, we expect these older stars to have few of them, and even those they have should be depleted in other, rare elements essential for Earth-like life. It should be noted that we do not actually know much about the rate of terrestrial planets around these metal-poor stars[6] because such stars are so rare in our part of the Galaxy, and so we do not know the degree to which life has difficulty arising around them.

The disk and bulge of the Galaxy formed shortly after the halo. The bulge is dense with stars, having witnessed many generations of star formation, and these stars can have very high metal content. The high density of stars poses several potential problems for the long-term development of life, most especially the high density of supernovae and other cataclysmic events that have the potential to sterilize large swaths space around them.

The bulge today has a lower rate of star formation than in the Galaxy's past, and so the rate of supernovae today is perhaps low enough to allow life to form. A second claimed problem is that the metallicity[7] of stars can be *too* high to support life. This is because there is a clear correlation between the metallicity of a star and the number of giant planets orbiting it. Metal-rich stars often host multiple giant planets like Jupiter, which could interfere with there being rocky planets in a star's Habitable Zone. The Galactic Habitable Zone posits a "sweet spot" of metallicity where only a single, Jupiter-like planet forms, and that the metallicity in the bulge today is too high for this.

The rest of the disk sports stars of various ages and metal contents, but in general the rate of star formation (and, thus, the rate of metal formation) decreases with Galactocentric distance (i.e., radius). In the outer reaches of the disk, there have been fewer generations of fewer stars, and so stellar metallicities are generally lower, perhaps too low for many rocky planets to form.

The punchline of the GHZ is that only in a range of radii are all of these considerations balanced in such a way that one expects to find a significant number of stars that are old enough to have evolved complex life (which Lineweaver et al. 2004 take to be at least 4 Gyr), formed in a place where there have been not too many supernovae, at a time and place with appropriate metal content. The balance of these effects is illustrated in Figure 7.1.

---

[5] See Section 2.9.

[6] These metal-poor stars are "population II" stars in the jargon of astronomy. "Population I" stars populate most of the Galaxy and have a wide range of metal content. "Population III" refers to the very first generation of primordial stars, none of which (are known to) exist today, and which contained zero metals.

[7] As much as I wish this referred to how hard stars thrash on electric guitars, it is just jargon for "amount of metal content," i.e., the abundance of elements heavier than helium.

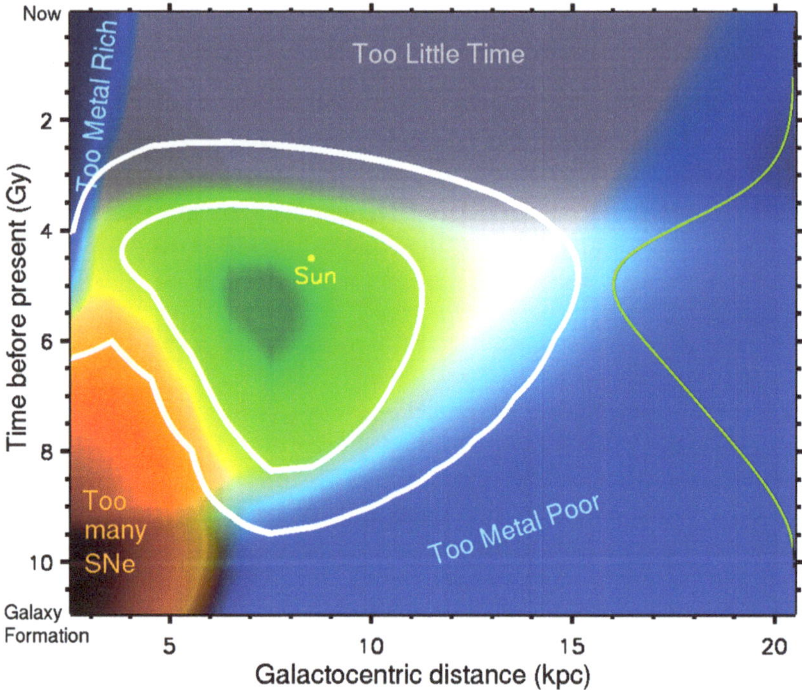

Figure 7.1. From Lineweaver et al. (2004), illustrating the evolution of the Galactic Habitable Zone with time. The y-axis refers to the time before present that a given star and its planetary system formed—today, at any given Galactocentric distance stars from a range of y-axis values will exist. Reprinted with permission from AAAS.

It should be noted that there are many uncertainties in this calculation. We do not know the frequency of finding rocky planets among metal-poor stars; we do not know the difficulties life encounters on such planets; we do not know the challenges life finds on in systems with different planetary architectures from ours; and we do not know the supernova rate in different parts of the Milky Way through cosmic time. Unlike the stellar habitable zone, where we can be sure liquid surface water is absent far beyond its boundaries, we cannot put hard limits on life outside the Galactic Habtibale Zone (Prantzos 2008).

7.3 Timing, Hard Steps, and the Great Filter

7.3.1 Anthropic Reasoning and Hard Steps

The Anthropic Principle (which we will encounter later in Section 10.2) states that we must experience the Universe in a manner consistent with that existence, because our ability to experience the Universe at all is predicated on that existence.

Carter (1983) applied the Anthropic Principle to the development of life on Earth to argue that the development of "intelligent" life is likely to be extremely rare using a timescale argument: the time it took intelligent life to arise on Earth (of order 4 billion years) is of the same order of magnitude as the lifetime of the Sun (around

10 billion years). Since, Carter argued, there is no physical reason these two timescales should be similar at all, it is likely that the former timescale is typically many orders of magnitude larger than the age of the Sun, and we only find it to be of similar order on Earth for anthropic reasons: we would not be here to measure it if it had taken the usual amount of time, because the Sun would have gone out by then.

In other words, the Anthropic Principle answers a puzzle: why should the time it took life to arise on Earth be similar to the time Earth has been habitable? This is hard to understand if intelligent life arises easily throughout the Galaxy, because if it were "easy" it would have occurred much earlier in Earth's history. It is possible that technological life generally arises throughout the Galaxy on any given habitable planet in around a few billion years, but that seems like a strange coincidence.

But if life takes generally an extremely long time to develop, then on those lucky few planets where life does arise precociously (i.e., before its star runs out of hydrogen) it will generally find itself around a middle-aged or old star.

Carter then suggested a physical reason that it typically takes so long for intelligent life to form: that there must be a requirement for the "acquisition of intelligence" that life first pass through a "series of intermediate steps" that are unlikely, in the sense that they each have a very low probability of occurring in the lifetime of a star.

Carter noted that at early times, the likelihood of achieving n steps of low probability per unit time grows like t^n, which steeply increases with t for large n. It works out that the expected time to complete the steps *assuming the unlikely outcome that they are completed before time* T, is roughly $T(1 - 1/(n + 2))$. That is, for no hard steps (uniform probability of intelligent life arising per unit time) we expect it to arise about halfway through its window on average ($T/2$ for $n = 0$), but if there were a large number of hard steps (say, n = 1000) then we would not expect to be here until the last 0.01% of the Earth's habitability.

Carter noted that the fact that the Sun still has billions of years left on the Main Sequence means that we did not arrive just at the very end of Earth's habitability "window." Today we understand that the Sun's luminosity increases at a rate such that liquid water will cease to exist on the Earth's surface in less than 1 billion years, meaning that we have arrived in the last quarter or so of Earth's habitability, implying that $n \approx 2$. Apparently, Carter concluded, there are a very small number of "hard steps."

Szathmáry & Maynard Smith (1995), attacking the problem of why complexity in some lineages increases with time at all, identified eight discrete "major evolutionary transitions" in which previously free-living organisms become integrated as units in larger wholes, accompanied by division of labor among these lower-level units and shifts in information storage and processing.

In an unpublished manuscript "Must Early Life Be Easy? The Rhythm of Major Evolutionary Transitions"[8] Hanson (1998) merged these ideas, identifying

[8] Available at https://web.archive.org/web/20250401012933/https://mason.gmu.edu/rhanson/hardstep.pdf.

Szathmáry & Maynard Smith (1995)'s major evolutionary transitions as the small number of "hard steps" that combine to make the emergence of complex life rare under the Carter argument.

Watson (2008) developed these ideas more formally, attempting to identify which of the major evolutionary transitions were truly "difficult" by determining which of them did not occur more than once, settling on seven of Szathmáry & Maynard Smith (1995)'s transitions that were truly difficult. Watson calls these transitions "critical," somewhat confusing the difference between steps that are low probability with those that were essential, and argued that the timings of each of the seven steps were consistent with each of them being approximately equally "hard."

Watson discarded an alternative model where the emergence of life was due to a very large number of slow but high probability steps, arguing that then "an explanation would be needed for the "Fermi Paradox"—why we see no evidence for abundant intelligent activity elsewhere in the cosmos" (but see Chapter 6 for more on this).

Carter's argument is contingent on there being a puzzle to solve in terms of the coincidence in the timescale of the emergence of intelligent (or animal) life and the evolution of the Sun. But Mills et al. (2025) argued that there is another possibility: billions of years is also the timescale of the evolution of the *Earth's surface*, which is intimately tied to the evolution of life itself. For instance, Salles et al. (2023) find that the evolution of macroscopic life on Earth was closely tied to plate tectonics and the emergence of certain kinds of land.

In other words, the "hard steps" model is predicated on explaining a coincidence that may need no explanation at all.

7.3.2 The Great Filter

A closely related argument to the "hard steps" model is that of the "Great Filter" (also from an unpublished Hanson manuscript in 1998, "The Great Filter—Are We Almost Past It?"[9])

The Great Filter argument is a proposed solution to the Fermi Paradox, arguing that there must be at least one very hard step to life becoming prevalent throughout the Galaxy, or we would see strong evidence of life everywhere. Hanson calls this the "great filter between death and expanding lasting life."

The philosophical twist in this explanation is that if the Great Filter is "behind" us—if it is some "hard step" toward the development of life—then humanity's lucky destiny is to be the first to settle the Galaxy because we have passed it and there is no reason to think another lies ahead. If, however, life like ours is common but there is some mechanism we have not yet encountered that destroys intelligent life before it can spread—if the filter is ahead of us—then humanity is doomed to suffer the same fate as the potentially large number of other prior intelligent species that failed to dominate the Galaxy.

[9] Available at https://web.archive.org/web/20250331020157/https://mason.gmu.edu/rhanson/greatfilter.html.

Hanson proffered self-destruction as the future filter, but since then others have suggested the filter might be a single previous intelligent species that periodically "weeds" the Galaxy of younger intelligent species.

It is important to remember that the Great Filter may not exist at all. It is a purely hypothetical challenge common to all life in the Universe that prevents it from being so pervasive in the Galaxy we could not have missed it by now. There may be many filters, none, or each kind of life might encounter its own filters.

7.4 Counterargument: Probabilities

The Rare Earth hypothesis is susceptible to two major counterarguments: that the astronomy does not support the small probabilities, and that the role of contingencies is overstated.

In the case of the Galactic Habitable Zone portion of the argument, our understanding of the abundance of Earth-like planets throughout the Galaxy is rather sketchy, with most of the concrete data we have coming from stars in our part of the Galaxy and with metallicities similar to the Sun. Also, while other parts of the Galaxy may be less hospitable for life, we do not yet know how to quantify this reliably. Prantzos (2008) provides a good description of these uncertainties, and finds that under reasonable assumptions, most of the Galaxy can be considered to be within the Galactic Habitable Zone.

We also do not understand the evolution of life on Earth well enough to know how long complex life takes to develop or how inevitable that progression is. Some major transitions, such as eukaryogenesis[10] appear so complex and essential for life as we know it that it seems likely they represent an unlikely event. However, this may reflect more about our (lack of) understanding of the event than its actual likelihood.

So while the timing of major evolutionary transitions on Earth can be interpreted as evidence that some of them are extremely rare, such arguments are based on a sample size of one, and so even to the degree they are not subject to the Anthropic Principle, they provide only limited evidence for how biology would evolve on *other* planets.

7.5 Counterargument: Contingency and Convergent Evolution

> For instance, on the planet Earth, man had always assumed that he was more intelligent than dolphins because he had achieved so much—the wheel, New York, wars and so on—while all the dolphins had ever done was muck about in the water having a good time. But conversely, the dolphins had always believed that they were far more intelligent than man—for precisely the same reasons.
>
> —Douglas Adams, *The Hitchhiker's Guide to the Galaxy*

[10] All complex multicellular organisms on Earth (animals, plants, mushroom-forming fungi, etc.) are eukaryotes. Eukaryotes package their genomes into membrane-bound nuclei and feature sub-cellular organs (or organelles) like mitochondria.

7.5.1 Contingency versus Convergence

The role of contingency in the development of Earth's biosphere is difficult to gauge, and it is important not to put too much stock in to what may turn out to be a just-so story.

As an example, consider a very small, very remote town with one medical doctor near retirement. Let's say the residents are hoping to find a new doctor to replace them. They might ask the doctor how they became interested in medicine, where they did their undergraduate, graduate, and other training, how they paid for school, and what life choices they made that led to them taking the job in such a small, remote town decades earlier.

They might calculate the probability that another person had followed that path, and, there being many steps each of low probability, conclude that no one on Earth would fit the bill, and they will never find another town doctor. But this would be a logical error: there are many paths to becoming a small-town doctor, with many steps along the way. We can identify some of the steps as being necessary, such as earning a medical degree, needing or being open to a job when one opens, and being willing to practice in a small town. These are *contingent* factors: things that may or may not be true for any given person, but necessary for them to serve as a small-town doctor.

Others aspects of the doctor's life may have led to them becoming a doctor, but are not universal factors: exactly which schools they attended, which scholarships they earned, the specific reason they became interested in medicine, and other aspects of their specific path to professional practice. These details may have been necessary for *them* given their circumstances, but other doctors will have arrived at the same place by different paths. These variable aspects of the path to an MD are *convergent* factors.

To give another example more relevant to SETI, one might look at the early development of civilization in Mesopotamia and Egypt and conclude that only the specific set of environmental and social factors there (the flooding of the Nile, the presence of the Fertile Crescent) were responsible for social structures that generated the Great Pyramids. But one would be wrong: large pyramid-building civilizations also arose in Mesoamerica and South America (for instance) completely independently and under very different environmental conditions.

To bring the analogy back to the Rare Earth Hypothesis, while the biosphere is certainly dependent on the size and structure of the Earth, it is not clear to what degree Earth's particular mix of plate tectonics, magnetic field strength, and climate stability are actually necessary for life. Venus and Mars may have hosted biospheres for billions of years, for instance, despite having none of those features.

The specific development of complex life on Earth involved billions of years of innovation at the molecular and microscopic level within cells, a story we understand only the outlines of. While it is true that biologists have identified major evolutionary transitions that seem to have been necessary steps toward how animal life functions, such as the development of eukaryotes, it is unclear whether there might be alternative routes that accomplish the same thing. Perhaps the development of multicellular life on Earth was uncharacteristically *late*.

The specific steps toward the development of animal life also may not need to be replicated elsewhere. Certain features of life are so advantageous that they tend to evolve independently in multiple lineages, a feature called *convergent evolution*. A famous example is the eye, which appears across the animal kingdom in species whose common ancestors had no eye.[11]

7.5.2 Monophyly and Polyphyly as Markers of Contingence and Convergence in Biological Evolution

A hallmark of convergent evolution is illuminated from our doctor example: the presence of a trait shared among those with different histories. Among creatures, this would be clear if a trait appears in multiple species whose last common ancestor lacked that trait. A classic example of convergent evolution is wings, which have evolved independently in bats, birds, and insects.

In the jargon of evolutionary biology we would say that wings are a *polyphyletic* trait. *Polyphyly*[12] is the property of a taxonomic group (i.e., a group of species grouped in a taxonomy for sharing some trait) that they do not include all species descended from some common ancestor. The converse is *monophlyly*, which arises when all creatures in a given taxon, and only those species, share a common ancestor.

An example of a trait with a monophyletic origin would be vertebrae, which exist only in the clade[13] Vertebrata.

Importantly, we only have access to traits of *extant* species that we have studied, so we can never be sure a given trait truly has a monophyletic origin: there could be undiscovered or extinct species from other clades that had the trait. Also, because we do not have access to last common ancestors, is also not always clear whether a trait is truly of polyphyletic origin: some descendants of a given ancestor may have lost that trait, making it appear inconsistently among its ancestors.

For instance, consider Figure 7.2, which shows a *cladogram* or map of the lineages of the primates. In this diagram, evolutionary descent goes upwards. Extant kids of primates are shown at the top, and extinct groups are not shown. Lines connect these groups to their last common ancestor with the other groups, at vertices.

The simians (i.e., all monkeys, apes, and humans) form a monophyletic group, meaning they are all descended from a unique common ancestor (that is, the most recent common ancestor of all living simians) that itself was a simian. Traits shared by all of these lineages (but not the lemurs, lorises, and tarsiers, such as their facial structures) were very likely shared with and inherited from their most recent common ancestor. An unlikely alternative scenario is that they all independently evolved these traits but all other lineages did not. The primates active at night, the

---

[11] Another is carcinization, the apparent tendency of crustaceans to evolve toward a common body type reminiscent of a crab.

[12] Pronounced PAH-lee-file-ee.

[13] A *clade* is all descendants of a common ancestor. Today, we can identify many clades with *taxa* (singular *taxon*) that we use to name and group species based on their traits, such as the class Mammalia or the phylum Chordata in Linnaean taxonomy.

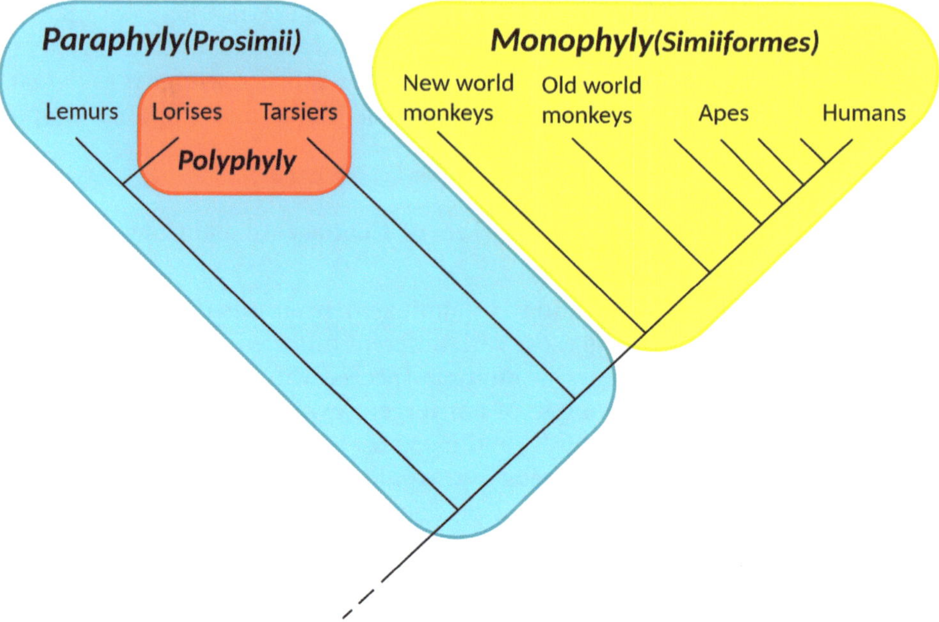

Figure 7.2. Cladogram of the primates showing descent (going upwards) from a common ancestor. The author obtained this image from the Wikimedia website, where it is attributed to and made available by Chiswick Chap under a (CC BY-SA 4.0) license.

lorises and tarsiers, are polyphyletic, implying that night activity evolved independently in those two lineages. An unlikely alternative scenario is that night activity is a trait that was shared by the last common ancestor to all primates, but that all other lineages lost this trait.

Evolutionary biologists can use genetic analysis of extant species to attempt to reconstruct cladograms like this, and to distinguish convergent from contingent evolution. For instance, if two species use different molecular pathways or genes to express some trait, then despite appearances they are likely to have evolved that trait independently.

7.5.3 The Convergent and Contingent Aspects of Intelligence

Another important example for SETI is the development of intelligence, which has occurred many times across the animal kingdom in many forms, arising from the brains of mammals such as humans, to the more distributed nervous systems of gastropods, to the collective intelligence and tool use of ant colonies. We may even have unrecognized forms of cognition on Earth—some have even suggested that interactions among trees or other plants might express itself as a kind of intelligence across very long timescales.

The octopus is a particularly good case study for the convergent and contingent components of the evolution of intelligence because octopi are molluscs, so only distantly related to chordates like humans. Octopi have large brains and, being predators, exhibit many aspects of intelligence, including problem solving and tool

Figure 7.3. Two panels from Figure 1 of Finn et al. (2009) illustrating the use of coconut halves as tools by the veined octopus *Amphioctopus marginatus*, including as a defensive shelter (B), and "stilt walking" (C) to move them from place to place for future use. © 2009 with permission from Elsevier.

use (Figure 7.3). Many other molluscs, such as barnacles or clams, and other chordates, such as sea squirts (ascidians) are clearly unintelligent, so the aspects of intelligence shared by octopi and humans clearly originated polyphyletically, i.e., they evolved convergently.

We can investigate which aspects of intelligence are contingent by looking at the last common ancestor of all clearly intelligent creatures. All obviously intelligent creatures on Earth are members of the clade Nephrozoa, which compose most of the clade Bilateria, the group of bilaterally symmetric animals, shown in Figure 7.4.[14] Nephrozoans are divided between the protostomes and the deuterostomes, distinguished by the manner in which the digestive track forms in embryos. Deuterostomes include echinoderms (like starfish) and chordates (which include vertebrates such as mammals, birds, and reptiles); protostomes include arthropods (like crabs and insects) and molluscs (like barnacles and octopi).

The last common ancestor of these creatures—and so the last common ancestor of all obviously intelligent creatures on Earth—is called the urnephrozoan or the protostome-deuterostome ancestor,[15] about which not much is known, but we can infer some things. All nephrozoans have nervous systems so we can infer that the urnephrozoan also had one, meaning it already had a complex network of neurons. It almost certainly existed by the Ediacaran Period 539–635 million years ago, when essentially all animal life was marine. It likely had genes common in photoreceptors, so possibly eyes, and possibly appendages, cilia, and segmentation. Its size was uncertain but it was likely small, perhaps 1 cm to 1 mm across. From this humble beginning, intelligence arose multiple times in convergent ways.

So what aspects of intelligence are contingent? We can look deeper into the evolutionary history of animals for some clues. In particular, we can ask which

[14] These clades are today confirmed with a combination of genetics and morphology, especially embryonic morphology, and some bilaterians to do not as adults exhibit obvious bilateral symmetry. Nephrozoa is also sometimes called Eubilateria.

[15] This creature was called the urbilaterian until it was recognized that Bilateria is a broader clade that includes xenacoelomorphs, which are neither deuterostomes nor protostomes. Since Nephrozoa is sometimes called Eubilateria, one might also call this creature the ureubilaterian. See Figure 7.4.

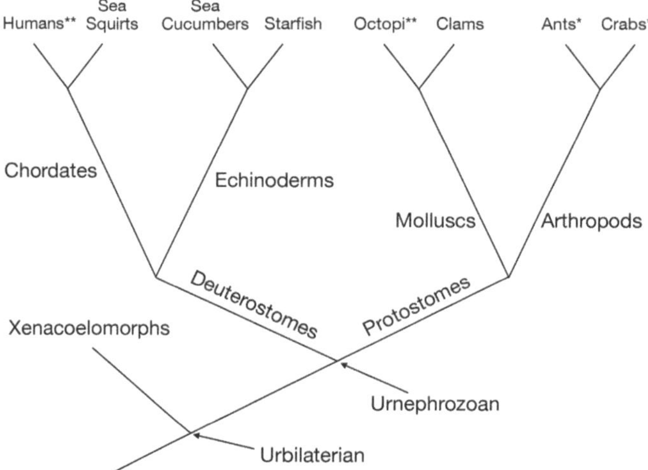

Figure 7.4. Cladogram of the clade Bilateria, showing some of the clades mentioned in the text (many branches of deuterostomes and protostomes are not shown). The urbilaterian is the last common ancestor of the entire clade, and the urnephrozoan is the last common ancestor of the clade Nephrozoa, within which all intelligent creatures reside. Humans and octopi are clearly intelligent (indicated by two asterisks) but other descendants of the urnephrozoan show less (or at least very different) intelligence (one asterisk) and others show essentially none (no asterisks). Intelligence thus appears to have evolved polyphyletically, and so is a convergent feature of whatever genetic toolkit the urnephrozoan shared with its descendants. The urnephrozoan itself had neurons, but was small and almost certainly unintelligent.

aspects of the urnephrozoan's nervous system evolved monophyletically or polyphyletically, that is, whether other ancient creatures had independently evolved similar neural structures.

At this point the topology of the tree of life becomes unclear, making answers murky and generating great controversy. There are five known major lineages of animals: two (Porifera (sponges) and Placozoa[16]) lack any kind of nerves, and three others show complex networks or central nervous systems synapses, and muscles: Ctenophora (comb jellies), Bilatera, and Cnidaria (jellies or "jellyfish").

Of these lineages, only the bilaterians have brains, however the jellies and comb jellies do have complex neural networks that allow for complex processing. Brains are clearly of higher complexity than these networks, however we know of no reason in principle that consciousness and cognition could not arise in a sufficiently complex neural network, even if it does not appear to have happened (yet) on Earth. Thinking about how cognition could arise on other worlds, we might require only that a structure like a neuron evolve.

Some authors, such as Moroz et al. (2021), have argued that genetic analysis shows that neurons, synapses, and muscles evolved at least twice, once in the comb jellies and once in the jellies and bilateria. Others, expressing the perhaps more conventional view, argue that the three neuron-bearing clades are monophyletic,

[16] Plocozoans are microscopic animals that lack any body plan or organs at all.

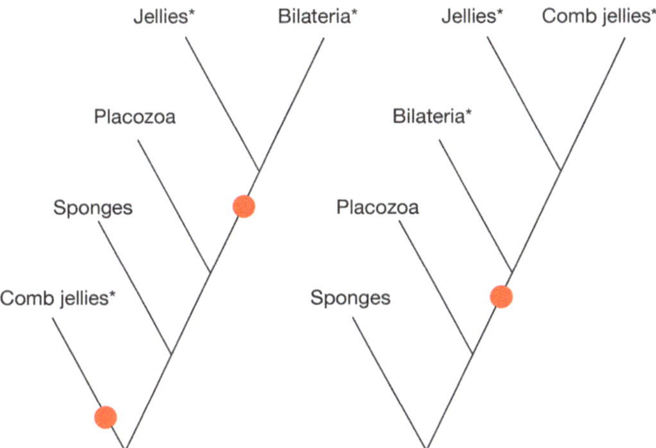

Figure 7.5. Two possible cladograms of the five extant clades of animals (metazoans), with asterisks indicating those clades with neurons, synapses, and muscles. The left is favored by Moroz et al. (2021) and argues that neurons evolved polyphyletically, and so is probably a convergent trait. That is, neurons evolved at least twice, at places shown by the red dots. The right is cited as a possibility by Redfield (2002), and shows neuron-bearing species being monophyletic, implying that the evolution of neurons was contingent (that is, having only a single origin, given by the red dot). The proper phylogeny of the animals is a matter of great contention and subject to rapid developments in evolutionary biology.

sharing a single neuron-bearing ancestor, with the sponges and Placozoans having diverged from this line before the neuron evolved. Figure 7.5 shows two possible phylogenies of animals.

This is an area of rapid progress in biology, however, and we may have a better sense of the landscape soon. If neurons indeed evolved polyphyletically, this may mean that not just intelligence but the underlying substrate of neurons is a convergent feature of evolution, and we might expect f_i to be quite high.

7.5.4 Other Aspects of Convergence and Contingency in Life

We have deeply explored what we can and cannot say about the evolution of intelligence in complex life, and the story is equivocal: only among the animals have brains and intelligence emerged, so it is perhaps not convergent, but within the animals it is unclear what underlying aspect of intelligence was so unlikely that it arose even once: once neurons evolved, brains and intelligence emerged multiple times, and it is not even clear if neurons emerged more than once. If intelligence is a "hard step" it is unclear what is hard about it, or if it is even fairly described as a "step."

Going back farther into Earth life's evolutionary past, things become even murkier. We can be sure that the path Earth life took to intelligence is unlikely, but we cannot be sure it is the only path at any step along the way. Even things as fundamental as water-based chemistry may be unnecessary for life beyond Earth: Saturn's moon Titan is far too cold for liquid water, but has a complex hydrological cycle of liquid ethane and methane, including rain, lakes, and rivers. The Huygens probe found its atmosphere to have an extremely rich organic chemical network with large molecules. Benner et al. (2004) has explored the possibility that Titan could actually host a biosphere based on this chemistry instead of water.

The bottom line is that if we include in our definition of life all of the varieties of life-as-we-don't-know-it, and accept that convergent evolution can lead to macroscopic organisms capable of intelligence and tool use, then it is not at all clear how to count necessary steps or how to compute their probabilities. The Rare Earth Hypothesis may turn out to be true, but much more work awaits before we can have confidence in that conclusion.

Box 7.1: Jargon Review: The Case against SETI

- **Contingency; contingent evolution:** Describing a trait or biological property that is peculiar or specific to a single group of animals, implying it may be unlikely to have evolved elsewhere.
- **Convergence; convergent evolution:** Describing a trait or biological property that arises independently in multiple lineages, such as wings or eyes. Convergent traits will plausibly arise elsewhere.
- **The Galactic Habitable Zone (GHZ)**: A proposed region within the Galaxy which has a sufficiently clement environment and metal content to support planets with an old biosphere on which complex life could evolve. The narrowness of the GHZ has been challenged as being based on several unproven assumptions about the necessary ingredients for such life to develop.
- **Rare Earth**: Name of the hypothesis that animal-like life is probably unique to Earth in the Milky Way because it is based on a series of individually unlikely contingencies, the product of whose probabilities is too small to expect all of them to have occurred more than once in every 100 billion stars.

  Counterarguments are that we do not actually know that these probabilities are all small, and that animal-like life might arise through a range of different paths via alternative biologies, geologies, and convergent evolution.
- **The Great Filter**: The hypothesis that there is some universal mechanism or set of mechanisms that prevent a species from becoming sufficiently spacefaring that it populates the entire Galaxy. The future of humanity would then depend on whether the filter is in front of us or behind us.
- **The Hard Steps model**: Model positing that there are a very small number of unlikely events in the history of the development of life on Earth that explain the arrival of large animals and human-like intelligence relatively late in the Earth's habitable lifetime.
- **Monophyly; monophyletic**: The property of a set of species being descended from a common ancestor, or of a unique trait common to all those species. A trait that appears only monophyletically (i.e., in all species of common descent and only in those species) is usually suspected to be the product of contingent evolution.
- **Polyphyly; polyphyletic:** The property of a set of species that does not include their common ancestor. A trait that appears polyphyletically may be suspected of being convergent. **Paraphyly** is an intermediate concept, describing a subset of lineages from and including a common ancestor. See Figure 7.2.

7.6 Further Reading

An Alternative to the Hard Steps Model: Mills et al. (2025).

References

Benner, S. A., Ricardo, A., & Carrigan, M. A. 2004, COCB, 8, 672
Carter, B. 1983, RSPTA, 310, 347
Finn, J. K., Tregenza, T., & Norman, M. D. 2009, CBio, 19, R1069
Gonzalez, G., Brownlee, D., & Ward, P. 2001, Icar, 152, 185
Lineweaver, C. H., Fenner, Y., & Gibson, B. K. 2004, Sci, 303, 59
Mills, D. B., Macalady, J. L., Frank, A., & Wright, J. T. 2025, SciA, 11, eads5698
Moroz, L. L., Romanova, D. Y., & Kohn, A. B. 2021, RSPTB, 376, 20190762
Prantzos, N. 2008, SSRv, 135, 313
Redfield, P. 2002, SSS, 32, 791
Salles, T., Husson, L., Lorcery, M., & Boggianiet, B. H. 2023, Natur, 624, 115
Szathmáry, E., & Maynard Smith, J. 1995, Natur, 374, 227
Ward, P., & Brownlee, D. 2000, Rare Earth: Why Complex Life is Uncommon in the Universe (New York: Copernicus)
Watson, A. J. 2008, AsBio, 8, 175

AAS | IOP Astronomy

The Search for Extraterrestrial Intelligence
Theory and practice
Jason T Wright

Chapter 8

The Case for SETI: The Drake Equation

> The reader may seek to consign these speculations wholly to the domain of science-fiction. We submit, rather, that the foregoing line of argument demonstrates that the presence of interstellar signals is entirely consistent with all we now know, and that if signals are present the means of detecting them is now at hand. Few will deny the profound importance, practical and philosophical, which the detection of interstellar communications would have. We therefore feel that a discriminating search for signals deserves a considerable effort. The probability of success is difficult to estimate; but if we never search the chance of success is zero.
>
> —Cocconi & Morrison (1959)

8.1 The Origins of the Drake Equation

On 1961 November 1, Frank Drake convened what would become the Order of the Dolphin at Green Bank, West Virginia to discuss the question of SETI. In a room in the dormitory today called the Drake Lounge, he wrote down his now famous equation like this:

$$N = R_* f_p n_e f_l f_i f_c L \tag{8.1}$$

and defined the terms this way:[1]

R_* = the mean rate of star formation over the period in which the stars now possessing communicative civilizations were being formed. If we are typical, R_* is about the mean rate of star formation 5 billion years ago;

[1] Or, at any rate, this is the version that appears in Drake (1965), the first citable publication by Drake with the equation in it, and an earlier paper by Sagan (1963). See Section 3.2.3.

doi:10.1088/2514-3433/adf5aech8

f_p = the fractions of stars which formed at that epoch with planetary systems;
n_e = the mean number of planets in each planetary system with environments permitting the development of life;
f_l = the fraction of such planets on which life actually develops;
f_i = the fraction of life-bearing planets on which intelligent life evolves;
f_c = the fraction of planets bearing intelligent life which give rise to a communicative civilization;
L = the mean lifetime in the communicative state of such civilizations.

Occasionally some of the terms are given different symbols and used differently from this. Most notably, the product of $f_p n_e$ is a variation of $\eta_\oplus$, widely used as the symbol for the measured number of "Earth-sized" or "Earth-massed" planets in the Habitable Zone per star (e.g., Howard et al. 2009). This relationship is imprecise, however, because Drake's n_e refers to planetary *environments*, which we are not usually able to determine.

8.2 Role of the Drake Equation in SETI

8.2.1 Purpose and Value of the Drake Equation

The purpose of the equation at the workshop was to organize the discussion on this new astronomical approach: what is the likelihood that there are signs of technology to be found? If so, how many are there and how nearby might they be? Is a search for technological life worthwhile?

Quantitatively, it is best thought of as an elegant solution to a Fermi Problem (Box 6.1): how many signs of technology are there in the Galaxy? Like any good Fermi Problem, the answer at first seems to be almost unknowable, but once broken down into its components becomes more amenable to estimates, or at least to being bounded.

The popularity of the equation after the workshop is a testament to the clarity with which it illustrates the plausibility of the success of SETI, and this is its primary value: demonstrating that reasonable, even likely values for its parameters yield a large value for N, making the search worthwhile.

It need not have turned out this way. Our understanding of stars and planets and life then or now could have been such that no reasonable values would generate $N > 1$, and it would in that case have been hard to justify the endeavor of SETI.

The equation can be used to argue for more than that as well, since for extreme but still plausible values for its parameters one can find $N \gg 1$, which implies that the nearest such technosignature could be quite nearby, and easily detectable.

Another great virtue of the equation is its simplicity, which allows it to be used as a tool for public engagement, as it provides a quantitative but accessible entry point for scientists to discuss a corner of science of great popular interest.

Box 8.1: Frank Drake (1930–2022)

Drake during his time at Green Bank Observatory. Credit: NRAO/AUI/NSF.

Prof. Frank Drake was an American astronomer and the preeminent pioneer of SETI. Born in Chicago, Illinois, Drake did his early education in engineering physics at Cornell and briefly served as a reserve officer in the US Navy where he received training in electronics. He did his graduate work in astronomy at Harvard under Cecilia Payne-Gaposchkin, where he worked with Bart Bok to build the 60 foot radio telescope. His thesis was on neutral hydrogen in Galactic clusters.

In 1958 Drake joined NRAO at Green Bank, West Virginia conducting radio observations in the solar system. He discovered the ionosphere and magnetosphere of Jupiter, and made some of the first measurements of the extremely high surface temperatures of Venus. There, he began Project Ozma, and hosted the Green Bank Workshop where he presented the Drake Equation.

After a brief stint at the Jet Propulsion Laboratory heading lunar and planetary sciences, he became a professor at Cornell, his alma mater. There, he became director of Arecibo Observatory, which he transformed from a primarily ionospheric science instrument into a premier instrument for astronomy and planetary science, both as a receiving telescope and as a powerful radar transmitter. In 1974 he used the newly commissioned radar to send the Arecibo message, which he helped design. He also helped design the Pioneer Plaques and Voyager Records.

In 1984 he joined the University of California, Santa Cruz as dean of Natural Sciences, where he remained on the faculty until his retirement in 1996. He also served as the Chairman of the Board of Trustees of the SETI Institute, and director of its Carl Sagan Center for the Study of Life in the Universe. He served as President of the Astronomical Society of the Pacific, vice president of the American Association for the Advancement of Science, and chairman of the Physics and Astronomy Board at the National Research Council. He was also elected as a fellow of the American Academy of Arts and Sciences, and a member of the National Academy of Sciences.

Drake was the inaugural recipient of the SETI Institute's Drake Award in 2001 for contributions to SETI or astrobiology through scientific research and space exploration.

For more on Frank Drake and his role in SETI, see Swift (1993) and Dick (1996).

The equation also helps illustrate how interdisciplinary SETI is, and situates it as a natural subset of astrobiology, the scope of which it nicely delineates. Each term in the equation could be the focus of a unit in a semester-long Life in the Universe course, covering the history of life on Earth through astronomy, planetary science, biology, and anthropology.

It is important to remember that at the time of the workshop, the study of life in the Universe had not yet been formalized. The Drake Equation helped frame the problem firmly in terms of questions that could plausibly be answered (or at lest studied), from the prevalence of Earth-like planets in the Galaxy to the mechanisms of the origin of life. For the next few decades, largely championed by Carl Sagan, NASA and scientists worldwide would begin to address the problem scientifically from many angles, most recently under the aegis of NASA's astrobiology initiatives. The Drake Equation ended up being the roadmap for decades of study.

8.2.2 What the Drake Equation is *Not*

Because of its centrality to the field, it is tempting to give the Drake Equation foundational status, akin to Newton's Laws of Motion in classical mechanics, or the Schrödinger equation in quantum mechanics, or Einstein's field equations in general relativity.

But this is an error, and one that leads to misguided efforts in the field. Unlike those equations, the Drake Equation is not an axiom upon which a formal mathematical system of physics can be built. One cannot use it to derive anything about alien behavior, the likeliest communication channels, or how to distinguish technosignatures from natural sources. Its main value is in the way it organizes and justifies the field, not in our ability to use it to perform calculations.

And, as we shall see, because its terms are so uncertain, one can do little better than generate estimates for N to within *several* orders of magnitude, and even then these estimates will require significant subjectivity. It seems more likely that we will constrain N experimentally via searches for technosignatures more quickly than we will constrain the product of the terms on the right-hand-side via studies of planets, life, and sociology. This does not mean the Drake Equation is useless or flawed—as we have seen, it has had great utility—it just means its value does not lie in robust calculation.

One sometimes hears critiques of the field of SETI that use the supposed flaws in the Drake Equation as a starting point. Such attacks miss the mark, misunderstanding the role the equation plays. If one could show that Maxwell's Equations did not hold in the classical limit, one would indeed have made a powerful argument against the reigning theory of electricity and magnetism because those equations are truly foundational. As we will see throughout this text, the foundations of SETI are much broader than this.

One also sees a lot of work by authors who rearrange, recompute, modify, add to, or reinterpret its various terms. This is valuable to the degree that the equation (and its modifications) can be used to illustrate assumptions in the field and suggest how we might supplement or adjust our expectations for what there is to find. Many authors also spend much time attempting to properly compute N with rigorous error bars, incorporating both statistical errors in our measurements of some terms and more epistemic uncertainty in terms for which we have little no data (like the fraction of planets on which life evolves). Others discuss the degree to which the history of life on Earth allows us to say anything about its terms (Section 10.2).

While such work has merit, one should always keep in mind that the Drake Equation's terms are fundamentally too uncertain and its assumptions to narrow for it to truly estimate the number of technological species in the Galaxy (or beyond), except to reiterate its original message: it is plausible that there could be a very large number of technological species in the Galaxy, which is a necessary and sufficient condition for SETI. Such a conclusion was profound in 1961; today it hardly needs further reiteration or refinement.

8.3 Assumptions and Structure of the Drake Equation

8.3.1 R_* and L

The first term in the equation, R_* strikes many as counterintuitive because it is a *rate*, where one might expect instead simply the number of stars in the galaxy. Indeed, in popular descriptions of the calculation, one begins with the roughly 200 billion stars in the galaxy and contemplates what fraction have technological life.

Drake's powerful insight was that it was important to consider that technological signals would have some duty cycle: some fraction of time they were "on," and that N must be proportional to this term. Over the lifetime of the star, this duty cycle could not be greater than that of technological life itself. That is, it's incorrect to just count the planets that will have technological life *at some point*; one must multiply that by the fraction of them that are emitting technosignatures *right now*.

One could imagine that he might have instead written the equation as

$$N = N_* f_\mathrm{p} n_\mathrm{e} f_\mathrm{l} f_\mathrm{i} f_\mathrm{c} f_\mathrm{t} \tag{8.2}$$

where N_* is the number of stars in the Galaxy, and f_t is the fraction of those with technological life emitting right now.

But what is f_t? The many stars in the Galaxy have a wide range of ages; among those with technological species at some point, some will have hosted technology that is long extinct, some will host technology in the future, and some have it right now. The fraction of time these stars spend with such a species emitting technosignatures is roughly the amount of time they emit, L, divided by the age of the Galaxy, let's call it T. So we have

$$N = N_* f_\mathrm{p} n_\mathrm{e} f_\mathrm{l} f_\mathrm{i} f_\mathrm{c} L/T. \tag{8.3}$$

But we can rearrange this to put the T beneath N_* and write $R_* = N_*/T$: the number of stars divided by their typical age, which is, roughly speaking, just the rate at which stars we see today have been forming. Conveniently for Drake, this number was actually rather well known at the time, at least to within a factor of a few.[2]

Phrasing the Drake Equation this way thus replaces an unintuitive parameter—our duty cycle of technology around a star over the life of the Galaxy, f_t—with a far more concrete and evocative one: L, the "lifetime" of a technosignature.

[2] If this all seems too hand-wavey to you, recall that the Drake Equation is a Fermi Problem: it does not purport to give precise answers. We need not be too precise here.

8.3.2 Steady State and Expectation Values

The presence of R_* and L in the Drake Equation indicate that it is meant to be a steady-state equation, essentially the solution to a balance between a process with a birth and death rate. That is, N grows every time a new technological species arises and starts transmitting, and shrinks every time those transmissions stop. If these processes balance out, then there will be some average lifetime L of those technosignatures, which the equation captures.

The true number should vary, of course. We do not expect the early Galaxy to have had many species, when it still did not have many stars, and the star formation rate has likely varied with time in the past. Drake asserted that the value should be the "mean rate…5 billion years ago" reflecting, presumably, his expectation that, as on Earth, other planets would take around 5 billion years to form technological life.

We also might wonder if the processes describing the way technosignatures arise and cease do not balance out—if, for instance, many technological species do not die out at all. In that case, we would expect N to steadily rise with time—we will examine this and related possibilities in Chapter 9.

We also expect the number to vary stochastically, especially if technologies come and go rapidly on a Galactic timescale. In that case, we should think of N as an expectation value, with the true value scattering above and below this number.

We might also worry about lookback time. Essentially, the time a species transmits, L, determines the width of a shell, Lc, within which transmissions can be detected. Only if Earth is, today, within that shell, could we detect those transmissions. But because the light travel time across the Galaxy is much shorter than the timescales in the steady-state assumption, we do not need to worry about this effect. Essentially, the Drake Equation assumes N is constant in time, so was the same value, statistically, when any signals we might receive are sent. This means that the Drake Equation is equivalent to our thought experiment about the shells (as shown by Grimaldi et al. 2018). This is consistent with astronomers' general attitude toward lookback time in the local Universe (Section 2.7), which is that we may as well ignore it and consider everything nearby as happening in the present.

8.3.3 The Meaning of L

Drake's L represents the total amount of time a species spends exhibiting a particular technosignature, but it is often conflated with the "lifetime of the civilization." Indeed, Drake, Sagan, and other proponents of the Drake Equation explicitly connected upper limits on L with fears of self-imposed human extinction from nuclear war.

We should be careful when making such connections. As we will discuss in Section 9.1.3, L refers to an integrated time across an entire species' existence, whereas the "lifetime of a civilization" is generally going to be much shorter than this. That is, civilizations have risen and fallen across human history, but this has had little impact on the total human population or the long-run development and accumulation of human culture and technology. Studies that use the historical approximate dates of the beginning and ends of various human civilizations in history are thus not really relevant to the question of what we should expect L to be for alien species (or humanity!)

8.3.4 Other Assumptions

The Drake Equation assumes that technological life will arise on an Earth-like planet, but as studies of Europa in the solar system have revealed, one might expect life to form anywhere liquid water exists, including in the vicinity of cold gas giant planets. We might then need to expand our assumptions about what goes into n_e.

Perhaps the most contested assumption the equation makes is that technological life stays put. Indeed, the logic of the Fermi Paradox argues that it will certainly *not* stay put, but will spread throughout the Galaxy. We will explore this more in the next chapter but for now we just note that the Fermi Paradox argument and the Drake Equation are based on mutually incompatible assumptions about alien behavior.

Neither of these assumptions actually weakens the argument that SETI is worthwhile because lifting them will tend to *increase* N, making the search even more worthwhile.

8.4 What Does N Represent?

The most nuanced part of the Drake Equation is what, exactly, N is measuring (or, equivalently, what $f_c L$ measures).

Drake defined it as "the number of communicative civilizations," but this is not the same thing as "detectable." We are a communicative species, but much or most of our communications are not of the sort that would be detectable at interstellar distances.

This cuts both ways with respect to the case for SETI. For the Drake Equation to justify SETI, L must refer not just to communication, but to remotely detectable communication. This has the odd effect of being a function not just of the nature of alien species, but of our ability to detect them. This is implicit in Drake's choice to restrict his equation to the star formation rate *in the Galaxy*—he was assuming species in other galaxies would be too far away to detect. At any rate, such a restriction should lower N, and requires us to measure L in terms of things we can detect.

On the other hand, Drake at the time was considering only radio technosignatures of the sort he sought with Project Ozma. As we have seen, within two years several other approaches were suggested, and so if the Drake Equation is going to justify SETI we should consider f_c to be the number of detectable technological activities of all sorts, and L how long they last.

This broadening of N is often seen in the literature, where the Drake Equation is taken to mean the total number of technological species., the total number of technosignatures, or some other thing. Variations on the Drake Equation have been developed for probes in the Galaxy and other technosignatures that might not naturally follow from the equation's structure.

In general, though, expanding the logic of the Drake Equation to all technosignatures increases N, and strengthens the case for SETI.

8.5 Drake's Values for the Equation Parameters

At the time of the workshop, only the first parameter, R_* had any firm value, and Drake estimated it at around 1 star per year, or a bit higher. This value, averaged over the past 5 Gyr, is still held to be approximately correct.

The next two parameters were a challenge because no exoplanets were known at the time (they would not be discovered for another 30 years). Planet formation and stellar rotation rates were not well understood at the time, and it was believed that the reason many stars like the Sun spin slowly was because they had transferred the angular momentum they should have collected during contraction to a planetary system.[3] Drake thus estimated f_p to be 0.5 from the statistics of stellar rotation periods. Today we understand the number to be a bit higher—the vast majority of stars appear to have planetary systems—approaching 1.

The mean number of planets per planet-bearing star, n_e, was even more challenging to estimate, and for this Drake looked to the solar system. The planets are approximately geometrically spaced[4] and as tightly packed as they can be dynamically. Venus, Earth, and Mars could all have liquid surface water given the correct atmospheric conditions.[5] Drake thus estimated n_e at 2–3. Today, we have estimates that are only slightly lower than this. We have rather weak constraints on planetary systems like the solar system, but considering lower mass stars and planets Earth size and larger, it appears that $\eta_\oplus \approx f_p n_e \approx$ 0.1–0.25, depending on how one defines an "Earth-like planet." This number does not include other sites where life might arise, such as subsurface oceans on moons like Europa, so Drake's number is still plausibly correct.

The next few terms Drake set to 1. Based on the discussions of the workshop, Drake claimed there was "near unanimity" that f_l should be 1. This attitude was heavily influenced by the Miller–Urey experiment of 1952, which showed that complex organic molecules that form the basis of life can be synthesized from what at the time was thought to be a good representation of Earth's early atmosphere and sparks meant to represent the effects of lightning. If we generate complex organic chemistry from simple gases in just a few days in the lab, the thinking went, surely a planet with billions of years to work would take the additional steps to form life. Today, most biologists would likely admit that f_l could *plausibly* be high but could

[3] Today we understand that cool stars spin down as they age because of coupling between their interiors and their ionized winds via their global magnetic fields.

[4] A celebrated rule of thumb is Bode's law, which says the semimajor axes of the planets roughly follow the rule $a = 0.4 + 0.32^n$ au, where $n = 1$ for Venus, and increases by one for each planet. This formula was credited with "predicting" the existence of Ceres in the asteroid belt (explaining the "missing" planet at $n = 4$) and Uranus at their correct orbital radii. It does not correctly describe the orbits of Mercury or Neptune. Today, the law is widely considered to be a coincidence, reflecting little more than the fact that the planets are "densely packed," and that the stability requirement that planets be more widely spaced as you move outwards in order to have stable orbits.

[5] Venus's surface is so hot not because of intense sunlight (its bright clouds reflect most of that away) but its intense greenhouse warming from its thick CO_2 atmosphere. The Earth would not have liquid water but for its much weaker greenhouse effect that keeps its surface warm. Mars apparently did once host liquid water, likely a consequence of greenhouse warming from a thick atmosphere it has since lost due to its low surface gravity and lack of magnetic field.

also be quite low, because we have not identified the path to the first life on Earth, and because the conditions for life to form might not be universally present on "Earth-like" planets. On the other hand, perhaps there are many paths to forming life, and even bodies like Titan with very different surface and atmospheric chemistries readily form life, and f_l truly is near 1.

Once life forms, evolutionary processes take over, and in 1960 Drake believed the consensus again was that complex life was inevitable, and intelligence was such a powerful survival trait that it would surely be selected by some species on any planet so he set $f_i \approx 1$. Today, many evolutionary biologists would likely point out that it took billions of years on Earth for neurons to evolve, and hundreds of millions of years after that for sufficiently neurologically complex creatures to evolve that might be intelligent. This long time might indicate that it was an unlikely event (Section 7.3.1), and the Anthropic Principle (Section 10.2) might argue that it would be almost arbitrarily unlikely. Modern values would thus be somewhere between very small and 1.

Drake separated out the next term, f_c, in light of the work of John Lilly and others demonstrating that dolphins exhibited extremely high, perhaps human levels of intelligence. He thus felt one should account for the possibility that a planet would evolve many intelligent species but none of them would generate technosignatures. Here, Drake invoked the strong evolutionary advantage of technology and "Terrestrial history" and set the value to "nearly 1." We discussed many of these issues in Section 7.5.3, and today we might again more plausibly put these numbers somewhere between very small and 1.

Multiplying these together, Drake concluded that $N \approx L$ when L is in units of years.[6] This extremely convenient distillation of the problem (which further motivates why he chose to use R_* and L) allows the problem to be easily parameterized.

The final term, L, Drake declined to estimate, preferring instead to bound the it in the style of a Fermi Problem. Drake found values of 10,000 yr to be on the low end, but even such pessimistic values led to there being 10,000 technosignatures waiting to be found in the Milky Way, or one in every 10 million stars (assuming the Milky Way has around 100 billion stars). Since stars have a local number density of approximately $n = 0.1$ stars pc^{-3}, this means the nearest technosignature would be a distance $d \approx (10^7/n)^{1/3} \approx 450$ pc distant (Drake estimated 300 pc). This is quite close—roughly the thickness of the Milky Way disk.

As for the high end of L, Drake cites estimates as high as 10^6 from Phillip Morrison (although as we will see next chapter, there is really no upper limit we can establish). Such a high number would make detection all the more likely, making the nearest star with a detectable technosignature only 100 pc away.

8.5.1 Plausible Values of N and the Great Silence

Drake's estimates are about as optimistic as one can get, so we can put $N = 10^6$ for the Milky Way as an upper limit (within the assumptions of the Drake Equation, which we will lift in the next chapter). We do not have firm reasons to bound f_l, f_i, or

[6] NEQLS L was Frank Drake's personalized license plate.

f_c on the low end, and indeed they could in principle be quite small as argued in Rare Earth Scenarios (Chapter 7). If we were to find evidence of extinct or extant life on Mars or Europa (for instance), that would strongly imply that f_l is not extremely small, but we would still not have lower bounds on f_i or f_c.

We thus have no real lower bound on N from the Drake Equation. Large values of N are certainly plausible, but it could very easily be less than 1.[7]

As such, we cannot use the Drake Equation to argue that SETI "should have succeeded by now" and that the Great Silence is therefore a puzzle that demands a solution. The Drake Equation admits a huge range of possible values for N, including those consistent with SETI having no targets in the Milky Way where it could be successful.

But as we have argued, Drake's purpose was not to precisely estimate the chances SETI could succeed, it was to establish that the search was plausible and worthwhile. For this, his values, which are still plausible today, are sufficient.

8.6 Further Reading

A deep dive into the Drake Equation: Vakoch & Dowd (2015).
Steady State and the Drake Equation: Kipping (2021).
Joint Analyses of the Fermi Paradox and Drake Equation: Prantzos (2013, 2020).

Exercises

1. Putting numbers in the Drake Equation
   (a) Put in your own best-guess numbers for the Drake Equation. How many communicative species are in the Galaxy?
   (b) Put some error bars on your numbers. What are your most pessimistic and optimistic estimates?
2. Literature on the Drake Equation
   Use ADS and the SETI bibgroup to explore the literature on the Drake Equation, focusing on papers primarily about it. Sort them into some categories like:
   - Modifications of the Drake Equation.
   - More rigorous uncertainties of the Drake Equation.
   - Estimates of the parameters of the Drake Equation based on what we know of life on Earth.

   What fraction of the literature is concerned with the Drake Equation and how is that fraction distributed? Do these papers get a lot of citations?
3. Investigating a Drake Equation calculation
   Use the ADS SETI bibgroup to find a paper modifying or doing a calculation with the Drake Equation. Answer one of the following sets of questions.

[7] Since N is an expectation value, this is not inconsistent with our existence. It simply means technosignature-producing species such as ours are very rare.

(a) If it modifies the Drake Equation: Is the modification warranted? Does the modification help improve one's estimate of the number of communicative species?
(b) If it does a calculation: How does their estimate for N differ from yours? What assumptions do they make that differ greatly from yours? What does the paper do that makes it novel?

References

Cocconi, G., & Morrison, P. 1959, Natur, 184, 844

Dick, S. J. 1996, The Biological Universe: The Twentieth Century Extraterrestrial Life Debate and the Limits of Science (Cambridge: Cambridge Univ. Press)

Drake, F. D. 1965, Current Aspects of Exobiology, ed. G. Mamikunian, & M. H. Briggs (Oxford: Pergamon) 323

Grimaldi, C., Marcy, G. W., Tellis, N. K., & Drake, F. 2018, PASP, 130, 054101

Howard, A. W., Johnson, J. A., Marcy, G. W., et al. 2009, ApJ, 696, 75

Kipping, D. 2021, RNAAS, 5, 44

Prantzos, N. 2020, MNRAS, 493, 3464

Prantzos, N. 2013, IJAsB, 12, 246

Sagan, C. 1963, P&SS, 11, 485

Swift, D. W. 1993, Seti Pioneers: Scientists Talk about Their Search for Extraterrestrial Intelligence (Tucson, AZ: Univ. Arizona Press)

Vakoch, D. A., & Dowd, M. F. 2015, The Drake Equation: Estimating the Prevalence of Extraterrestrial Life through the Ages (Cambridge: Cambridge Univ. Press)

Chapter 9

The Case for SETI: Beyond the Drake Equation

> The universe is a pretty big place. If it's just us, seems like an awful waste of space.
> —*Contact* (Box 18.1)

While the Drake Equation and reasoning around it provides a sufficient justification for SETI, it can actually understate the potential for the detection of technosignatures by a significant fraction.

In this chapter, we explore the maximal case for SETI, considering by what fraction the Drake Equation might understate the number of technosignatures in the Galaxy. As a baseline, we'll first consider how N compares to a similar calculation for the number of sites of biosignatures.

9.1 Comparing The Number of Biosignatures and Technosignatures

One conclusion one can draw from the Drake Equation too seriously is that it seems to show that searching for non-technological biosignatures—signs of "unintelligent life," might be a significantly better search strategy than SETI. Let's take a look at that reasoning, and what it misses.

If we think of technosignatures more broadly and do not just focus on communication, we can write a pair of Drake-like equations for biosignatures and technosignatures generally. The first

$$N_{\rm bio} = R_* f_{\rm p} n_{\rm p} f_{\rm l} L_{\rm b} \tag{9.1}$$

captures the number of planets with detectable biosignatures, where $f_{\rm l}$ here is interpreted as the fraction of planets that develop sizable biospheres of the sort that

doi:10.1088/2514-3433/adf5aech9

give rise to remotely detectable biosignatures (and, potentially, technological life) and L_b is the average length of time that biosphere remains detectable.

Similarly, we write

$$N_{tech} = R_* f_p n_p f_l f_t L_t \tag{9.2}$$

which captures the number of planets with detectable technosignatures, where here f_t is the fraction of planets with life with detectable technosignatures of any sort and corresponds to $f_i f_c$ in the equation's canonical form, and L_t is the length of time that such technosignatures remain detectable.

9.1.1 The Case for $N_{bio} \gg N_{tech}$, and the Flaw in That Argument

A naive conclusion from the Drake Equation is that $N_{bio} \gg N_{tech}$, based on at least two seemingly sound assumptions. The first assumption is that $f_t \ll 1$, since the development of technology is presumably a rare event. The second assumption is that any technological life that *does* arise will exist for only a small fraction of the time that life of any kind exists on the planet.

In other words, these two assumptions would lead to the expectation that

$$\frac{N_{tech}}{N_{bio}} = f_t \frac{L_t}{L_b} \ll 1. \tag{9.3}$$

Both of these assumptions seem sound but don't hold up to scrutiny, because they are based on the sole example of the evolutionary history of Earth.

The primary flaw in the reasoning from the Drake Equation is that it restricts technology to a small subset of biospheres in time and space, and that it does so *tacitly* and *by assumption*. In truth, while technology may originate from biology, it may also exist *independently* of it in many ways that the Drake Equation does not capture.

To give some examples:

1. Technology can long outlive the biology that created it. If humans died out, many of our technologies, such as the Great Pyramids, would live on for thousands of years or longer, and many of our more subtle technosignatures might last millions of years. One can thus have technosignatures without technological life, and even without a biosphere.
2. Technological life can spread to other worlds, and thus create sites of technosignatures that outnumber biospheres. We have already done this in the solar system.
3. Once created, technology could self-replicate and spread without any further association with life at all. We have not developed such technology, but it is "just engineering," not something that might be forbidden by physics or practical matters.
4. Technosignatures can exist beyond planets, for instance arising from spacecraft. Life is (presumably) restricted to planetary surfaces.

Let's take a look at two ways in which technosignatures might outnumber biospheres. In what follows we will use humanity's future in the solar system as an example to demonstrate the plausibility of these arguments, but one should not read too much into this. What matters is not how humanity's technology persists or spreads or grows in the future, but whether it is possible that another species' technology has done so.

9.1.2 Abundance

Consider the solar system, which at the moment hosts only a single planet with remotely-observable biosignatures (at least, any obviously discernible ones) but has *four* major planets with potentially detectable technosignatures. In addition to Earth, Jupiter and Venus are both orbited by probes actively transmitting radio waves, and Mars hosts several orbiters, landers, and rovers.

Thus, in our solar system, we have $N_{\rm bio} = 1$, $N_{\rm tech} = 4$ (and this does not count interplanetary probes between the planets). The radio technosignatures of these other planets are obviously weak at the moment, but there is no reason they might not grow by huge factors. There are substantial plans to increase humanity's presence on Mars, and regardless of whether that happens or not, or whether it takes decades or millennia, the prospect for Mars having strong technosignatures but no biosphere is clear.

Looking further, there is no reason that future humans or an alien species could not choose to use some uninhabitable worlds as sites for significant technology, whether for industry, energy generation, or even waste processing, creating entire *technospheres* spatially independent of the biosphere which spawned them. Such *service worlds*[1] may not host any biology at all, and may outnumber sites of biology from their originating species.

And of course, as we have seen in Chapter 6, the reasoning behind the Fermi Paradox argues that life can spread beyond its planetary system, potentially filling the entire Galaxy.

Finally, technology might exist at many sites beyond the worlds counted by the $n_{\rm e}$ term in the Drake Equation. In addition to planets outside the Habitable Zone, technology might exist between planets, between stars, or in abundance around stars with no planets (as Dyson suggested with Dyson spheres, Chapter 15). In terms of the Drake Equation, we can write that it is possible that $f_{\rm t} \gg 1$.

Thus while the Drake Equation may underestimate $N_{\rm bio}$ to some degree, there are reasons inherent to technology that permit us to argue that the attendant degree of underestimation will be much larger for $N_{\rm tech}$, and by potentially enormous factors.

Interestingly, technology might make searches for biosignatures more likely. Science fiction stories about Mars includes future visions where the entire surface and atmosphere is engineered to support Earth life (a process called *terraforming*). A species that spreads throughout the Galaxy that terraforms planets might therefore cause $N_{\rm bio}$ to be quite large.

[1] The term was coined by Manasvi Lingam in as-yet-unpublished work.

9.1.3 Longevity

The next place the Drake Equation implies low values of $N_{\mathrm{tech}}/N_{\mathrm{bio}}$ is in L_{t}, which one might think must be much smaller than L_{b}. In particular, L_{b} for Earth is of order 4 Gyr, and L_{t} seems to be of order decades to millennia, based on human technosignatures.

But that value for L_{t} only looks at humanity's *past*. What matters, of course, is the total value for L_{t} including our future, and its value is highly uncertain.

Much ink has also been spilled tying the L_{t} in the Drake Equation to fears of human catastrophes, for instance from nuclear or biological warfare or climate change, and arguing it would be optimistic to think it could be orders of magnitude larger than a thousand years or so; indeed, such arguments are as old as the Drake Equation itself (e.g., Drake 1965; Mamikunian & Briggs 1965; Shklovski & Sagan 1966; Gott 1993; Charbonneau 2022). But human's technology's survival is essentially unlimited in theory, even in the face of inevitable natural catastrophes. Apart from these modern examples, Earth-analogs from human history teach us that a technological downshift—to temporarily become less technological until circumstances improve—is a common and healthy adaptation to catastrophe in human history, and that technology and longevity are in this way inextricably linked (Denning 2011).

Further, L_{b} in the past is of order a few Gyr, but Earth's future probably has $\lesssim 1$ Gyr of habitability remaining, putting a hard limit on L_{b}.[2] Earth's technosignatures, however, have no such limits since even an uninhabitable Earth can still host technology (as Mars does). It is therefore plausible that $L_{\mathrm{t}} > L_{\mathrm{b}}$ might be achievable for the solar system.

Note that this conclusion is somewhat independent of what we imagine becomes of the life that created this long-lived technology. We might imagine this life spreads with the technology, or that the technology itself is self-sustaining (as purely mechanical or "post-biological life"). It is also possible that the technology would be derelict but still detectable for very long times (much like the Great Pyramids of Giza, for instance, would last for a long time, even if humanity went extinct). In particular, service worlds with planet-wide technology or large orbital structures might be detectable as technological for Gyr or longer.

In short, our expectations that $L_{\mathrm{t}} \ll L_{\mathrm{b}}$ should not be based on Earth's past, but on its future, and that the probability distribution of $L_{\mathrm{t}}/L_{\mathrm{b}}$ in this future *even for human technology* has a long tail extending to large values.

9.1.4 Other Species and L_{t} for Earth Life

Even if we accept that Earth life is a good guide to extraterrestrial life *and* that humanity's future is short, we can *still* argue that L_{t} could be quite large because

[2] This is because as the Sun consumes hydrogen in its core to produce energy, helium builds up in the core, which counter-intuitively causes its output to increase. As the Sun becomes more luminous, the Earth will heat up until it is uninhabitable.

Earth has a plausible chance of giving rise to *another* technological species after humanity's extinction.[3]

As we saw in Section 7.5.3, few of the most striking traits of our species are altogether unique in the animal world: intelligence, tool-use, communication, and social behavior are all common. Given that it has already happened (at least) once, there is no inexorable reason to think that as long as animal life exists at all, another species cannot develop an even longer-lived technological society.

Pessimism about humanity's future is thus not sufficient to argue that L_t for Earth is short, especially if one believes that our propensity for self-destruction or environmental degradation is a particularly human failure.

9.1.5 We Do Not Know L_t for Earth's Past

In a similar vein, if we grant that other terrestrial species could be capable of technology, then we cannot even use Earth's past to precisely constrain L_t empirically because we do not know it! Our assumption that $L_t \ll L_b$ is usually based on what we *think* we know about Earth, namely, that remotely detectable technosignatures on Earth have only existed for decades or, optimistically, perhaps in the millennia since humans developed agriculture—that is, $L_t/L_b \sim 10^{-7} - 10^{-6}$.

This assumption seems well justified, but a little examination reveals that we actually have very little to base it on. Schmidt & Frank (2019) discussed the *Silurian Hypothesis* that we are not the first technological species on Earth.

They have argued that, while humanity's impact on the geological record (marking the Anthropocene epoch) will be indelible, the fact that it is due to technology will be indiscernible on a timescale of millions of years. That is to say, the Anthropocene will be obvious over geological timescales as an important event in Earth's past, but there may be little to no evidence that it was the result of technology, as opposed to being a purely natural phenomenon. The conclusion is that if there have been prior episodes of technological life on Earth, it is possible that *we would not know it.*

So while Earth's past certainly teaches that $L_t < L_b$, the magnitude of the difference could easily be much smaller than we might guess. If there have been tens of millions of years of prior technology, and if this technology could be conceivably detectable, then $L_t/L_b \sim 10^{-2}$. We just don't know.

In short, using Earth as a guide for our expectations for L_t/L_b is probably unreliable because we do not know L_t for Earth's past; and even if we did we should not use it to predict L_t for Earth's future; and even if we did, we should not expect Earth to be a good guide for alien life; and even if we did, we should expect a broad distribution of longevities across alien species.

[3] This argument goes back at least as far as Drake (1965), p. 328.

9.2 The Case For SETI: The Bottom Line

Above, we have compared $N_{\rm bio}$ and $N_{\rm tech}$, but the lesson is not that one search strategy is superior to the other (at best we can say it is very unclear which is better). Rather, we have used non-technological life as a baseline to illustrate where the Drake Equation potentially dramatically underestimates $N_{\rm tech}$.

Most importantly, the Drake Equation counts only planets where technology *arises* from life, but technology might exist at many more sites than this because it can spread, and it might last much longer than life, because it can avoid the ultimate fate of planetary biospheres.

This is, essentially, the more appropriate conclusion of Fermi Paradox arguments. As we have seen, those arguments make the dubious assertions that life *will* spread exponentially through the Galaxy, and that we *would definitely* have noticed them by now if they had. If we demote both assertions from certain to possible, then we arrive at the conclusion that the Galaxy *could very plausibly* be filled with technology that is *easily* detectable.

So in the end, by employing a more nuanced and careful version of the Fermi Paradox argument, we are left at a far stronger place than where we began in 1961 at Green Bank when Drake wrote down his equation: for reasonable assumptions about planets and life, there will be a large number of potentially detectable technologies in the Galaxy. The prospect of finding it is sufficiently tantalizing and important, that we should look for it.

9.3 Searches for Biosignatures versus Technosignatures

In this chapter we've looked at how the number of biosignatures and technosignatures compare to illustrate the maximal case for SETI, using searches for biosignatures as a baseline, finding that it is plausible that $N_{\rm tech} \gg N_{\rm bio}$.

The conclusion here is not that SETI is an obviously better search strategy than searching for biosignatures (though that is certainly a plausible conclusion). Nonetheless, because so much more effort goes into finding non-technological life than SETI, we should finish the argument of Wright et al. (2022) in order fully rebut the opposite conclusion, that searching for biosignatures is a clearly superior strategy.

Just counting the number of sites of biology and technology is not enough—the signatures must be detectable to be useful. To that end, we must consider both the detectability of the signatures, and their ambiguity.

On the matter of detectability, we can bound the problem by starting with Earth today. It is unclear whether Earth's technosignatures are more detectable than its biosignatures. If we had access to our current technology on α Centauri, it would be extremely challenging to detect Earth's biosignatures (one would presumably need a large space coronagraph, and want to detect reflection spectra of chlorophyll, or perhaps methane and ozone in Earth's atmosphere). Earth's technosignatures, however, might be more easily detectable via our leaked radio transmissions, such as radar. And as Kardashev showed, the upper limit for the detectability of technosignatures is quite high.

On balance, we might call the relative detectability of technosignatures and biosignatures somewhere between a tie and strongly favoring technosignatures.

On the matter of ambiguity, we see in Chapter 11 that some technosignatures have no ambiguity at all—nature does not, indeed cannot, produce monochromatic radio signals or light pulses. Other forms of technosignatures—infrared heat signatures for instance—are extremely ambiguous.

For biosignatures, the situation is similar. Seeing an obvious set of fossils of macroscopic life in Mars rocks or a fish-like creature swimming in the Europan ocean would be rather unambiguous, but most work on remote biosignatures, like simultaneous methane and ozone in an atmosphere, will be inherently ambiguous.

On balance, we might call it a draw between the two approaches, with both looking for both ambiguous and clearly dispositive signatures of life.

We then have four ways to compare the two searches: abundance, longevity, detectability, and ambiguity. For the first three, either search could have the upper hand, but technosignatures have a much higher ceiling, being potentially many orders of magnitude more abundant, long-lived, and detectable than biosignatures. Either could be less ambiguous, but when restricting to remote detection outside the solar system, technosignatures can be absolutely unambiguous, while biosignatures will likely not be.

In short: it's not a competition, and both strategies should be pursued because the Drake Equation and extensions on its reasoning can only show plausibility, and both approaches are plausible.

Box 9.1: Jargon Review: Beyond The Drake Equation

- **Technosphere**: By analogy with *biosphere*, the technology on or near the surface of a planet.
- **Terraforming**: The process of engineering the atmosphere and surface of another planet to give it Earth-like climate and make it habitable for humans. By extension, the process of any such planetary engineering, including by other species to adapt to their own preferred climate.
- **Service world**: A planet with an extensive technosphere but no (or practically no) biosphere, for instance because it is the site of an interstellar settlement but has not been terraformed. If interstellar settlement is common, most sites of technology might be on service worlds.
- **Silurian hypothesis**: The possibility that there existed a previous terrestrial technological species. The degree this is allowed by paleontological evidence is unclear.

9.4 Further Reading

A deep drive into the Drake Equation: Vakoch and Dowd (2015)
Steady State and the Drake Equation: Kipping (2021)
Joint Analyses of the Fermi Paradox and Drake Equation: Prantzos (2013, 2020).

References

Carroll-Nellenback, J., Frank, A., Wright, J., & Scharf, C. 2019, AJ, 158, 117

Charbonneau, R. 2022, PhD thesis, Univ. Cambridge

Denning, K. 2011, Civilizations Beyond Earth: Extraterrestrial Life and Society, eds. D. A. Vakoch and A. A Harrison (New York: Berghahn Books) 74

Drake, F. D. 1965, Current Aspects of Exobiology, ed. G. Mamikunian, & M. H. Briggs (Oxford: Pergamon) 323

Gott, J. R. 1993, Natur, 363, 315

Kipping, D. 2021, RNAAS, 5, 44

Lingam, M., & Loeb, A. 2019, AsBio, 19, 28

Mamikunian, G., & Briggs, M. H. 1965, Current Aspects of Exobiology (Oxford: Pergamon)

Prantzos, N. 2013, JAsB, 12, 246

Prantzos, N. 2020, MNRAS, 493, 3464

Schmidt, G. A., & Frank, A. 2019, IJAsB, 18, 142

Seager, S. 2018, IJAsB, 17, 294

Sheikh, S. Z., Huston, M. J., Fan, P., et al. 2025, AJ, 169, 118

Shklovskiĭ, I. S., & Sagan, C. 1966, Intelligent Life in the Universe (Delta-Books) (San Francisco, CA: Holden-Day)

Vakoch, D. A., & Dowd, M. F. 2015, The Drake Equation: Estimating the Prevalence of Extraterrestrial Life through the Ages (Cambridge: Cambridge Univ. Press)

Wright, J. T., Haqq-Misra, J., Frank, A., et al. 2022, ApJL, 927, L30

Chapter 10

Philosophical Principles and Puzzles of SETI

> We are unanimous in our conviction that the only significant test of the existence of extraterrestrial intelligence is an experimental one. No *a priori* arguments on this subject can be compelling or should be used as a substitute for an observational program.
>
> —An international petition authored by Sagan (1982)

> [T]he now-elaborate and extensive discourse concerning the Fermi Paradox…is, quite literally, a substantial body of analysis about nothing, which is now evolving into metaanalysis of nothing.
>
> —Denning (2013)

Many *a priori* arguments for and against SETI have been offered, and we might count Fermi–Hart–Tipler style arguments among them. In this chapter we explore some of the more philosophical arguments for and against SETI. As we head into this labyrinth, it is good to keep a ball of thread attached to Sagan's international petition of 1982, co-signed by dozens of SETI supporters and skeptics, emphasizing that we cannot really know what is out there except by looking.

10.1 The Copernican Principle

A major shift in Western thought about the place and role of humanity in the cosmos occurred in the 16th century CE with the posthumous publication of *De revolutionibus orbium coelestium* by the Polish astronomer Nicolaus Copernicus,

doi:10.1088/2514-3433/adf5aech10

who constructed a heliocentric model of the solar system that rivaled the older, geocentric model of Ptolemy in accuracy.

This triggered the (punningly named?) *Copernican Revolution*, part of a series of transformational events in Western philosophy that gave rise to the Renaissance and the formulation of modern science. Further work by Galileo, Kepler, Newton, and many others would build on Copernicus's ideas to establish not just the physical truth of heliocentrism and the foundations of modern astronomy, but the universal character of physical law, and the scientific worldview.

Part of this shift was away from the view that the Earth was the (literal and figurative) center of the Universe, and that humans were privileged within it, to the point that the Universe itself was designed specifically for the benefit of humanity. By reducing the Earth's status from the central object of the cosmos to just one of several planets orbiting the Sun, Copernicus contributed to a broader shift in explaining the world and the cosmos in purely mechanical terms with no appeal to God or miracles.

With the development of modern cosmology in the 20th century came a figurative extension of this aspect of the Copernican Revolution called the *Copernican Principle*,[1] which states that our vantage in the Universe is not special. This is a component of the Cosmological Principle, which states that the Universe is isotropic and homogeneous on its largest scales and which forms part of the mathematical basis for the equations that underlie cosmology.

The statement that "Earth and humans are not special" can also be applied to the possibility of life elsewhere in the Universe. After all, in an infinite universe[2] governed by uniform physical laws, even the unlikeliest of events will repeat somewhere. There are thus only two numbers of things in the Universe: zero, and infinity.

So, the Copernican Principle would seem to imply that if life arose on Earth, it will have also arisen elsewhere—we are not special. This is the fundamental argument for SETI: humanity serves as an "existence proof" that the search is worthwhile (as we saw in Chapter 8).

10.2 The Anthropic Principle

The Copernican Principle comes with an important caveat: it only applies on the largest of scales. On small scales it obviously fails, and in important ways. For instance, humans certainly *do* occupy a very specific and particular part of the Universe: we live on the surface of a rocky planet with an atmosphere, not in a random spot in deep space. The Copernican Principle states that such privileged vantages exist throughout the cosmos, not that every spot in the cosmos is identical.

[1] So named by Austrian-British cosmologist Hermann Bondi.

[2] Although the Concordance Model of modern cosmology predicts that the Universe is truly infinite in physical extent, this is technically an open question, and may be ultimately unknowable. For the purposes of the instant argument, however, the enormous size of the Universe, including those parts beyond our horizons, is sufficient to make our point.

We should expect, then, that by virtue of our own existence certain aspects of our vantage and our surroundings will be special, perhaps even unique out to very large distances. This is the *Anthropic Principle*[3] and it tempers many conclusions that would seem to follow from the Copernican Principle.

10.3 The Sleeping Beauty Paradox

Anthropic arguments can be philosophically brain-twisting. Consider the so-called Sleeping Beauty paradox. In this puzzle, a person is (voluntarily, one would hope) subjected to a repeated experiment, about which they are fully informed: on Sunday, they are put to sleep, and a fair coin is tossed. On Monday, they are awakened, interviewed, given an amnesia drug that makes them forget Monday, and put back to sleep. If the Sunday coin came up tails, then on Tuesday the Monday ritual is repeated. On Wednesday they are awakened, until the next Sunday when the experiment repeats.

During the interviews, the participant is asked what the probability is that the fair coin tossed on the previous Sunday had come up heads. The answer is not obvious: the answer to that question should naively be 1/2 because the coin is fair. Since this would have been the participant's position on Sunday, and since this would be an outside observer's answer any time, and since having the participant waking up at a later date adds no information, the probability must remain the same. This is the "halfer" position.

However, the participant will be awakened twice if the coin is tails, and only once if it is heads, and so is interviewed twice as often with a coin showing tails as heads, suggesting that only 1/3 of the time they are interviewed is the coin heads. This is the "thirder" position.

The Sleeping Beauty paradox illustrates the role that actually being *in* a situation complicates what otherwise appears to be a straightforward exercise in probability. The connection to the Anthropic Principle is that the participant's consciousness changes the probabilities: during an interview, it cannot be Tuesday if the coin is heads, whereas outside observers are always awake on Tuesday, and so the coin could be heads for them.

The resolution of the paradox involves subtle reasoning about the distinction between a credence (your degree of confidence in a conclusion) and a probability (a more well-defined concept), and about what assumptions you are willing to make about how probabilities should be constructed when they are predicated on your existence as a conscious being.

In particular, the Anthropic Principle underlies the Rare Earth and Hard Steps arguments of Section 7.3.1, which state that our existence might be extraordinarily unlikely.

---

[3] So named by Australian cosmologist Brandon Carter. It is sometimes called the *weak* Anthropic Principle to distinguish it from the *strong* Anthropic Principle, which states that the laws of physics themselves must, by virtue of the fact that we can contemplate them, be such that we can exist. The strong version argues that the laws of physics could not, for instance, be such that there were no matter in the Universe or that it be entirely a dense plasma forever. This form of the Principle is more controversial, since it has been interpreted variously to imply that the Universe was created so that observers would exist, or that other universes with other laws of physics must exist.

10.4 Doomsday Arguments

One controversial extension of the Copernican and Anthropic Principles is the so-called Doomsday Argument, which has been used to argue that there must not be many intelligent species in the Galaxy, and that humans are unlikely to last more than tens of thousands of years.

A simplified version of a true story helps illustrate the logic of the Doomsday Argument. During World War II, Allied forces captured many German tanks, and noticed that they were all imprinted with serial numbers. Under the assumption that these numbers were sequential and that tanks were captured randomly, the Allies could determine how many tanks the Germans had constructed.

For instance, if the highest serial number on any captured tank read 3000, then clearly the Germans had made at least 3000 tanks. If this was the highest number among 100 captured tanks, then it would seem very unlikely that the Germans had constructed, say 30,000 tanks, since the odds that all of the captured ones would just happen to be among the first 3000 would be astronomically small. It is far more likely that the true number would be only a bit larger than 3000. From the distribution of serial numbers, then, the Allies could estimate the number of tanks remaining in the field.

The Doomsday Argument asks us to consider all humans that have existed *or will ever exist* as having a sequential serial number, and to imagine that your individual consciousness is that of a random human among that set. With these assumptions, one can conclude it is very likely that one's serial number is "typical" of all numbers. By the Copernican Principle, it seems unlikely that one would be, say, among the very last or very first humans to ever exist, past or future. Indeed, by definition there is only a 1% chance that one's serial number would be among the first 1% of humans to ever exist.

Let us assume for concreteness that 100 billion humans have existed so far on Earth, so one's serial number is around 100 billion. The Doomsday Argument then says there is only a 1% chance that the total number of humans that *will* ever exist is 100 times larger than this, or 10 trillion. Given that around 150 million people are born each year, this implies that there is only a 1% chance that humans will maintain that birth rate for another 70,000 years. Indeed, the longer one projects human populations into the future, the larger the implied total population of humanity and the smaller the chance that one would just happen to be among the very first humans.

In other words, there cannot be a future with quadrillions of humans in it, because it is extraordinarily unlikely that one would randomly find oneself to be in the present and not among those quadrillions.

Gott (1993) has extended this argument to include not just humans but all conscious creatures in the cosmos. Consider the possibility that there is a Galactic supercivilization of quadrillions of intelligent beings. If one assumes one's consciousness is that of a random being, then the chances that one would be among the 100 billion humans instead of the quadrillions of aliens is vanishingly small. Gott then concludes that humanity's total population must be near the median (to within one or two orders of magnitude) of all intelligent species in the Universe.

In other words, there cannot be species with vast numbers of members throughout the Galaxy, because it would be extremely unlikely in such a case to find ourselves among humans, and not those larger species.

The Doomsday Argument has a fascinating, almost perverse appeal to it, but it seems to have some obvious counterexamples. For instance, if one indexes every bit transmitted via WiFi in the history of the world, then a similar argument would consider the index of the bits currently being transmitted as typical. Since WiFi is a relatively recent technology, one could conclude it will only last of order decades. But this is clearly silly: the number of bits transmitted per day continues to increase as WiFi technology improves, and it is easy to imagine that the technology will still be used 50 years from now when data rates are orders of magnitude higher. Indeed, the parallel to future humans is clear here: humans have a capacity for continued (even exponential) growth, especially if we eventually populate other stars. If the Doomsday Argument fails for WiFi bits, why not people?

Unsurprisingly, there is a considerable debate around the validity of arguments invoking the Copernican and Anthropic Principles, especially those in the spirit of the Doomsday Argument and Sleeping Beauty Paradox. Brandon Carter was an early advocate of such reasoning, and important additions to these arguments were made by Canadian philosopher John Leslie and Swedish–British philosopher Nick Bostrom.

These primarily revolve around examining the assumption that an observer is a random member of a set of observers. The primary issue is the *reference class* chosen: who counts as an observer? For instance, how should we count humans? Is it fair to include aliens in the count? Do all members of *Homo sapiens* count, or should we include other members of the genus Homo? Should we consider only those who contemplate this problem at all? Garisto (2020) and Lacki (2023) provide a nice overview of these arguments, and their complexity.

The Doomsday Argument has also been attacked as fallacious on other grounds, including the conflation of future duration with total duration, confusion over the meaning of confidence intervals, and that the argument ignores relevant information about the nature of humans.

One way to see the challenges behind using a Doomsday Argument is to imagine one is testing the hypothesis that we live in a universe that will contain 200 trillion humans versus one that will contain 200 billion humans. Given that we are, say, human number 100 billion, the logic of the Doomsday Argument says that the chances that we would have such a number is of order 1/2 in the Universe with 200 billion humans, and of order 1/2000 in the other, meaning the chances humanity has a short future are favored 1000:1.

A Bayesian, however, will insist that you specify your prior belief on which universe you are in. One might say that since we don't know any better, the two universes have equal probability. But if we are a random observer in a "multiverse," then we would *a priori* expect that we are 1000 times more likely to live in a universe that contains 200 trillion humans than 200 billion humans, which exactly cancels out the 1000:1 odds that come from knowing *which* numbered human you are, meaning you cannot favor either scenario. Another way of seeing this is that each universe contains one 100 billionth human: it's a coin flip which one you actually are.

The bottom line is just to reinforce the message at the beginning of our chapter: these sorts of philosophical explorations can be fascinating, but ultimately we cannot determine how likely SETI is to succeed solely by looking at ourselves.

Box 10.1: Jargon Review: Philosophical Puzzles

- **The Copernican Principle**:
  1. An axiom of modern cosmology, stating that our vantage in the Universe is typical. Part of the Cosmological Principle, stating that the Universe is homogeneous and isotropic on the largest scales.
  2. By figurative extension, the axiom that Earth and humans "are not special," either in terms of location or the laws of nature. Implies that throughout the Universe we should expect there to be many other stars like the Sun, planets like the Earth, and even biospheres with technological species like ours.
- **The (weak) Anthropic Principle**: Rule that we can draw few or no conclusions on the likelihood of something occurring from a sample size of one when that one is ourselves. The reason is that our own existence is a precondition of our being able to calculate probabilities in the first place, and so must be satisfied no matter how unlikely our existence is. The Anthropic Principle tempers conclusions about human-like creatures based on the Copernican Principle.
- **The Strong Anthropic Principle**: The position that the laws of the Universe must be consistent with the existence of observers in order for there to be observers (like us) to measure them. Used variously to argue that the Universe is finely tuned for observers, or that other universes exist with different laws of physics.
- **Doomsday Argument**: The reasoning that humanity is very unlikely to have a long future based on the supposedly low probability that we would find ourselves at the very beginning of that long future.

10.5 Further Reading

Gott's Doomsday Argument: Gott (1993).
Anthropic Bias: Bostrom (2013).
How to select observers in light of the anthropic principle: Garisto (2020).
The "noonday" argument rebuttal to the Doomsday Argument: Lacki (2023).

References

Bostrom, N. 2013, Anthropic Bias: Observation Selection Effects in Science and Philosophy Studies in Philosophy (London: Taylor and Francis)

Denning, K. 2013, Astrobiology, History, and Society, Advances in Astrobiology and Biogeophysics, ed. D. A. Vakoch (Berlin: Springer) 301

Garisto, R. 2020, PhRvR, 2, 033464

Gott, J. R. III 1993, Natur, 363, 315

Lacki, B. C. 2023, IJAsB, 22, 354

Sagan, C. 1982, Sci, 218, 426

Part III

Practice

Chapter 11

Strategies of SETI

> Many people have an emotional investment in the outcome [of SETI]. The question touches on religious and political matters where predispositions have traditionally played important rôles. But it is abundantly clear from the history of science that no convincing resolution of this issue is likely to come from protracted debates carried on with great passion and sparse data. We have an alternative denied to the medieval scholastics: we are able to experiment. We can organize a scientifically rigorous systematic search for extraterrestrial intelligence using the technology of modern radioastronomy. That is where the energies should be focused of those concerned with the great issue of the existence of other technical civilizations in the cosmos.
>
> —Sagan & Newman (1983)

11.1 Principles of Communication SETI

The first and most common form of modern SETI is searches for communicative activities. Most broadly, this includes the sending of any signal meant to convey a message, by any means, to anyone. Most excitingly, it includes communications intended for, and to be interpreted by, humans. The implications of the detection of such a signal have driven much of the interest and activity in the field since it began.

Two challenges of communication SETI are detecting the photons (or other carriers) from a signal at all, and distinguishing them from other (e.g., natural) photons.

11.1.1 The Inverse Square Law

11.1.1.1 Omnidirectional Beacons

As radiation from a small source travels great distances through space, it spreads out, making it harder to detect.

doi:10.1088/2514-3433/adf5aech11

Consider the Sun, which emits roughly $N_{\text{emit}} = 10^{44}$ photons every second. The photons it emits in one second leave the surface at the speed of light in a shell one light-second thick. Far from the Sun at distance d, this thickness is extremely small compared to its radius, so we can treat it as a surface with area $4\pi d^2$. The shell then has $N_{\text{emit}}/(4\pi d^2)$ photons per unit area. This falloff with d^2 is called the *inverse square law*.

Imagine that a distant observer is attempting to detect the Sun's light by collecting light for one second in a telescope with area $A = 1$ m^2. The fraction of the photons from the Sun detected by the telescope is just the fraction of the surface area of the shell subtended by the telescope, given at large distances by $A/(4\pi d^2)$. This means our telescope collects

$$N_{\text{collect}} = N_{\text{emit}} \frac{A}{4\pi d^2} \tag{11.1}$$

photons.

If the observer is $d = 1$ pc away, then they will receive around 600 million photons. This makes sense: 1 pc is the distance to the nearest star, like α Centauri, which is easily detected by the naked eye as a bright star in a tiny fraction of a second with less than 1/1000 of the collecting area of that telescope.

Things quickly become challenging, however, for more distant stars. At 100 pc, less than the thickness of the Galactic disk, one detects $d^2 = 100^2$ times fewer photons, or 60,000. This is still enough for our telescope to make a clean detection in just one second, but the eye would receive only 60 in a second, which is too few for a human to register as a distinct star.

Far across the Galaxy, at, say, 10 kpc, d is another 100 times larger, and so we now detect just 6 photons per second from the star with our telescope. Stars like the Sun in the far reaches of the Galaxy require large telescopes and long integration times to detect.

Applied to communication, this makes things quite daunting. Consider sending a signal by flashing an optical light for one second with a power of $P = 1$ MW, equivalent to the most powerful continuous lasers on Earth. Since photons contain energy hc/λ, for monochromatic optical light at $\lambda = 0.5\mu$ we have

$$N_{\text{emit}} = \frac{P}{hc/\lambda} \approx 10^{24} \text{ photons per second} \tag{11.2}$$

If sent out omnidirectionally, this is 20 orders of magnitude lower than our previous example for a Sun-like star, meaning even at 1 pc the laser would not be detectable with our telescope, having an arrival rate of only $N_{\text{collect}} = 6 \times 10^{-12}$ photons s^{-1}.

11.1.2 Directed Transmissions

Communication can be enhanced by increasing the gain of the transmitter (Section 2.3.2). By placing a paraboloidal dish of diameter D behind the transmitter, or sending the light through a lens, one can collect light that would have gone in other directions and focus them in a narrow range of angles toward an intended receiver.

Despite the popular image of a laser beam being a perfectly collimated cylinder, there is an ultimate limit to the amplification we can get, governed by diffraction at the edges of the dish, given by Equation (2.15):

$$G = \varepsilon\left(\frac{\pi D}{\lambda}\right)^2 \tag{11.3}$$

where ε is of order unity and depends on the dish properties. This restricts the opening angle of the outgoing beam at large distances to be no smaller than $\sim\lambda/D$ in radians. At large distances from the transmitter, all outgoing transmissions from such a transmitter travel in a beam that is roughly described as a cone.

In our laser example, the gain from putting a paraboloidal mirror with diameter 1 m behind our 1 MW laser is $G \sim 2{,}000{,}000$. Applying this to the example of our laser detected at distance d by a 1 m^2 telescope, we have from the inverse square law (ignoring the ε term for now):

$$N_{\text{collect}} = \frac{P\lambda}{hc}\left(\frac{\pi D}{\lambda}\right)^2 \frac{A}{4\pi d^2} \tag{11.4}$$

$$= \frac{P\pi D^2 A}{4hc\lambda d^2} \tag{11.5}$$

$$\approx 8300 \text{ photons s}^{-1} \tag{11.6}$$

which is quite a large number! Clearly, it is possible to get photons through interstellar space, at least to the nearest stars, with a sufficiently tight beam. Nonetheless, the inverse square law will eventually catch up with us: at 100 pc, one only receives ~8 photons per second, and at 10 kpc it requires long integrations to detect even a single photon.

11.1.3 Distinguishing Photons

The second major problem with interstellar communication is that even if you generate enough photons to be detected at your destination, they are liable to get lost among all of the other photons in the Universe.

At optical and infrared wavelengths, the primary confounder will be the host star of the transmitter.

For instance, in our example above, we determined that a 1 MW laser launched from a 1 m telescope would send 8300 photons s^{-1} to a collector on Earth—but the number for the host star, if it is like the Sun, would be 600 million photons s^{-1}. The laser will get entirely lost in the noise.

The transmitter will need to manipulate the photons in some way to make them cleanly distinguishable from these "background" photons. There are two common ways to overcome background noise often employed in human communications: time and frequency compression.

11.1.3.1 Time Compression

If one has enough energy to transmit, say, one million photons and one wants them to be detected, it is a poor strategy to dribble them uniformly out over many days. Better to send them packed in pulses, which will stand out better against the steady hum of background noise.

This strategy is quite robust, because there are only a few kinds of natural sources that can pulse or appear to pulse; in general, sources have a minimum timescale for getting brighter or dimmer based on their size.

Astrophysical sources of radiation typically have large sizes, because one needs a lot of material to generate enough photons to be detected all the way at Earth. A cloud of hydrogen generating 21 cm radiation, for instance, can be many light-years across, and even stars have sizes measured in light-seconds.

Consider a source that, for whatever reason, suddenly got significantly brighter. Since we generally see light emitted from many parts of an object (the surface of a star, or throughout a cloud) we will receive the new photons first from the nearest portion of the object, then later from the other parts, and finally from the most distant part. This means that no matter how well coordinated the brightening of the object is, we will always see a slow ramp up in brightness that takes at least as long as a typical light crossing time of the object. We can write, roughly, that the shortest timescale τ for an object of size L to change brightness is

$$\tau \approx \frac{L}{c}. \tag{11.7}$$

Receiving a pulse of 100 photons in the span of, say 1 ns, thus implies either deliberate engineering or a source that is no larger than around 1 foot across—and such a small source would have to be very bright indeed to generate enough photons for us to detect.

How does technology produce such short pulses? By generating pulses that are highly directional from an organized source. As a purely heuristic example of how organization and directionality helps, reconsider our example above, but now imagine that the cloud emits its pulse of photons in an organized way: first the back, then in a wave moving toward the front at the speed of light. Any observer in the direction of the wave will see all the photons arrive at once from the entire cloud.

There are some caveats to this argument: pulsars famously have narrow pulse widths because they are both highly directional and *rotating*. The beam of radiation sweeps a path across the sky, and we only see it for the brief time it is pointed toward us, set by the beam width and rotation period of the pulsar.

Also, a front of material moving near the speed of light (for instance in the ejecta of a supernova or the material in a jet from an active galactic nucleus) will emit light that is relativistically beamed in the direction it travels. This makes the component of the material moving directly at us appear much brighter than material moving in other directions, shrinking the effective size of the emitting area and creating brightness variations on timescales smaller than the true light crossing time of the entire emitting region.

But such sources are generally rare, and characterized by other spectral features that reveal what is really going on. For the vast majority of photons in the Universe, time compression is an effective mechanism for making photons stand out against the noise, provided the receiver has the equipment necessary to detect them at sufficient time resolution.

One issue is how to convey content beyond "this is an artificial signal" if all of one's photons arrive at once. One solution is to vary the frequency of the photons in each pulse.

Another is to allow the arrival time of the pulse to itself convey information. Consider a nanosecond wide signal that pulses once per second. This means that there are one billion possible times of arrival, each second. Which billionth of a second contains the pulse is then a new piece of information that arrives each second, corresponding to $\log_2 10^9 \approx 30$ bits per second of information. More sophisticated schemes are also possible (Section 13.3).

11.1.3.2 Frequency Compression

The flip side of time compression is frequency compression. Most astronomical noise sources are broadband, i.e., they emit at a wide range of frequencies. By giving all of the photons in a signal very similar frequencies, they can stand out to receiving equipment that is sensitive to photon frequency.

In astronomy terms, we would write that a spectrograph can segregate photons by frequency, and that a communicative signal would show up as a spectral line in just one or a small number of spectral channels. In traditional radio communication, this is associated with the carrier wave: a strong, monochromatic signal that is modulated to encode information. In frequency space (that is, in a spectrum of the signal) these modulations appear as "sidebands" at the sum and difference frequencies of the carrier wave and the encoded information.

There are, of course, natural astronomical sources that emit spectral lines, but as with time compression there is a limit to how frequency compressed such emission can be. The reason is that photons are almost always emitted from gas or solids that have finite temperature, meaning that the emitting ions or atoms or molecules have some intrinsic motion. This thermal motion imparts their emitted photons with Doppler shifts which we witness as a spread in frequencies.

For instance, in order for something to emit narrowband optical radiation thermally, it must be warm enough to excite electrons in atoms or molecules to energy levels corresponding to optical light, let's say with wavelength $\lambda = 0.5\mu$, so having energy $E = hc/\lambda$. This implies the material is at a temperature characteristic of these energies $E = kT$, and so the components are moving at characteristic velocities $E = \frac{1}{2}mv^2$. We thus have

$$\frac{hc}{\lambda} \approx 2.5 \text{ eV} = \frac{1}{2}mv^2. \tag{11.8}$$

For a characteristic mass of a proton, $m = m_p$, this works out to velocities of approximately 20,000 km s^{-1}. This implies the internal motions of the atoms will produce random Doppler shifts of

$$\frac{\Delta\lambda}{\lambda} = \frac{v}{c} \approx 7 \times 10^{-5}. \tag{11.9}$$

Laser line widths are often expressed in terms of $\Delta\nu$, which in this case works out to about 30 GHz. We thus don't expect thermal sources to show line widths narrower than this. For reference, typical commercial laser pointers have bandwidths of about 1 GHz, and narrower lasers are readily available. Lasers can thus easily distinguish themselves even from thermally excited line emissions. Note that natural optical lines can also be generated via fluorescence in a colder gas with less Doppler broadening.

At much colder temperatures Doppler broadening can be quite small, but in this case the thermal radiation will be in the radio spectrum. Even a very cold gas (let's say 30 K) has atoms that are moving at $\sim$100 m s^{-1}, or one one millionth of the speed of light. This imparts a range of Doppler shifts of around 1 kHz in radio waves at 1 GHz. Typical human communication uses carrier waves with widths much less than 1 Hz, or 1000 times narrower. The gas cloud itself will also typically have internal motions of order km s^{-1}, and so there is no way for a gas cloud (or any source at reasonable temperature) to generate signals as frequency compressed as a technological signal.

Frequency compression thus serves the dual purpose of concentrating photons into a narrow range of frequencies that allow them to be distinguished from the background, and signal that the photons are probably artificially generated.

In both the time and frequency cases, the degree to which a signal stands out depends on the degree of compression. For arbitrarily high levels of compression, the discrimination depends on the resolution of the receiver: as one's time or frequency resolution increases, the number of background photons in any time or frequency bin drops, but the number of communicative photons is constant.

Figures 11.1 and 11.2 show examples of how signals can stand out when compressed in the time and frequency domains.

11.1.4 The Uncertainty Principle

Taking advantage of compression requires that the receiver match their detection method to the transmission method. This in general could require the receiver to know or correctly guess the method of transmission compression: one cannot have signals arbitrarily compressed in *both* dimensions, because of the *Gabor Uncertainty Principle* or *Gabor limit*.[1]

[1] Here we are discussing only the classical relationship between the widths of functions in the time and frequency domains under a Fourier transform. The Heisenberg Uncertainty Principle in quantum mechanics is closely related (indeed the whole idea comes from Heisenberg), but we are not here describing an inherently quantum effect, just a property of waves in general that all quantum particles, having a wave nature, also share. The fact that a particle's energy is proportional to its frequency, and that the uncertainty principle also applies to position and momentum, both further deepen the physical meaning of Heisenberg's version. The Gabor limit is sometimes represented as $\sigma_f \sigma_t \geqslant 1/4\pi$, which is appropriate for multidimensional signals, but a constant of 1 is appropriate for 1D signals in the time domain (see MacLennan 2024).

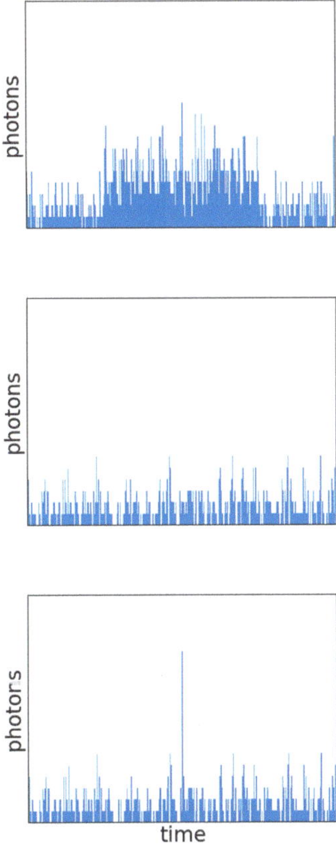

Figure 11.1. Example of time compression. In the top panel, 500 background photons arrive randomly throughout the time period and 500 signal photons arrive in the central 50%. The signal is clear. In the middle panel, only 15 signal photons arrive, and are not detectable against the background photons. In the bottom panel, the same 15 signal photons are concentrated in a single time bin, and stand out cleanly. If the detector can resolve the time of arrival of photons precisely enough, the signal can easily be detected.

The Gabor limit says that the product of a pulse's width in the time and frequency domains must be no smaller than 1, that is

$$\Delta\nu\Delta t \geqslant 1. \tag{11.10}$$

For instance, a 1 μs pulse of radio emission has a minimum spread in frequencies of 1 MHz. A narrowband signal that is only 1 Hz wide takes at least 1 s to send.

One way to understand this is that the frequency of a signal is not well defined if the wave does not contain a full oscillation. That is, the fact that the signal started and stopped means it is not an infinitely long train of sinusoidal variation at a single frequency, and the change in the strength of the electromagnetic field in this starting and ending are characterized by a range of other frequencies. The longer the train of sinusoids, the more these edge effects are diluted, and the more precisely the frequency of the whole train can be defined.

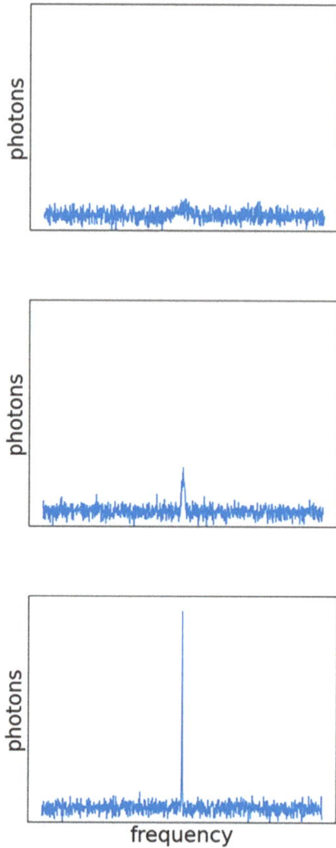

Figure 11.2. Example of how frequency compression allows a signal to stand out against noise. In all three cases, a Gaussian signal is present with the same total area against the same noise. When the signal is compressed in frequency, it stands out cleanly, provided the spectrograph measuring the signal has sufficient resolution to resolve the line.

This has consequences for compressed signal transmission and reception. For transmission, the maximum one can compress a signal to maximize its detectability over noise is set by its duration (in the case of frequency compression) or bandwidth (in the case of time compression). Figures 11.3 and 11.4 illustrate these tradeoffs, and shows that one need not necessarily choose between time and frequency compression. One could, for instance, compress modestly in both time and frequency and still achieve optimum discrimination from background noise.

The limits of the Gabor Uncertainty Principle also apply to detection. That is, we cannot sort photons into time–frequency bins spanning an area less than one. We often must choose when we collect data what "shape" our detections will be in the time–frequency domain.

Ideally, we would sort photons into the smallest possible channels to maximize our ability to distinguish signal photons from noise (in the optical, this noise typically comes from the stellar background photons; in the radio, it typically comes from the system itself, or the noise associated with linear amplification). If signals

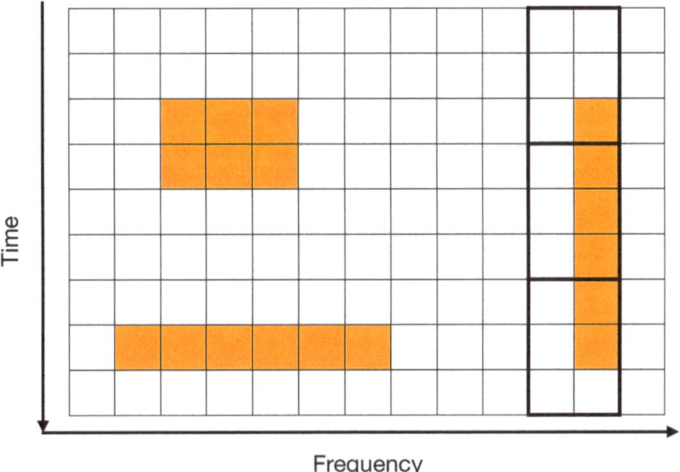

Figure 11.3. Schematic showing maximally compressed signals in a dynamic spectrum (Section 12.3). Time proceeds vertically down the plot, and signals arrive at the times and frequencies shown in orange. Each cell has a $\Delta\nu\Delta t$ product (area) of 1/6, so by the Gabor uncertainty principle, all signals must occupy at least 6 cells. The right hand signal is a long duration narrowband transmission. The bottom signal is a very brief broadband transmission. The upper left signal is a compromise between these cases. When detecting signals, there will generally be both background noise from astrophysical sources and instrumental noise in each cell; compression helps a small number of photons stand out by concentrating them in as few cells as possible. Note that there is no limit to how compressed in time or frequency these cells could be, so long as no signal is compressed in an area less than 1. The black rectangles are an example of a choice of detection channels that, although at the Gabor limit, fail to maximize detectability because they are not the same shape as the signal. The receiver has guessed incorrectly at the optimum detection scheme.

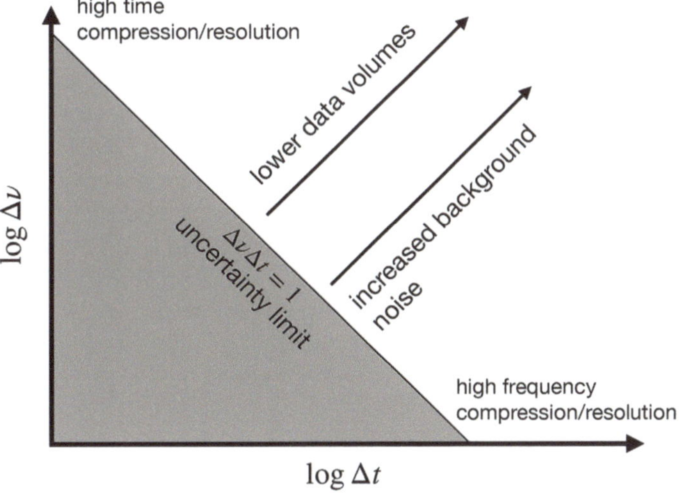

Figure 11.4. Illustration of the frequency–time resolution space tradeoff. Having larger "pixels" in time–frequency space moves one to the upper right of the diagram, requiring less storage/computation, but including more noise in each measurement against which the signal must compete.

are not maximally compressed, we can then always "bin up" to coarser resolutions to detect them.

One challenge of data taken at this limit is that there is a lot of it. Typical digital hardware can operate at speeds of MHz to GHz, so for concreteness let's say we check for a signal every microsecond (i.e.,, we'll scale things so that one box in Figure 11.3 has a height of 1 μs). The maximally compressed bandwidth of such a signal is roughly $\Delta\nu = 1/\Delta t = 1$ MHz (setting the scale of the x-axis). If we are operating in the radio at a frequency near 1 GHz, then to cover an octave (factor of 2) in frequency requires approximately 1000 frequency channels. We thus have a data rate of 1000 measurements per microsecond, which is of order GB s^{-1}. This is just manageable with modern equipment.

As we will see in Chapter 12, at radio frequencies, one can measure "raw voltages" and perform a Fourier transform to produce *dynamic spectra* with arbitrary "pixels" in time–frequency space, in principle allowing one to search for maximally compressed signals of all kinds. In practice, storage and computation limits mean that voltage data are typically reduced to dynamic spectra with specific, pre-chosen pixels at resolutions significantly coarser than the Gabor limit. Breakthrough Listen, for instance, typically uses roughly 3 Hz frequency resolution in 18 s bins, so $\Delta\nu\Delta t \sim 40$ saving a factor of $\sim$40 in computation and storage space over work at the Gabor limit.

In the optical or infrared spectrum the situation is more challenging because the frequencies are about 10^5 times higher and one cannot typically measure the electric field voltage directly, instead relying on photon counting (Section 13.2). The bottom line is that for higher energy photons, one must generally decide in advance what "shape" to search for signals for in time–frequency space and use specialized hardware for that purpose.

For instance, one may physically sort photons by frequency in a spectrograph, which generally will not achieve high resolution in both time and frequency. Typical astronomical high-resolution spectrographs are quite inefficient, and achieve typical resolutions of $R = \nu/\Delta\nu = 10^5$ (Section 2.5). Such instruments are generally not built for high time resolution work, and so can only take data every minute or so.

Fast detection of photons is possible in principle (PanoSETI achieves this resolution; Section 13.7.1), but the detectors are expensive and not as mature as technologies such as CCDs or CMOS detectors, and they have not been used as a detector in a high-resolution spectrograph. Optical/infrared communication SETI thus typically chooses either long exposures on spectrographs at $R \sim 10^5$, leading to $\Delta\nu\Delta t \approx 10^{11}$, or broadband work at micro- to nanosecond time resolution, for $\Delta\nu\Delta t \approx 10^5$.

At high energies such as X-rays, one typically has time resolution of around 1 ms and modest frequency resolution $R < 100$ simultaneously, yielding extremely high values of $\Delta\nu\Delta t \approx 10^{15}$.

11.1.5 Other Methods

Photons have properties beyond their frequency and their time of arrival that can, in principle, be used to carry information.

The most obvious is polarization, either circular or linear (Figure 2.4). Indeed, in our radio communications with spacecraft, we generally use purely circularly polarized radiation to improve detection signal-to-noise ratios.[2] This makes polarization a potential way to distinguish communication photons from natural ones, but not as powerful as time or frequency compression: natural circular polarization is generally produced by the presence of magnetic fields by a variety of mechanisms, and linear polarization occurs whenever photons are scattered. Most natural sources have low levels of polarization, perhaps $<0.1\%$, but some can be highly, nearly purely polarized.

There are various modes of quantum communication with photons that can be exploited to increase the number of bits encoded per photon, to make communications secure, or to serve as part of a distributed quantum computing scheme. It is possible that photons used in such communications would show significant spatial or temporal correlations that would reveal them to be artificial (see Hippke 2021 for some examples).

We will return to these ideas in Chapter 13.

11.1.6 Natural Power Sources

Since a major challenge of interstellar communication is generating enough photons to survive the trip across space, one alternative is to use stars or other naturally bright objects as power sources and somehow modulate their light in a communicative way.

This idea has long been used in human communications. As we saw in Section 3.1.3, in the 18th century Gauss invented the heliotrope, which reflected sunlight to mark a position at great distances. This was later developed into the heliograph, which used a pivoting mirror to flash signals such as Morse code, and such devices were mused to be ways to communicate with Venus, Mars, and the Moon.

We will explore examples of this technique in SETI in Sections 15.7.1, 17.2.4, and 17.7.

11.2 Kinds of Technosignatures and Searches

There are a variety of approaches to technosignature search. In this book we will look at several of the more mature approaches, but in this section we will explore some of the aspects that distinguish them.

11.2.1 Non-communication or Artifact SETI

Wright et al. (2018b) define "artifact SETI" as "the search for physical manifestations of technology, exclusive of communicative transmissions." This is then a broad category, being virtually any technosignature that is not a communicative signal.

---

[2] This is for two reasons. One is that radio telescopes typically are sensitive to both polarizations in different channels, so working only in one channel halves the noise of detection. The choice of circular over linear polarization is because the sense of circular polarization is insensitive to the relative orientations of the source and receiver, while linear polarization does not have this property.

The most literal form of artifact SETI is the search for physical artifacts in the solar system, and this is indeed how Freitas (1983) introduced the (now deprecated) term SETA (The Search for Extraterrestrial Artifacts). This part of the search is motivated by the same assumptions behind the Fermi Paradox: that if technological life exists, it will be spacefaring and eventually explore the entire Galaxy, including the solar system. We should therefore look for evidence of such exploration, either in the form of free-floating probes or artifacts on or under the surfaces of planets.

At the other extreme, Cirkovic (2006) and Bradbury et al. (2011) describe "Dysonian SETI" as the search for megastructures or similar mega-engineering projects in the Galaxy and beyond, such as Dyson spheres. Such artifacts might be detected via their infrared heat signatures, for instance, as Dyson (1960) originally suggested.

Physical artifacts could generate a variety of other technosignatures, including artificial pollutants in atmospheres, high energy particles, quickly moving objects or stars, or bursts of radiation designed for propulsion. Indeed, since we presume any technosignature will involve a physical underlying technology, it can be difficult to circumscribe the limits of what it means to search for "artifacts."

The primary difference between communication and artifact SETI comes then from the rationale for the search. In communication SETI, the target of the search is the signal itself, and the nature of the transmitter is incidental. The technosignature in this case is the target of the search. In artifact SETI, the target of our search is the underlying technology itself, and the technosignature merely the means to that end.

11.2.2 Searches for Extant Versus Derelict Technology

Technology might remain detectable long after it has fallen into disrepair. For instance, the long-term stability of the swarm elements of a Dyson sphere is probably poor unless they are actively managed to prevent collisions (Lacki 2025), but if they are self-repairing and autonomous they might last as a generator of infrared heat signatures long after their creators are gone. Artifacts in the solar system might be active, producing communicative signals, generating heat, maintaining attitude, and maneuvering; or they might be long dead, subject to collisions with asteroids or meteorites, or lying beneath the regolith or sands of a terrestrial body.

We do not know the limits of technology's ability to self-maintain. If we detect a communicative signal, it will likely have been transmitting for a long time (Drake's L must be large in order for $N > 1$). But this does not necessarily imply that its creators have been around that long; it may have simply been built to last. Indeed, technology itself might even be able to adapt, replicate and grow much like life does, completely blurring the line between bio- and technosignatures.

11.2.3 Scales and Kinds of Artifact Technosignatures

One can, at least roughly, map the artifact technosignature landscape by the form of the technosignature and the scale of the technology itself. In general, the more distant a technology is, the grander its scale need be for us to detect it.

At the smallest scales/distances we would have searches within the solar system (Chapter 14), for instance for physical artifacts such as probes, but also for things like city lights (Loeb & Turner 2012) or reflections from solar collectors (Dyson 2003). We might also look for chemical evidence of past technology, perhaps recorded in the Martian ice caps or in planetary atmospheres. If there is active technology using energy, it could be detected by its thermal emission. Such forms of SETI have many synergies with work in planetary science.

On larger scales we could look for technosignatures associated with exoplanets (Chapter 16), such as their satellite belts (Socas-Navarro 2018; Sallmen et al. 2019), structures at their Lagrange points (Korpela et al. 2015; Gaidos 2017; Moores & Welch 2018), or even directly imaged on their surfaces (for instance with the Solar gravitational lens; Section 14.4.2). The planets themselves might have atmospheric technosignatures (Section 16.3; Lin et al. 2014; Kopparapu et al. 2021) akin to NO_xs or chlorofluorocarbons. Kuhn & Berdyugina (2015), Berdyugina et al. (2018), and Berdyugina & Kuhn (2019) have explored how one might identify rotationally modulated hotspots, for instance due to cities (Section 16.4). Much work in this direction might occur in concert with other work to study exoplanets in general. These technosignatures roughly correspond to Kardashev 's Type I species (Section 4.2.2).

Circumstellar material might be on a scale even larger than planets, such as a Dyson sphere (Chapter 15, Wright 2020), either around an ordinary star or potentially around other objects as well (Semiz & Oğur 2015; Osmanov 2016; Imara & Di Stefano 2018). Such material can be detected via its infrared heat signatures, or by the starlight it blocks (Wright et al. 2016). Other suggestions for stellar-scale technosignatures include stellar atmospheric pollution (Section 16.3; Whitmire & Wright 1980) or "stellar engines" that move the stars themselves (Badescu & Cathcart 2000; Svoronos 2020). These technosignatures roughly correspond to Kardashev 's Type II species.

Finally, we can follow the original spirit of the Fermi Paradox and imagine pan-galactic technologies, considering Kardashev's Type III species. Indeed, as Wright et al. (2014b) pointed out, if the SETI pessimists are correct and the lack of any obvious galaxy-scale engineering proves that we are alone in the Milky Way, then unless we are alone in the *Universe, other* galaxies should show signs of such engineering. We might search for infrared heat signatures across whole galaxies (Wright et al. 2014a; Griffith et al. 2015), stellar population management (for instance, to suppress cataclysmic events such as gamma-ray bursts or supernovae), or other forms of galactic engineering (Section 17.5). Certain very luminous sources might even be visible at such distances (Lingam & Loeb 2017).

We might then group these various artifact technosignatures into three broad camps: detecting artifacts themselves via the light they emit or block, detecting the changes they have on their environment, and via the excess heat they generate from energy use. This grouping is expressed in Figure 11.5.

11.2.4 Searches for Ambiguous versus Dispositive Technosignatures

As with biosignatures, the issue of how one can know that one has succeeded in detecting life after detecting what appears to be a technosignature is complex.

Spatial Scale / Distance	Solar System	Exoplanetary	Circumstellar	Extragalactic
Structures	•Free floating •Surface / subsurface •Lights / collectors	•Satellite belts •Lagrange point megastructures •Surface features	•In transit •Directly imaged	•Direct energy propulsion
Environmental changes	•Martian ice cores •Venusian atmosphere •Ancient Earth	•Atmospheric spectroscopy •Surface changes •Commonalities among exoplanets	•Stellar atmosphere pollution •Circumstellar gases •Stellar motions	•Stellar population management •Stellar motions
Excess Heat	•Exothermic "asteroids" •Heat islands	•Rotationally modulated surface hotspots	•Dyson spheres	•Dyson spheres

Figure 11.5. Some potential technologies and artifact technosignatures grouped according to their scale and their kind. This table is not intended to be comprehensive.

A great virtue of searches for communicative signals is their prospect for being dispositive: any signal from beyond the Earth that is sufficiently compressed in time or frequency must be artificial. Such a detection would scale the entire "Ladder of Life Detection" at once (Neveu et al. 2018), showing that complex life that has produced detectable technology exists elsewhere in the Universe.

Such a detection also offers the prospect of more detailed study than an ambiguous detection. In the most extreme case, the signal may contain decipherable information. Such a signal could then include anything as grand as plans for powerful technology or the "Encyclopedia Galactica," or as intimate as the instant thoughts of a single being at the controls of the transmitter.

But even absent such content (either because it is undecipherable, or because it is absent), the mere presence and form of an unambiguous signal can be studied in the same way as biosignatures to deduce the properties of the technology itself, and the nature of its environment. Knowing its form would also motivate additional searches for similar phenomena elsewhere (either because the technology could be from a spacefaring species that spread it far and wide, or because it represents an engineering optimum for a particular problem that is appreciated by many species).

Of course, there is no guarantee that unambiguous technosignatures are detectable, and there may be many more and more easily discovered ambiguous technosignatures. Dyson spheres, for instance, will share many (but not all) of the characteristics of natural circumstellar material (Wright et al. 2014a, 2016), and the spectral signatures of atmospheric technosignatures like CFCs might be similar to that of combinations of purely natural gases.

Searches for ambiguous technosignatures such as these will have a longer road to prove the discovery of alien life, one that will largely proceed as natural explanations are slowly ruled out (Wright et al. 2014b). Still, it can serve to generate candidate technosignatures for follow up by more dispositive forms of SETI.

Searches for unambiguous technosignatures may be less dispositive than they seem at first glance, however. When a novel anomaly is first discovered it may seem to be so strange at first as to be inexplicable except as technology. Theorists are a clever bunch, however, and what seems inexplicable at first may soon have more natural explanations than data. The discovery of the interstellar comet 1I/'Oumuamua, for instance, at first generated much puzzlement and, naturally, some speculation that it could be artificial (Bialy & Loeb 2018). Rather quickly, however, planetary scientists were able to show that it fit within the bounds of expectations for natural objects ('Oumuamua ISSI Team et al. 2019; see Section 14.3.2).

A third approach is to search for extremely ambiguous technosignatures, by pursuing a SETI interpretation of objects that *do* seem to have satisfactory natural explanation. Vidal (2016), Vidal (2019), and Haliki (2019) have explored the possibility that certain kinds of binary stars or pulsars might be technological (or, even, living things themselves). This approach can have the disadvantage that not only might it be challenging to test the hypothesis, but the motivation for the hypothesis in the first place may be unpersuasive to many. On the other hand, one would hardly want to miss a technosignature just because it had been misinterpreted as natural, so such work still has some merit, provided it is well grounded.

11.2.5 Commensal/Archival versus Dedicated Searches

Because SETI has historically been funding-starved, it has often had to "piggyback" on other projects, as, for instance, with the SERENDIP project at Arecibo, which observed simultaneously with other instruments (Lampton et al. 1992). Such projects often sacrifice being able to point a telescope in exchange for having access to its focal plane at all times.

Such work has historically been called "parasitic," however this term is now discouraged because it implies that the observatory (or the rest of astronomy) suffers from such projects. A better term borrowed from biology, and more commonly used today, is "commensal" implying that SETI benefits from such work while doing, at worst, no harm. Indeed, even this understates the case: having SETI work for "free" at an observatory often expands its science case, its publication record, and its public profile, and so the truth is something a bit closer to "symbiotic."

Closely related is archival work, which makes use of data collected for other purposes. Albrecht & Balazs (1988) and Djorgovski (2000) were early proponents of big-data and algorithm-based approaches in sky surveys, a strategy followed by, for instance, Carrigan (2009) and Griffith et al. (2015) for Dyson spheres, and the VASCO project for disappearing stars (Villarroel et al. 2020). More targeted data is also useful, as with the Tellis & Marcy (2015) and Tellis & Marcy (2017) searches for continuous wave lasers near nearby stars and potential Habitable Zone exoplanet host stars.

A dedicated search has the advantage that it can be designed as a careful survey, and so more easily focus its efforts on likely targets, and more easily calculate upper limits. Such work can also make use of dedicated hardware that is more sensitive to the technosignature being searched for.

Dedicated SETI searches using competitive time on public telescopes can be challenging because it has not happened much in the past, so such work can seem foreign to many time allocation committees. This may change if such proposals become more common, and such work begins to have a positive effect on observatories' paper publication rates.

11.2.6 Searching for Beacons versus Passive Technosignatures

Many of the earliest searches for technosignatures were searches for beacons (Section 5.1)—signals intended to be discovered, perhaps specifically targeted at Earth itself. Such technosignatures should be strong, obvious, long-lived, and unambiguous, by design, but they may seem contrived to many.

Such searches are contrasted with *passive technosignatures*, or the search for technosignatures not intended to broadcast one's existence. Such technosignatures form a broad category from the consequences of industry (for instance thermal radiation from infrared heat signatures or atmospheric pollutants), to relics and artifacts (such as probes or artifacts in the solar system.)

Searches for beacons would seem to be much more likely to succeed than other kinds of searches *if* such signals exist. The primary difficulty in detecting them will likely lie not in achieving the requisite sensitivity (since the signals should, by design, be quite strong), but in identifying the appropriate Schelling points.

An intermediate case is *eavesdropping*: searches for communicative signals not intended for us (Section 16.5). For instance, it is easy to imagine that a spacefaring species would want to communicate with its brethren across the stars, or among the planets in its planetary system, or even just across its planetary surface using radio or laser communications, as we do. These transmissions inevitably leak into space, and especially if Earth happens to be in the beam of a powerful interstellar transmission we might detect these obviously artificial transmissions.

There is a misperception that early SETI programs, especially in radio SETI, were motivated entirely on the optimistic assumption that there exist beacons to be found. In fact, the pioneers in the field recognized that success might come not from the detection of a bright, narrowband signal meant for us, but from indistinct "clamor" or "chatter" of leaked communications or other sources of radio transmissions. Drake (1965), for instance, described a method for identifying sub-threshold signals in radio data via cross-correlation of signals taken at different times, and Dyson (1966) described looking for the "stray electromagnetic fields and radio noise produced by large-scale electrical operations."

It is important to recognize that searches for many or most technosignatures are sensitive to *both* beacons and other technosignatures, such as leaked emission. For instance, Arnold (2005)'s suggestion of looking for transiting megastrutures in *Kepler* data (Section 15.7.1) works fine as a perfectly general suggestion, but his specific application was to use such megastrutures as beacons. The Breakthrough Listen radio search is sensitive to not just radio beacons, but also signals not meant for us that we might intercept, or the radio emission of non-communicative radiation such as that used for directed energy propulsion systems (Lingam & Loeb 2017). The strategy to search for beacons thus need not be *exclusive* to such technosignatures.

11.2.7 Model-based versus Anomaly-based Searches

11.2.7.1 Anomaly-Based Searching: "Generalized SETI"

"Model-based" searches that propose a particular technosignature and search for it have the disadvantage that they require us to guess or deduce the form of alien technology—if not technology-as-we-know-it then technology-as-we-can-imagine-it. It is common to the point of cliché in such discussions to invoke Clarke's Third Law, that "any sufficiently advanced technology is indistinguishable from magic,"[3] meaning that we should not model alien technology on things we understand, but instead look for apparent violations of the laws of physics, or at least physics that does not account for technology.

An alternative to model-based searching, therefore, is to assume that technosignatures are rare but may be evident as highly unnatural phenomena, and look for outliers along any arbitrary dimension in astronomical catalogs. This approach was articulated well by Djorgovski (2000), who referred to it as *generalized SETI.*

Such searches might seem at first to stretch the very definition of SETI: if one is merely looking for that which one does not understand, then one is not explicitly searching for alien technology, but is merely looking for the unusual, not necessarily the technological. But this approach is well-motivated by the fact that any persuasive detection of alien technology will necessarily have to be so different from any plausible natural source, that only technological origins make sense. Any such signal should, to a sufficiently sensitive anomaly-detection algorithm, stand out quite cleanly.

Indeed, one can then proceed in much the same way as described in the next subsection, examining the strongest outliers among many axes to find both extreme natural objects and good SETI candidates. The primary difficulty is that it may be hard or impossible to model exactly what the "strength" of a technosignature parameterizes. But because it presumes little or nothing about the nature of alien technology, this generalized approach may be our strongest tool in avoiding human biases in what alien technology will be like.

A great advantage to this approach is that it is essentially guaranteed to lead to interesting science: astrophysical anomalies are often inherently interesting (Box 3.1), and so searching for them explicitly will lead to extreme objects and potentially new astrophysical phenomena.

11.2.7.2 Model-based Searching

The most straightforward way to find something is to have a model for what it looks like and look for that. This approach has the disadvantage that it requires "xenopsychology"[4] or "xenosociology": we must imagine why and/or how the engineers of the technology built it in order to model it, and in order to justify its existence to motivate the search. But a great strength of this approach is that it straightforwardly allows for upper limits on the technosignature to be calculated.

---

[3] And Gehm's corollary: "Any technology distinguishable from magic is insufficiently advanced."

[4] The term "xenopsychology" appears to originate with Robert A. Freitas, defined as "the higher mental processes of" alien life. I use the term to refer generally to needing to "get into the heads" of alien life to understand how we might find them, for instance when finding Schelling points (Chapter 5).

Presuming that one can express the strength of such a technosignature along a single axis, one can schematically illustrate that strength among all objects in a catalog, for instance from an all-sky survey, as in Figure 11.6.

Even before any analysis of individual objects is undertaken, the most extreme object in the catalog represents a first upper limit on the technosignature. Of course, this upper limit may be extremely weak, but that is the nature of first upper limits. As a referee complained on an early, weak upper limit on neutrinos in a manuscript by Davis (1955):

> Any experiment such as this, which does not have the requisite sensitivity, really has no bearing on the question of the existence of neutrinos. To illustrate my point, one would not write a scientific paper describing an experiment in which an experimenter stood on a mountain and reached for the Moon, and concluded that the Moon was more than eight feet from the top of the mountain. (Bahcall & Davis 1982, p 245)

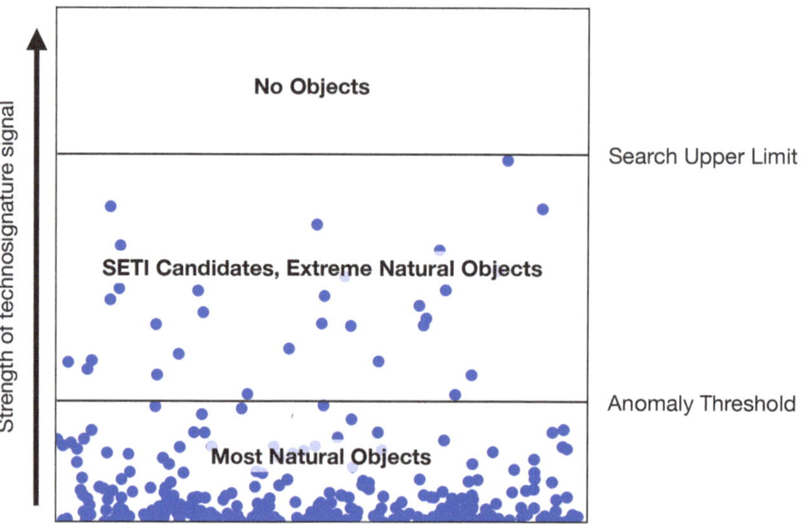

Figure 11.6. Heuristic illustration of how upper limits can be set in SETI for a model-based search with natural confounders, such as a search for Dyson spheres. One makes measurements of many objects provisionally interpreting all signals as potential technosignatures, and ranks them by their strength. Above some threshold no objects exist, and this provides a first, weak upper limit on the prevalence of that technosignature. An anomaly threshold is set by either the sensitivity of one's search, or at a higher level that provides a manageable number of candidates. One then explores the candidates between the threshold and the upper limit with complementary measurements that can be expected to distinguish natural from artificial causes for the signal, or at least confirm that the object is a member of a known, natural class and that the hypothesis of artifice is not warranted. As the most extreme natural objects become identified as such, one's upper limit drops. More thorough searches can then improve on this limit through better sensitivity or better follow-up. Ancillary science can then be performed on the extreme natural objects, which are likely to be inherently interesting simply because they are extreme.

Davis would go on to win the Nobel Prize for the discovery of cosmic neutrinos and the ultimate success of that project shows that the referee had undervalued the setting of weak limits as part of the process of discovery.

Likewise, the first upper limits on technosignatures, especially those that have not been explored before, may at first seem uselessly weak. We should welcome such upper limits as important first steps, and as challenges to others to improve upon them.

Once this upper limit has been established, it can be improved upon by examining the most extreme objects just below this limit. In general, such objects will be rare and interesting exactly because they are extreme (whatever their nature). As their nature becomes understood, they can be removed as SETI candidates from the list (or the model that measures technosignature strength can be modified to accommodate them as a source of "background") and the upper limit can be improved.

Of course, upper limits are only useful to others if they are published. For many years, very few SETI searches published quantitative upper limits, and these must therefore be inferred and estimated based on the known properties of the instrument and the published target lists. Fortunately, the community has become much better about publishing its results, and several schemes for announcing upper limits have been developed (Howard et al. 2004; Enriquez et al. 2017; Wright et al. 2018a). Technoscarch, a resource developed by Jill Tarter and others, tracks both inferred and explicit upper limits in SETI and can be found at technosearch.seti.org.

11.3 The Cosmic Haystack

> To seke out one lyne in all hys bookes wer to go looke a nedle in a medow.
>
> —Thomas More, 1557

> ...the task is far greater than many of us have the strength to acknowledge, let alone tackle. We are searching for a few needles in a fantastic haystack of inconceivable size.
>
> —Frank Drake (1976)

11.3.1 A Multidimensional "Haystack"

Figure 11.6 shows the strength of all transmissions in a single axis, but this greatly oversimplifies the search for most technosignatures, essentially assuming that one has already performed an all-sky survey sensitive to some persistent technosignature of interest.

If one has only searched part of the sky, one cannot rule out anything (what you are looking for may simply be in another part of the sky), rather one can put a limit on the *abundance* of the search target. In radio SETI, for instance, there is also the matter of choosing the right frequency, looking at the right time, and so on. Dixon (1973) defined six such dimensions one must search in: distance, direction, time, "signaling rate, bandwidth, and modulation, all interrelated," and frequency.

In one early visualization of the search space for radio communication SETI, Wolfe et al. (1981) identified several relevant parameters to consider in preparation for the NASA SETI program: polarization, repetition rate of transmission, signal modulation, three spatial dimensions, central transmission frequency, and receiver sensitivity. They then "compressed" the three dimensions of polarization, frequency of transmission, and signal modulation by considering only obvious, continuous signals searched for with receivers sensitive to orthogonal polarizations. Next, they compressed the three spatial dimensions into a single axis in terms of the number of targets or beams surveyed. They then graphically illustrated a three-dimensional "Cosmic Haystack" through which various SETI programs had hunted, looking for "needles" (see Figure 11.7). The term and visualization also appear in Tarter (1985).

Similarly, Papagiannis (1985) described the "SETI search space" of two NASA SETI search programs for narrowband transmission in terms of the frequencies searched, fraction of sky surveyed, and instrument sensitivity. He distinguished between the *Targeted* search that parametrized the haystack in terms of the number of targets surveyed (discrete sources at good sensitivity) and a *Sky Survey* program parameterized in terms of sky coverage (i.e., a shallower search for any transmitters in the beam).

Tarter (2007) and Tarter et al. (2010) described the search space as a "nine-dimensional haystack" in terms of three spatial dimensions, one temporal dimension (when the signal is "on"), two polarization dimensions, central frequency, sensitivity, and modulation.

11.3.2 Extending the Haystack Metaphor

This "needle in a haystack" metaphor is especially appropriate in a SETI context because it emphasizes the vastness of the space to be searched, and it nicely captures how we seek an obvious product of intelligence and technology amidst a much larger set of purely natural products. SETI optimists hope that there are many alien needles to be found, presumably reducing the time to find the first one. Note that in this metaphor the needles are the *detectable signatures* of alien technology, meaning that a single alien species might be represented by many needles.

Much of the SETI literature is dedicated to guessing or reasoning which haystacks might contain many needles. For instance, if we posit that extraterrestrial species are trying to get our attention and so have created obvious "beacons," (e.g., Dixon 1973), then we can focus our efforts on those haystacks or parts of haystacks where needles would most obviously be left for us to find, (i.e., we can identify which haystacks represent Schelling points in a game of mutual search, for instance by guessing at "magic frequencies"; Wright 2017).

There is also the issue of recognizing a needle when one finds it, and not mistaking it for a particularly odd piece of straw. Communication SETI typically seeks signals

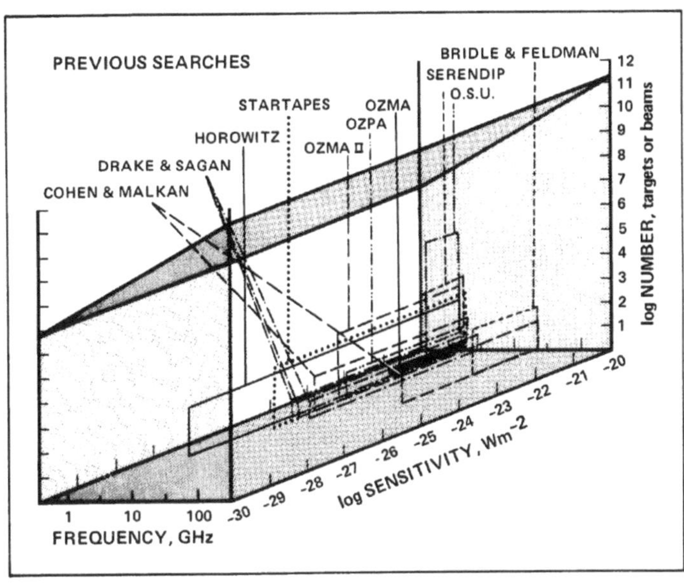

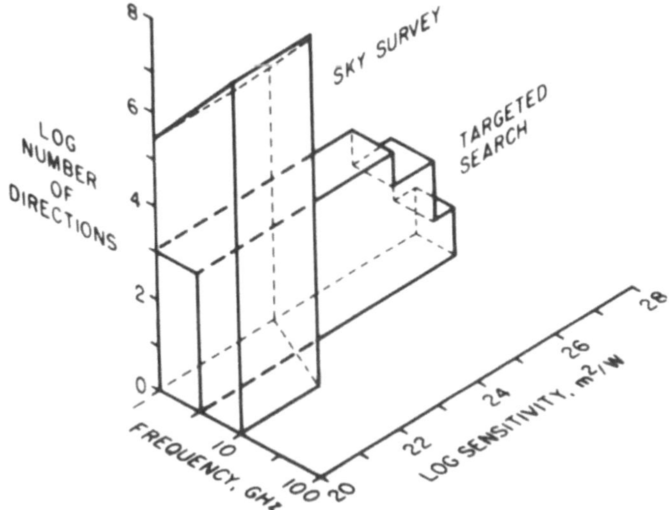

Fig. 7 SETI MOP search volume
(NASA Ames Research Center)

Figure 11.7. Figure 2 from Wolfe et al. (1981) (top) and Figure 7 from Billingham & Tarter (1993) (bottom) showing the "Cosmic Haystack" and the parts of it searched by various previous radio searches. Highly sensitive searches capable of detecting transmissions from more distant or less powerful transmitters push deeper to the lower left in these diagrams. Wide searches that search many targets push upward, and the frequency range surveyed spans the last dimension. Most of these early searches were extremely narrowband, relying on "magic frequencies." The NASA-SETI Microwave Observing Project (MOP) had two searches, a deep, targeted search, and a shallow, all-sky survey (Section 3.3.3). Credit: NASA.

so obviously artificial that there will be no doubt that the search has succeeded. Artifact SETI seeks astrophysical anomalies and exotica that are consistent with predictions for alien technology, but may not be dispositive of such technology (Wright et al. 2014b). Here the metaphor breaks down somewhat, but we might interpret the needles and straw in these haystacks to be hard to distinguish. Finding such maybe-needles nonetheless helps communication SETI efforts to narrow down which parts of their haystack to focus on to find more unambiguous needles.

11.3.3 Uniform Distribution of Bright Beacons

One way to compress the search space that also serves to compare searches is to choose a fiducial transmitter and follow the Drake figure of merit to determine the upper limit on the uniform space density of such transmitters. That is, we express a search in terms of the number density n of transmitters uniformly or randomly distributed in space and continuously emitting a narrowband signal at fiducial frequency and power. This is not a well-motivated haystack, in the sense that we have no reason to think such a network of uniformly distributed interstellar transmitters would exist,[5] but it provides an easy-to-understand and one-dimensional metric of search completeness. The number of such transmitters that a search will detect after surveying a solid angle Ω at sensitivity ${\phi_{\rm min}}^{-1}$ is given by

$$\frac{n\Omega}{3}d_{\rm max}^{3} \tag{11.11}$$

where $d_{\rm max}$ is the distance of the most distant transmitter that can be detected,

$$d_{\rm max} = \left(\frac{\rm EIRP}{4\pi\phi_{\rm min}}\right)^{\frac{1}{2}}. \tag{11.12}$$

For instance, the Enriquez et al. (2017) L-band search had a sensitivity of 17 Jy in 2.7 Hz channels, and a signal-to-noise ratio detection threshold of 25. For a transmitter with power 1 MW (because our fiducial transmitters are isotropic, this is also the EIRP), this corresponds to a distance limit of $d_{\rm max} = 2.7$ pc. The search covered 10.6 square degrees, or $\Omega = 3.23 \times 10^{-3}$ sr. Since it found no such transmitter, we can put a rough lower limit on the mean separation between any transmitters in L band as

$$l = n^{-\frac{1}{3}} \gtrapprox \left(\frac{\Omega}{3}\right)^{\frac{1}{3}} d_{\rm max} = 0.27\ {\rm pc}. \tag{11.13}$$

This is not a very tight limit, but that is not surprising because Breakthrough Listen is a targeted survey performed with a high-gain antenna that measures completeness by the number of nearby stars it observes, not the volume of space it explores. It does underscore a point about the amount of searching done to date, however, in that a

[5] But see Gertz (2018).

null result such as that of Enriquez et al. (2017) cannot rule out even very dense networks of continuous megawatt omnidirectional transmitters.[6]

11.3.4 Haystack Boundaries

Calculating search completeness requires defining the boundaries of the search; in many cases, one can always look harder for ever rarer or more subtle or more extreme events. The huge range of potential technosignatures also means that one survey cannot hope to be sensitive to all of them, requiring searchers to hypothesize a particular form of signature that they can put interesting limits on with their survey.

For instance, the Drake figure of merit in Section 12.7.1 implicitly defined a haystack with a boundary at a certain distance; while searches would be sensitive to signals from beyond that distance (indeed, out to almost *any* distance if they were strong enough) survey speed (or, in our case, completeness) would only be calculated with respect to signals up to that threshold distance. Similarly, with respect to sensitivity: while searches would be extremely sensitive to nearby sources, the figure of merit only considered sensitivity to signals stronger than some threshold.

Examples of haystack boundaries that might define a "complete" survey would be:

- All transmissions above a threshold power, or all signals above a certain received flux.
- All transmissions in a range of frequencies or (more practically) all frequencies to which a particular receiver or instrument is sensitive. In the radio, anthropogenic interference may also pose a practical limit here, meaning one must be content to search a noncontiguous set of frequencies.
- All transmissions that originate within a given distance, or in a given fraction of the sky, or from a set of sources (e.g., all stars within 10 pc, all Sun-like stars in the Northern Hemisphere brighter than $V = 10$, etc.)
- Transmissions narrower than some bandwidth.
- Transmissions with characteristic repetition rates below some threshold.
- Signals of any polarization strength or form (including unpolarized signals).
- All signals of a given modulation.

11.3.5 How Much SETI Has Been Done? Haystack Fractions

> Extraterrestrial civilizations have not yet been found, because in effect they have not yet been searched for.
>
> —Kardashev (1985)

When defining their Cosmic Haystack and calculating the fraction searched to date, Tarter et al. (2010) concluded that the amount of searching that had actually

[6] The haystack that was completely searched by Enriquez et al. (2017) is continuous transmitters near their 692 target stars emitting narrowband signals arriving at Earth at the time of their observations in the 1.1–1.9 GHz region with $\gtrsim 5.2 \times 10^{12}$ W.

been done was quite small compared to the total volume of the haystack, equivalent to the ratio of the volume of the glass of water to that of the world's oceans. "If you were looking for fish instead of extraterrestrial intelligence," they concluded "I don't think that you would conclude that there are no fish in the ocean after this meager sampling!"

This calculation was repeated by Wright et al. (2018a). Their haystack for radio searches included these dimensions:

1. Sensitivity to transmitted or received power
2. Transmission central frequency
3. Distance and position (three spatial dimensions)
4. Transmission bandwidth
5. Time/repetition rate (i.e., when the signal arrives)
6. Polarization
7. Modulation

where "modulation" was actually a hard to quantify catch-all parameter for the nature of the signal, acknowledging that power from a signal might be detected but not recognized because it was not of the sort the search algorithm was trained to recognize.

Wright et al. (2018a) rigorously computed a volume for the other eight dimensions, using extremely generous haystack boundaries. Their updated fractional volume in terms of the volume of Earth's oceans was a hot tub's worth of water. Progress! But still not enough to claim anything like a thorough refutation of the hypothesis that strong signals are waiting to be found.

11.4 The Nine Axes of Merit

> Every search for alien civilizations should be planned to give interesting results even when no aliens are discovered.
> —Freeman Dyson's self-described "First Law of SETI Investigations," as told to the author.

11.4.1 What Makes a Good Technosignature?

A great deal of subjectivity enters SETI when discussing the most likely technosignatures to search for. To some, the likelihood that someone, somewhere has built an obvious beacon or ostentatious display of technology is high enough that a search for technosignature is likely to succeed. To others, the best use of efforts is looking for the presumably more numerous but harder-to-detect passive signs of technology like pollution. Many proposed technosignatures track the proposers' expertise: Charlie Townes, inventor of the laser, thought laser communication was the best bet, radio astronomer Frank Drake thought radio signals were more likely, and examples since then abound.

Socas-Navarro et al. (2021) described the "technology balance" between senders and receivers that must be met before contact can be made: only a technology with a large ichnoscale can be detected by technology like ours, but with much better detection capabilities we could detect less obvious technology. The best technosignatures are then the ones that we are very sensitive to with our technology, and have a low value on the ichnoscale. This balance between receiving and transmitting capabilities is also described in Klein & Gulkis (1985).

Socas-Navarro et al. (2021) also note that in addition to our being sensitive to the technosignature, we need to be able to search a large number of targets to have a high chance of success. Searches that can target the whole sky are thus more likely to succeed than star-by-star searches.

At the Houston NASA Technosignature Workshop in 2018 (Section 3.5.2), many of these ideas were discussed. David Kipping proposed that the best technosignatures would be the ones that not only required the least intentionality or the least contrived activities on the part of the ETIs, but also were least ambiguous (unlike, say, heat signatures from a Dyson sphere, which could just as easily be from a disk of dust). Technosignatures that were either unambiguous or not too contrived were "OK", those that were both were ideal.

Also at the workshop Adam Frank suggested a different balance, between technosignatures that can't be avoided, like heat signatures from industry, and those that can be easily detected.

11.4.2 The Nine Axes

At the conclusion of the workshop, Sofia Sheikh presented a compilation of these and other ideas, including Freeman Dyson's First Law of SETI Investigations, in the form of nine qualitative and sometimes subjective qualities a technosignature search would need to be ideal. No proposed technosignature meets all of the criteria, but considering them all in concert helps identify which might hold the most merit.

Published in Sheikh (2020) the Nine Axes of Merit are divided into two parts: three axes are about humanity, describing how *practical* the search is given our current capabilities, and the other six are about the technosignature itself, describing how abundant or prominent the technosignature might be:

Box 11.1: Sheikh's Nine Axes of Merit

Describing us:

1. Observing Capability
2. Cost
3. Ancillary Benefits

Describing them:

4. Detectability
5. Duration
6. Ambiguity
7. Extrapolation
8. Inevitability
9. Information

11.4.2.1 Observing Capability

This axis captures our ability to conduct a broad survey of a given technosignature, for instance by surveying a large number of stars, a large number of frequencies, or sustaining a search for a long time. A high observing capability speaks to a thorough search for all of the variety and locations of a given technosignature.

11.4.2.2 Cost

This axis captures how much developing or using our observing capability will cost. This axes has been especially important for SETI because funding for actual searches has been scarce, and favors projects that can be conducted commensally or with archival data.

11.4.2.3 Ancillary Benefits

Freeman Dyson's First Law of SETI Investigations (quoted at the top of this section) can be followed in at least two ways. One is that the data collected in a search will be dual purpose, allowing a researcher who has a null SETI result to publish interesting natural science. Indeed, this is often quite easily done if one is looking for anomalies, since the false alarms in such a search are, by definition, extreme in nature.

The second is that the conduct of the search might have spinoff benefits: the equipment developed for broadband radio SETI, for instance, has application in the studies of flares, fast radio bursts, and pulsars.

11.4.2.4 Detectability

This axis refers how well a proposed technosignature stands out, or how ostentatious it is. Nearly complete Dyson Spheres would be extremely obvious across the Galaxy, whereas a belt of satellites orbiting a nearby star would be extremely challenging to detect. This axis essentially tracks the ichnoscale of the technosignature, but is also a function of the natural sources it must compete against to be noticed at all.

11.4.2.5 Duration

This is equivalent to Drake's L: a long-lived technosignature will have a higher abundance in the Galaxy. At one extreme, trying to detect a brief event like a planet-wide cataclysm (from nuclear war, for instance) would require such events to be extremely common to be found; at the other end, a passive megastructure orbiting a star or a structure just below the lunar surface might last billions of years.

11.4.2.6 Ambiguity

A major advantage to communication SETI is the lack of ambiguity of a detection: a narrowband radio signal can only be the result of technology (Section 11.1.3). Other signals, like an infrared excess suggestive of a Dyson sphere, can only be the first step on the road to proving alien technology has been found.

Searches that "skip to the end" and provide the champagne-popping proof quickly are clearly more meritorious.

11.4.2.7 Extrapolation

This axis captures the degree to which we must extrapolate beyond what we understand how to engineer, or even what we understand is possible. Where the detectability axis describes the *scale* of the technology itself, this axis captures something about its sophistication.

For instance, sending powerful radio beacons into space requires no leaps of imagination or engineering over current practice: if we wanted to increase our radio output by many orders of magnitude and broadcast to millions of stars, it would simply be a matter of scale to build and power the radio dishes.

Sending probes to another star system or building a Dyson sphere—to say nothing of a neutrino beacon or gravitational wave transmitter—requires us to solve engineering challenges we have probably not even contemplated yet. Such leaps of imagination require that there be no unforeseen showstoppers that might make the project even more challenging than we imagine, and that an alien species has done far more than we have.

11.4.2.8 Inevitability

This axis captures the degree of contrivance or inevitability in a technosignature, or the amount of appeal to "xenopsychology" involved. For instance, leaked radio transmission from cross- and inter-planetary communications might be quite common among spacefaring species, since it sits at an engineering optimum for coordinating activities across a planetary surface or throughout a planetary system. Building powerful beacons pointed at Earth requires a specific set of motivations that we can hardly expect to be universal.

Infrared heat signatures score well on this axis, since they are a thermodynamically required result of any energy use, and so are truly inevitable. Building a Dyson sphere, however, requires that a species do more than just use energy, but that they repurpose significant amounts of mass in their planetary system, which is not inevitable.

11.4.2.9 Information

Like the ambiguity axis, this final axis captures not the probability of success, but its payoff. A communicative technosignature from a nearby star harbors both the possibility of containing decipherable content itself, and of allowing followup of the system for discovery of additional technosignatures. A probe in the solar system might score highest on this axis, being amenable to physical examination. At the other end of the axis is a "bare contact" scenario, in which we detect, say, unmodulated narrowband radio emission from a distant galaxy. We would then know (and could know) nothing more about the transmitters. (A communicative signal which contains no information is sometimes in SETI called a *dial tone* or *doorbell*.)

11.4.3 Example

Sheikh (2020) provides a tool for generating graphical representations of where a given technosignature lies. Figure 11.8 shows one example for communication SETI, such as with radio or laser technosignatures.

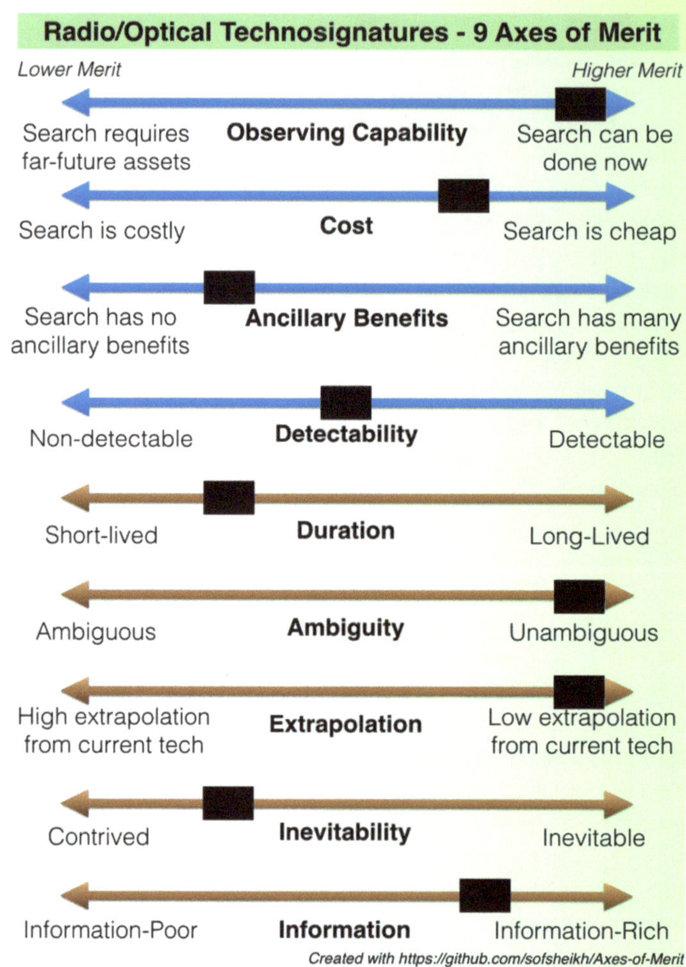

Figure 11.8. From Sheikh (2020), the Nine Axes of Merit for communication SETI as with radio or laser technosignatures. Reproduced with permission, © Cambridge University Press 2019.

11.4.4 Subjectivity and Degeneracies

It is important to note that there are degeneracies along these axes (they are not, if you will, orthogonal). For instance, the detectability axis captures the *strength* of the signal, but strong signals require less observing capability and cost to detect, and might imply large extrapolations from human technology, and more contrivance (i.e., why would they do something so ostentatious?).

Where technosignatures rank on these axes will also be subjective, especially along the extrapolation, inevitability, and ancillary benefits axes.

For these reasons, as with the Drake Equation, it is best to consider the nine axes as a list of things to consider when thinking about the problem, or a rough map of the landscape within with technosignatures reside, and not as an easily quantified and rigorous statement of the problem.

Box 11.2: Jargon Review: Strategies

- **Passive technosignature**: A technosignature arising incidentally and not from an attempt to communicate or be noticed, such as infrared heat signatures or spectroscopic absorption by atmospheric pollution.
- **Doorbell** or **Dial tone**: A communicative signal devoid of any modulation or information beyond its own existence.
- **Axes of merit**: A set of nine desirable qualities of a proposed technosignature search.
- **Cosmic haystack**: The multidimensional space that describes the search space of possible technosignatures, for instance spanning frequency, location on the sky, and strength.
- **Commensal** or **Piggyback SETI**: A SETI search done with equipment being used for other purposes, for instance by tapping into the data stream of a radio telescope doing other science. Sometimes called "parasitic" SETI.
- **Generalized SETI**: Searching for generic astrophysical anomalies and outliers that might be indicative of technosignatures.
- **Xenopsychology/xenosociology**: Terms used in this book whimsically to refer to efforts to imagine the motivations or actions of ETIs.

Exercises

1. Time Compression

   Imagine you receive regular pulses every second that are 1 ns wide. Let's explore what they could be from, assuming no engineering is involved:

   (a) What is the physical size of the largest region that could have generated the pulses if it's an isotropic emitter?

   (b) Could it be a pulsar? Let's see. First, draw a picture to determine the relationship between pulse width and the opening angle of the beam from the pulsar.

      i. What would the opening angle of the pulsar's beam need to be to generate pulses this narrow?

      ii. Typical pulsar opening angles are around 10° or so. Is that consistent with the previous part of this problem?

      iii. What would a pulsar's period need to be with a 10° opening angle to generate a 1 ns pulse?

      iv. Look up the radius of a neutron star. What would its rotational velocity need to be to have such a short period?

      v. Look up the shortest period for pulsars. Could they generate a 1 ns pulse?

2. Frequency Compression

   Now let's say you want to go find some laser lines at optical frequencies. Let's see what kind of optical spectroscopy we'll need.

   (a) Gas that emits in the optical region needs to have its atoms or ions put into the appropriate upper energy level so they can de-excite and emit the photon. Assume a simple two-level atom or ion. Roughly how much energy separates the upper and lower levels?

(b) Assume the atoms are excited by collisions with electrons (which is very often the case). How fast do the electrons have to be moving to give enough energy to excite the atoms?
(c) What temperature gas does this correspond to?
(d) Now, assume the atoms are the same temperature as the electrons. How fast are they moving? (You'll have to make a rough guess at their mass.)
(e) Calculate the Doppler shift of the emitted photon from an atom moving this fast. How wide do you expect the emission lines from the gas to be?
(f) The resolution of an optical spectrograph is measured as $R = \lambda/\Delta\lambda$. This means narrow lines at wavelength λ are always observed to be at least $\Delta\lambda$ wide because of instrumental limits, even if they are intrinsically more narrow than this. What R value would you need to measure the width of an emission line for the gas in the problem above?
(g) Typical high-resolution optical spectrographs have $R \sim 10^5$. Is this enough to distinguish an artificial laser from a natural emission line?

3. Telescope Size and Detectable Transmissions

   Consider a pair of telescopes on different planets across the galaxy communicating via optical lasers with a wavelength of λ and transmission power P. They both have diameter D and are diffraction limited, and wish to ensure that they can detect r photons per second from the other telescope (ignore backgrounds and the problem of distinguishing photons for now).
   (a) Derive an expression for their maximum distance d over which they can achieve this. How does d depend on D?
   (b) Now put in some numbers: assume λ= 500 nm, $P = 1$ MW, and $r = 10$ photons s^{-1}. Approximately what D is required to communicate with the nearest stars? With stars out to 100 pc? Across the galaxy?
   (c) Interpret these numbers in terms of humanity's ability to participate in interstellar communication via optical light

4. Let's imagine a "maximal" optical SETI experiment that searched the entire sky over the entire optical and near-infrared region of the spectrum (lets call it 0.3–3 μ) at the Gabor limit in either the time or frequency domain. Let's say that each measurement of each patch of sky at each Δt is one byte of information.

   Imagine that this experiment involved telescopes of diameter $D = 3$ m collectively observing every patch of sky at their diffraction limit.
   (a) What is the data rate of such an experiment?

      Hint: This is an order of magnitude problem, so don't sweat small corrections.

      First calculate the range of frequencies covered by the experiment.

      Then calculate the number of "pixels" on the sky the telescopes will have at the diffraction limit.

      Then, calculate the range of frequencies covered in each data collection interval Δt. How many frequency bins do you need to cover your entire bandwidth?

Then the total data rate is bits per measurement × number of frequency bins × number of pixels × measurements per second.

(b) How does this data rate compare to all of the Internet traffic in the world? (In 2017 the Internet carried about 5×10^{14} bits per second.)

(c) Repeat the exercise with a Gabor-limit all-sky survey from 1–10 GHz with a $D = 30$ m telescope.

References

Albrecht, R., & Balazs, B. 1988, Proc. 99th IAU Colloq.: Bioastronomy - The Next Steps ed. G. Marx (Dordrecht: Kluwer Academic Publishers) 315

Arnold, L. F. A. 2005, ApJ, 627, 534

Badescu, V., & Cathcart, R. B. 2000, JBIS, 53, 297

Bahcall, J. N., & Davis, R. 1982, Essays in Nuclear Astrophysics, ed. N. Schramm, et al. (Cambridge: Cambridge Univ. Press) 243

Berdyugina, S. V., & Kuhn, J. R. 2019, AJ, 158, 246

Berdyugina, S. V., Kuhn, J. R., Langlois, M., et al. 2018, Proc. SPIE, 10700, 107004I

Bialy, S., & Loeb, A. 2018, ApJL, 868, L1

Billingham, J., & Tarter, J. 1993, Space Biology and Medicine: Space and Its Exploration—Volume I, ed. A. E. Nicogossian, et al. (Reston: VA) 247

Bradbury, R. J., Cirkovic, M. M., & Dvorsky, G. 2011, JBIS, 64, 156

Carrigan, R. A. 2009, ApJ, 698, 2075

Cirkovic, M. M. 2006, arXiv:astro-ph/0606102

Davis, R. 1955, PhRv, 97, 766

Dixon, R. S. 1973, Icar, 20, 187

Djorgovski, S. 2000, ASP Conf. Ser. 213, Bioastronomy 99, ed. G. Lemarchand, & K. Meech (San Francisco, CA: ASP) 519

Drake, F. D. 1965, Current Aspects of Exobiology, ed. G. Mamikunian, & M. H. Briggs (Oxford: Pergamon) 323

Drake, F. 1976, TecRv, 78, 22

Dyson, F. J. 1960, Sci, 131, 1667

Dyson, F. J. 1966, Perspectives in Modern Physics: Essays in Honor of Hans A. Bethe on the Occasion of His 60th Birthday, July, 1966, ed. R. E. Marshak, J. W. Blaker, & H. A. Bethe (New York: Wiley) 641

Dyson, F. J. 2003, IJAsB, 2, 103

Enriquez, J. E., Siemion, A., Foster, G., et al. 2017, ApJ, 849, 104

Freitas, R. A. 1983, JBIS, 36, 501

Gaidos, E. 2017, MNRAS, 469, 4455

Gertz, J. 2018, JBIS, 69, 263

Griffith, R. L., Wright, J. T., Maldonado, J., et al. 2015, ApJS, 217, 25

Haliki, E. 2019, IJAsB, 18, 455

Hippke, M. 2021, AJ, 162, 1

Howard, A. W., Horowitz, P., Wilkinson, D. T., et al. 2004, ApJ, 613, 1270

Imara, N., & Di Stefano, R. 2018, ApJ, 859, 40

Kardashev, N. S. 1964, SvA, 8, 217

Kardashev, N. S. 1985, The Search for Extraterrestrial Life: Recent Developments ed. M. D. Papagiannis (Dordrecht: Reidel Publishing Co.) 497

Klein, M. J., & Gulkis, S. 1985, Proc. Symp. The Search for Extraterrestrial Life: Recent Developments (Doredrect: Reidel Publishing Co.) 397
Kopparapu, R., Arney, G., Haqq-Misra, J., Lustig-Yaeger, J., & Villanueva, G. 2021, ApJ, 908, 164
Korpela, E. J., Sallmen, S. M., & Leystra Greene, D. 2015, ApJ, 809, 139
Kuhn, J. R., & Berdyugina, S. V. 2015, IJAsB, 14, 401
Lacki, B. C. 2025, ApJ, 985, 191
Lampton, M., Bowyer, S., Werthimer, D., Donnelly, C., & Herrick, W. 1992, AcAau, 26, 189
Lin, H. W., Gonzalez Abad, G., & Loeb, A. 2014, ApL, 792, L7
Lingam, M., & Loeb, A. 2017, ApJL, 837, L23
Loeb, A., & Turner, E. L. 2012, AsBio, 12, 290
MacLennan, B. 2024, Foundations of Field Computation, Ch. 6, Unpublished Manuscript
Moores, J. E., & Welch, D. 2018, RNAAS, 2, 34
Neveu, M., Hays, L. E., Voytek, M. A., New, H., & Schulte, M. T. 2018, AsBio, 18, 1375
Osmanov, Z. 2016, IJAsB, 15, 127
'Oumuamua ISSI TeamBannister, M. T., Bhandare, A., et al. 2019, NatAs, 3, 594
Papagiannis, M. D. 1985, Natur, 318, 135
Sagan, C., & Newman, W. I. 1983, QJRAS, 24, 113
Sallmen, S., Korpela, E. J., & Crawford-Taylor, K. 2019, AJ, 158, 258
Semiz, I., & Oğur, S. 2015, arXiv:1503.04376
Sheikh, S. Z. 2020, IJAsB, 19, 237
Socas-Navarro, H. 2018, ApJ, 855, 110
Socas-Navarro, H., Haqq-Misra, J., Wright, J. T., et al. 2021, AcAau, 182, 446
Svoronos, A. A. 2020, AcAau, 176, 306
Tarter, J. C. 1985, Interstellar Migration and The Human Experience. Conference Held at Los Alamos, NM, May 1983, ed. B. R. Finney, & E. M. Jones (Berkeley, CA: Univ. California Press)
Tarter, J. C. 2007, HiA, 14, 14
Tarter, J. C., Agrawal, A., Ackermann, R., et al. 2010, Proc. SPIE, 7819, 781902
Tellis, N. K., & Marcy, G. W. 2015, PASP, 127, 540
Tellis, N. K., & Marcy, G. W. 2017, AJ, 153, 251
Vidal, C. 2016, AcAau, 128, 251
Vidal, C. 2019, IJAsB, 18, 213
Villarroel, B., Pelckmans, K., Solano, E., et al. 2020, arXiv:2009.10813
Whitmire, D. P., & Wright, D. P. 1980, Icar, 42, 149
Wolfe, J. H., Billingham, J., Edelson, R. E., et al. 1981, NASA Conf. Publication, Vol. 2156, ed. J. Billingham (Washington, DC: NASA) 391
Wright, J. T. 2017, Handbook of Exoplanets, ed. H. J. Deeg, & J. A. Belmonte (Berlin: Springer) 186
Wright, J. T., Griffith, R. L., Sigurdsson, S., Povich, M., & Mullan, B. 2014a, ApJ, 792, 27
Wright, J. T., Mullan, B., Sigurdsson, S., & Povich, M. S. 2014b, ApJ, 792, 26
Wright, J. T., Cartier, K. M. S., Zhao, M., Jontof-Hutter, D., & Ford, E. B. 2016, ApJ, 816, 17
Wright, J. T. 2020, SerAJ, 200, 1
Wright, J. T., Kanodia, S., & Lubar, E. 2018a, AJ, 156, 260
Wright, J. T., Sheikh, S., án Almár, I., et al. 2018b, arXiv:1809.06857

Chapter 12

Radio Technosignatures

> It has been suggested that the apparent absence of a major reworking of the Galaxy by very advanced beings, or the apparent absence of extraterrestrial colonists in the solar system, demonstrates that there are no extraterrestrial intelligent beings anywhere. At the very least, this argument depends on a major extrapolation from the circumstances on Earth, here and now. The radio search, on the other hand, assumes nothing about other civilizations that has not transpired in ours.
>
> —Sagan (1982)

12.1 Radio Astronomy Basics

Radio telescopes work very differently from those at near infrared or higher frequencies. Rather than collecting and counting photons as they are destroyed and converted to energy, a radio telescope uses an antenna to probe electric field voltage, and uses linear amplifiers to measure the variations of that voltage due to light passing by. The heritage of such work comes from electrical engineering, and so much of the language and mathematics is different from that in work in, say, the optical or X-ray regions.

One consequence of using an antenna instead of a photon-counter is that multiple instruments can "share" the input signal. In the optical (for instance), once a photon has struck the detector it is destroyed and only the properties measured by that instrument can be known (for instance, the spatial or spectral properties of the light involved.) With a radio telescope, the antenna accepts power from some large range of frequencies dictated by the properties of the receiver hardware at the telescope focus, and the voltage is sampled at some rate and amplified. This information can then be shared and processed across multiple instruments which might be designed to study the light gathered in many different ways.

doi:10.1088/2514-3433/adf5aech12

12.1.1 The Mathematics of the Fourier Transform

The fundamental observable at a radio telescope is the voltage as a function of time. Converting these voltages into the power received at given frequency is most naturally performed via a *Fourier transform*, which recasts data from the time domain into amplitudes and phases in the frequency domain. The square of these amplitudes is the power at each frequency or, in the language of astronomy, the spectrum.

Consider a series of voltage measurements starting at time $t = 0$ and spanning some integration time T. Having measured this series $V(t)$ we wish to compute the power present at each frequency. For light at a single frequency, the voltage at the antenna is given by a pure sine wave

$$V(t) = A \sin(\omega t + \phi) \tag{12.1}$$

where A is the (half-)amplitude of the variation in the electric field strength, ϕ is the phase at $t = 0$, and $\omega = 2\pi\nu$ where ν is the frequency of the light. The power[1] at this frequency is proportional to A^2: if one quadruples the power one puts into a radio transmitter, the voltage variations it induces in distant antennae is doubled.

The true voltage one measures is a combination of power at many frequencies from many sources, including noise from the instrument electronics. A theorem of Fourier theory says that one can represent *any* finite, real time series of N evenly-sampled measurements as the sum of sine waves with a unique set of phases and amplitudes:[2]

$$V(t) = A_0 + \sum_{j=1}^{\lfloor N/2 \rfloor} A_j \sin\left(\frac{2\pi jt}{T} + \phi_j\right) \tag{12.2}$$

where $\lfloor . \rfloor$ represents the floor function that rounds non-integers down, and for even N we always take $\phi_{N/2} = \pi/2$.

Note that the term on the left, $V(t)$, is just a collection of N voltages, ordered by t the time they were taken, and the right-hand side is a collection of $N/2$ semi-amplitudes A_j and phases ϕ_j, (so also N pieces of information, ordered by frequencies $\omega_j = 2\pi j/T$). Because powers are measured from A_j^2, the phases generally do not have physical significance in an absolute sense (although the

[1] Here we are using the term "power" somewhat loosely, and not strictly in terms of erg s^{-1}. The reason is that this is the term for these quantities in the jargon of Fourier transforms and electrical engineering, and the expressions we use for this "power" are *proportional* to physical fluxes and powers. While in principle one can rigorously compute the correspondence between the voltage at the antenna and the power of radio waves falling on the telescope dish, in practice one simply finds the constant of proportionality by calibrating ones' measurements by measuring the signal generated when observing an astrophysical standard source of known strength.

[2] When two sine waves add, we get a new sine wave at the same frequency but different amplitude and phase. That is, we can always write $A \sin(\omega t + \phi_1) + B \sin(\omega t + \phi_2)$ as $C \sin(\omega t + \phi_3)$. This means that no matter how many sources are contributing to varying the voltage, we can always write the resulting time series as the sum of the same set of sine waves.

differences in phases between frequencies or telescopes are important in the measurement of radio waves, especially for an interferometer).

Also note that this representation of $V(t)$ is of a sum of a finite number of sine waves, having frequencies

$$\omega_j = \frac{2\pi j}{T} \tag{12.3}$$

set by the number of evenly-spaced samples N taken in time T. We will explore the consequences of this restricted set of frequencies (which need not exactly correspond to the actual frequencies contributing to $V(t)$) below.

It is a great mathematical convenience and perennial headache for students new to the field to recast this equation as:

$$V(t) = \sum_{j=-\lfloor (N-1)/2 \rfloor}^{\lfloor N/2 \rfloor} C_j e^{2\pi i j t/T} \tag{12.4}$$

where C_j is now a set of N *complex* numbers.[3]

These C_j are related to our sine waves as:

$$C_j = \begin{cases} \frac{A_j}{2} e^{i(\phi_j - \pi/2)} & j > 0 \\ A_0 \ (=\langle V(t) \rangle) & j = 0 \\ C_{-j}^{\dagger} & j < 0 \\ A_j & j = N/2 \text{ when } N \text{ is even} \end{cases} \tag{12.5}$$

where $\langle \cdot \rangle$ represents an average and $\dagger$ represents the complex conjugate.

These coefficients are unique and can be found via the *discrete Fourier transform* (DFT):[4]

$$C_j = \frac{1}{N} \sum_{k=-\lfloor (N-1)/2 \rfloor}^{\lfloor N/2 \rfloor} V\left(\frac{kT}{N}\right) e^{-2\pi i j k/N} \tag{12.6}$$

where we have discretized the time variable with $t = kT/N$, such that the first measurement is at $t = 0$ and the last is at $t = (N-1)T/N$. We can determine our amplitudes and phases from these C_j with:

$$A_j = \begin{cases} 2|C_j| & j > 0 \\ C_j & j = 0, N/2 \end{cases} \tag{12.7}$$

[3] That is, they have real and imaginary parts, $i = \sqrt{-1}$, and $\exp(ix) = \cos(x) + i \sin(x)$. If this is all Greek to you, just keep in mind that complex numbers have both an amplitude and a phase, and $C \exp(i\omega t)$ is a fancy way to represent something oscillating at such an amplitude and phase with frequency ω. This is all a mathematical convenience to describe real things; there is nothing "imaginary" or even particularly "complex" about this approach.

[4] There are many conventions for the DFT, particularly whether the indices should start from $-\lfloor (N-1)/2 \rfloor$ or 0, and whether the $1/N$ normalization factor belongs in the DFT, its inverse (Equation (12.4)), or divided between them as $1/\sqrt{N}$ each, for symmetry.

where |.| represents the norm or absolute value operator, such that $\left|C_j\right| = \sqrt{C_j C_j^{\dagger}}$, and

$$\phi_j = \arctan \frac{\mathrm{Re}[C_j]}{-\mathrm{Im}[C_j]} \tag{12.8}$$

where the Re[] and Im[] operators extract the real and imaginary parts of the coefficient. These phases are just the complex phases of C_j plus $\pi/2$ (because we chose to express things as sine waves instead of cosine waves).

Note that by switching from N real numbers to almost N complex numbers, we have introduced twice as many parameters into our description of $V(t)$ in terms of frequencies (because these complex numbers have both real and imaginary components), and we now also have a set of coefficients at what appear to be *negative* frequencies. Both of these complications will pay off, however. For now, we can recognize that the information for negative j values (the "negative frequencies") is actually completely redundant with that at the ordinary frequencies because $C_{-j} = C_j^{\dagger}$—these extra parameters are just (conjugate) duplicates of the more easily interpreted information at positive frequencies. That is, the "negative frequencies" are mathematical fictions introduced by our choice of variables.

The reason the Fourier transform works this way is that it is applicable to any complex function—not just our real voltages—and so these coefficients at negative j values are in some sense there to give us "room" to encode information about the imaginary parts of $V(t)$. The imaginary parts of our complex voltages $V(t)$ are all zero, meaning we do not need this "room," and so the C_j's at negative frequencies have predictable values whose information is fully redundant with the coeffiecents for the positive frequencies.

One convenience of moving to complex numbers is that it gets the ϕ_j terms out of the nonlinear sine function and into the linear coefficients C_j, where they can be more quickly computed from the data using linear algebra via Equation (12.6). Another is that the discrete Fourier transform can be computed extremely quickly via methods like the *fast Fourier transform* (FFT), an algorithm that can convert between $V(t)$ and C_j values for a time series of length N in order $\mathcal{O}(N \log N)$ time (instead of the naïve $\mathcal{O}(N^2)$ scaling from computing each coefficient separately, as in Equation (12.6)).

A final benefit is that it makes the math of heterodyning extremely simple and transparent, without the need to do trigonometric gymnastics to calculate things.[5]

[5] This is, indeed, one of the primary benefits of complex analysis in physics: trigonometric identities and relations which must normally be memorized or derived with many lines of algebra can often be expressed in just a few lines of ordinary algebra using complex numbers and e^{ix}. Since sine functions describe the behavior of waves, whenever working with waves it is worth the effort to get over the initial challenges of working with complex numbers to represent real quantities to greatly simplify complicated calculations down the road.

12.1.2 Practical Limitations of the Fourier Transform

We described above the math of going from $V(t)$ to power as a function of frequency, but the method cannot be used for arbitrary frequencies. Theorems of Fourier theory dictate which frequencies we can measure power at rather strictly.

Inspection of Equations (12.2) and (12.3) shows that the frequency resolution of our spectrum, $\Delta\nu$, is dictated by T, the total duration of our time series by:

$$\Delta\nu = \frac{1}{T} \tag{12.9}$$

When using a DFT, if we want 1 Hz spectral resolution, we must integrate for at least 1 s, which, as promised, is consistent with the uncertainty principle from Section 11.1.4 and the Gabor limit.

The highest frequency we can measure power in is called the *Nyquist frequency*, and is dictated by the number of measurements we make and the total integration time. Inspection of Equations (12.2) and (12.3) shows this is

$$\nu_{\max} = \frac{N}{2T}. \tag{12.10}$$

Since N/T is the *sampling rate* of our instrument, we can write that we cannot access frequencies higher than half of our sampling rate. This makes sense: to tell something has changed, we much sample it faster than its rate of change. If we sampled a signal at exactly its frequency, we would never see it change, but by sampling at twice this rate we will be able to measure it at each crest and trough.

Because we are sensitive to frequencies from zero all the way up to the Nyquist frequency, the sampling rate also determines the *bandwidth* B of our observations, that is the range of frequencies we can measure (but we will actually get around this limitation by a factor of 2 later via heterodyning.)

While it is true that we can *uniquely* describe our sampled function $V(t)$ via powers at the frequencies prescribed by Equation (12.3), this does not mean that we have *accurately* represented the voltage as a function of time at the telescope by those frequencies. This is because the frequencies we have chosen are dictated by our sampling and integration time, not by how the electric field is actually varying.

For instance, what if a signal is at a frequency near but not exactly at the frequencies in Equation (12.3)? In this case most of it will actually show up at the nearest two frequencies, but not all of it. We only perfectly capture all of the power from frequencies whose phases change by exactly a multiple of 2π in time T.

This is because formally, a Fourier transform applies to strictly periodic functions that continue on infinitely. In fact, we do not have access to the entire signal we seek, just samples in some time window, which breaks the perfect connection between Fourier amplitudes and the actual amplitudes at the telescope.

Figure 12.1 shows how the *window function* describes the parts of the signal we record, and is typically 1 between times $t = 0$ and $t = T$ (we record all data equally) and 0 outside of it. If the signal repeated itself exactly before and after this window, there would be no problem (the blue case in Figure 12.1), but some frequencies are

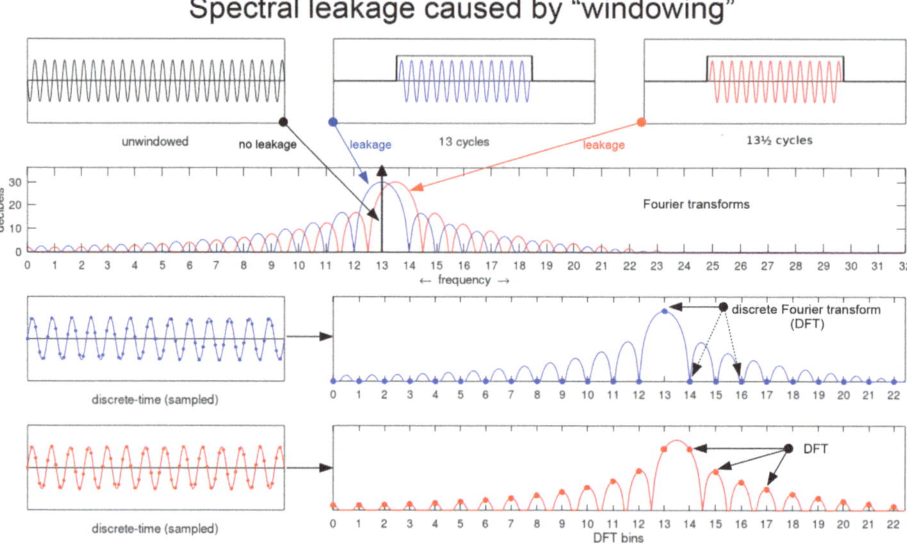

Figure 12.1. Illustration of how a real signal is represented by a discrete Fourier Transform. *Top:* three examples of a sinusoidal signal: an infinite train (black, top left), and a train observed for some finite time T (red and blue, top center and right). The *window function* describes this time, shown as a black tophat function. *Second row:* the corresponding Fourier transforms (note the y-axis is in log space, measured in decibels (dB)). The infinite train shows up at a single frequency, but the truncated trains produce power at many frequencies, described by a sinc function. *Bottom rows:* The discrete Fourier transform of the truncated signals sampled regularly for a time T, shown as circles atop the (continuous) Fourier transforms. For frequencies that are an exact multiple of the window function (blue), the DFT measures power at exactly that frequency, and zero at all other frequencies (because the sinc function is zero there). For other frequencies (red), power is mostly split between the two closest frequencies to the true frequency, but also leaks into neighboring frequencies. This is called *DFT leakage*. This leakage can be minimized by methods such as tapering. The author obtained this image from the Wikimedia website, where it is attributed to and made available by Bob_K under a CC0 license.

not integer multiples of $1/T$, so the signal gets "cut off" abruptly: the next voltage we would have measured would not have been identical to the first voltage we measured (the red case). This sharp discontinuity is interpreted by the DFT as the voltage signal having a sharp discontinuity or "edge," which results in extra power appearing at many frequencies across the spectrum. This phenomenon is called *DFT leakage*.

A theorem of Fourier theory states that pure signals show up in the spectrum in a pattern that is the Fourier transform of the window function. The Fourier transform of a "tophat" window function is a sinc function (second row of Figure 12.1). For signals at exactly the frequencies of the DFT, this function is 1 at that frequency and zero everywhere else, so those frequencies have all of their power at the appropriate frequency and nowhere else (third row). but for other frequencies this will not be the case, and their power "leaks" into frequencies across the spectrum. In practice, most of the power does not leak far from the true frequency (the y-axis in Figure 12.1 is logarithmic, so exaggerates the power far from the peak for clarity).

DFT leakage can be ameliorated by tapering the window function, i.e., multiplying the data in the time domain by the inverse Fourier transform of a function that keeps the power confined to one or two frequency bins. A common choice is a wide sinc function, whose Fourier transform is a narrow rectangle. This largely mitigates the problem of leakage, but somewhat degrades the resolution of the spectrum.

Another much bigger problem comes from frequencies above the Nyquist frequency. Power from these signals are present in $V(t)$ and so they will be represented in the spectrum, but where they show up depends on a complex interaction between them and the window function that typically causes them to show up in sharply defined frequencies where they actually have no power. This is called *aliasing*. Aliasing can be ameliorated in hardware by rejecting frequencies higher than the Nyquist frequency with a low-pass filter before recording them for analysis.

Radio engineers employ these and many other techniques to ensure that the spectrum produced by a telescope is as faithful a representation of the detected radio emission as possible.

12.1.3 Heterodyning in Radio Astronomy

A practical limitation to the scheme described in Section 12.1.1 is the rate at which the voltages from the telescope can be sampled. Looking at the Nyquist frequency, if one wishes to measure the power at the frequency of the Water Hole, around 1.4 GHz, then one must sample the voltages at a rate of at least $N/T = 2 \times 1.4$ GHz, or 2.8 billion times per second. This is not, in general, practical.

Instead, radio astronomers use *heterodyning* to transfer the electrical signals from the voltage at the antenna to lower frequencies that can be more easily sampled. The essence of heterodyning comes from the fact that the product of two sine waves at similar frequencies is a pair of (co)sine waves at the sum and difference of those frequencies:

$$A \sin(\omega_1 t) \times B \sin(\omega_2 t) = \frac{AB}{2}(\cos((\omega_1 + \omega_2)t) - \cos((\omega_1 - \omega_2)t)) \quad (12.11)$$

Radio telescopes employ analog electronics called *mixers*, which effectively perform a product on two input voltages,[6] and a *local oscillator* (LO), which is a tone generator that produces an extremely precise, accurate, and pure electrical signal at

[6] A mixer is a funny electrical component: voltages have physical units of volts (V), so it is strange to think that one can multiply them together, since their product would have units of V^2, which is not a thing. The solution is that mixers employ nonlinear amplifiers, meaning the factor by which they pass current due to an input voltage depends on that input voltage (as with a diode). There is some characteristic voltage, V_D, above which this nonlinearity becomes significant. For instance, if the input signal is v, the output might have the form $v(1 + v/V_D)$, where the second term in parentheses is the small nonlinear correction to linearity. If one adds two input signals $v = v_1 + v_2$, then in this case the output would include a $v_1 v_2/V_D$ term, which has the frequencies resulting from the multiplication we seek.

a tuneable frequency $\omega_{\rm LO}$. The mixer multiplies the LO output and the amplified input from the radio antenna, which generates sum and difference *beat frequencies* in addition to the original signals.

If the LO frequency is very close to the frequency of interest, then the power of the difference beat frequency will be shifted to a very low frequency, where it can be practically sampled.

We can see this mathematically with a heuristic model of how the process works.[7]

If we could multiply our actual signal at the telescope $V(t)$, by a *complex* oscillation near the frequency of interest, we would end up with heterodyned voltages

$$V_h(t) = V(t)e^{i\omega_{\rm LO}t} = \sum_{j=-\lfloor (N-1)/2 \rfloor}^{\lfloor N/2 \rfloor} C_j e^{(\omega_{\rm LO} - \omega_j)t} \tag{12.12}$$

which has the C_j values we want at the frequencies of interest near $\omega_{\rm LO}$, but now oscillating at very low frequencies we can measure. We could then sample this complex-valued set of voltages at some more reasonable sampling rate across a bandwidth N/T around the frequency of interest.

Of course, there is no such unit as "imaginary voltage"; what we are doing is keeping track of two separate products that are real voltages, and associating them with the real and imaginary parts of a complex voltage. That is, because we can "share" the input signal, we can split it into two components and use a pair of mixers to multiply one stream by $\cos(\omega_{\rm LO}t)$ and the other by $\sin(\omega_{\rm LO}t)$, and we label these the real and imaginary components of a complex signal.

We can then perform the Fourier transform on these two data streams interpreted as a single stream of complex numbers, yielding the C_j values we seek, representing the power near this frequency. Importantly, because our voltages are now complex--valued, we will not have a redundant set of C_j's, and so our bandwidth will be for frequencies across the full set of coeffcents (including the "negative" frequencies where $\omega_j < \omega_{\rm LO}$, which are really just positive frequencies that have been heterodyned down to where we can sample them.)

This trick allows an instrument to sample some finite bandwidth given by its sampling rate around any frequency, not just around zero frequency, allowing radio telescopes to access essentially arbitrarily high frequencies in the radio and microwave regions of the electromagnetic spectrum. This is illustrated in Figure 12.2.

The bandwidth can be increased further by multiplexing the problem, splitting the signal across many spectrographs each tuned to a slightly different $\omega_{\rm LO}$. The bandwidth of the spectrograph is then limited only by the amount of computing power and storage available to perform the analysis.

[7] This explanation illustrates the principles of heterodyning; the actual implementaiton in hardware is more sophisticated than this.

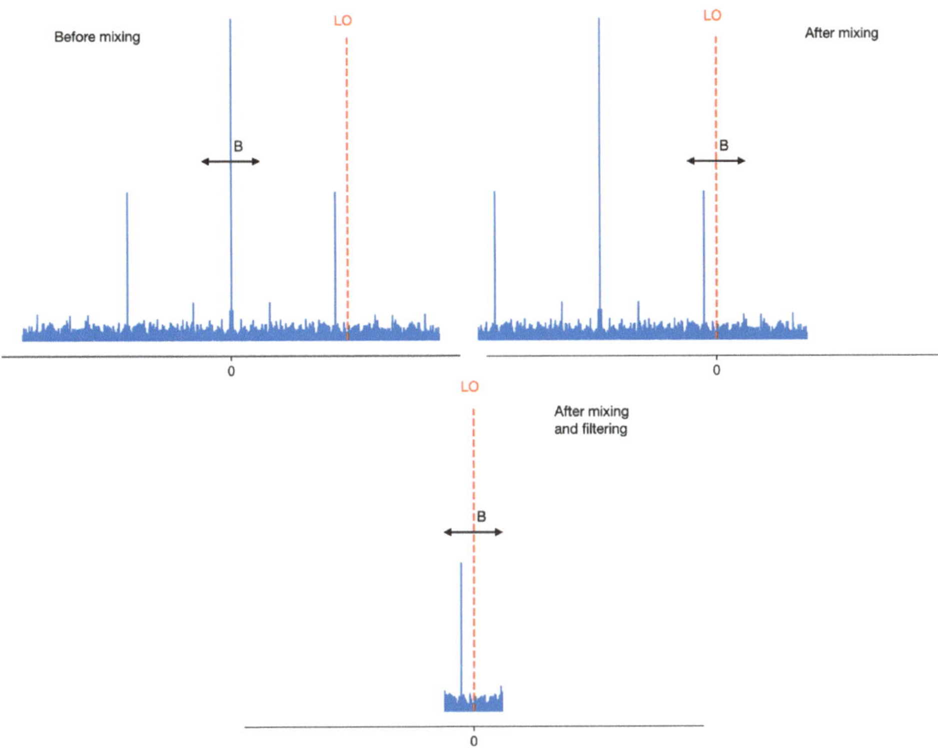

Figure 12.2. A schematic representation of heterodyning. *Top left*: The true underlying spectrum (a Fourier Transform, generating power vs. frequency) of a real signal consisting of a pure tone at a single frequency, plus noise. The total span of frequencies detected is set by the receiver and other equipment, and in practice would have some lower cutoff. The zero frequency here represents the overall voltage level in the signal and is not astrophysically significant. The other two spikes each contain half of the power of the signal, representing the positive and negative frequency components (here we plot the square of their amplitudes; they are actually complex numbers and complex conjugates of each other, because this is the Fourier transform of a real signal). The range indicated by B is the bandwidth of the spectrograph, equal to its sampling rate. The frequency of the signal we wish to measure is much higher than our sampling rate, so we need to resort to heterodyning. To do this, we introduce a signal near the frequency of interest with a local oscillator (LO). *Top right*: We mix the voltage stream with the complex voltages from the LO signal, which has the effect of shifting all of the frequencies to more negative values, and the LO signal appears at zero frequency. Our signal of interest is now within the bandwidth of our instrument. Note that since we multiplied by a complex voltage, the amplitudes at negative and positive frequencies are no longer complex conjugates of each other, but have exactly the amplitudes of the higher-frequency powers we seek. *Bottom*: A low-pass filter removes other frequencies, ensuring that there will be no aliasing of other powers when we perform our DFT, recovering the spectrum. This process can be repeated in parallel at many LO frequencies to recover as much of the spectrum as we can afford to compute.

In practice, this general scheme is performed in a more practical and optimized manner. A popular choice is the *polyphase filter bank*, which multipliexes the problem across many filter banks (i.e., individual spectrographs that are given different parts of the spectrum to analyze) and mitigates variations in sensitivity across each band and DFT leakage issues described in Section 12.1.2.

Box 12.1: The Arecibo Telescope (1963–2020)

Image courtesy of the NAIC, Arecibo Observatory, a facility of the NSF.

The National Astronomy and Ionosphere Center (NAIC) in Puerto Rico, more commonly called Arecibo Observatory after the nearest city, was at the forefront of ionospheric and radio astronomy research and was host to several iconic SETI projects.

Built in 1963 to study the Earth's ionosphere, develop missile tracking technology, and conduct radio astronomy, it hosted the Arecibo Telesocpe, the largest and most sensitive radio telescope in the world. Frank Drake (Box 8.1) was its director from 1966–1968, and in 1969 it became a civilian project overseen by the NSF. It was managed for decades by Cornell University.

The telescope had a 305 m spherical primary dish that rested in a natural depression. It pointed vertically, bringing radio waves to a linear focus. Instrumentation was mounted on a receiving platform suspended by cables connected to three towers surrounding the main dish. Astronomical sources passing nearly overhead could be tracked by moving the receiving platform, which provided for observations up to 20° from zenith.

It was equipped with a powerful interplanetary radar, which was used to help track and map near-Earth asteroids, and which could transmit a 1 MW continuous wave beam at 2.4 GHz, with an EIRP (Section 2.3.2) of order 60 TW. This radar has become a standard unit for transmission power in SETI (representing $\iota = 1$, see Section 4.2.1), although the precise value used varies from source to source.

The observatory made many major advances in astrophysics, especially in the study of pulsars and radar tracking and mapping of solar system planets and asteroids. It played a key role in the discovery of the first planets outside the solar system, around a pulsar, by Alex Wolszczan and Dale Frail in 1992. The telescope became a well recognized symbol of astronomy and SETI thanks to appearances in popular culture such as the James Bond film *GoldenEye*, and its central role in the first act of *Contact* (Box 18.1).

In 1973, Frank Drake and others used it to send the Arecibo Message (Section 19.2.3) as a demonstration of its new radar capabilities, and for many years it hosted the SERENDIP projects (Section 3.3.3), which fed their data to the worldwide SETI@Home project (Section 3.10).

The observatory suffered severe structural failure in 2020 October, and while mitigation efforts were underway the receiving platform suspension failed, crashing the equipment into the dish and destroying the telescope. Today efforts are underway to preserve the site for research and education purposes.

12.2 Sensitivity to Narrowband Signals

12.2.1 Radio Telescope Sensitivities

The sensitivity of a radio telescope depends on the bandwidth of the signal and the receiver.

In the absence of background signals (either astrophysical or terrestrial), a radio spectrometer is limited by its internal noise, expressed in terms of the *system equivalent flux density* (SEFD). Being a flux density,[8] it is typically measured in Janskys, 10^{-26} W m^{-1} Hz^{-1}. It can be thought of as the power density of signal that must fall on a square meter of the telescope aperture for it to be detected against the noise of the system itself in a single measurement. All other things being equal, larger telescopes have lower SEFDs.

The SEFD of a telescope is sometimes expressed in terms of the fictitious *system temperature* of the equipment and effective[9] collecting area A of the telescope by:

$$\text{SEFD} = \frac{2kT_{\text{sys}}}{A}. \tag{12.13}$$

This is equal to the flux density the telescope *would* receive from the sky if its beam were filled by a uniform blackbody source at temperature T_{sys}. Larger telescopes have narrower beams, so would receive less flux density from this (fictitious) background. The power from this background noise is proportional to the bandwidth involved, which is why frequency compression (Section 11.1.3) is so powerful: putting all one's power into a sufficiently narrow signal means it need compete only with a small amount of instrumental or background power.

Radio telescopes can integrate on a source for a long time to improve their sensitivity. The power from the instrumental background is constant and can be calibrated away, and the residual noise decreases as the square root of the number of times it is independently measured. The number N of independent[10] samples (measurements) one can take is given by $N = \tau\delta\nu$ where τ is the integration time and $\delta\nu$ is the range of frequencies sampled.[11]

The sensitivity S_ν, measured as a flux density, of the telescope to a signal across the range of frequencies $\delta\nu$ is thus given by[12]

$$S_\nu = \frac{\text{SEFD}}{\sqrt{\tau\delta\nu}} = \frac{2kT_{\text{sys}}}{A\sqrt{\tau\delta\nu}}. \tag{12.14}$$

---

[8] The reader unfamiliar with this term may wish to consult Section 2.2.2 before proceeding.

[9] This differs from the literal area of the dish because of projection effects onto the sky, obstruction of the aperture, imperfect reflective properties, and other practical effects.

[10] I am eliding a subtlety about Nyquist sampling here that involves a factor of two that cancels later on; one should consider this a purely heuristic derivation of the radiometer equation.

[11] This is from the uncertainty principle, Section 11.1.4. To make an *independent* sample one must wait some amount of time for the electric field to change, which is limited by the range of frequencies you are considering.

[12] In radio astronomy this is usually expressed as the *radiometer equation* in terms of the noise in the measured system temperature, $\sigma_T = T_{\text{sys}}/\sqrt{\tau\delta\nu n_{\text{pol}}}$, where n_{pol} accounts for whether both polarizations are measured. SETI signals may be highly polarized, so present in only one channel, although many SETI programs combine them, leading to an additional factor of $\sqrt{2}$ from our calculations.

Note that this quantity represents the flux density of a signal with a strength equal to that of the noise in the telescope—a large value of S_ν implies large amounts of noise, and thus low "sensitivity" in a vernacular sense.

In practice detections are actually made in terms of flux, not flux density. A modern spectrograph has many channels of individual width $\delta\nu$, and a signal of width $\Delta\nu$ might be narrower or broader than such a channel. In the case that the signal is narrower than the channel, sensitivity to flux in a single channel is

$$S = S_\nu \delta\nu = \mathrm{SEFD}\sqrt{\frac{\delta\nu}{\tau}} = \frac{2kT_{\mathrm{sys}}\sqrt{\delta\nu}}{A\sqrt{\tau}}. \tag{12.15}$$

For signals wider than a single channel, we must consider the total contribution to this noise across the entire bandwidth $\Delta\nu$, which spans $\Delta\nu/\delta\nu$ channels, when determining the sensitivity. Each frequency channel contributes to this noise in quadrature, meaning the total amount of noise grows only as the square root of the number of channels, so we have

$$S = S_\nu \delta\nu\sqrt{\frac{\Delta\nu}{\delta\nu}} = \mathrm{SEFD}\sqrt{\frac{\Delta\nu}{\tau}} = \frac{2kT_{\mathrm{sys}}\sqrt{\Delta\nu}}{A\sqrt{\tau}}. \tag{12.16}$$

The sensitivity to a signal thus can be expressed as a flux as

$$S = \mathrm{SEFD}\sqrt{\frac{\Delta f}{\tau}} = \frac{2kT_{\mathrm{sys}}\sqrt{\Delta f}}{A\sqrt{\tau}}. \tag{12.17}$$

where Δf is the *larger* of $\Delta\nu$ and $\delta\nu$.

12.2.2 Formal Sensitivities to Narrowband Radio Transmitters

The flux F received from a spatially *unresolved* source[13] is related to the power P and gain G of the transmitter (Section 2.3.2) by:

$$F = \frac{PG}{4\pi d^2} \tag{12.18}$$

and d is the distance to the target.

Radio SETI programs often set some signal to noise ratio threshold (SNR) for their detection algorithms to detect the total flux received to a single channel for narrow-band signals, or across a wider band for resolved signals. In other words, the sensitivity of the radio telescope is to fluxes

$$F > (\mathrm{SNR})\ S \tag{12.19}$$

Radio SETI programs typically report their sensitivities in terms of the EIRP (equivalent isotropic radiated power, equal to PG, see Section 2.3.2) of the

[13] That is, one that does not fill the beam, as with a distant radio transmitter. Signals that fill the beam have flux densities that are independent of distance because the surface brightness of an object does not change as it gets farther away.

transmitter they could detect (since it is only this product that is constrained by a measurement).[14]

Putting this together, we find that the sensitivity of a radio survey to artificial signals is

$$\mathrm{EIRP}_{\min} = PG = (\mathrm{SNR})(\mathrm{SEFD})(4\pi d^2)\sqrt{\frac{\Delta f}{\tau}} \quad (12.20)$$

$$= (\mathrm{SNR})(8\pi k d^2)\frac{T_{\mathrm{sys}}}{A}\sqrt{\frac{\Delta f}{\tau}} \quad (12.21)$$

where Δf is the *larger* of $\delta\nu$, the width of a single channel, and $\Delta\nu$, the width of the signal.

Equation (12.21) illuminates the drive toward certain designs of radio SETI instrumentation. The first two terms cannot be optimized in hardware. The T_{sys} term is the product of the receivers, which at higher frequencies are literally cooled to keep their noise levels low.[15] The effective collecting area A drives astronomers to using larger telescopes, such as the 100 m Green Bank Telescope, the (now defunct) 300 m Arecibo telescope, and the 500 m FAST telescope.[16]

The $\sqrt{\tau}$ term in the denominator drives one to longer integration times, but only up to a point: as we will see in Section 12.4.3, the motion of the telescope produces Doppler shifts that will move any narrow signal across multiple channels, complicating matters.

Finally, the $\sqrt{\Delta f}$ term is in the numerator. Since one wishes to minimize EIRP, this drives searches for narrowband signals to develop specialized hardware to build the narrowest spectral channels practical, up to the presumed resolution of a communicative signal. Once $\delta\nu < \Delta\nu$, this benefit disappears and the sensitivity is fixed. Most SETI searches report their EIRP limits in this regime, as their sensitivity to narrow or unresolved signals.

12.3 Dynamic Spectra

As discussed in Section 11.1.3 (and as will be elaborated on in Section 13.5), there are good reasons to expect that communicative radio signals will be highly compressed in frequency. Human communicative radio signals tend to employ carrier waves that are extremely narrow—less than 1 Hz wide for a several GHz signal—and much narrower than any natural astrophysical source.

[14] Use of this term is sometimes misinterpreted as an *assumption* by SETI astronomers that the transmitters they seek are isotropic, and so must be extremely powerful. The term is actually inherited from radio engineering and implies exactly the opposite: if one were really assuming an isotropic transmitter, there would be no need to qualify the power measured as being "equivalent" or "effective," it would simply be the power.

[15] The system temperature is not a literal temperature of any component in the instrument, but the radiation from warm elements in the optical path does contribute to raising it.

[16] The F stands for "five hundred meter" and the T for "telescope," but as the Los Angeles Angels baseball team will tell you, sometimes the structure and context of a sentence demand a bit of redundancy.

Because radio SETI works at much higher spectral resolutions than typical for radio astronomy, it typically employs specialized hardware producing extremely fine channelization of the data. That is, it collects data over some integration time, and then checks many different narrow frequency channels to see if they contain more power than neighboring channels.

Because the signals radio SETI seeks are expected to drift (Section 12.4.1), it is important to track any such signal as it changes with time so the signal does not get smeared out and lose contrast against the background noise. This requires acquiring not just a single spectrum, but a series of spectra separated by some amount of time characteristic of the signal drift. The result is a *dynamic spectrum* or *waterfall plot* of the radiation from a source, which is represented as an image composed of stacked rows of high resolution spectra taken at a series of times (as shown in the bottom panels of Figure 12.3).

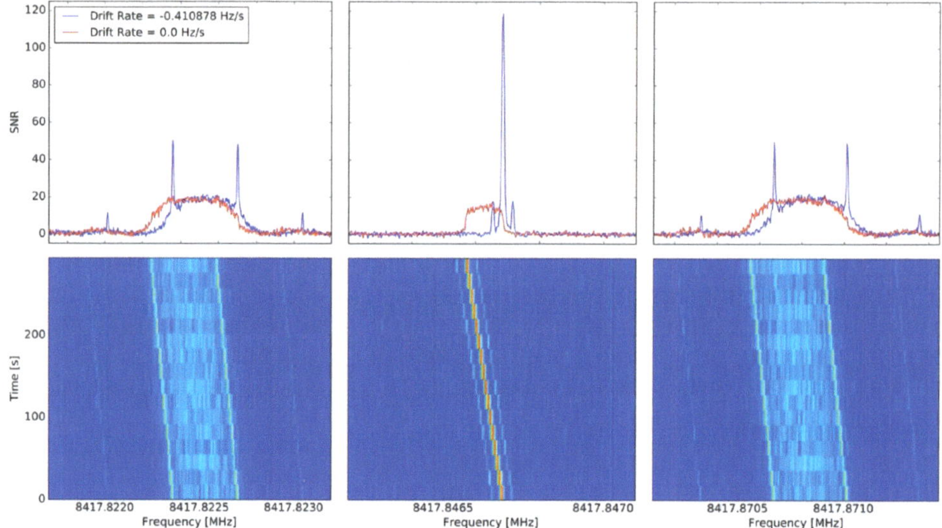

Figure 12.3. From Isaacson et al. (2017). Dynamic spectrum (or waterfall plot) of the detection of the Voyager 1 downlink by Breakthrough Listen using the Green Bank 100 m telescope, illustrating how narrowband signals are identified in dynamic spectra. The temporal resolution is 18 s for a total integration time of 5 minutes, and the spectral resolution is 3 Hz. The central column shows the narrow, unresolved carrier wave with two nearby sidebands, and the left and right columns show the more distant sidebands. The shape of these sidebands is particular to the form of modulation of the carrier wave NASA uses to transmit binary data, and is evidence that the signal is communicative and rich with information. These signals are transmitted at constant frequency in the frame of Voyager 1 centered around 8415 MHz. They are Doppler shifted, and show significant drift in 5 minutes due to mutual acceleration, primarily due to the rotation of the Earth. In red, the top plots show the total power for the full 5 minute integration, i.e., the sum of the columns in the dynamic spectra below them. Those plots show in blue the total power if the power is instead summed along slants through the dynamic spectra to account for the drift and recover the true spectrum of the transmitted signal. To realize this benefit on an unknown, alien signal, one must try a range of drift rates, and one must have sufficient time resolution that the signal does not drift significantly in a single integration.

The resolution of a dynamic spectrum is limited by fundamental and practical effects. The uncertainty principle (Section 11.1.4) sets the minimum area of a "pixel" on such a spectrum, such that $\Delta\nu\Delta t \geqslant 1$. The choice of this tradeoff—whether to favor higher frequency or temporal resolution—must be made when one is working with the voltage data measured at the telescope.

Practically, the data volumes in radio SETI are extremely large and one must employ significant data reduction and compression. For scale: if one wanted to maintain maximum temporal resolution at 1 Hz frequency resolution, then one would collect one spectrum every second. Modern SETI backends have bandwidths in excess of 1 GHz, meaning that if each element of each spectrum is represented by a single byte, the data rate for storage is 1 GB s^{-1}, a dynamic spectrum from a single 5 minute integration would be 0.6 TB, and data from a single hours-long observing session of many targets would generate over 10 PB.

Programs typically, therefore, do not save their voltage data or their full resolution dynamic spectra, instead binning the dynamic spectra down to larger time-bandwidth "pixels" as a concession to data storage realities. For instance, the Breakthrough Listen project at Green Bank Observatory typically stores data for work with SETI at 18.25 s time resolution and 2.7 Hz channelization, resulting in a savings of a factor of $\sim$40 over the maximum resolution data product, while still preserving the information necessary for signal detection.

Still, one would prefer to preserve whatever data one could for interesting signals for further diagnostics, such as the polarization or fine time and frequency structure of a signal that is not preserved in the default dynamic spectrum data product. One solution is to employ a ring buffer or similar temporary storage space where the raw voltage data is retained just long enough for a search for signals to be performed. One can then save some small subset of the voltages just at the frequencies and times of particularly interesting signals.

Such a scheme is in some sense optimal, but can be technically challenging to implement, especially given the extremely high computational demands of broadband SETI. In practice, most searches to date are performed long after the observations themselves are conducted, meaning verification must rely on followup observations.

12.4 Barycentric Drift

12.4.1 The Relativistic Doppler Formula

A challenge of narrowband radio communication in space is the mutual motion of the sender and receiver. On Earth, this is not a problem for receivers and transmitters fixed to the surface, but in space the problem can be severe.

For searches at a single, "magic" frequency one must hope that the transmitter has corrected for their own motion (relative to some mutually obvious frame of reference, see Section 5.2.4). In this case one must only correct for one's own motion.

This is accomplished for terrestrial transmitters through the use of a precise *ephemeris*, a list of the positions and orientations of all of the bodies in a planetary system. The NASA Jet Propulsion Laboratory maintains the best ephemerides of the solar system, and combined with a precise time, location on the Earth, and Earth orientation parameters can be translated to a position and velocity ($\vec{v}$) of the Earth in the frame of the solar system barycenter (Section 2.14.4).

If the target is in direction $\hat{r}$, then the Doppler shift[17] the receiver experiences compared to the solar system barycenter is

$$\frac{\Delta\nu}{\nu} = \frac{1 + \vec{v}\cdot\hat{r}/c}{\sqrt{1 - |v|^2/c^2}} - 1 \tag{12.22}$$

or, in terms of redshift z and wavelength,

$$z = \frac{\Delta\lambda}{\lambda} = \frac{\sqrt{1 - |v|^2/c^2}}{1 + \vec{v}\cdot\hat{r}/c} - 1. \tag{12.23}$$

12.4.2 Barycentric Corrections in Dynamic Spectra

Importantly, because terrestrial receivers accelerate as the Earth rotates, this shift can change rapidly enough that signals will change their frequencies noticeably during an observation.

Figure 12.3, from Breakthrough Listen observations of Voyager 1 (Isaacson et al. 2017) shows the effect. During a 5 minute integration, the frequency of the signal changed by around 50 Hz as the Earth rotated.

This figure also illustrates one way to correct for this effect, which is to "shift and stack" each integration (each row of the dynamic spectrum) to counteract this drift. This requires that one retain data with a sufficiently high time resolution that the signal is not smeared within a single time bin.

The maximum acceleration experienced by a terrestrial transmitter toward a source due to the rotation of the Earth (the dominant source) is given by

$$a/c = \frac{4\pi^2 R_\oplus}{c}\cos(\text{lat})\frac{1}{\text{day}^2} \tag{12.24}$$

$$= 1.1 \times 10^{-10}\ \text{s}^{-1}\cos(\text{lat}) \tag{12.25}$$

$$= 0.11\ \frac{\text{Hz s}^{-1}}{\text{GHz}}\cos(\text{lat}) \tag{12.26}$$

$$= 0.11\ \text{nHz}\cos(\text{lat}) \tag{12.27}$$

[17] This formula is relativistic and accounts for nearly all effects. There are a few additional effects, such as the Shapiro delay and secular acceleration, which operate below the 10^{-10} level. See Wright & Eastman (2014) for more details. On should never use the 1D relativistic Doppler formula for this purpose, since the Earth moves in 3D (see Wright 2019).

where lat is the latitude of the receiver. Note that the drift in Hz s^{-1} depends on the frequency of the light involved. Since radio/microwave work is at frequencies of order GHz, this can be expressed equivalently as Hz s^{-1} GHz^{-1} or nHz.

For instance, Breakthrough Listen typically uses fine channels 3 Hz wide. With 18.25 s time resolution, they can be sure that an observation at 2 GHz will not drift across more than one channel boundary in a single time bin, but may need to worry about smearing at higher frequencies. A challenge of working at higher time resolution, however, is that data rates, and computational and storage requirements are correspondingly larger.

Another approach available to radio astronomers is *coherent de-Doppler* shifting, which occurs prior to the generation of dynamic spectra. In this scheme, one multiplies the complex voltages from heterodyning by an additional factor of

$$\exp(-\mathrm{i}\omega t^2 a/c)$$

where a is the acceleration one is correcting for. (Note that this factor is frequency dependent, so each section of the spectrum will need its own correction.) This has the effect of "un-slanting" the dynamic spectrum before it is made, and so before the voltages are converted to powers and put into time–frequency bins. As a result, there is no sensitivity lost to smearing signals at this drift rate across bins, although all zero-drift rate RFI (Section 12.5) will now be drifting.

12.4.3 Accounting for Accelerations of the Transmitter

If a transmitter is not correcting for its own motion in some frame—perhaps because it expects a particular receiver to do this correction or to match its physical accelerations, for instance because it sits on the same surface—then we should expect its signal to drift even in the solar system barycentric frame (or any other initial frame). This is in essence the test Hewish et al. (1968) performed (on pulse frequency as opposed to light frequency) to check whether the first pulsar, PSR B1919+21, might be an alien transmitter (Box 3.1).

In this case, one must check many possible drift rates, and it is impractical to employ coherent de-Doppler techniques, which would require performing the computationally expensive step of reducing voltage data to dynamic spectra many times. Rather, one must shift-and-stack the dynamic spectrum, as shown in Figure 12.3, at very possible slope of interest.

There are several methods for doing this, the goal being to sum power in a dynamic spectrum along a drifting signal so that its signal-to-noise ratio can be maximized despite the drift (as shown in Figure 12.3). A naïve approach would scale as the number of time–frequency "pixels" in the image times the number of slopes checked. However, performing this summation for two different but very similar slopes involves a large number of identical calculations, which means the computational complexity can be lower than this.

The most popular de-Doppler method is the "tree" de-Doppler method, first used in the context of pulsars by Taylor (1974) and described in a SETI context by Siemion et al. (2013), Enriquez et al. (2017), and Houston (2023). Another related method is the

Table 12.1. Some Drift Rates for Transmitters on Illustrative Surfaces

	Drift Rate (Hz s^{-1} GHz^{-1})
Earth rotation	0.11
Io orbit	2.39
Jupiter rotation	7.2
Maximum theoretical rotation for terrsetrial planets	87
Maximum observed for exoplanet orbit (Kepler-78 *b*)	191

doubling accumulation drift detector (DADD; Cullers & Deans 1992). Care must be taken for drift rates larger than the "one-to-one" slope, when signals begin to drift into a neighboring channel during a single integration, as described by Margot et al. (2021).

But how much might we expect the signal to drift? Our own terrestrial transmitters (transmitting to other terrestrial transmitters or doing interplanetary radar) are accelerating up to 0.11 Hz s^{-1} GHz^{-1}, but this limit is set by the particularities of Earth's radius, rotation period, and orbital properties.

Sheikh et al. (2019) explored the range of accelerations one would expect from transmitters on a variety of planetary surfaces. Some extreme values they found are listed in Table 12.1.

Because the time it takes to search drift rates grows with the number of drift rates checked, there is a tradeoff between computational cost and thoroughness (Table 12.1). Sheikh et al. (2019) recommend searching up to a conservative upper limit of 200 Hz s^{-1} GHz^{-1}, when practical. Note that this will in general be far above the one-to-one line of one's dynamic spectrum, so signals at these drift rates will be badly smeared across many pixels, resulting in lower sensitivities. Li et al. (2023), looking at the drift rates one might expect from real distributions of exoplanets, recommend a more practical limit of 0.4 Hz s^{-1} GHz^{-1}, which they show should account for more than 99% of uncorrected drifts from the surfaces of exoplanets.

12.5 Radio Frequency Interference

> ...because of the growing problem of radiofrequency interference by civilian and military transmitters, the search program will become more difficult the longer we wait. This is the time to begin.
>
> —Sagan (1982)

The primary challenge of finding an alien radio signal in search data is radio frequency interference (RFI) from anthropogenic sources. The equipment built by radio SETI engineers is capable of detecting extraordinarily weak signals from artificial transmitters—and they routinely do to the tune of millions of detections per observation.

The radio spectrum of the Earth is already quite crowded, so much so that it must be carefully regulated and allocated to enable its usefulness for wireless communication of all sorts. A few, narrow regions of the spectrum are ostensibly protected for radio astronomy, but these are generally only around commonly observed lines such as the 21 cm line of hydrogen. We do not know what spectrum to expect an alien signal from, and if they exist they are almost certainly at a frequency where ubiquitous terrestrial equipment is already transmitting.

Radio telescopes, of course, are highly directional, and isolate signals coming from small fractions the sky, but they are not perfect. The wave nature of light means that they must accept radiation from a range of angles with varying sensitivity (Section 2.3.1). Far from the direction they are pointed, their sensitivity is very low but it is not zero, and in a few directions called *sidelobes* can be as high as tens of dB[18] below the sensitivity of the main beam (see Figure 12.4). These sidelobes

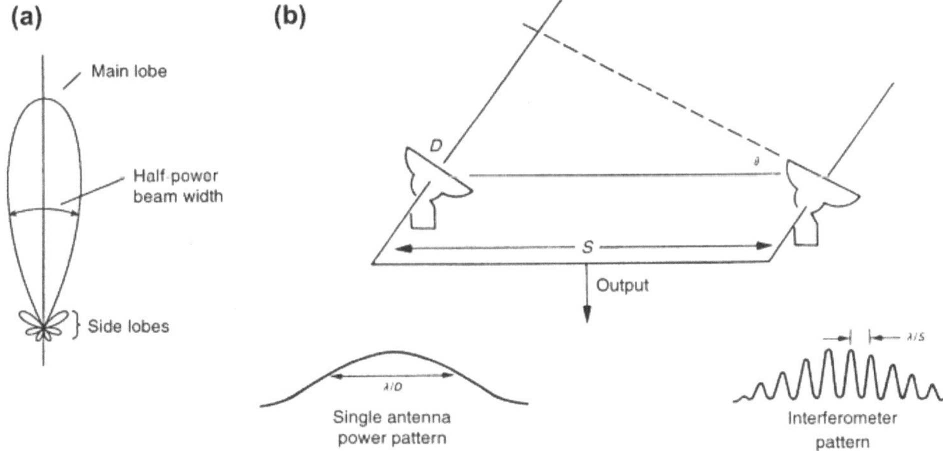

Figure 12.4. (a) A schematic of a beam pattern from a radio telescope, showing the many lobes where the telescope is sensitive to radiation. Sensitivity is maximized in the direction the telescope is pointed (indicated by the vertical line), and drops off quickly with angular distance from this direction (indicated as distance from the telescope on a log scale). The side lobes are much less sensitive, typically being down by a few tens of dB (a few factors of ten). (b) A schematic of an interferometric beam pattern. Bottom left is the primary beam pattern on the sky, here with angle on the x-axis and linear sensitivity on the y-axis. The half-power width of the beam is roughly λ/D. When the signal from two telescopes pointed in the same direction is combined, light from their overlapping beams interferes constructively and destructively depending sensitively on its location on the sky, producing an interference pattern, called *fringes*, with wavelength $\sim\lambda/S$, where S is the projected telescope separation. Combining signals from N telescopes allows radio astronomers to create $N(N-1)/2$ overlapping sets of fringes, and tease out a complete map of the radiation within the main beam. Reprinted from de Pater & Kurth (2007) © 2014 with permission from Elsevier.

[18] Radio and sound engineers measure signal strength in decibels (dB), or tenths of a bel. Bels are a logarithmic unit representing a factor of 10, so a decibel is a factor of $10^{1/10} \approx 1.26$. The bel is named after Alexander Graham Bell, and it's just one of those things that the name of the unit drops a single letter from the proper spelling of its eponym when it's just going to be abbreviated anyway (especially since the watt does not!)

invariably include objects on the ground (like radio stations, home WiFi routers, or even microwave ovens) and in the sky (like satellites and aircraft). Because these transmitters are perhaps 10^{15} times closer than nearby stars, even a telescope's very small sensitivity to them is enough for them to swamp the weak signals SETI astronomers seek to detect.

In addition, the Earth's orbit is increasingly filled with satellites broadcasting to the ground, and which have passive electronics generating low levels of radio waves easily detected by large radio telescopes.

The challenge of radio SETI is thus not just picking up a weak signal; it is in distinguishing it from the huge number of actual artificial signals detected with every observation.

12.5.1 RFI Mitigation: Drift Rates

The single most powerful discriminant between terrestrial and other transmitters is the drift rate (Section 12.4.3). Sources fixed to the surface of the Earth typically transmit carrier waves or other signals at fixed frequencies with zero Doppler shift (Section 2.5) with respect to a radio telescope, and so will have zero-drift rate, while the changing motion of the Earth (and potentially an alien transmitter) will cause any signal from space to change with time. When searching for radio signals, SETI astronomers thus assess the degree to which a signal is drifting and reject any signal without such a drift.

If the frequency of a signal from space is not changing due to the motion of the transmitter, then in general one expects to detect a negative drift rate—that is, one expects the frequency to drop with time. This is for the same reason that the pitch of car engines or sirens drops when a car passes you: as things pass by, they go from approaching you (the Doppler shift is toward higher frequencies) to moving away from you (the Doppler shift is toward lower frequencies).[19]

One might wonder if alien transmissions intended for Earthly receivers might not adjust their transmission frequencies dynamically to account for the Earth's motion to improve their detectability, but remember that different parts of the Earth rotate at different speeds and in different directions. Unless the alien transmitter had advance knowledge of the precise location of the receiving telescope, a nearly perfect ephemeris of the Earth's motion, and a precise distance to the Earth, it could not completely correct for this and the signal would appear to drift.

This filter is not perfect. Many terrestrial signals have complex structure in frequency space that includes changing frequencies, and some transmitters, such as

[19] This asymmetry may seem strange since it is not time reversible—shouldn't an object going "the other way" have its pitch go up? And the pitch of the engines of cars on a racetrack cannot continuously drop; there must be some time when they are rising again. The answers to the first puzzle is that a drift rate is due to an acceleration, which is a *second* derivative in time, and so preserves its sign under time reversal (it involves dt^2, so the substitution $dt \rightarrow -dt$ changes nothing). The puzzle for the racetrack is that the pitch goes up on the parts of the oval where the cars are turning back toward the observer. The analogous situation for transmitters in space is for stars "below the pole" (at hour angles greater than 6 hr). For most observatories this is a small region of the sky generally avoided by optical astronomers because it is low on the horizon, but radio astronomers occasionally make use of it, and there we expect positive drift rates.

aircraft, satellites, and our deep space probes exhibit large drift rates because they are moving quickly (or, more precisely, quickly accelerating with respect to the telescope). Others have poor control of the frequency at which they transmit, and so their signals change in frequency in manners that mimic the behavior of a steady, non-terrestrial source. Finally, a transmitter (terrestrial or alien) might have a complex frequency structure that is not well characterized as a single steady or drifting frequency.

12.5.2 RFI Mitigation: Spatial Filtering

One feature of RFI that should cleanly distinguish it from a real source on the sky is that real sources are tightly localized on the sky and are centered in the primary beam of the telescope, and so the flux one receives from them is extremely sensitive to the pointing of the telescope. RFI, since it enters through sidelobes, will have a more complex dependence on telescope pointing. Using this fact to distinguish RFI and real sources is called *spatial filtering*. This is typically done in two ways: on–off pointings, and interferometry.

12.5.2.1 On–Off Pointings

On–off pointings are used for single-dish telescopes. The idea is that RFI tends to be detected at a wide range of telescope pointings, and so is insensitive to small adjustments in the pointing direction of the telescope, while a weak signal on the sky will become undetectable if one points even slightly away from it.

A common strategy, then, is to "nod" the telescope back and forth and check if a signal is only present when the telescope is pointed directly at the target of interest. The "off pointings" must be at a small enough distance from the target of interest that the sidelobes are in largely the same locations, but far enough from the target of interest that it will be essentially undetectable in the off pointings. One then checks for signals that are present in all of the on pointings but none of the off pointings.

A typical strategy is to spend half of one's time on target and half in the "off pointing" in a series of perhaps three on–off pairs. An efficient use of this strategy is to point the telescope at a second target of interest during the off pointing, and so use each target as the other's control pointing: it is essentially impossible that two true signals of interest would come from two nearby locations on the sky with identical frequencies and drift rates.

12.5.2.2 Interferometry

When it comes to RFI rejection, interferometers (Section 2.3.3) have an enormous advantage over single-dish telescopes in that they have many more methods for identifying the direction of incoming signals, and so can perform RFI rejection without spending time pointed away from the source of interest.

Interferometry can be performed in many modes, and there is a tradeoff between the computational effort required to correlate signals among the many radio telescopes and the sophistication one can use to reconstruct the localization of the signals detected. In general, interferometers collect light from their *primary beams*, or the region of the sky probed by each element in the array. For the Allen Telescope Array (composed of $D = 6.1$ m dishes) operating in the Water Hole (at 1.42 GHz, or $\lambda = 21$ cm) the primary beam has a Gaussian shape about $\lambda/D \approx 2°$ across (Section 2.3.1). When all of its elements are pointed in the same direction, the array can then reconstruct signals coming from specific locations within the primary beam with resolution corresponding to the widest separation between array elements, which is around $D = 300$ m, or to $0\overset{\circ}{.}04 \approx 3'$.

That is, the ATA can reconstruct an image with $0\overset{\circ}{.}04$ pixels, set by the separation of the telescopes, in a circle with roughly 2° diameter set by an individual telescope's size (see Figure 12.4). The more elements an interferometer has, the more sensitivity it has to weak signals, and the cleaner and sharper the underlying image it creates is.

The simplest and fastest method of interferometry is called *beamforming*, wherein one combines the signals from many telescopes linearly to reconstruct the signal isolated in a single synthesized beam within the primary beam (one "pixel" in our previous example). The computational complexity of this mode is low, scaling linearly with the number of telescopes and the number of beams one wants to synthesize. One can then synthesize two beams—one on the target of interest and another at a different pointing of interest where sensitivity to signals from that beam of interest is extremely low—and use these as simultaneous "on–off" pointings. One then constructs spectra in each of these beams and searches for signals present in the "on beam" but not in (or much, much weaker in) the "off beam."

A more sophisticated but computationally intensive method is to construct an image of all of the sources in the primary beam (essentially, get a spectrum for every pixel in the primary beam) and search for signals at each frequency that are highly localized to a single direction. This is more comprehensive in that it provides powerful diagnostics that a detected signal is truly coming from a single point on the sky, and this is how searches on low-frequency arrays like LOFAR are routinely performed. But at microwave frequencies, this can be computationally prohibitive for large arrays, since the complexity scales as the *square* of the number of telescopes.

12.5.3 RFI Mitigation: Multi-site Followup

Because RFI is generally quite local, a robust method of RFI management is coincidence rejection using very distant telescopes. In this method, telescopes in entirely different RFI environments (say, Green Bank, the VLA, and the ATA) observe a target simultaneously, and only signals exactly common to the multiple sites are considered (after accounting for the different drift rates due to the motion of the Earth). This is generally challenging because it requires redundant observations, restricts the region on the sky that can be observed, and requires coordinating telescope schedules. It is also limited by the degree to which the telescope instrumentation is matched in sensitivity and bandwidth: only the brightest signals at the VLA, for instance, will be visible at the much smaller ATA.

A less demanding variation on this scheme is to have a somewhat less sensitive telescope with narrow bandwidth capabilities "on call" to the west of the primary search instrument. If the search can be performed in real time, then especially promising candidate signals can trigger an interruption of the ordinary tasks of the follow up instrument, which can quickly be re-tasked with verifying the signal. The followup instrument can then do a longer integration at just the frequencies of interest to make up for its lower sensitivity and bandwidth. The telescope should be to the west to ensure the source is visible even after a brief delay (of order an hour or two) at the primary instrument to search the data and trigger the alert.

12.5.4 RFI False Alarms

Figure 12.5 shows the RFI situation at Green Bank Observatory, where an L band observation will in general detect *millions* of real narrowband signals from human technology (and this only counts those whose morphology triggered the particular algorithm, `TurboSETI`, used in that work). There, a combination of on–off filtering and drift rate filtering reduced the number of hits to a manageable number that could be examined by eye.

Figure 12.6 shows a candidate signal detected by the Breakthrough Listen team at Green Bank Observatory toward the star HIP 4436. This candidate signal passed all tests: its components are too narrow to have a natural origin; it is drifting, so it is unlikely to originate from a fixed terrestrial transmitter; it is drifting in the negative

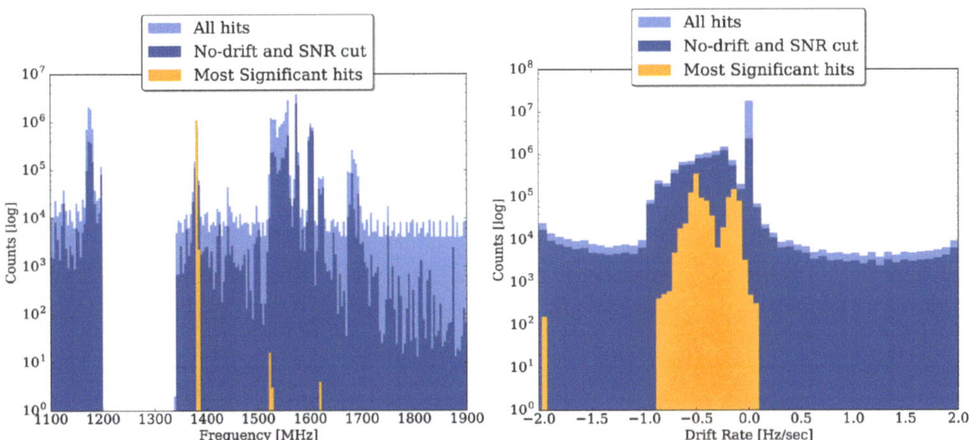

Figure 12.5. Figures from Enriquez et al. (2017) showing the problem of RFI at Green Bank Observatory, which is located in a National Radio Quiet Zone. These histrograms show the total number of "hits," or detections of narrowband radio signals due to technology, as a function of frequency (*left*) and drift rate (*right*). The various colors represent successive steps in the RFI-rejection algorithm, such as requiring a nonzero drift rate, a signal-to-noise ratio above a certain threshold, and that the signal appears in all of the on-pointings. Note the logarithmic y-axis: algorithms must search through *millions* of real technological signals from human technology. In this L band search, an entire region from 1.2 to 1.34 GHz is ignored because there is a filter on the telescope receiver that suppresses RFI from aircraft radar at these frequencies. This signal is so strong, that without very strong suppression, it would contaminate observations at all other frequencies in the band.

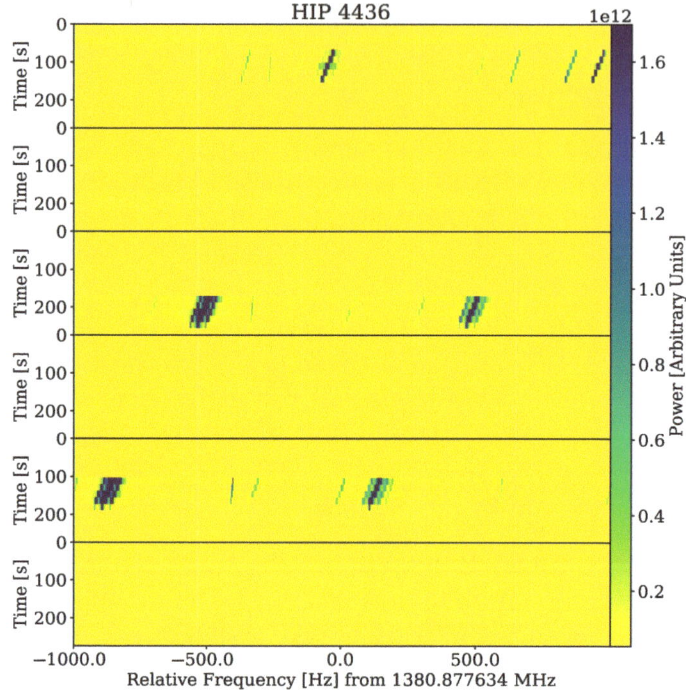

Figure 12.6. Figure from Enriquez et al. (2017) showing the dynamic spectrum near a detection passing all RFI rejection filters employed by Breakthrough Listen at Green Bank Observatory at the time. Time reads from top to bottom, with each block corresponding to a different pointing. This observation employed an ABACAD pattern, where A is the on pointing and B, C, and D were different off pointings. This source was ultimately identified by the Breakthrough Listen team based on similarities with known patterns and frequencies of RFI generated by GPS satellites and air traffic control. © 2017 the American Astronomical Society, all rights reserved.

direction, consistent with an object in space; the signal is clearly strong and not noise; and it appears in all of the on pointings while being completely absent from all of the off pointings, indicating it has been localized on the sky well over the course of half an hour.

The Breakthrough Listen team ultimately determined this source is due to RFI based on similarities with known patterns and frequencies of RFI generated by GPS satellites and air traffic control. The likely explanation is that the signal would have appeared in the off pointings if it had been persistent, but that its repetition every 10 minutes or so just happened to phase up with the nodding cadence.

These sorts of pathologies are all but inevitable when millions of signals are detected, and so radio SETI searches often require significant person-power and detective work to sort through the RFI "noise."

In addition to generating false positives, RFI can create false *negatives* in several ways. Perhaps the worst is when the overwhelming number of RFI hits leads SETI researchers to build overeager filters that reject alien signals as well as RFI. Strong

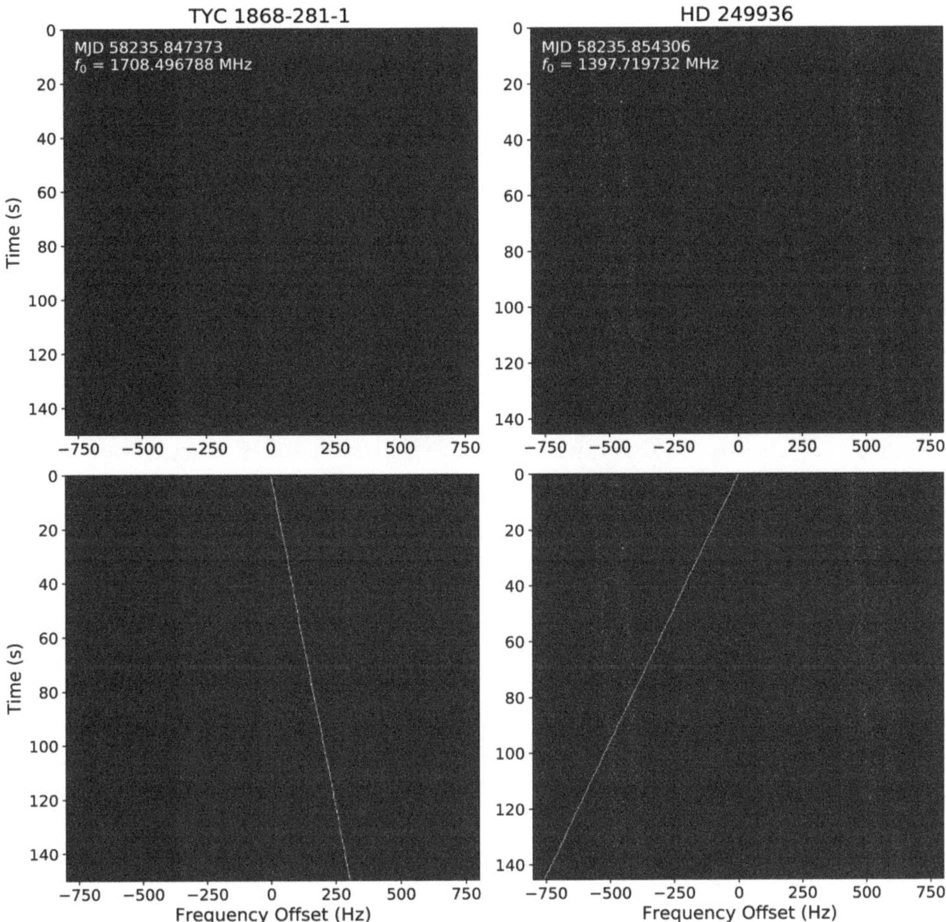

Figure 12.7. Figure from Margot et al. (2021) illustrating an injection-recovery test for real signals on top of RFI. The top two figures show a "clean" region of the radio spectrum (*left*) and one contaminated with RFI (*right*). The bottom figures show the same regions with artificially-injected signals present as a test of the automated detection algorithm's sensitivity to signals in the presence of RFI. When the signals were injected at a level 20 times weaker than shown here, the algorithm correctly identified the signal on the left, but was unable to distinguish the signals on the right.

RFI can also blind us to weak signals at the same frequencies: an algorithm might reject power detected at a given frequency because it is clearly RFI, and so miss a weaker alien signal superimposed on it (see Figure 12.7).

Finally, there is the problem that not all RFI can be easily classified and identified, meaning it is not always possible to distinguish RFI from real alien signals. The classic case of this is the Wow! Signal (Box 3.4), but this occurs regularly in the sense that much RFI can only be identified as such with some finite confidence, meaning only clear alien signals matching a particular expectation can be considered strong candidates.

Mitigating this requires understanding the RFI environment well, which is an ongoing project at radio observatories around the world. Understanding the RFI environment can improve SETI astronomers' confidence that they can reject certain signals as RFI, so they can focus on unusual signals.

But ultimately as with the Wow! Signal, the only true test of a signal of ambiguous origin will be verification observations from another site or at another time.

12.5.5 blc1

The most thoroughly studied recent false alarm from RFI is Breakthrough Listen Candidate one (blc1), detected at the Parkes telescope during observations of Proxima, the closest star to Earth (and host to at least one planet).

Figure 12.8 shows the dynamic spectrum, which includes several long stares at Proxima and a few off pointings in which the signal is not seen. The signal is very faint, at the limits of detection, but has many of the hallmarks of a signal from space: it is drifting, and it is not present in the off pointings.

This signal was the first for the Breakthrough Listen team that passed its initial internal checks, and so generated some excitement and the development of the next round of verification to determine if it was RFI. The fact that it was apparently from the direction of Proxima was especially interesting, since the required transmitter power to generate the signal would be quite modest and plausible. Unfortunately, the signal was leaked to the press, generating an enormous amount of interest and distraction for the team while it conducted its analysis (see Section 20.7.1).

Several aspects of the signal were immediately suspicious. One is its transient nature, since it is not present in all of the "on" pointings. Another is that these data did not follow the usual on–off cadence because the data were actually taken for another purpose, and the SETI analysis was being done commensally. As such, the on–off test lacked its usual discriminatory power.

A deep investigation eventually revealed that there were several signals with nearly identical morphologies seen at the same time that were not initially flagged because they also appear in the off pointings, indicating they are certainly RFI.

More damningly, these "lookalike" signals were separated from each other and from blc1 in frequency space by exact integer multiples of 2.000004 MHz. This is a hallmark of a nonlinear interaction among radio signals called *intermodulation* in which weak copies of a signal are generated at harmonic multiples of the sum and difference frequencies of two strong signals, similar to the action of a mixer. Also telling is the suspiciously precise base frequency of 2 MHz, whose value and precision are common among human-made oscillators, but which alien radio engineers have no reason to choose (since Hz is not a natural unit).

Since all of the "lookalike" signals were RFI, and since blc1 was clearly a member of the family, and since its relationship to the other signals was tellingly anthropocentric, blc1 was ultimately rejected as all but certainly coming from a poorly-understood source of RFI.

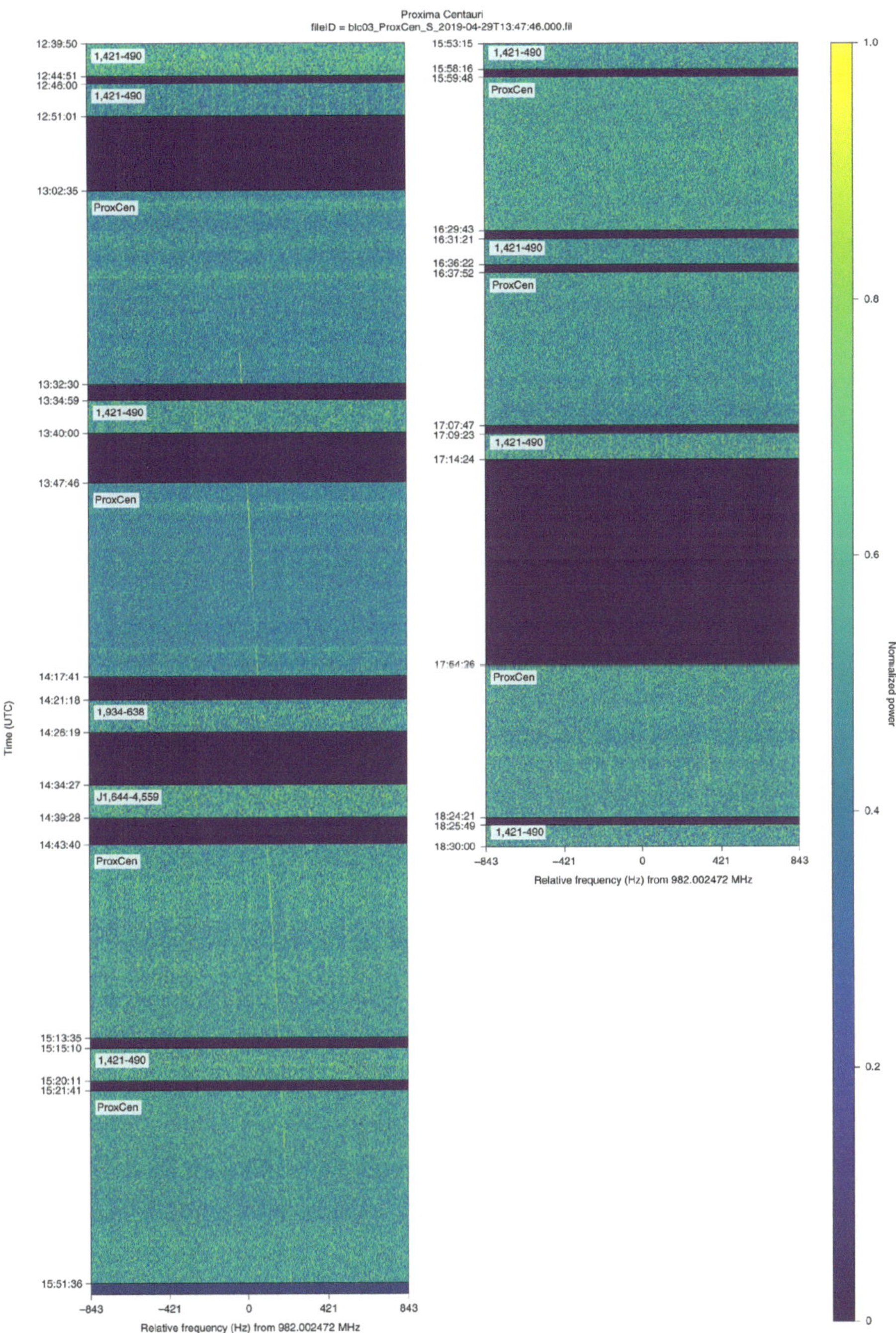

Figure 12.8. Figure from Sheikh et al. (2021) showing the dynamic spectrum of an hours-long stare at Proxima (labeled "ProxCen" for "Proxima Centauri") and several off pointings, including the candidate signal blc1. No data were collected in the blank regions, which are mostly slews of the telescope. The blc1 signal is the thin slanted line of variable slope, most cleanly seen around UTC 13:55 or so. CC BY 4.0.

12.5.6 Lessons of blc1

Despite being a false positive, the blc1 experience was a large step forward for radio SETI. Unlike with the Wow! Signal, there was enough contextual information to rule the candidate out, and there was enough positional information to perform long followup observations on Proxima to check for a repeated signal.

The fact that such a weak signal was found and flagged—in data not even optimized for SETI at that—is proof that the Breakthrough Listen pipeline indeed had the sensitivity necessary to perform such detections, and the algorithms were working as intended. Finally, the experience was an important dress rehearsal for the day that a real signal is detected, both in terms of post-detection protocols with the public and for signal verification.

The importance of blc1 was emphasized by the publication of the false alarm as the cover articles in the highly selective journal *Nature Astronomy* (Smith et al. 2021; Sheikh et al. 2021). The latter article laid out a plan for future signal verification:

1. Rule out instrumental artifacts.
2. Confirm the signal is not in the off-pointings at a low level that might have escaped automated detection.
3. Check other observations for clear RFI at the same frequency.
4. Compare its drift rate to that of human technology such as space probes and Earth orbiting satellites.
5. Compare its drift rate to that expected for signals constant in the barycentric frame, or from transmitters fixed to planets in the target system.
6. Search for similar signals in archival data of the target.
7. Search for similar signals at other frequencies in the same data set.
8. Compare any signals from the previous two steps to see if they might be of the same origin as the candidate signal.
9. Consider if any related signals are clearly RFI.
10. Re-observe the target from the same site and other sites to confirm its persistent nature.

12.6 Some Historic, Current, and Future Searches

Radio SETI is the most mature of the branches of the field, as reflected by the many radio searches that have been performed. Figure 12.9 shows the sky coverage and sensitivity of many such searches, including several that are or will soon be performing an order of magnitude more searching than has even been performed.

Early searches were by modern standards extremely narrow band and had limited sky coverage, although the long-running program at Arecibo covered an impressive area at modest sensitivity.

The privately funded Project Phoenix in the 1990s and 2000s was revolutionary in its broad spectral coverage and large sky coverage, which would not be matched for another 10 years. The Allen Telescope Array had (and continues to have) great promise with its unprecedented coverage from 1 to 9 GHz, although prior to 2020 or

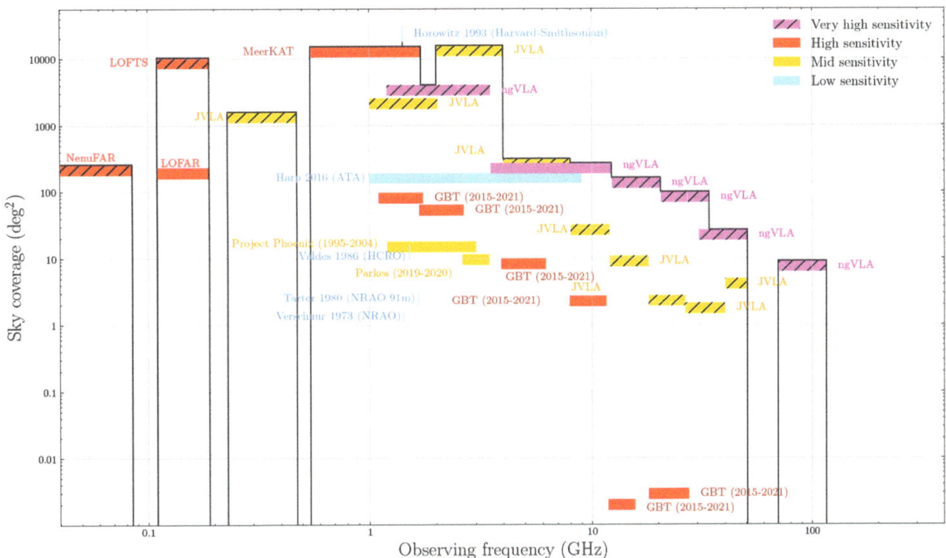

Figure 12.9. Some historical (dates given), many current, and some near-future (hatched) radio SETI programs across the radio spectrum. Color indicates the survey sensitivity in terms of the distance at which the Arecibo interplanetary radar could be detected, with thresholds of 25, 75, and 250 pc dividing the four categories. Note the logarithmic y-axis. "JVLA" is the current VLA, "ngVLA" refers to the planned upgrade to the "next generation" VLA. Figure courtesy Owen Johnson, after Ng et al. (2022) Figure 5 based on code written by Leo Rizk.

so its modest number of small dishes and limited staffing limited its sensitivity and duty cycle (Harp et al. 2016).

The Breakthrough Listen project, which began at Green Bank Observatory in 2015, ushered in a new era of extremely sensitive pointed observations, including some work up past 10 GHz. Not shown is complementary work in the southern hemisphere by the Parkes telescope (Price et al. 2020; Gajjar et al. 2021). Other work at Green Bank includes surveys performed as part of the UCLA SETI course by Jean-Luc Margot (Margot et al. 2018, 2021, 2023).

Today, work by Breakthrough Listen at the VLA on the COSMIC SETI backend (Ng et al. 2022) allows for commensal SETI work at high sensitivities across a very large range of frequencies, by performing beamforming on nearby stars within the primary beam of the telescopes when they are used for nearly any other science. Upgrades to the next generation VLA (ngVLA) will lead to a dramatic increase in sensitivity.

At lower frequencies, the MeerKAT array is already being used by Breakthrough Listen in a manner similar to COSMIC at the VLA to survey the southern sky (Czech et al. 2021). This extremely large interferometer is a precursor to the square kilometer array (SKA), an ambitious telescope proposal that harkens back to the promise of Project Cyclops (Section 3.3.2). At ever lower frequencies, LOFAR and the Murcheson Wide Field Array (MWA) offer extremely large sky coverage and survey speeds, thanks to the lower computational cost at working at lower frequencies (Garrett et al. 2017; Johnson et al. 2023).

12.7 Quantifying Searches

12.7.1 Radio Survey Figures of Merit

A given telescope and instrumentation will take some amount of time to complete a survey for a given sort of technosignature down to a given sensitivity. Intuitively, one would like to define a *survey speed*—i.e., haystack volume (Section 11.3) surveyed per unit of time—that described the haystack volume searched as a linearly increasing function of integration time or telescope time.[20]

Drake et al. (1984) and Enriquez et al. (2017) provide good discussions of some proposed figures of merit for "survey speed" in radio SETI work, and describe the difficulties in constructing one that allows fair comparison among surveys with different strategies and assumptions. Such figures of merit are usually of smaller dimensionality than haystacks because many dimensions are common across surveys, and so add no new information when comparing speeds of surveys. For instance: one rarely considers the sensitivity to polarization of the instrument because modern radio instrumentation is equally sensitive to orthogonal polarizations; one would not consider the repetition rate of the transmitter because searching for lower repetition rates simply requires more telescope time regardless of the instrumentation used.

As a more specific example, Drake et al. (1984) described a commonly used figure of merit for narrowband radio searches:

$$\mathrm{BW}_i \Omega \phi_{\mathrm{min}}^{-\frac{3}{2}} \tag{12.28}$$

where BW_i is the instrument bandwidth; Ω is the fraction (solid angle) of the sky surveyed, and ϕ_{min} is the minimum detectable flux (W m^{-2}) in a single observation. The $-3/2$ dependence arises because the distance d out to which one can detect a signal of a given intrinsic strength scales as $d \propto \phi_{\mathrm{min}}^{-\frac{1}{2}}$, and the volume one probes in a single beam scales as Ωd^3. The "haystack" implied by the Drake figure of merit is thus a literal 3D volume times a frequency dimension, hence 4D. Lesh & Tarter (2015) extend this haystack in two more dimensions, incorporating the polarization and number of stars surveyed as additional parameters.

As pointed out by Enriquez et al. (2017), such haystacks have some curious properties, as a search speed, that seem counter-intuitive (but which we will see later

[20] In practice, this will not be strictly possible, because different parts of a haystack will be searched at different powers of time; also, some survey speeds are calculated per pointing, ignoring the integration time required.

can be virtues.) For instance, it treats all points in space equally and does not give a higher weight for regions (directions) more likely to host signals, such as (for instance) nearby stars. It also considers signal strength in a rough way: within a distance d it gives equal weight to all space within the beam (even though one's sensitivity to very nearby sources is much better, so one's survey speed there should be in some sense higher), and zero weight to the space beyond d (even though one's sensitivity to signals beyond this distance is nonzero). The primary value of the Drake figure of merit, then, is that it offers a scaling relation useful for comparing the speeds of radio surveys hunting for narrowband emission anywhere in the sky down to a certain *transmission* strength.

Tarter (2001) describes the figure of merit proposed by Dreher & Cullers (1997), which is a function of the Equivalent Isotropically Radiated Power (EIRP) of a hypothetical transmitter, given by

$$\ln\left(\frac{\nu_{\rm high}}{\nu_{\rm low}}\right)N_{\rm stars}({\rm EIRP})\eta_{\rm pol}N_{\rm looks} \tag{12.29}$$

where $N_{\rm stars}$ is the number of stars in a given pointing to which one can detect a signal of strength EIRP, $N_{\rm looks}$ accounts for the number of visits to the same sky position, $\eta_{\rm pol}$ is a number between 0 and 1 that describes the survey sensitivity to polarization, and the ν terms describe the top and bottom of the bandwidth surveyed. We note that, in our terms and notation, two surveys surveying the same parts of the sky at relatively narrow bandwidths will compare by this metric roughly as

$$\frac{{\rm BW}_i}{\nu}\Omega\phi_{\rm min}^{-\frac{3}{2}}N_{\rm looks} \tag{12.30}$$

where we have approximated the number of stars surveyed by the volume of space explored out to the limiting distance given by the survey sensitivity. Note that this recasting breaks down for surveys so sensitive or complete that they "run out" of stars in the Galaxy (see Figure 4 of Tarter 2001).

Enriquez et al. (2017; Equation (9) therein) prefer a metric proportional to the inverse square of the minimum detectable flux of an unresolved narrowband signal, which they call the *Survey Speed Figure of Merit*:

$${\rm BW}_i\phi_{\rm min}^{-2} \tag{12.31}$$

appropriate for the Breakthrough Listen survey. The motivation here is that, for observations of fixed channel bandwidth (i.e., the bandwidth of the widest resolvable signal), sensitivity to signals of a given strength scales as the usual $\sqrt{\tau}$ where τ is the observation or integration time, so this metric is linear in time. That is, a project with a figure of merit twice as large will complete the same survey in half the time. Solid angle does not appear in this metric because Breakthrough Listen is a targeted survey performing observations down to a given *instrumental* sensitivity, and so the survey is defined by the number of stars it surveys, independent of their distance, not by a physical volume. To that end, Enriquez et al. (2017) also define the *Continuous Wave Transmitter Figure of Merit* of an entire survey, inversely proportional to

$$\frac{\mathrm{BW}_i}{\nu} N_{\mathrm{stars}} \mathrm{EIRP}_{\mathrm{min}}{}^{-1} \tag{12.32}$$

where ν is the central frequency of the instrument bandpass and $\mathrm{EIRP}_{\mathrm{min}}$ is the EIRP of the weakest transmitter that could be detected. Because it is based on the minimum detectable EIRP and not detectable flux, this figure considers surveys of nearby stars more meritorious than surveys of an equal number of more distant stars (because the former can detect weaker signals).

The figure of merit one chooses thus depends on the haystack one is searching and how one is searching it; because different searches historically have had very different search methodologies—even among narrowband radio searches at the same wavelengths—it can be difficult to compare their relative search speeds. This will also be true with haystacks in general: the size, shape, and dimensionality of the haystack—and one's completeness in searching it—will necessarily depend on the sorts of needles one is searching for.

These figures of merit are often used in a way that assumes that the search is effectively complete across the instrumental bandpass down to the sensitivity given by Equation (12.21). In fact, radio spectrometers have variable sensitivities, and the search algorithms may depend on signal parameters in ways that yield much lower sensitivities than these equations would suggest. Good practice is to estimate one's sensitivity as a function of these parameters, for instance via injection-recovery techniques, and to provide an appropriately averaged summary value.

For instance, Margot et al. (2023) points out that the sensitivity of an algorithm to signals at higher drift rates depends crucially on the time–frequency resolution of the dynamic spectrum (as we discussed in Section 12.4.3). RFI also severely compromises one's detection efficiency and sensitivity in many parts of the spectrum, reducing the effective bandwidth of the observations.

12.7.2 Setting Upper Limits from Radio Surveys

A closely related concept to defining survey speed is defining upper limits set by a completed search. A common strategy is to summarize a survey in just two dimensions, following the Enriquez et al. (2017) continuous wave transmitter figure of merit, in terms of its two components: the *transmitter rate*, given by

$$\left(\frac{\mathrm{BW}_i}{\nu} N_{\mathrm{stars}}\right)^{-1}$$

and the EIRP threshold (often conservatively estimated in terms of the most distant star in the survey, but also perhaps using the mean or median star).

For instance, Price et al. (2020) compiled many surveys including those by Breakthrough Listen as shown in Figure 12.10. The most constraining surveys—the ones that say the most about the lack of detected signals—are the ones that search the most bandwidth across the most targets at the highest sensitivity.

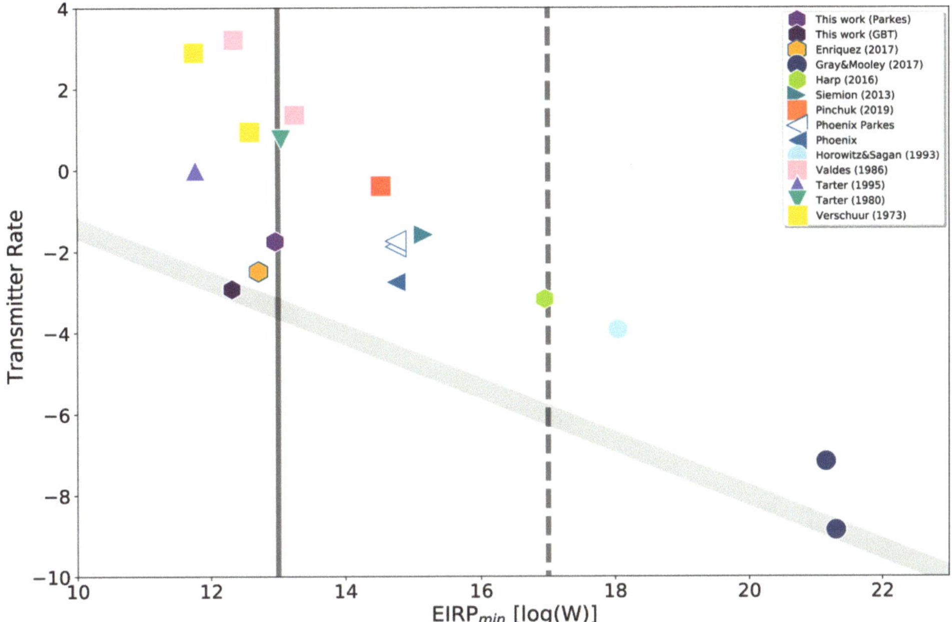

Figure 12.10. Upper limits from Price et al. (2020) for several radio surveys. The y-axis is the base-10 logarithm of the transmitter rate $\left(\frac{\mathrm{BW}_i}{\nu}N_{\mathrm{stars}}\right)^{-1}$, or the number of targets times the fractional bandwidth searched. The x-axis is the EIRP of the weakest transmitter that could be detected in the survey (at some characteristic distance of the targets). The gray line is simply the highest line that can be drawn beneath all of the points, representing where the most constraining surveys are on the plot. The true luminosity function of transmitters pointed at Earth, if it is linear in this space, must be below this line.

The heuristic here is that there is some distribution of transmitters with various strengths and frequencies, which are presumed to be randomly distributed in log(ν) and according to some luminosity function in EIRP, where more powerful transmitters are probably rarer.

After checking N stars, one weights each star by the fractional bandwidth, corresponding roughly to the fraction of the radio spectrum covered. A null result then rules out all luminosity functions that would pass above that point in the survey. In this figure, the Parkes and GBT surveys in the upper left represent pointed searches of nearby stars, which were very sensitive to relatively weak transmitters. In the lower right are searches of the Great Galaxy in Andromeda and its satellite: so a search of billions of stars, but one that is sensitive only to their very brightest transmitters.

Of course, a survey that does not improve upon previous surveys in this space may still have value: it may target places more likely to have life or use strategies that make them more likely to succeed than blind searches.

An extension of this strategy is to consider that the beams of radio telescopes are typically quite wide and contain many more stars than just the target star. They thus

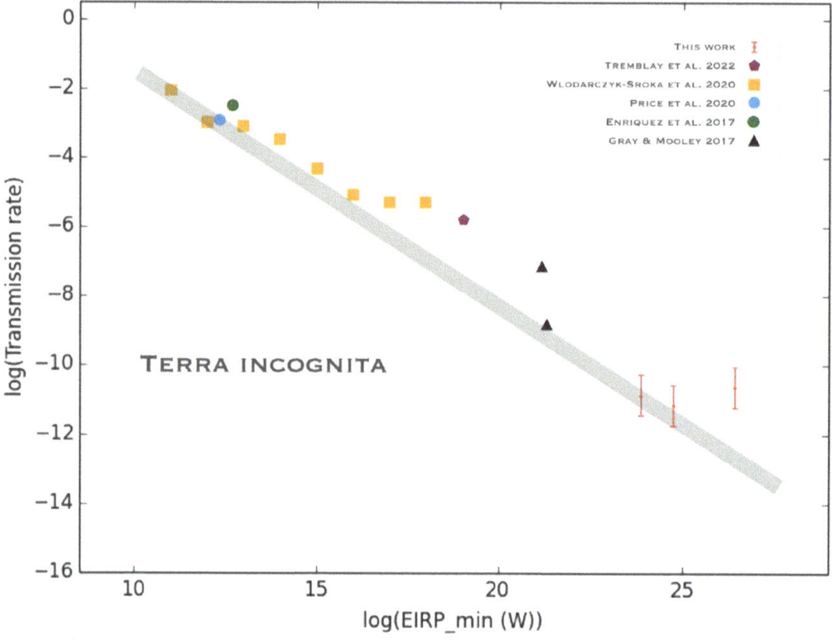

Figure 12.11. From Garrett & Siemion (2023), an extension of the radio transmitter rate upper limit diagram accounting for background galaxies and stars in the Breakthrough Listen survey beams.

perform a sensitive search on a nearby star but also a less sensitive search on a large number of background stars, including an enormous number in background galaxies.

Wlodarczyk-Sroka et al. (2020), Garrett & Siemion (2023), and others have worked to include these background, inadvertent search targets, and their extensions appear in Figure 12.11.

Other reductions in dimensionality are also possible, of course. One can also drop the assumption of continuous transmission and ask what duty cycles one would be sensitive to.

Margot et al. (2023) attempted to account for the effects of RFI and loss of sensitivity at high drift rates by using injection-recovery tests to determine the true sensitivity of various surveys, and to include the duty cycle dimension, producing the upper limit plot seen in Figure 12.12. Here, the y-axis is the 95% confidence interval on the number of stars with detectable transmitters, plotted against how often they transmit to Earth. This metric accounts for stars in the Milky Way in the beam of the Green Bank Telescope search but not background galaxies. "Detectable" here is different for each star, since stars are at different typical distances across and within each survey, and the sensitivity also depends on the searches' bandwidth and detection algorithms.

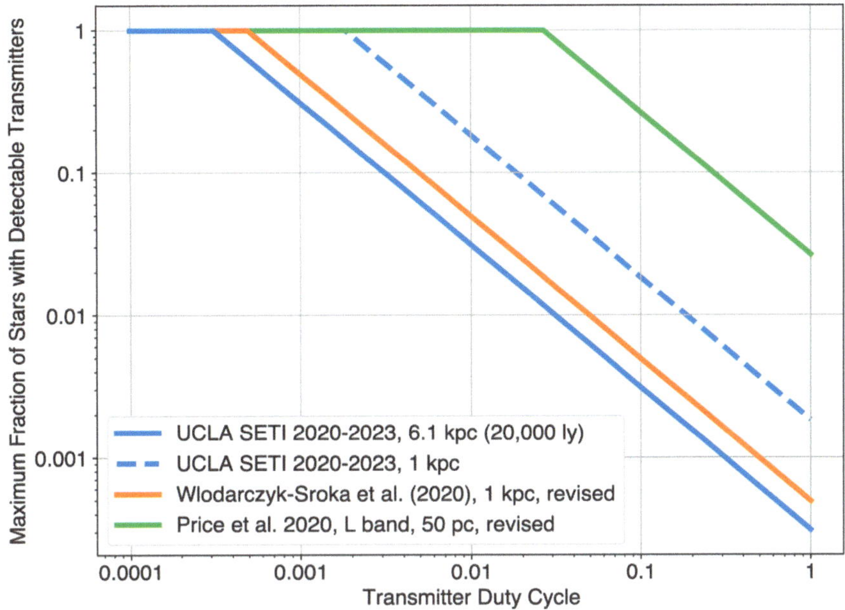

Figure 12.12. Upper limits from Margot et al. (2023) for the fraction of stars emitting signals strong enough for us to detect as a function of those star's duty cycles. Longer total dwell times on stars lead to better limits on their transmission duty cycles.

Box 12.2: Jargon Review: Radio Signal Processing

- **Dynamic spectrum/waterfall plot**: A sequential set of spectra stacked vertically in time showing the evolution of signals as a function of time and frequency.
- **Fourier transform (FT)**: A mathematical transformation from a signal in the time domain to amplitudes and phases in the frequency domain, and a step in the translation of voltages into power spectra.
- **Discrete Fourier transform (DFT)**: A special case of a Fourier transform on discrete measurements of regularly sampled data, appropriate for digital processing.
- **Fast Fourier transform (FFT)**: A common algorithm for efficiently computing a DFT.
- **Sampling rate/bandwidth**: The rate at which voltages are measured. It is equal to the bandwidth, or range of frequencies, over which one can measure a spectrum.
- **Nyquist frequency**: The highest frequency one can measure powers for, equal to half of the sampling rate.
- **Beat frequency**: New frequencies introduced when two signals are multiplied, equal to the sum and difference of the input frequencies.
- **DFT leakage**: The appearance of powers in a spectrum at frequencies far from their true frequency because of the discrete nature of the DFT.
- **Aliasing**: The appearance of powers in a spectrum from frequencies outside the nominal bandwidth of that spectrum because of the finite integration time.

Box 12.3: Jargon Review: Radio Telescopes

- **Primary beam**: The diffraction limited set of directions a radio telescope is sensitive to in the direction it is pointed, with angular width $\sim\lambda/D$. Equivalently, the angular pattern of power it can transmit.
- **Side lobes**: Other directions with nonzero sensitivity, into which unwanted signals can enter.
- **Beamforming**: Combining light from many telescopes in an interferometer to generate a narrow, synthetic beam within their common primary beam.
- **Local oscillator**: A device that produces an extremely precise, accurate, and pure sinusoidal electrical signal at a tunable frequency.
- **Heterodyning**: Shifting an electrical signal detected at the telescope to a lower frequency where it can be efficiently sampled, by mixing it with the signal from a local oscillator.
- **Mixer**: An analog electrical component that generates an output signal that includes the product of two input signals.
- **(Coherent) De-doppler:** A technique that corrects for the drift in frequency of a signal in time. Coherent methods act on voltages before a dynamic spectrum is generated.
- **Polyphase filter bank**: A set of many parallel computers that use particular methods to efficiently heterodyne and process incoming voltages, each on working on a piece of the spectrum to collectively produce a broadband spectrum corrected for DFT leakage and other artifacts.
- **System equivalent flux density (SEFD)**: A measure of the noise present in a radio measurement due to the entire receiving system, expressed as the flux density of a real source that would produce the same amount of noise as the system.
- **System temperature**: The SEFD expressed as the temperature of a blackbody that filled the primary beam with the same flux density.

Exercises

1.

   (a) The $D = 100$ m Green Bank Telescope is extremely sensitive, with an SEFD of around 15 Jy ($= 15 \times 10^{-26}$ W m^{-2} Hz^{-1}) in Ku band (around 14 GHz). There also exist plans to outfit it with a 500 kW Ku band interplanetary radar transmitter.

   If the radar transmitter packed all of its power into a 1 Hz-wide signal that lasted 15 s, at what distance would it be detectable to itself at a signal to noise ratio of 10?

   (b) The Galactic disk is about 150 pc thick, and the stars within it near the Sun have a number density of roughly 0.1 per cubic parsec. How many stars are within range of this transmitter?

   (c) How does the distance out to which we can detect the signal scale with the diameter of the telescopes involved if they are identical? Assume that the system temperature is the same in both telescopes.

   (d) How far could the signal be detected if $D = 300$ m? How far is this compared to the size of the Galaxy?

2. * For those with an astronomy, especially a radio astronomy, background. Check Cocconi and Morrison's math. Notes:
   - The authors play a bit fast and loose with factors like π. You may include them for a more rigorous calculation.
   - The equation at the top of column 1 on p. 845 has a typo. The second term in parentheses should be squared (i.e., it should read $\left(\frac{\lambda}{l_d}\right)^2$).
   - c./s. means Hz, and their τ is the integration time of their receiver. The signal-to-noise ratio of a detection should grow as $\sqrt{\tau}$.
   - F is the total band searched, in steps of the receiver bandwidth, Δf_{d}.
   - They use W to represent a power, the subscript b to represent the background, and the subscript s to represent the sender. They express the background flux in terms of the specific surface brightness or specific intensity $(dW_{\mathrm{b}}/dSd\Omega df)$ (Section 2.2.2).
   - Their version of the radiometer equation is equivalent to our Equation (12.14), except that it is expressed as variations in the total flux, not the flux density, so $\Delta B = S_\nu \delta\nu$ and $B \propto kT_{\mathrm{background}}\delta\nu/A$.

   (a) Solve for f_{min} and reproduce their expression.

   (b) Derive the authors' expression for the power required to send a signal detectable above the Galactic background (you will want to compute the flux density of the signal and the background and set them equal).

   The background received by the detector is proportional to its collecting area and the size of its beam on the sky. The signal from the transmitter received by the detector is proportional only to the detector's area (not its beam size, because the source is effectively a point, and does not "fill the beam").

   Set the detected transmission power in a channel equal to the detected background power in that channel to solve for the necessary transmission power to stand out above the background.

   (c) Confirm that the time one would observe a range of frequencies F by making sequential measurements accros F, each with channel width Δf_{d} at some sensitivity ΔB does not depend on the bandwidth Δf_{d} chosen.

References

Cullers, D. K., & Deans, S. R. 1992, SETI Signal Detection Algorithms, NASA Technical Report, (Moffett Field, CA: SETI Program Office, NASA Ames Research Center)

Czech, D., Isaacson, H., Pearce, L., et al. 2021, PASP, 133, 064502

de Pater, I., & Kurth, W. 2007, Encyclopedia of the Solar System, ed. L. A. A. McFadden, P. R. Weissman, & T. V. Johnson (Amsterdam: Elsevier) 695

Drake, F. D., Wolfe, J., & Seeger, C. L.SETI Science Working Group 1984, SETI Science Working Group Report NASA Technical Paper 2244, *National Aeronautics and Space Administration, Scientific and Technical Information Office*

Dreher, J. W., & Cullers, D. K. 1997, IAU Colloq. 161: Astronomical and Biochemical Origins and the Search for Life in the Universe, ed. C. Batalli Cosmovici, S. Bowyer, & D. Werthimer (Cambridge: Cambridge Univ. Press) 711
Enriquez, J. E., Siemion, A., & Foster, G. 2017, ApJ, 849, 104
Gajjar, V., Perez, K. I., Siemion, A. P. V., et al. 2021, AJ, 162, 33
Garrett, M. A., & Siemion, A. P. V. 2023, MNRAS, 519, 4581
Garrett, M., Siemion, A., & van Cappellen, W. 2017, arXiv:1709.01338
Harp, G. R., Richards, J., Tarter, J. C., et al. 2016, AJ, 152, 181
Hewish, A., Bell, S. J., Pilkington, J. D. H., Scott, P. F., & Collins, R. A. 1968, Natur, 217, 709
Houston, K. M. 2023, AcAau, 212, 505
Isaacson, H., Siemion, A. P. V., Marcy, G. W., et al. 2017, PASP, 129, 054501
Johnson, O. A., Gajjar, V., Keane, E. F., et al. 2023, AJ, 166, 193
Lesh, L., & Tarter, J. C. 2015, AAS Meeting Abstracts #225, 260.02
Li, M. G., Sheikh, S. Z., Gilbertson, C., et al. 2023, AJ, 166, 182
Margot, J.-L., Greenberg, A. H., Pinchuk, P., et al. 2018, AJ, 155, 209
Margot, J.-L., Pinchuk, P., Geil, R., et al. 2021, AJ, 161, 55
Margot, J.-L., Li, M. G., Pinchuk, P., et al. 2023, AJ, 166, 206
Ng, C., Rizk, L., Mannion, C., & Keane, E. F. 2022, AJ, 164, 205
Price, D. C., Enriquez, J. E., Brzycki, B., et al. 2020, AJ, 159, 86
Sagan, C. 1982, Sci, 218, 426
Sheikh, S. Z., Wright, J. T., Siemion, A., & Enriquez, E. 2019, ApJ, 884, 14
Sheikh, S. Z., Smith, S., Price, D. C., et al. 2021, NatAs, 5, 1153
Siemion, A. P. V., Demorest, P., Korpela, E., et al. 2013, ApJ, 767, 94
Smith, S., Price, D. C., Sheikh, S. Z., et al. 2021, NatAs, 5, 1148
Tarter, J. C. 2001, ARA&A, 39, 511
Taylor, J. H. 1974, A&AS, 15, 367
Wlodarczyk-Sroka, B. S., Garrett, M. A., & Siemion, A. P. V. 2020, MNRAS, 498, 5720
Wright, J. T., & Eastman, J. D. 2014, PASP, 126, 838
Wright, J. T. 2019, RNAAS, 3, 6

AAS | IOP Astronomy

The Search for Extraterrestrial Intelligence

Theory and practice

Jason T Wright

Chapter 13

Optical Communication Technosignatures

> There is every reason to believe that Mars and other planets are inhabited...But if intelligent creatures do exist, as we may assume they do, elsewhere in the universe, I should not expect them to try to communicate with the Earth by wireless. Light rays, the direction of which can be controlled much more easily, would more probably be the first method attempted.
>
> —Albert Einstein, 1921, (quoted in Einstein et al. 2009, 430)

13.1 Schwartz and Townes

Shortly after Cocconi & Morrison (1959) argued for radio waves as the optimum communication channel for interstellar communication, Schwartz & Townes (1961) argued that optical[1] light was superior. Just the previous year, "optical masers" (what today we would call lasers) had been invented, opening up a world of possibilities for the manipulation of light.

Schwartz & Townes (1961) imagined that light would be frequency compressed to overcome the noise from the starlight of their host star, and be transmitted at frequencies where stars have strong absorption lines, such as in the Balmer lines. In retrospect this might be a poor choice: these lines are darker by a factor of a few, which provides a modest improvement in signal to noise, but also frequenty show

[1] In this chapter we consider frequencies of light significantly higher than the GHz regime where radio SETI operates. For succinctness I will often refer to this regime as "optical" or "optical/near-infrared" light, but the arguments generalize to nearly any higher frequency. Far- and mid-infrared astronomy is largely impractical from the ground, as is work in the ultraviolet and beyond (Section 2.1). Practically speaking, then, the content of this chapter is largely relevant for optical (and near-infrared) methods.

doi:10.1088/2514-3433/adf5aech13

emission features due to stellar flares and chromospheric activity. A insufficiently-high resolution spectrograph thus might mistake a signal for stellar activity.

Over 20 years later, Townes (1983) would make a strong and detailed case for optical or near-infrared work over radio work in SETI. He there discussed the merits of transmitting signals via radio and optical methods assuming practical constraints on the sizes of optical and radio telescopes. Many of the assumptions in that work today seem somewhat dated, but the part that endures is his identification of two very strong advantages higher wavelengths offered, despite the assumptions that had been made by radio astronomers about the optimality of the Water Hole.

The first major advantage (which is actually far more important) is that the geometric gain associated with optical telescopes is so much higher than radio telescopes. Since gain scales as $(D/\lambda)^2$, and since optical photons are $\sim 10^4$ times smaller than radio photons, one can focus an optical beam millions or billions of times more precisely than a radio beam with the same size primary mirror. Townes (1983) argued it would thus be more energy efficient to send tight optical beams to many stars than to waste energy transmitting omnidirectionally.

The second advantage comes from the observation that many of the early arguments for radio waves assumed equations for noise in a communication channel appropriate for antennae using linear amplification, like radio telescopes, and not for photon-counting devices as used in the optical. As a result, radio astronomers assumed a level of "quantum noise" arising from the discrete nature of photons being amplified by their equipment to be fundamental (see the "quantum limit" line in Figure 5.1, implying all higher frequencies are worse than the terrestrial microwave window). In fact, as we will see below, this is not the case.

13.2 Linear Amplification Noise versus Photon Counting Noise

Here we will take a bit of a deep dive into photon statistics to explain Townes's point about why the physics of radio telescopes lead to the wrong conclusion about the limits of information conveyed by photons at higher frequencies. This will also help us understand encoding schemes and why optical communication can be more efficient than radio in terms of bits communicated per photon (if not per erg).

13.2.1 Modes

Photons are quantum objects, having both wave and particle nature. In some applications it makes sense to think of them as classical particles, and in other cases waves. We can always, though, describe light as exciting certain "modes" of a system like a detector or telescope.

Photons have certain properties as they arrive at our detector: their position on the detector, their time of arrival, their angle of arrival, and their frequency. There is a limit to how well one can distinguish these things in ways that trade off on each other: one cannot distinguish their frequencies very finely *and* their arrival times; one cannot distinguish where they arrived *and* their angle of arrival to arbitrary precision. Roughly speaking, the limits of these tradeoffs (which are all governed by the uncertainty principle) are described by *modes*, which we can think of as the smallest meaningful volumes in this arrival-time/frequency/position/angle space.

Just as we have the Gabor uncertainty principle for time and frequency (Equation (11.10)):

$$\Delta\nu\Delta t = 1$$

we also have for the position and incoming angle of the photons:

$$\Delta A\Delta\Omega = \lambda^2 = \frac{c^2}{\nu^2}$$

where $\Delta\Omega$ is the range of solid angle from which the photons arrive and ΔA is area over which they arrive.

13.2.2 Occupation Numbers

We refer to the number of photons sharing a common set of these properties subject to these limits with n, which we call the *occupation number* of the radiation.

For faint astronomical sources like stars giving off natural radiation from a great distance, this number is usually much less than 1: photons arrive much less frequently than one every period of the oscillation of their electric fields. For blackbody radiation (like the thermal infrared photons inside of an oven) this number is given by

$$n = \frac{1}{e^{h\nu/kT} - 1} \tag{13.1}$$

For very long wavelengths on the Rayleigh–Jeans tail (low frequencies) of full intensity blackbody radiation this is roughly $n = kT/h\nu$, and the occupation number can be much greater than 1.

When $n \gtrsim 1$, the photons are no longer arriving randomly at one's detector: all of the photons arriving in the same mode essentially arrive together. This means the arrivals are not statistically independent, which means normal Poisson statistics do not apply. This might be the case for lasers, or for photons transmitted by an artificial source using quantum communication. This essentially never comes up in optical or infrared astronomy because we do not observe undiluted blackbody radiation that fills our telescope's beam (it would come up if we observed, say, a spot on the Sun). It can come up in radio astronomy, however, when observing extended thermal sources, or in the case of system noise.

13.2.3 Single-mode and Multi-mode Optics

For instance, a "single-mode optical fiber" is an extremely thin fiber optic cable whose core transmits only a single mode of light at most optical frequencies. One transmits energy through the fiber by exciting this mode with an electric field, much like one can excite standing waves on a string. The product of the core area A and its solid angle of acceptance of light Ω is equal to λ^2. n then describes how many photons pass through that aperture from that range of angles per period of the light.

A telescope accepts photons from space over a range of angles across its entire aperture; it thus accepts photons across many modes. We can choose how to organize these modes when we detect the photons, but the nature of a telescope is such that we usually try to make sharp images—at the diffraction limit if possible. In this case, light striking the breadth of the mirror interferes to create a beam.[2] In the diffraction limit, the beam has width $\sim\lambda/D$ where D is the telescope diameter and this is light from a single mode. If the telescope forms an image, the image is in some sense described in terms of "pixels," each representing a separate mode and being as small as the diffraction limit.

A diffraction limited telescope can thus focus light from a star into a single-mode fiber: it couples one mode of light coming from the sky to one mode of light in the fiber. The larger the telescope (the larger A is) the smaller the range of angles on the sky (Ω) it does this for. If the telescope is not diffraction limited its image of the star will be too large for a single mode fiber. In this case one must use a multi-mode fiber, which is wider than a single-mode fiber with the same acceptance solid angle.

13.2.4 Photon Counting

In optical astronomy, we use photon-destroying detectors like CCDs, which register a "count" after a photon deposits its energy in the detector. When counting photons on a detector, we usually assume that the arrival of photons in each mode are independent events, and that they are described by Poisson statistics, but this is wrong in cases where $n \gtrsim 1$. The general expression that accounts for this bunching is that the variance in the number of photons that arrives is given by

$$\sigma^2 = n(n+1)\frac{A\Omega\Delta\nu\Delta t}{\lambda^2} \tag{13.2}$$

(where we've assumed n is constant across all of these modes). In the limit where $n \ll 1$, this reduces to the normal rule that $\sigma \propto \sqrt{N}$. But when the occupation number is high, the variance increases: the arrival of photons is not independent, and so one cannot "beat down the noise" as efficiently. This is rarely the case in optical astronomy, where our sources are faint.

13.2.5 Linear Amplifiers

Linear amplification, as is done at radio frequencies, works differently. Radio detectors need not destroy the photons they detect: they actually measure the electric field of the radiation which they amplify. As a result, multiple detectors can "share" the signal detected by an antenna. A consequence of this amplification process (which is also inherent in any heterodyning process) is additional noise: the variance in such a system is

[2] The term comes from radar where the beam is generally *outgoing* from radiation generated an the focus; for a telescope the light is generally *incoming*, and light from the beam forms a point at the focus.

$$\sigma^2 = (n(n+1)+1)\frac{A\Omega\Delta\nu\Delta t}{\lambda^2} \tag{13.3}$$

where the extra "+1" is significant when $n \ll 1$. In radio work, there is virtually always a source of high-occupation number noise, either from the instrument itself or from a uniform background which fills the beam of the telescope, so $n \gg 1$ and this extra source of noise is not important. At optical wavelengths, it can be very important.

This is the source of the extra "quantum noise" in Figure 5.1 at higher frequencies. As Townes (1983) points out, however, this extra noise can be avoided if one uses photon counters instead of linear amplifiers, as is possible for higher energy photons.

13.3 Efficient Encoding Schemes

13.3.1 Encoding Schemes

We are often interested in conveying digital information, and even when conveying analog information it is useful to think in terms of bits. Perhaps the simplest digital encoding is an on–off scheme in which in a given time interval ΔT a signal either arrives or it does not, corresponding to a 1 or 0 bit. This is called *on–off keying*.

We could also allow the time of the signals' arrival itself be the information we seek. This is called *pulse position modulation* (PPM) and is generally considered optimal for present-day optical interplanetary communications.

Many other encoding schemes are possible, and even superior, but these two will serve as case studies below.

13.3.2 Accommodating Noise: Error Correction

The noise associated with communication leads to the challenge that one must devise a way to correct the errors associated with that noise. For instance, in a photon-counting detector with on–off keying, we will typically expect the actual number of detected photons in a time slot to be $n \pm \sqrt{n}$. If one is very unlucky or the signal is weak, one might by chance detect none and infer the wrong value for the bit. Similarly, if there is a background source of noise in the channel, then one might mistake the noise for signal and make the opposite mistake.

To account for this, one needs to build some redundancy into one's encoding, and there are many encoding schemes to do this. Typically, redundant information is provided with a summary statistic of the information, such as the parity of the bits in the main message. The receiver can then compare their own calculation of this statistic from the main message with the transmitted value to see if an error has occurred.

13.3.3 The Classical Limits of Error Correction

When errors are encountered, there are two solutions: in low-latency two-way communication, the receiver can request that information be re-sent, but this is presumably not practical for interstellar communication. Instead, the transmitter can provide forward error correction, in which all information is multiply redundant, providing not just error *detection* but error *correction.*[3]

A common expression for the maximum information rate R (in bits per time) that can be conveyed in a noisy channel is the Shannon–Hartley limit

$$R = B \log_2\left(\frac{S}{N} + 1\right) \tag{13.4}$$

where B is the bandwidth of the transmission, and S/N is the signal-to-noise ratio.

This makes intuitive sense: the higher the noise, the more redundancy one must build into one's encoding to account for this, so the lower the bit rate. In the limit of zero signal, one can convey zero information. The rate is proportional to the bandwidth, which makes sense because high bandwidth implies faster modulation or, equivalently, more frequency channels through which information can flow.

At high signal-to-noise ratios, one can confidently transmit many bits of information at once, for instance by going beyond the single bit conveyed through on–off keying and making the number of photons that arrives itself a piece of information (i.e., by having four possible transmission strengths per slot, one conveys two bits per slot).

Put more heuristically: the bandwidth B can be thought of as the number of symbols sent per second. At large values, the signal-to-noise ratio is proportional to the number of distinct amplitudes that can be distinguished for each symbol, x. Since it requires $\log_2 x$ bits to distinguish x possible symbols, we have that the total bit rate is $B \log_2(S/N)$. The extra $+1$ in Equation (13.4) ensures positivity at low S/N.

13.4 The Information Content of Photons

13.4.1 Photon Information Efficiency

In the radio, one typically modulates the classical properties of a carrier wave—its amplitude (AM), frequency (FM), or phase—to convey information. The Shannon–Hartely limit applies to such "classical" communication schemes that do not make use of the quantum nature of light. An example of such a manipulation is the preparation of a Fock state in which n photons all occupy the same mode and so will be detected as one. Such a state is ideal for a PPM scheme of communication, and cannot be described as a classical EM wave.

The amount of information conveyed per photon (i.e., per erg) is an interesting qunatity when considering the efficency of communication. For instance, in on–off keying, if one expects a signal containing either 0 or n photons to arrive every nanosecond, then one could expect in one second to receive $\sim 5n \times 10^8$ photons corresponding to $1/\Delta T = 10^9$ bits of information: this is a gigabit (Gb) channel

[3] There is an old saying along the lines of "Never take two chronometers to sea—take one or three." A variation is "A person with a watch knows what time it is. A person with two watches is never sure." The lesson is that singly redundant information tells you only when a piece of information might be wrong; you need multiple redundancies to correct the error.

with, on average, a *photon information efficiency* (PIE) of $b = 2/n$ bits per photon in the absence of noise.

In the case of our example above of PPM, we have 10^9 possible arrival times per second, so the arrival time one measures corresponds to $\log_2 10^9 \approx 30$ bits of information. If one uses n photons to send this information in one second, then one has then transmitted with a PIE of $b = 30/n$ bits per photon, which is more efficient than on–off keying on a per-photon basis in the absence of noise.

13.4.2 Classical Limits

Adding noise understandably lowers the information content on a per-photon basis, since one must use some photons for redundancy. For classical detection of light, as in the radio via linear amplification, we can express the Shannon–Hartley limit in terms of the number of photons conveyed as

$$R = B \log_2\left(\frac{n_s}{1 + n_n} + 1\right) \tag{13.5}$$

where n_s is the number of signal photons per mode arriving in time ΔT, B is the bandwidth, n_n the number of noise photons, and the "+1" in the denominator comes from the linear amplification process[4] (Banaszek et al. 2020). In the limit where $n_n \gg 1$, we arrive at the usual form of this limit in terms of the signal-to-noise ratio.

In interstellar communication where one is photon starved, $n_s \ll 1$, and in the corresponding limit with low noise ($n_n \ll 1$) we have[5] $R = Bn_s \log_2 e$. Since the number of photons one receives per unit time is Bn_s, we have that the maximum information light can carry when noise is minimized is PIE $= R/Bn_s = \log_2 e \approx 2.88$ bits per photon. This is much less than the $30/n$ in our example above because of the noise inherent to linear amplification and the shot noise associated with light in a purely classical state.

13.4.3 Photon-counting Limits

Interestingly, this limit can be overcome by exploiting the non-classical features of light, for instance by using photon counting detectors detecting light prepared in a non-classical state.

A pulse position modulation scheme that counts the arrival of individual photons, for instance, yields a data rate superior to the Shannon–Hartley limit in the low-noise case (Hippke 2021; Banaszek et al. 2020):

$$R = Bn\left(\log_2\frac{1}{n} - \log_2 \ln\frac{1}{n}\right) \tag{13.6}$$

[4] A single mode in integration time ΔT has width $\Delta\nu = 1/\Delta T$. The number of modes N in bandwidth B is thus $B\Delta T$, and the rate photons arrive $r = Bn_s$.

[5] I've rewritten $\log_2(1 + x)$ as $\ln(1 + x)/\ln(2) = \log_2(e)\ln(1 + x)$ then taken the Maclaurin approximation $\ln(1 + x) \approx x$.

The simplicity and efficiency of PPM, the fact that it is used in our own optical communications, and the fact that it can be used in photon-counting systems to surpass the Shannon–Hartley limit suggests that something like it that is robust to noise might be used for interstellar communication. This motivates the search for *pulsed* optical transmissions instead of the continuous wave transmissions envisioned by Townes (1983).

13.4.4 Quantum Communication

There is an emerging field of *quantum communication* which seeks to improve the efficiency, robustness, and security of communications by using non-classical states of light in sophisticated ways, including the PPM scheme above. Because the field is still young, it is unclear what the implications are for SETI, or what likely modulation and compression we might expect for interstellar communication by other species.

In the absence of noise, the ultimate limit on communication efficiency is given by the Holevo limit, which in the photon-starved case $n \ll 1$ is given by:

$$R = Bn\left(\log_2\frac{1}{n} + \log_2 e\right). \tag{13.7}$$

This is higher than the limit for a PPM scheme by 3–6 photons per bit. Interestingly, in these cases the PIE $= R/Bn$ grows without bound as n shrinks. This suggests that for photon-starved communication, as with deep space communication, there is a potentially great advantage on a per-photon basis to using photon-counting detectors and encoding schemes like PPM that rely on the quantum nature of light.

These considerations motivate searches for light that has been prepared in ways consistent with quantum communication (Hippke 2021). One possibility is to carefully examine the statistics of the arrival times of photons, which requires very fast detectors that can record individual photons. For incoherent light, as from most natural sources, these arrival times follow a characteristic pattern, such that photons tend to arrive "bunched up." Coherent laser light will show essentially random arrival times, and certain non-classical states characteristic of quantum communications may show "anti-bunching" (a steady arrival rate). Other tests for the coherence of light are also possible (Benton 2019).

We should perhaps not put too much stock in the above considerations, however. If communicative optical signals exist they may not optimize bits per photon, or bits per erg, or even detectability. Any transmissions from ETIs likely balance a variety of practical and technical matters, and so might employ a signal as simple as classical FM radio, despite its theoretical inefficiency.

13.5 Time versus Frequency Compression

Classically, and purely in terms of standing out against background noise, a communication scheme can be equally efficient when employing pure time compression/modulation with tightly pulsed broadband light, or pure frequency compression with an always-on signal (as with AM or FM radio). However, practical considerations with respect to the generation of signals, sources of noise, and interstellar transmission of light argue for working with light pulses at optical and near-infrared frequencies, and with monochromatic carrier waves in the radio and microwave.

Any transmission through interstellar space must contend with the interstellar medium (ISM). Dispersion by the ISM will broaden broadband pulses temporally which makes pulsed radio communication impractical. For instance, consider radio transmissions that wish to transmit pulses to communicate with $\Delta T \approx 1$ ns. By the uncertainty principle, such pulses much have a broad bandwidth of around 1/1 ns = 1 GHz. But variations in the refractive index of the ISM with frequency mean the different parts of these pulses will arrive at different times, with variations of order *milliseconds*, or 1000 times larger than the timing resolution needed to decode the message (Section 2.2.4). Such pulses can be de-dispersed and recovered (indeed, this is done for pulsars and fast radio bursts to recover their true time profiles), but doing so requires taking data at high time and frequency resolution, defeating the purpose of generating a purely time-compressed signal in the first place.

In the optical or infrared the situation is much better, because the frequencies are higher. Since the index of refraction of the ISM scales here as $1 - n \propto \nu^{-2}$, the effects on arrival time are smaller by at least a factor of 10^8 than in the radio. In addition, a 1 ns pulse of $\lambda = 1\mu$ near-infrared light has a minimum bandwidth of 1 GHz, but since this light has frequency $c\lambda = 3 \times 10^5$ GHz this is only a broadening of 3 parts per million, so the difference in arrival times will be much smaller than in the radio. Pulsed optical emissions are thus quite practical up to very short pulses (Hippke 2018).

For compression in frequency, however, the situation is reversed for more practical reasons.

Optical spectroscopy at lower resolutions is generally done with gratings and prisms, and these have hard limits on their resolving powers set by their sizes. A good rule of thumb is that a grating or prism has a minimum size $R\lambda$ where R is the resolving power $R = \lambda/\Delta\lambda$.[6] For optical light, this means for $R \sim 10^6$ one needs a grating at least ~10 cm large. An $R = 10^{10}$ optical spectrograph would need elements ~1 km across! This is why techniques that perform extremely high resolution spectroscopy (frequency resolution) generally involve heterodyning and linear amplification (Figure 12.2), instead.

[6] This is not a strict, fundamental limit. Immersing a grating within a prism can combine the power of both, in which case the limiting size is $R\lambda/n$ where n is the index of refraction of the medium, having typical values of 1.1–4 for glasses but which in principle could be higher. It might also be possible to multiplex many narrowband spectrographs (which have minimum size RB where B is their bandwidth in units of wavelength) to achieve a compact, broadband, high-resolution spectrograph. Other solutions include a virtual image phased array (VIPA) which can exceed the $R\lambda$ limit by perhaps a factor of 100. Nonetheless, it seems unlikely that these techniques would be able to generate many order-of-magnitude improvements over the $R\lambda$ limit that would permit the construction of a compact $R \sim 10^9$ broadband spectrograph. The problem may be worth investigating rigorously, however.

Precise timing, by contrast, puts no such hard constraints on detector size, and so very high temporal compression can be searched for with a compact instrument, today up to ns time resolution but in principle much faster.

We thus might expect radio transmissions to be primarily frequency-compressed to avoid problems with the ISM, and optical transmissions to be time-compressed because it makes detection easier.

Note that the frequency of optical/infrared light is high enough that one does not necessarily have to choose between frequency and time compression: even if one compresses, say, 3μ light in frequency at $R \sim 10^6$, one can still additionally compress it in time to a resolution of 10 ns and obey the Gabor limit at $\Delta t \Delta \nu = \Delta t c \Delta \lambda / \lambda^2 \geqslant 1$. It is thus possible that a pulsed signal could also be narrowband, and so detected in either the time or frequency domains.

13.6 Continuous-wave (Frequency Compressed) Searches

13.6.1 Principles of Optical Spectroscopy

We covered heterodynining in Chapter 12. Such techniques are also possible in the infrared and even the optical using lasers as local oscillators, but they have very narrow relative bandwidths (because the range $\Delta \nu$ is the same as in the radio but ν is much higher at infrared wavelengths).

There is another technique, Fourier Transform Spectroscopy, that is generally used on bright, constant sources. In this technique light from the source is sent through an interferometer whose path length is slowly changed over a large throw ΔL (of order meters) at a steady rate. The resulting fringe pattern's Fourier transform is then the spectrum of the light, measured at resolution $\Delta L / \lambda$, which can be of order 10^6 or even higher. Such spectroscopy has proven very useful for the Sun but is usually not practical for stars, especially faint ones.

Most astronomical spectroscopy at optical and infrared wavelengths is usually accomplished with either a prism or a grating. A prism is a purely transmissive element with a high index of refraction and angled surfaces. Prisms provide relatively low dispersion, and so are good for low resolution spectroscopy. Prisms are generally made of some sort of glass, and so can be very large and heavy when the beam size is large, as with a large telescope.

Another option is to use a *diffraction grating*, which can be transmissive or reflective. A grating has many evenly spaced grooves or ridges, and light reflecting or diffracting off of these grooves will constructively or destructively interfere according to the angle of its reflection and its wavelength, resulting in it being dispersed by wavelength. Light constructively interferes when light scattered from neighboring grooves has a path length difference of $m\lambda$, where m is an integer *order*. This order is higher when the outgoing light is at a steep angle of reflection. The ultimate resolution of such a grating scales with the total number of grooves across the grating times the order of the interference.

The highest resolution optical and infrared spectrographs are generally accomplished with *echelle gratings*, whose grooves are angled such that light striking at a high angle of incidence is reflected back nearly in the direction it came. This allows the light to interfere at very high order (often $m \approx 80$–120) and very high resolutions to be achieved.

Because light from many orders in an echelle spectrograph emerges across a wide range of wavelengths, it must be dispersed again in an orthogonal direction to separate the orders. The result is an echelleogram, in which each of the orders, spanning a different range of wavelengths running in one dispersion direction, is stacked at a different place in an orthogonal dispersion direction.

This secondary dispersion is called cross-dispersion and is done at lower resolution, often by a large prism. An example of an echellogram is shown in Figure 13.1.

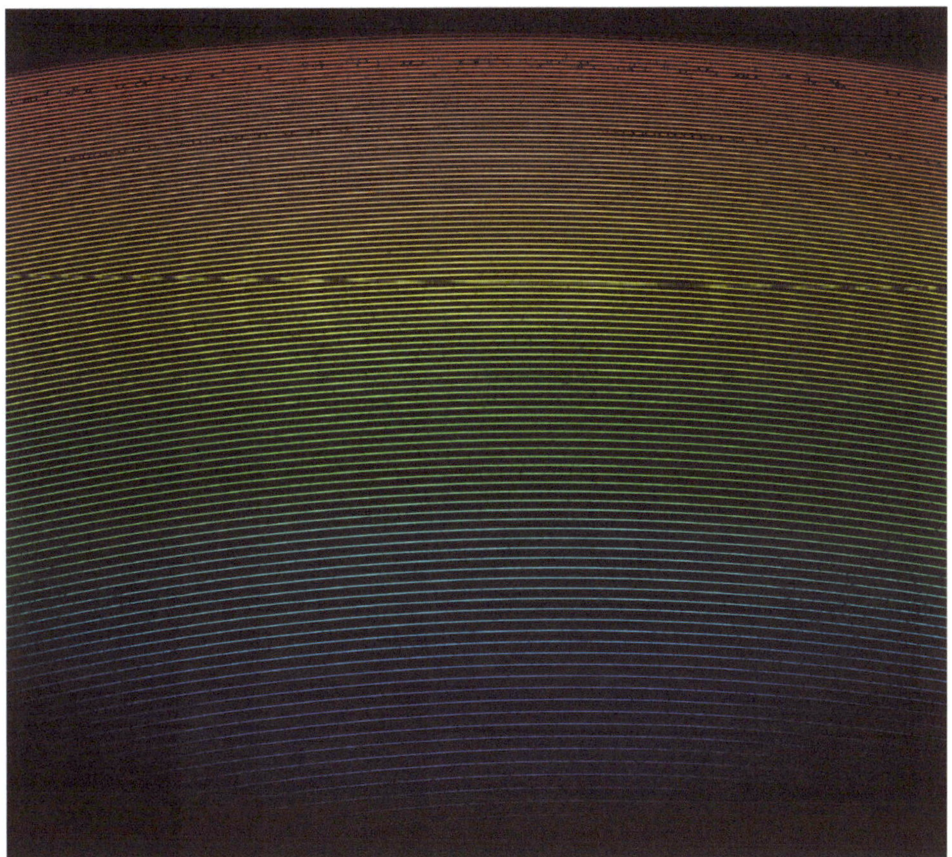

Figure 13.1. An echellogram from the NEID spectrograph of the star 51 Pegasi. The horizontal arcs or stripes are different orders of the spectrum. The spectrum can be read like a book, with frequency increasing along each stripe and then continuing on the next stripe down. The horizontal direction is the high dispersion direction, and the vertical direction is the cross-dispersion direction. The colors have been added for illustrative purposes and are roughly accurate. (This image may appear to have overlapping oval patterns—if you see these, they are an artifact of aliasing from pixelation of this high resolution image, and they should disappear if you zoom in.) Credit Guðmundur Kári Stefánsson/Princeton University/Penn State/NOIRLab/NSF/KPNO/AURA. CC BY 4.0.

13.6.2 Methods of Optical Spectroscopy

Optical telescopes generally are capable of imaging a relatively wide region of sky, but spectroscopists generally study only a single star or other source in that region. There are four general ways this is accomplished.

Perhaps the most common method is to employ an aperture in the image plane that allows light from only a single, small region of the sky to enter the spectrograph. This is usually done using a *slit*[7] or rectangular aperture, but recently has been done commonly with optical fibers, which generally have circular cross sections.

Figure 13.2 illustrates how the geometries of slits and fibers interact with the image and position of a star and a nearby laser.

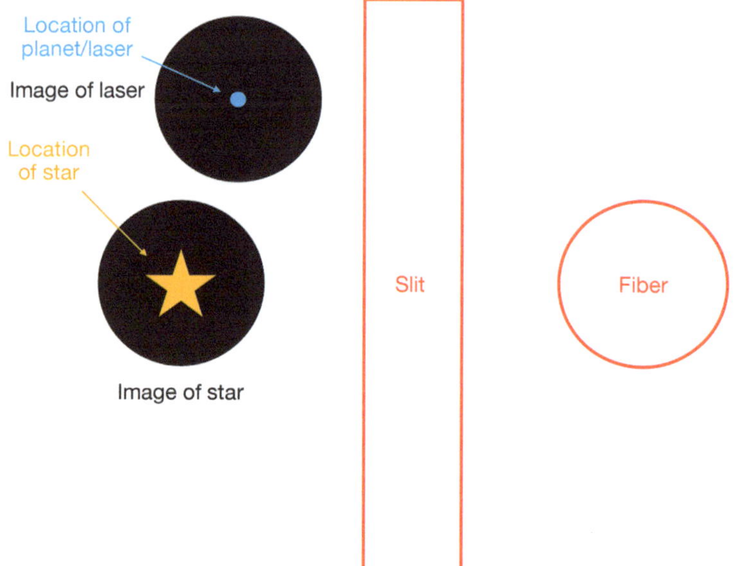

Figure 13.2. Schematic showing how slit and fiber geometries interact with images of stars and potential lasers near them. *Left:* The black circles represent the apparent angular size of a star and laser source due to the telescope's angular resolution and/or seeing, as seen from the Earth. The actual position of the star and laser are indicated, and their actual angular sizes are extremely small. *Right:* Slit widths and fiber diameters, shown for scale, are typically chosen to be similar to the seeing as a compromise between spectral resolution (favoring narrow slits) and efficiency (favoring slits wide enough to capture all of the starlight). For typical slit sizes matched to ground-based seeing (of order $1''$), the separation implied by this figure between the planet/laser and the star is quite large, corresponding to perhaps 3 AU around the nearest stars, and much more than that for more typical search targets.

[7] Although we often disperse light with diffraction gratings, and narrow slits can cause diffraction, spectrographs do not use slits to disperse light. Slits are typically many wavelengths wide and do not cause significant diffraction—the dispersion is done by the grating or prism, not the slit.

13.6.2.1 Slit-Based Spectroscopy

The camera of a spectrograph images the slit and disperses that image in one dimension. Figure 13.3 shows this schematically: the slit selects out part of the images of the laser and star, then the spectrograph images them and disperses them by wavelength in the horizontal direction. Since the laser is monochromatic, it appears only at a single place and the edge of the slit is visible. The starlight represents all wavelengths, and so one can think of the resulting rainbow spectrum as a series of overlapping images of the slit at different wavelengths. Wider slits thus result in more blurring in the spectral dimension. Information in the vertical dimension (corresponding to position on the sky) is preserved in a slit-based spectrograph.

An important way in which this figure is merely schematic is that an actual spectrograph's optics and pixels have finite resolution strike only a few times better than the width of the slit, while here the laser image is shown perfectly resolved. Another is that the apparent angular extent of a source on the sky is given not by a disk but by a Gaussian or Airy or similar function, resulting in no sharp edges in the spatial (vertical) dimension in the actual spectrum.

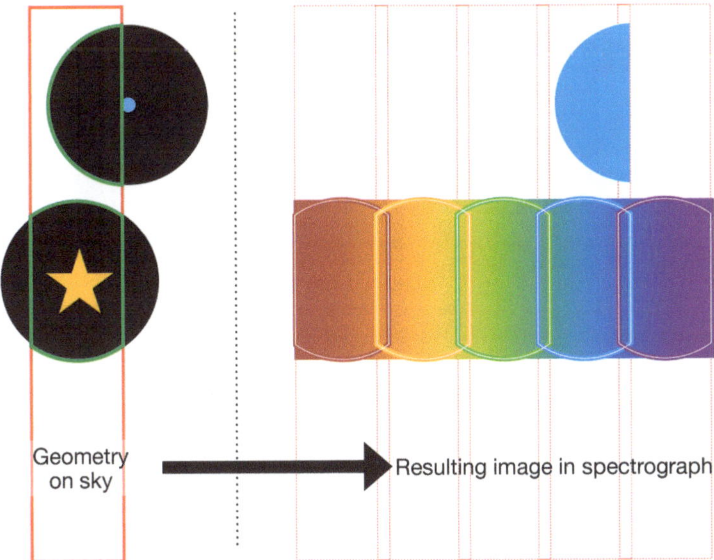

Figure 13.3. Schematic showing how a spectrograph treats light entering a slit when it is placed on the star from Figure 13.2. *Left:* the geometry of a slit aperture, the image of a star, and the image of a laser. Only light falling within the outline of the slit (green outlines) will arrive at the spectrograph. *Right:* How light entering the slit is arranged on an echellogram. One can think of the spectrum generated by the spectrograph as an infinite series of overlapping images of the slit at different wavelengths: some of these are shown as overlapping thin dashed red outlines. Where starlight fills the slit, it is dispersed into a linear spectrum by wavelength. Because the laser is monochromatic, its image on the slit is reproduced at just a single wavelength. Note that even though the laser is actually just outside the slit (at the center of the blue circle), atmospheric seeing nonetheless scatters about half of its light into the slit. If it is separated from the star by sufficient angular distance in the vertical direction, its light will be distinct from the star; otherwise it will be blended and might appear as a stellar emission line.

13.6.2.2 Fiber-based Spetroscopy
Optical fibers are today commonly used in high resolution spectroscopy to transmit light from the focal plane of the telescope to the entrance of the spectrograph. Optical fibers generally scramble the light that they transmit (this is in fact a reason they are often used) so they do not transmit spatial information on the sky. This means that if two sources are within the diameter of a fiber, their light will be combined. It also means the shape of the resulting image of a laser line on the echellogram will simply be that of the illuminated fiber tip.

13.6.2.3 Slitless Spectroscopy
Another option is to disperse light without the aid of the slit, for instance by introducing a prism or grating into the optical path of the light being imaged with an ordinary camera.

This has the result of turning every point of light into a small, low-resolution spectrum. A disadvantage to this method is that for crowded fields these spectra can overlap, and that it cannot be easily extended to high resolutions except over very narrow bandwidths (because the spectra would become very long). Under slitless spectroscopy, a laser would appear as a dot because it is monochromatic, whereas stars will appear as linear features.

13.6.2.4 Integral Field Units
Another option is to image the field onto a lenslet array, send the light that enters each lenslet to a dispersion element, and then reimage each lens's dispersed output onto a separate region of a detector. This allows one to then reconstruct a full image on a pixel-by-pixel basis at any wavelength one chooses. This is more complex, but avoids the problem of crowding and overlapping spectra from slitless spectroscopy.

This technique is used, for instance, on planet-imaging coronagraphs as a way to extract a low-resolution spectrum from an imaged planet. This generally can only be done at low spectral resolutions because of the very large number of pixels that would be needed to record light from a large number of distinct wavelengths across all of the pixels on an image.

13.6.3 Prior Searches

13.6.3.1 Betz and Townes
The first actual searches for optical or infrared communicative transmissions appear to be from Albert Betz, who developed (Betz 1986) and with Charlie Townes executed (Betz 1993) a search for 10μ CO_2 laser light using heterodyning spetroscopy. They planned to survey around 300 stars over a few years, with bandwidth sufficient to account for Doppler shifts of around 50 km s^{-1}. They planned to use a pair of truck-mounted 1.7 m telescopes ordinarily employed as an interferometer. It is unclear how much of the survey they completed, if any; they did not publish any results.

13.6.3.2 Use of Planet Search Spectra
The discovery of the planet 51 Pegasi *b* in 1995 via precise radial velocities started a revolution in astrophysics. The technique, which dominated exoplanet discovery for

a decade and remains an essential component of exoplanet characterization, requires taking many measurements over time of a star's radial velocity, deduced from the Doppler shift of the spectral features of its host star. Typical Doppler shifts induced by planets are generally less than 30 m s^{-1}, corresponding to shifts of less than one part in 10^7 (Section 2.5).

These measurements are generally made with high resolution optical and near-infrared spectrographs with resolving powers of 10^5—enough to cleanly resolve the stellar absorption features without dispersing the light too much, which increases noise and the number of pixels needed to make a measurement.

The upshot is that since 1995 many observatories around the world have spent significant fractions of their time repeatedly taking high resolution spectra of nearby stars—almost exactly the strategy one would employ in a large laser SETI search. This is thus an excellent opportunity for commensal SETI.

The first such search was by Reines & Marcy (2002), who searched for laser lines in in roughly 10,000 spectra taken as part of a planet search of over 500 stars using the slit-based HIRES instrument at Keck. Subsequent searches with similar methodologies include Tellis & Marcy (2015), who searched the regions of a tall slit far from the target star; Tellis & Marcy (2017) who searched at lower sensitivity for lasers from sources near or coincident with the target star; and Zuckerman et al. (2023) who applied a similar technique using the Automated Planet Finder Telescope at Lick Observatory.

13.6.4 Identification of Laser Lines

At high spectral resolutions, the most sensitive way to search for laser lines in spectra makes use of the two-dimensional data that preserve the spectral and spatial information from the slit or fiber.

Figure 13.4 shows a candidate laser signal from the search for lasers in exoplanet search data by Tellis & Marcy (2017). This search was for lasers spatially near or coincident to the target stars. They first searched the stellar spectra for unusual emission lines, then explored the two-dimensional data to see if the extra emission was consistent with the seeing at the telescope. In this case, the extra signal was clearly from a high energy particle event.[8]

Other common sources of false positives are stellar flares and emission in Earth's atmosphere, both of which cause emission lines at known wavelengths (so are easily identified and ignored). Other sources of spurious signals can include scattered light in the spectrograph, and even scattered light in the atmosphere from terrestrial lasers, for instance used for satellite-to-Earth optical communication or LIDAR measurements.

To rule out laser lines from Earth's atmosphere or other nearby source, one can look at the profile of the light in the spatial/cross-dispersion direction. It should follow the pattern of seeing for the star shared by neighboring wavelengths. If the light fills the slit (appearing as a tall rectangle and not isolated to just the stellar trail) it cannot be from a distant source.

[8] Astronomical detectors are exquisitely sensitive to high energy particles, and so it is very common in any observation to find a few or many pixels contaminated by them. Their typical source is weakly radioactive material surrounding the detector, such as uranium in the concrete of the walls of the spectrograph room. Astronomers colloquially call these in the spectra "cosmic rays" even though that term strictly refers to high energy particles from space, and these particles are generally local.

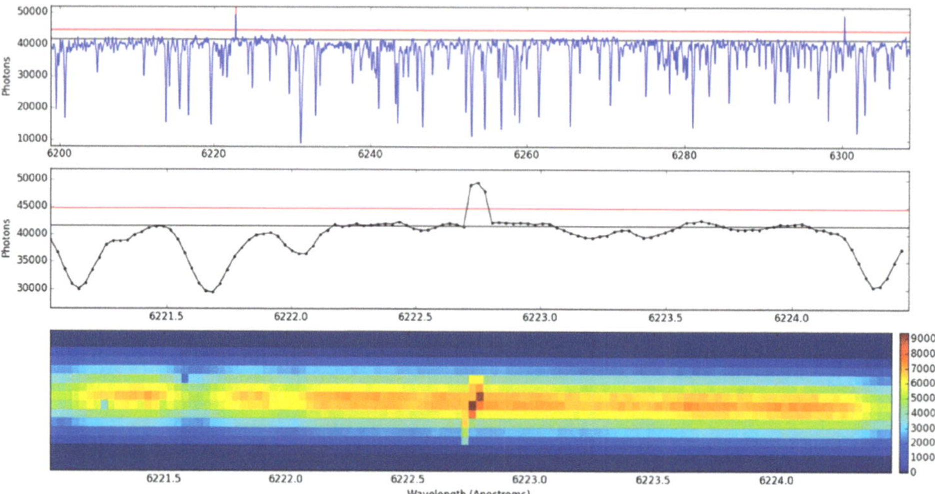

Figure 13.4. Figure 4 from Tellis & Marcy (2017) showing the diagnostics used to discount a candidate laser signal from HD 5319. *Top:* a small portion of the spectrum of the star in blue, with its many absorption features. The black and red lines indicate various thresholds used to detect unusual emission. *Middle:* a zoom-in on one of the emission features seen in the spectrum. *Bottom:* the region of the two-dimensional data that gave rise to the middle plot. Here the vertical dimension is along the long side of the slit, and preserves the spatial information of the star's extent on the sky, as in Figure 13.3. The emission feature does not share the same spatial structure as the image of the star, as would be expected if it had passed through the atmosphere. Its irregularity is characteristic of a high energy particle striking the CCD, a common occurrence.

For fiber-based searches, it is important to have a second fiber set far from the star that simultaneously captures light from the night sky. This is often, but not universally, done, especially for faint sources, to correct for contamination from atmospheric emission. Without such a simultaneous measurement of the sky emission, it is challenging or impossible to rule out a local cause for an observed emission line.

13.7 Pulsed (Time-compressed) Searches

13.7.1 Past Searches

An early suggestion to compress in time instead of frequency was made by Ross (1965), but the first pulsed laser searches would not occur until the 1990s. Possibly the first pulsed optical effort was that of Stuart Kingsley, who built a the Optical SETI Observatory in Columbus, Ohio in 1990 (see Ross & Kingsley 2011).

Shvartsman et al. (1997) and Beskin et al. (1997) developed the MANIA complex in Russia to search for monochromatic laser pulses from nearby sunlike stars with timescales as short as 100 ns, and later extended the search to the southern hemisphere (Beskin et al., 1998). An early American effort was that of the SEVENDIP project at the Leuchner 30 in telescope in California, sensitive to ns pulses (Werthimer et al. 2001).

The most thoroughly developed and extensive optical SETI program is probably that of Paul Horowitz and his students using Oak Ridge and Princeton observatories, a project which developed multiple degrees in the field (Box 13.1). The system developed by Andrew Howard et al. (2004) had 5 ns time resolution and was sensitive to fluxes of 100 photons m^{-2}. They made nearly 16,000 observations of nearly 6000 stars using the 1.6 m telescope at Oak Ridge, which was later supplemented with the 0.9 m telescope at Princeton for better coincidence rejection of false alarms.

Box 13.1: Paul Horowitz (1942–)

Photo by the subject in his natural habitat, Logitech webcam.

Prof. Horowitz is an American physicist and electrical engineer, and a pioneer in laser SETI.

Horowitz, who was an amateur radio operator at age 8, was taught electrical engineering by his older brother. He has spent his entire career at Harvard, where he earned undergraduate and graduate degrees in physics and served as a professor of Physics and Electrical Engineering. He is the co-author of the classic textbook *The Art of Electronics.*

Horowitz was inspired to work in SETI as an undergraduate by a professor of his, Ed Purcell, and his paper/lecture on radio SETI (Purcell 1960). He was further inspired in graduate school when his roommate took a course by Carl Sagan, and later by a lecture by Frank Drake.

As a faculty member, for his sabbatical he asked Drake if he could "look for life in Puerto Rico," and in 1977 he spent three or four months looking for extremely narrowband radio signals at Arecibo. In 1981 Jill Tarter asked him to pick the work up again as a fellow at NASA Ames and Stanford. There, with colleagues he built "Suitcase SETI," a 128,000 channel spectral analyzer that fit in three suitcase-sized boxes. These formed the basis for the META (8 million channel) and BETA (billion channel) projects at Oak Ridge Observatory, and laid the foundations for many radio SETI searches since then.

Horowitz championed graduate education in SETI. Working with a series of PhD students, Horowitz continued his radio work and pioneered pulsed laser SETI at Oak Ridge Observatory, supervising the first PhD thesis in the field (Darren Leigh, 1988), three of the first six theses, and four of the first nine (see Table 1.1). Horowitz is also a co-I of the PanoSETI project with Shelley Wright of UC San Diego.

Horowitz won the Drake award from the SETI Institute in 2021 for his contributions to radio and laser SETI. He also serves as a member of the JASON Defense Advisory Group.

For more on Horowitiz's group's work in SETI, see Leigh & Horowitz (2011).

Howard et al. (2007) later extended the methodology to wide fields, a project extended by Mead (2013, for the eighth PhD thesis in SETI). More recently, the VERITAS telescope array (designed for detections of Cherenkov radiation, see below) has taken up the search.

In parallel to these efforts, work at Lick Observatory led by Remington Stone, Dan Werthimer, Shelley Wright, and others used the Nickel 1 m to perform a targeted search in the optical and near-infrared (NIROSETI Stone et al. 2005; Maire et al. 2019b). This work has led to the planned PANOSETI instrument (Wright et al. 2018; Maire et al. 2022) and its extensive nearly-all-sky optical survey with a narrower infrared component.

13.7.2 False Positives

Unlike with radio SETI, there is no significant source of anthropogenic pulsed light in the night sky that might generate false positives (Howard & Horowitz 2001). The largest such source appears to be LIDAR measurements from Earth-observing satellites (Maire et al. 2022) which are rare and easily identified.

Instead, the dominant sources of signals are due to electronic glitches, and due to Cherenkov radiation (Sections 2.1, 2.6.1, and 2.6.3) from cosmic rays entering the Earth's atmosphere.

As charged particles from space enter the atmosphere at very high velocities, they find themselves in a medium in which their speed exceeds that of the local speed of light which causes them to emit radiation that is strongly beamed in their forward direction. The timescale for these flashes of light when they reach the the ground is 4–6 ns.

This presents a source of noise but also something of an opportunity because detecting these flashes to study cosmic rays is routinely done by projects such as VERITAS. The VERITAS array consists of very large (12 m) optical telescopes.

Because there are so few flashes and their precise location in the sky is not a crucial parameter, the VERITAS array does need to not have very good angular resolution. As a result, the quality of the figure of the mirrors need not be very good, allowing them to be composed of many segments only roughly aligned by astronomical standards.[9]

An example of detections of cosmic ray showers are shown in Figure 13.5. Cherenkov telescopes like VERITAS are in principle sensitive to similar flashes of light from extraterrestrial laser signals, which would appear as a point source with no detectable parallax. Cherenkov showers from cosmic rays show significant parallax even when observed from baselines of just a few tens of meters because they are relatively close to the telescopes (of order 10 km), readily distinguishing them.

[9] Because they observe atmospheric sources of light and do not produce sharp images, Cherenkov telescopes like VERITAS are often overlooked when identifying the world's largest optical telescopes. At 12 m they are larger than the twin Keck telescopes, which for many years were considered to be the largest in the world.

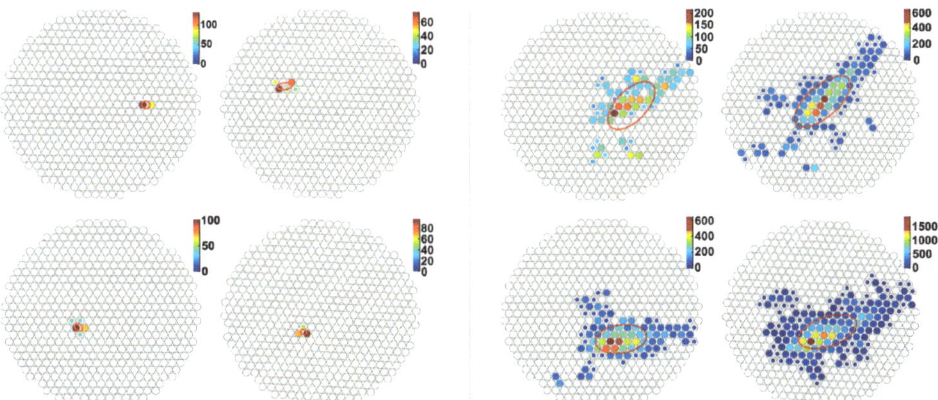

Figure 13.5. *Left:* Cherenkov light from a low-energy cosmic ray as observed from the four VERITAS telescopes. The smaller circles represent the individual "pixels" of the image formed by the telescopes. The signal is seen coming from significantly different parts of the sky because of parallax from the relatively nearby event. *Right*: The Cherenkov light from an "air shower" caused by the many particles caused by a cosmic ray's interaction with a nucleus in the Earth's atmosphere. The many charged particles produce light from a range of positions on the sky. A pulsed laser signal from deep space can be distinguished as a point of light showing no appreciable parallax. Modified version of Figure 2 of Abeysekara et al. (2016). © 2016 the American Astronomical Society, all rights reserved.

13.7.3 Rejecting False Positives and Data Practicalities

The data rates associated with ns resolution on even a single-pixel instrument aimed at a single star is of order GB s^{-1}, meaning that a night's worth of data is around 30 TB. Performing such an experiment across a wide field to distinguish Cherenkov light will quickly generate PB of data—an unmanagable amount.

The standard strategy then is to store data temporarily in a ring buffer, and only save data in the buffer when an unusually large number of photons in a single time slot triggers the system. The threshold above which one triggers then sets the data rate, and determines one's sensitivity.

For single-pixel detectors, one ideally performs coincidence detection to cut down on false positives. The simplest method, employed by Howard et al. (2004), is to use a beamsplitter to send light to two independent photon detectors. Here, one only triggers if *both* detect a threshold signal *simultaneously*. This removes instrumental noise downstream of the beamsplitter as a source of false positives. It leaves one open to upstream sources, however, like Cherenkov radiation.

To distinguish Cherenkov radiation, one ideally employs widely separated telescopes. If the separation is of order 10–100 m, then Cherenkov showers will still be coincidental, but parallax will mean that they appear in different parts of the sky and, if the time relsolution is sufficiently high, at slightly different times because of differential light travel time between the sites.

Howard et al. (2004) went a step further and placed a second site in Princeton, NJ, far from Oak Ridge in Harvard, MA, and where there should be no upstream coincident events. This required absolute timing precision at the two sites to ns precision, and careful consideration of light-travel time differences between the sites. The sensitivity of the combined system is then the lesser of the two sites.

13.8 Upper Limits

A haystack calculation along the lines of Section 11.3 has not yet been performed for laser work, as it has in the radio. Searches for continuous wave (frequency-compressed) signals are also in principle sensitive to pulsed lasers, provided they are also narrowband and their total power during the integration is above the continuous wave detection threshold. Searches for pulsed signals will generally have some range of pulse durations where they are maximally sensitive, requiring a haystack that includes repetition rate and pulse duration as dimensions.

13.8.1 Frequency-Compressed (Spectroscopic) Searches

When computing upper limits set by spectroscopic laser searches it is important to consider the area of sky covered by the search. As Figure 13.2 shows, the sensitivity of the search to a laser strongly depends on the seeing (i.e., the size of the image of the star), the relative separation of the star and source of the laser, and the shape of the slit or fiber.

The orientation of the slit on the sky in this sort of work is usually in the altitude direction, i.e., with the long axis pointed toward the zenith. This is done to ensure that atmospheric diffraction does not send the edges of the spectrum of the star out past the slit. The effect is that the orientation of the slit is somewhat random from observation to observation, depending on the position of the star in the sky.

Whether laser light enters the slit (or fiber) thus depends on the physical distance of the laser to the star, the orientation of it and the star with respect to our line of sight, our distance to the star, the happenstance of the orientation of the slit that observation, and the seeing. How bright the laser appears in the spectrum also depends on how much of its image falls within the slit.

Typical fibers and slit widths are matched to the seeing at a given site and so are of order $1''$. Typical targets of planet searches are stars with distances of 10–100 pc. A planet orbiting within 1 AU of one of these stars will thus generally be $0''.01-0''.1$ from the star, which will generally put it inside the fiber or slit. Even for very nearby stars, a laser mounted on a planet at a few AU will generally be within $1''$ of the slit or fiber, so the blurring of Earth's atmosphere will put at least some of the light into the spectrograph.

This means that for most lasers within a few AU of their stars, the laser light will be blended with the starlight, and details about seeing and slit orientation are not important. The relevant noise is then from the fact that the laser will have to compete against the noise from photons from the star to be detected.

The situation is somewhat different for telescopes using adaptive optics or space-based telescopes. There, image sizes are set by the diffraction limit of the telescope, which is generally of order 1/10 arcsec. There, unless a very wide slit is used the laser light might be expected to be missed for nearby stars.

Slit-based searches are also sensitive to lasers very far from the host star, as in the case of Figure 13.2. In these cases the laser light is well separated from the star on the echellogram, so could be detected from even just a few photons in principle. The space thus searched, and the sensitivity, is then a complicated function of how the slit was oriented during the observations.

The upper limits one computes would ideally be expressed in a manner similar to radio work, where one expresses the EIRP (gain times power) of the laser line one would have detected in a spectrum, if it were present. This is complicated because the underlying stellar spectrum is complex, so one's sensitivity is a complex function of wavelength and the algorithms used to flag potential candidates.

Typical sensitivities with the 10 m Keck telescope for targets well separated from their host star are 90 W launched from a diffraction limited 10 m telescope (Tellis & Marcy 2015). For laser light blended with the host starlight, typical exposure times are set to achieve a uniform signal-to-noise ratio on the star, meaning the detection thresholds for lasers are independent of distance but of order 3 kW–10 MW (Tellis & Marcy 2017).

Current typical upper limits for wide swaths of the sky from work with slitless spectroscopy are generally sensitive only to very powerful lasers, with powers of around 60 GW when launched from diffraction limited 10 m telescopes at a distance of 30 pc (Marcy et al. 2022). There is also a project at the SETI Institute run by Elliot Gillum called LaserSETI that uses slitless spectroscopy.

13.8.2 Time-compressed (Pulsed) Searches

Upper limits on pulsed laser searches depend largely on the level of background light in an angular resolution element. In an ideal situation, an instrument could monitor a star at sufficiently high temporal resolution that the chances of two photons arriving in the same time slot are so low that they rarely happen. In this case, any time such a pileup is seen is potentially significant, and observing only a few photons might constitute a detection. Photons prepared in quantum states with high arrival time correlation, like Fock states, are especially likely to be detected, even at lower average flux levels.

In practice, pulsed optical telescopes may have very large pixels on the sky, encompassing many stars resulting in a high enough background flux that one may need dozens or hundreds of photons in a single time bin to stand out against the noise.

The best upper limits on pulsed searches to date are those of the Harvard/ Princeton and VERITAS surveys. Figure 13.6 shows how such upper limits are often expressed: in terms of the number of transmitters per star that could be detected with the given equipment. When such upper limits are expressed in terms of the sensitivity of the receiving instrument, and not the power of the transmitter, they are useful for establishing what practical search space has been covered, but more information is needed to turn these into upper limits on transmitter capabilities.

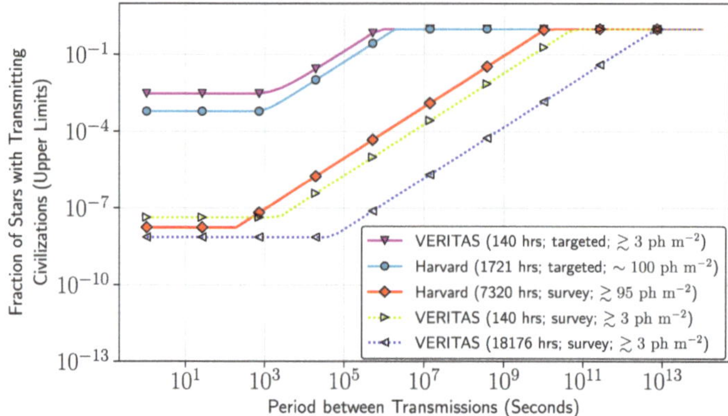

Figure 13.6. Figure 7 of Acharyya et al. (2023) showing the upper limits of the targeted and survey searches of the various Harvard and VERITAS surveys. CC BY 4.0

Roughly speaking, if one spends a time T integrating on a given target out of a list of N total survey targets then one can expect to have seen something if the typical time between pulses $t \ll T$, in which case a null detection implies that the fraction of all stars regularly transmitting pulses one could have detected must be less than roughly $1/N$. When $t \gtrsim T$, one's chance of having seen a flash goes roughly as T/t, so the fraction of stars transmitting in the result of a null detection is $t/(NT)$.

These numbers are rough, however, and elide many haystack dimensions, including the fact that these surveys operate at different wavelengths, and have different sensitivities to different pulse durations and directions on the sky. Maire et al. (2019a) for instance searched around 1300 stars with stares of around 300 s, yielding upper limits of around 10^{-2} transmitters per star for pulse durations less than 5 minutes with a sensitivity of around 60 photons m^{-2}.

The VERITAS limits are the tightest in terms of sensitivity and are for signals containing around 3 photons m^{-2} in pulses less than a few ns of duration. For scale, the photon flux density of a pulse at Earth goes as:

$$F = \frac{P\lambda}{4\pi d^2 hc}\left(\frac{\pi D}{\lambda}\right)^2 \frac{1}{r\Delta t} \tag{13.8}$$

where the first fraction captures the photons per time arriving per unit area from a transmitter of power P at distance d sending pulses at wavelength λ. The second fraction captures the geometric gain advantage of launching the beam from a diffraction limited telescope of diameter D. The final term captures the amount of concentration of the signal in pulses of duration Δt sent at a rate r.

For a 100 W signal sent omnidirectionally (so the geometric gain is 1) from the nearest star (distance 1 pc) toward Earth with 1 ns pulses every second in green light (500 nm), we have 2×10^{-5} photons m^{-2}, far below the ability of even VERITAS to detect it. If the beam were focused on the inner solar system specifically (spanning about 2 AU) it would have a width of approximately $2''$, requiring only a modest 5 cm launch telescope, and have a geometric gain of around 10^{10}. In this case we would detect a whopping 2×10^5 photons m^{-2}, easily detectable in these surveys.

We thus see that laser search limits can be quite tight for pulsed signals directed specifically at the Earth.

Box 13.2: Jargon Review: Optical SETI

- **Mode**: A manner in which the electromagnetic field can be excited. A single mode is limited in frequency-time space by $\Delta\nu\Delta t \leqslant 1$ and angle-area space by $\Delta A\Delta\Omega \leqslant \lambda^2$.
- **Occupation number**: The number of photons in a given mode.
- **On–off keying**: A simple encoding scheme in which bits are conveyed via whether a given time slot does or does not contain any photons.
- **Pulse position modulation**: An encoding scheme in which the arrival time of a pulse conveys information. This scheme is used in our interplanetary optical communications and is in some ways optimal for non-quantum communication.
- **Quantum communication**: Communication schemes that employ the preparation of non-classical states (those that cannot be described by the classical equations of electromagnetism, such as Fock states.)
- **Diffraction grating**: A particular optical element, which can be either transmissive or reflective, that disperses light, i.e., sends light of different wavelengths into different directions.
- **Echelle grating**: A reflective, high incidence angle diffraction grating used for very high resolution optical work.
- **Slit**: A small opening at the focal plane of a telescope for selecting out light for spectroscopy. Its width often determines the resolution of the spectrograph.

Exercises

1. The "Fundamental Theorem of Optical SETI"

   Let's demonstrate the "Fundamental Theorem of Optical SETI," which states that "Earth 2000 technology could generate a pulse that outshines the Sun by four orders of magnitude in white light, independent of range." For "Earth 2000" technology, let's assume we can use the Keck 10 m telescope. Howard et al. (2004) also based this on the Helios laser, which could generate pulses 3 ns wide, each containing 3.7 MJ at optical wavelengths, with a 10 Hz repetition rate.

   For this problem, express powers in the everyday units of Watts so we can get a better sense of the scale of the problem involved. Remember these calculations are rough, so you might need to choose nominal values for some parameters, and to be loose with factors of order unity.

   (a) First, let's do a calculation for a continuous wave optical (500 nm) laser (so, not pulsed). Here, the benefit is frequency compression, so we can use the calculations from Chapter 12. Assume that the bandwidth of the laser is 1 GHz. What is the flux density (i.e., specific flux, meaning per flux per Hz) at a distance d as a function of the power, P, of the laser?

(b) Next, let's compute the flux density of the Sun at this wavelength at distance *d*. *Hint: for this problem you can make a reasonable assumption about the solar flux density, like it's a blackbody. Be careful, though, that there are* two *versions of the Planck law: be sure to use the one for flux density, and not intensity. Also, remember to distinguish between the flux density at the surface of the Sun, and that at some distance away from it.*

(c) Now, compare these two: what power do you need to have the laser outshine the Sun? Do continuous wave lasers that powerful exist? (You might need Google for this one—make sure you're looking at continuous not pulsed lasers) How does this number depend on distance? Why does it have that dependence?

(d) Next, let's try a pulsed laser. Instead of computing the flux density, let's just count the photons in a 3 ns pulse. First, how many photons can the Helios laser pack in there?

(e) Next, launch those photons out of the Keck telescope. What is the number of photons per square meter per pulse, as a function of distance d? *Hint: remember that laser beams cannot be perfectly collimated—far from the telescope, what determines the opening angle of the beam?*

(f) What does this work out to at 10 ly in photons per square meter during a 3 ns pulse?

(g) Next, calculate the number of photons in a 3 ns time interval from the Sun per square meter at distance d.

(h) What does this work out to in terms of photons per square meter at 10 ly during the 3 ns interval?

(i) Compare the two. How much brighter is the pulse than the Sun during those 3 ns?

References

Abeysekara, A. U., Archambault, S., Archer, A., et al. 2016, ApJL, 818, L33

Acharyya, A., Adams, C. B., Archer, A., et al. 2023, AJ, 166, 84

Banaszek, K., Kunz, L., Jachura, M., & Jarzyna, M. 2020, JLwT, 38, 2741

Benton, D. M. 2019, PASP, 131, 074501

Beskin, G., Borisov, N., Komarova, V., et al. 1997, Ap&SS, 252, 51

Beskin, G., Plokhotnichenko, V., Bartolini, C., Guarnieri, A., & Piccioni, A. 1998, 19th Texas Symp. on Relativistic Astrophysics and Cosmology ed. J. Paul, T. Montmerle, & E. Aubourg (Amsterdam: North-Holland) 328

Betz, A. 1986, AcAau, 13, 623

Betz, A. 1993, ASP Conf. Ser. 47, Third Decennial US-USSR Conf. on SETI ed. G. S. Shostak (San Francisco, CA: ASP) 373

Cocconi, G., & Morrison, P. 1959, Natur, 184, 844

Einstein, A., Buchwald, D. K., Sauer, T., et al. 2009, The Collected Papers of Albert Einstein: Correspondence, January-December 1921, The Berlin Years, Vol. 12 (Princeton, NJ: Princeton Univ. Press)

Hippke, M. 2018, JApA, 39, 74
Hippke, M. 2021, AJ, 162, 1
Howard, A. W., Horowitz, P., Wilkinson, D. T., et al. 2004, ApJ, 613, 1270
Howard, A., Horowitz, P., Mead, C., et al. 2007, AcAau, 61, 78
Howard, A. B., & Horowitz, P. 2001, Proc. SPIE, 4273, 153
Leigh, D., & Horowitz, P. 2011, Millions and Billions of Channels (Berlin: Springer) 107
Maire, J., Wright, S. A., Barrett, C. T., et al. 2019a, AJ, 158, 203
Maire, J., Wright, S. A., Barrett, C. T., et al. 2019b, AJ, 158, 203
Maire, J., Wright, S. A., Holder, J., et al. 2022, Proc. SPIE, 12184, 121848B
Marcy, G. W., Tellis, N. K., & Wishnow, E. H. 2022, MNRAS, 515, 3898
Mead, C. C. 2013, A Configurable Terasample-per-Second Imaging System for Optical SETI (Cambridge, MA: Harvard Univ. Press)
Purcell, E. 1960, The Quest for Extraterrestrial Life: A Book of Readings, 1980, ed. D. Goldsmith (Mill Valley, CA: Univ. Science Books) 188
Reines, A. E., & Marcy, G. W. 2002, PASP, 114, 416
Ross, M. 1965, Proc. IEEE, 53, 1965
Ross, M., & Kingsley, S. 2011, Searching for Extraterrestrial Intelligence, ed. H. Paul Shuch (Berlin: Springer) 147
Schwartz, R. N., & Townes, C. H. 1961, Natur, 190, 205
Shvartsman, V. F., Bernstein, I. N., & Beskin, G. M. 1997, A&AT, 13, 13
Stone, R. P. S., Wright, S. A., Drake, F., et al. 2005, AsBio, 5, 604
Tellis, N. K., & Marcy, G. W. 2015, PASP, 127, 540
Tellis, N. K., & Marcy, G. W. 2017, AJ, 153, 251
Townes, C. H. 1983, PNAS, 80, 1147
Werthimer, D., Anderson, D., & Bowyer, C. S. 2001, Proc. SPIE, 4273, 104
Wright, S. A., Horowitz, P., Maire, J., et al. 2018, Proc. SPIE, 10702, 107025I
Zuckerman, A., Ko, Z., Isaacson, H., et al. 2023, AJ, 165, 114

AAS | IOP Astronomy

The Search for Extraterrestrial Intelligence
Theory and practice
Jason T Wright

Chapter 14

Solar System Technosignatures

> Are there artefacts or sites in the solar system which have been left behind by other intelligent species? Most of us think not…
>
> It would be best not to be overly smug and assume we do not need to propose protocols for recording artefacts or sites left by other intelligent species either directly or indirectly (e.g., as robotic probes). Some of us may wish to assume that we are totally alone in the Universe, or at least in the Milky Way (e.g., Ward & Brownlee, 2000), but if we treat that assumption as a "fact" then we will not be prepared for dealing with unexpected evidence.
>
> —Campbell (2004)

14.1 The Rise and Fall of Martian Civilization in the 20th Century

14.1.1 Claims of Life

With the launch of *Sputnik* in 1957, the Space Age dawned just two years before modern SETI began. 60 years earlier, Percival Lowell was confidently describing to the public the system of canals and locks he had observed across the Martian surface, and tracking its seasonal changes in vegetation.

In the 1920s, telescopic observations of Mars had begun to cast doubt on Mars's habitability, finding that it had very low temperatures (Pettit & Nicholson 1924) and in the 1930s astronomers concluded there was very little water or oxygen in the atmosphere on the planet (Adams & St. John 1926).

Still, who knew what sort of life might still thrive in such an environment? And although many or most astronomers doubted Lowell's interpretations and promotions that Mars hosted a network of artificial canals and abundant surface biota and an intelligent civilization, some, such as Earl Slipher (1962), maintained life might explain Martian surface features such as the linear features and regular color

doi:10.1088/2514-3433/adf5aech14

changes well into the 1960s (Section 3.1.3). Measurements of the pressure of the Martian atmosphere typically suggested it had around 1/10 the atmosphere of Earth's, which would allow liquid water on its surface, and Sinton (1959) claimed that infrared absorption features indicated the presence of chlorophyll-like substances on the surface as the origin of the darker surface features.

In his collaboration with Sagan (Shklovskiĭ & Sagan 1966), Shklovskiĭ described data that appeared to show the orbit of Phobos, one of the moons of Mars, was decaying at a rate far too fast to be explained by atmospheric drag (since Mars's atmosphere is too thin for that), by tides, or by radiation pressure if it were a natural moon. Shklovskiĭ suggested that atmospheric drag could be responsible if the Moon had very low mass. Perhaps, Shklovskiĭ suggested, it was in fact hollow—that is, not a natural moon at all, but a large artificial satellite. Did the Martians have spacecraft?

Indeed, one must imagine that many of the engineers planning the first interplanetary missions at NASA might have grown up reading in popular science texts or perhaps even being taught in school about the Martian civilizations Lowell described. The only way to tell for sure was to go there, and with the rockets of the Space Age, that is just what NASA did.

14.1.2 The *Mariner* Missions

In 1964 NASA launched the *Mariner 4* probe toward Mars to collect the first close-up images of the surface of another planet. To guide them, they turned to Earl Slipher's maps from Lowell Observatory, which included many of Lowell's canals (compare Figures 3.3 and 14.1).

Alas, *Mariner 4* found no forests or cities. Although it returned only a few images of part of the surface of Mars, what little they showed revealed a largely barren, crater-filled surface (see Figure 14.2). Subsequent missions, including *Mariner 9*, found the entire surface to be of this nature—fascinating in its own right, but nothing resembling the thoroughly inhabited surface of the Earth. The presence of so many craters revealed that the surface suffered erosion much more slowly than Earth's, and confirmed that the planet was extremely dry. Close-up photos by *Mariner 9* also showed that Phobos was, indeed, just a rock.

Whatever popular perception of widespread Martian life remained from Lowell's claims were dashed. While the possibility of microscopic or even macroscopic life on Mars remained, much of that hope seemed put to rest when the *Viking* landers showed that the surface was, indeed, a desert, and its life detection experiments were negative or inconclusive. A major scientific question, it seemed, had been answered (Murray 1972). Indeed, after the *Mariner* missions, descriptions in science fiction of Mars ceased to describe the possibility of Mars teeming with biota and technology.

And yet, the discovery that Mars lacked technology had been achieved with hardly any scholarship on the topic addressing it. One of the only (if not *the* only) serious, published scientific analyses of the problem was by Sagan & Lederberg

Figure 14.1. Work at the NASA Jet Propulsion Laboratory on *Mariner 4*. Note the annotated copy of Slipher's map of Mars on the wall for reference (Figure 3.3). This map was based on observations of Percival Lowell and others to track Mars's putative vegetation and canals, i.e., its biosignatures and technosignatures. Until the *Mariner* missions arrived at Mars, the question of a civilization on Mars was technically open. Courtesy of the JPL Archives CL#25-2234.

(1976), who conceded that there was no obvious technology on Mars: the "canals" had proven to be illusions; the surface patterns were attributable entirely to dust and wind; photos of Phobos showed that it indeed, appeared to be just a rock; and at 100 m resolution there were no signs of technology (for instance, checkerboard patterns characteristic of Earth agriculture.)

And yet Sagan & Lederberg (1976) cautioned against reading too much into the apparent lifelessness of the surface:

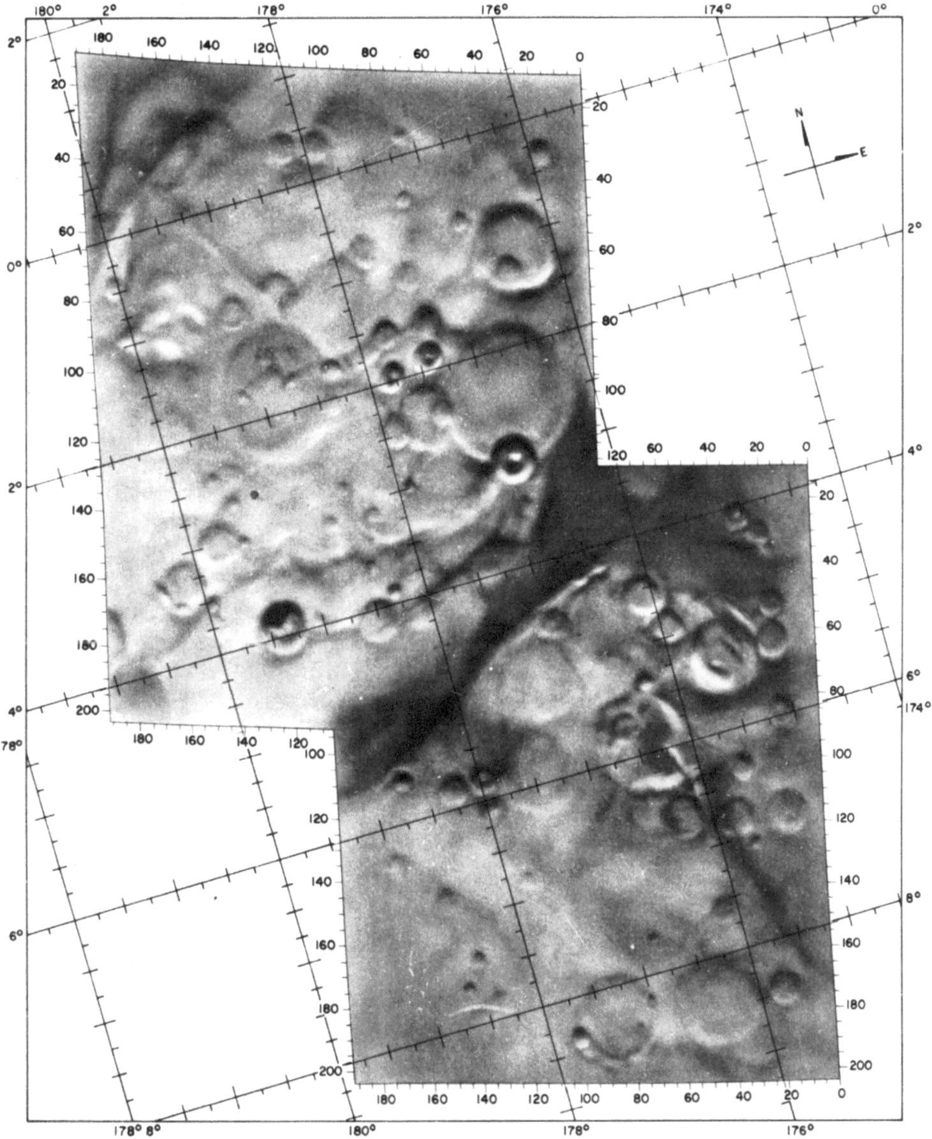

Figure 14.2. De-projected images from the *Mariner 4* mission showing the first detailed images of the Martian surface. There are no obvious technosignatures or biosignatures in this image. Credit: NASA.

> ...to surrender the search is to give too little credit to the adaptive versatility of living organisms, a versatility whose boundaries are understood very poorly. Biological evolution has a major stochastic component, which makes biology more akin to history than to planetary physics. Whether Viking 1976 succeeds or fails, however, subsequent

> landed spacecraft on Mars should have a wider range of biology experiments, able to search for a broader spectrum of conceivable Martian organisms.

Fifty years later, Sagan's optimism is still with us, as NASA makes the search for microbial life in the solar system—whether extinct or extant—a top priority, and a top reason to continue our study of Mars.[1] And if Mars once harbored water and teemed with life whose traces we might still detect today, it also might, like Earth, have hosted or even launched technology.

How small or damaged or eroded or covered by dust would such technology need to be to have escaped our notice? What is the time horizon for such studies on Mars, where, unlike Earth, life, rain, and plate tectonics have not been active for billions of years? Presumably the possibility that Mars once hosted technology becomes more remote as our imagery and understanding of the surface improves, but little proper scientific effort has been put into quantifying this.

14.1.3 The Face on Mars

The topic of solar system SETI provokes some discomfort among planetary scientists both because it comes close to the study of UFOs (Section 18.7) and the pseudoscience of "ancient" alien visitors to Earth (Section 18.6). At least some of this comes from an episode involving dubious claims of surface technosignatures on solar system bodies: the "Face on Mars."

In 1976 a NASA press release based on *Viking 1* orbital imagery casually noted in a figure caption regarding a particular mesa-like landform: "The huge rock formation in the center, which resembles a human head, is formed by shadows giving the illusion of eyes, nose and mouth."[2] Figure 14.3 shows the image, and indeed the shadows and black dots from missing data appear to show an eye socket, nose ridge, and mouth. The human brain is wired to identify human faces readily with only a few rough cues for eyes and a mouth, for instance in emoticons or US electrical outlets. This property, called *pareidolia* (Liu et al. 2014), is at play with this "face on Mars."

At least some people took NASA's imaginative identification seriously, and began interpreting the *Viking* imagery to be evidence of an artificial structure on the Martian surface designed to look like a human head. Note that the hypothesis here is quite contrived: the evidence of its artificiality is that it resembles a human head, but why would there be a giant structure carved to look human on Mars?

The suggestion began with Vincent DiPietro and Gregory Molenaar, two engineers at Goddard Space Flight Center, who in their personal capacities requested and obtained the *Viking* imagery and began an extensive project of data

---

[1] "Determine whether life ever arose on Mars" is at the top of the list of NASA Mars Exploration Science Goals retrieved 2024 September 2.

[2] JPL Viking Press Release P-17384.

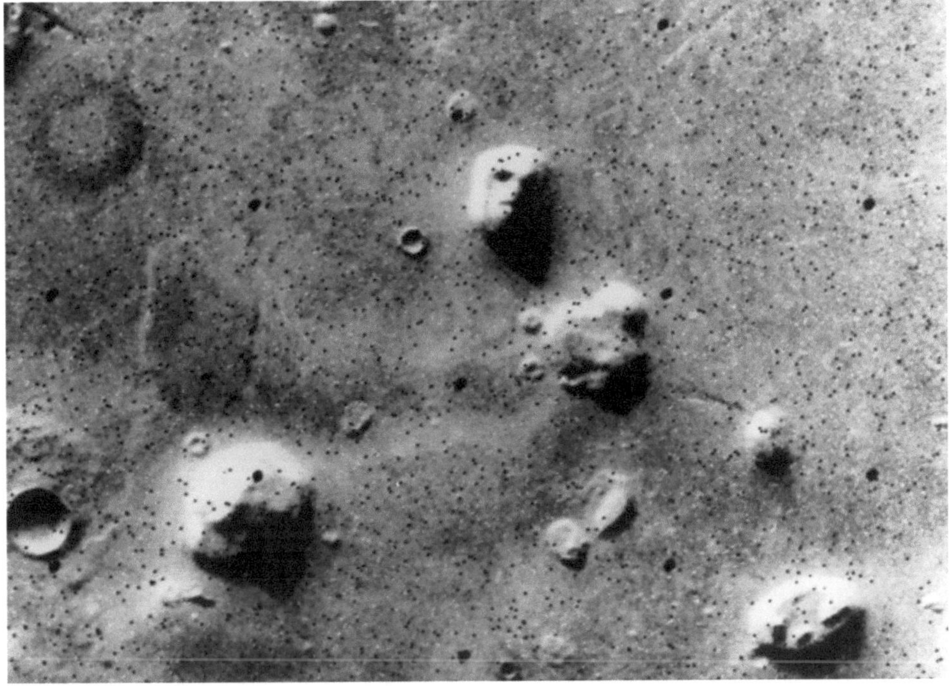

Figure 14.3. NASA's "Face on Mars" image. Black dots are missing data, not real features. Note how one of the black dots appears where a nostril would be, adding to the illusion. Image credit: NASA/JPL.

analysis on it. Their work on two different images of the feature convinced them it was unnaturally bilaterally symmetric and truly characteristic of a human head.

They presented their findings at the 1980 summer meeting of the American Astronomical Society (Dipietro & Molenaar 1980) and many subsequent scientific gatherings around the United States, at which point the ideas that the "head" might be artificial generated publicity in major newspapers and OMNI magazine. They discussed their findings with Carl Sagan, later writing, "in our opinion Dr. Sagan seemed very impressed." Publicly, Sagan expressed skepticism at the claims, but he echoed DiPietro and Molenaar's arguments among his reasons for humans to go to Mars.

To promote their work among scientists, DiPietro & Molenaar (1982) self-published their claims in a book *Unusual Martian Surface Features*. A few years later the cause was picked up by another engineer, Mark J. Carlotto, who began a program of extensive image analysis of the Cydonia region on Mars in which the "head" is located. This region contains many isolated buttes and other features, which in a series of papers and books Carlotto argued formed a complex of artificial structures (see Figure 14.4), including a "fortress," a pyramid (also identified by DiPietro and Molenaar), and more.

Following popular interest in the feature, NASA and ESA made sure to acquire high resolution imagery: the ESA *Mars Express* mission revealed the entire region in great detail (see Figure 14.5), and from 1998 to 2001, NASA's *Mars Global*

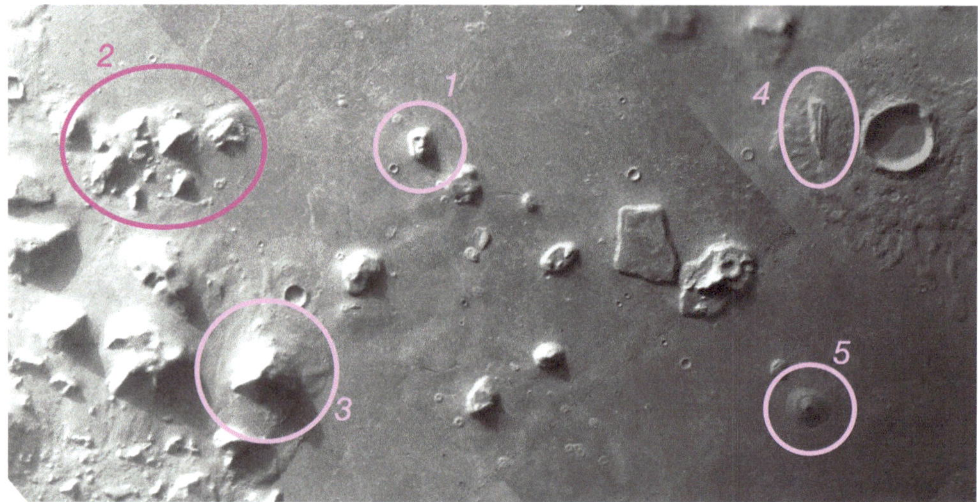

Figure 14.4. Mosaic of Viking 1 images of Mars showing Carlotto's labeling of various features, including the "Face on Mars" (1), the "city" (2), a "pyramid" identified by DiPietro et al. (1988) (3), a "cliff" (4), and the "tholus" (5), a purportedly unusual volcano-like mound. Compare Figure 14.5. Figure composed by the author from NASA imagery, after Figure 1 of Carlotto (1997).

Figure 14.5. The Cydonia region in color as derived from Mars Express imagery from multiple angles, with annotation by the author. Compare Figure 14.4, whose footprint is roughly indicated in magenta along with the corresponding features identified by Carlotto (1997). Image credit ESA. The author obtained this image from from the Wikimedia website, where it licensed under CC BY-SA 3.0 IGO.

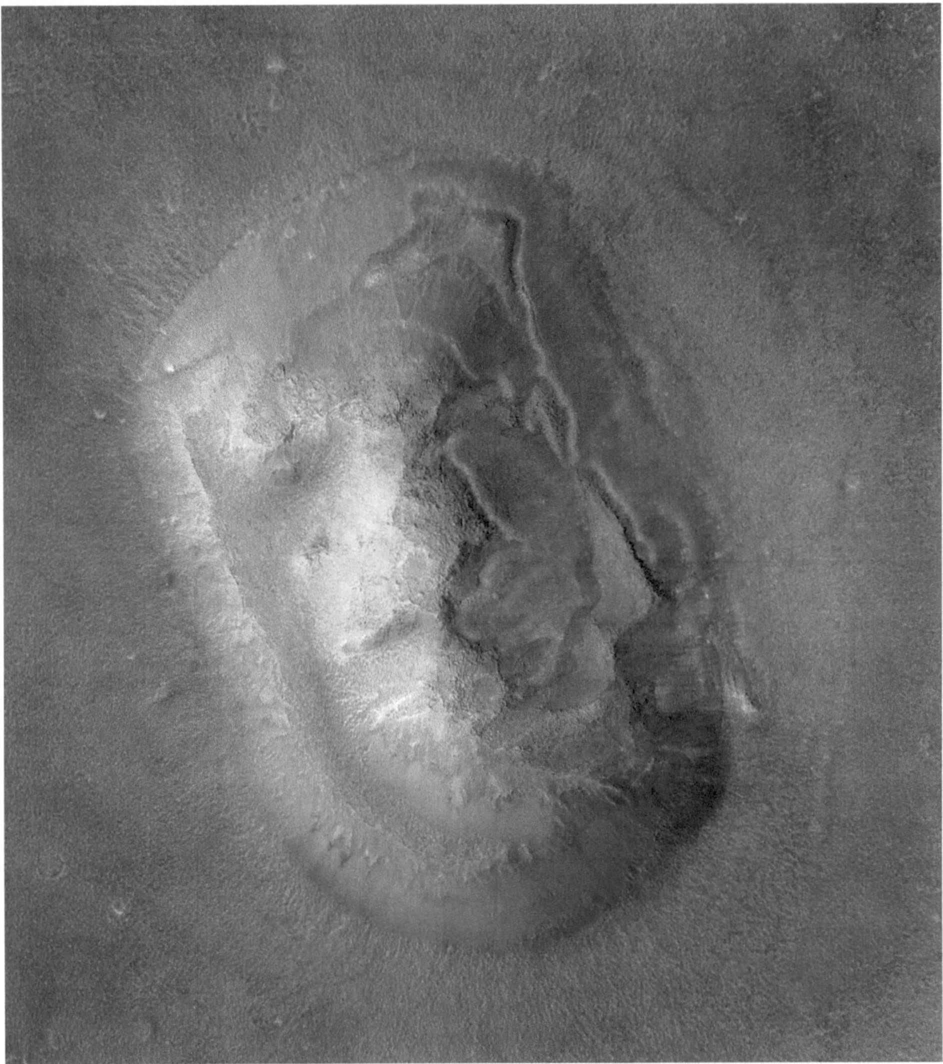

Figure 14.6. The "Face on Mars" as seen by the Mars Global Surveyor in 2001. The resolution in this image is about 2 m per pixel. Credit: NASA/JPL/MSSS.

Surveyor observed the region several times revealing a thoroughly natural looking, eroded, asymmetric landform (one such image can be seen in Figure 14.6).

Undeterred, Carlotto would continue to publish papers arguing the head shows evidence of symmetry that "strongly suggests that it is an artificial object."[3]

[3] Carlotto also self-published books on various terrestrial ruins being much older than archaeologists believe—suggesting Earth hosted another technological species in its recent past—and that UFOs represent manifestations of this ancient terrestrial technology. See Sections 18.6 and 18.7.

Figure 14.7. Montage of some of the "coverage" of the "Face on Mars" in the *Weekly World News* over the years. These made-up stories were loosely based on DiPietro, Molenaar, and Carlotto's claims about the "face," and helped maintain the feature's place in popular culture. Note the reference to Todd and Jenkins' 1924 radio experiment in the bottom right (Section 3.1.3). Clippings courtesy of and © Weekly World News and Bat Boy, LLC.

Carlotto's claims did not find any purchase in the scientific community, but did find wide attention in popular culture beyond the media splash made by DiPietro and Molenaar. One driver of this was occasional, sensational coverage of the "Face on Mars" by the fake-news tabloid *Weekly World News* that was a staple of supermarket checkout aisles in the United States in the 1980s and 1990s[4] (see Figure 14.7). The *News* originally attributed the claims to an anonymous Soviet source based on a fictional Soviet mission to Mars, and the "Face" became one of the tabloid's recurring "characters."

14.1.4 Lessons From the "Face on Mars"

It is possible that the relative lack of work on solar system SETI, especially involving surface features, is attributable to the strong "giggle factor" (Section 18.6) associated with such work because of the "Face on Mars."

[4] *Weekly World News* was an over-the-top parody of its tabloid neighbors in the checkout aisle (such as its sister publication *The National Enquirer*) with fictional celebrities like "Bat Boy." In addition to obviously fake news it offered some, true, offbeat news items, humorous horoscopes, parody Op-Ed columns, and other trappings of yellow journalism. It is unclear to what degree everyone who saw its covers featuring the Face on Mars were in on the joke, however.

Certainly, serious work on Martian, lunar, and other surface features searching for technosignatures must contend with this context and consider how to do better. Specifically: in addition to working in a scientifically credible way, one must navigate popular misconceptions about alien life in the solar system, the legacy of the Face on Mars episode in science and the media, and sensationalist journalism. In addition, the episode raises real and important concerns about solar system SETI: what, exactly is the path toward success of such work?

One lesson is that it or any search for solar system technosignatures needs to involve experts in planetary science. Experts in geomorphology and remote sensing of topographic features largely dismissed the Face on Mars claims, especially after subsequent imagery revealed it to be an illusion. Archaeologists that use spacecraft to identify ruins on Earth can also advise how to distinguish artificial structures from natural features. Incorporating such expertise early can ensure that work will withstand later scrutiny.

Following the outline in Chapter 11, ideally one would be able to set strong upper limits on the *lack* of technosignatures smaller or more subtle than some threshold, and have a followup plan for determining the nature of anomalies, which presumably are of at least some interest as extreme natural phenomena. The path here is unclear, but if it could be developed then there exists substantial imagery of solar system surfaces from interplanetary probes that might be scrutinized for technosignatures.

14.2 Bracewell Probes

14.2.1 Bracewell's Proposal

Solar System SETI is in some sense the oldest form of SETI, going back centuries to studies of technology on the Moon, Mars, and Venus (e.g., Section 3.1.3, where we discussed contemplation of and attempts to communicate with Martians). In the modern era, this subfield begins with the American physicist and electrical engineer Ron Bracewell.

In a paper published shortly after the foundational work of Cocconi & Morrison (1959) (Section 4.1), Bracewell (1960) presented a Drake Equation-style argument that unless communicative species are extremely abundant—essentially present at every Sun-like star—the nearest will be much farther away than the targets of Project Ozma (Section 18.2). Rather than set up megawatt transmitters pointing at thousands of nearby stars all the time, he argued it is much more natural to send autonomous probes to explore those stars. Such probes could be communicative, eliminating the need to overcome $1/r^2$ attenuation over great distances, and the need to guess magic frequencies (Chapter 5): one needs only find the probe, not the signal.

He suggested such a probe might store large amounts of information and be capable of autonomous communication with Earth. Bracewell's imagining of how contact would proceed is quite specific:

> Such a probe may be here now, in our solar system, trying to make its presence known to us. For this purpose a radio transmitter would seem

> essential. On what wave-length would it transmit, and how should we decode its signal? To ensure use of a wave-length that could both penetrate our ionosphere and be in a band certain to be in use, the probe could first listen for our signals and then repeat them back. To us, its signals would have the appearance of echoes having delays of seconds or minutes such as were reported thirty years ago by Størmer and van der Pol (Størmer (1928)) and never explained.
>
> To notify the probe that we had heard it, we would repeat back to it once again. It would then know that it was in touch with us. After some routine tests to guard against accident, and to test our sensitivity and bandwidth, it would begin its message, with further occasional interrogation to ensure that it had not set below our horizon. Should we be surprised if the beginning of its message were a television image of a constellation?

Even if we find his guess on the manner in which a probe would be detected or in which a probe would communicate to be extremely speculative, Bracewell's underlying logic that it is reasonable that probes could be in the solar system stands quite well today.

14.2.2 Benefits of Physical Media over Photons

Bracewell probes also have another major advantage, articulated by Rose & Wright (2004), which is that "inscribed matter" provides extremely high bandwidth and information content compared to photons.

This principle that the physical transport of matter has higher bandwidth than using photons is well known in computer science, and perhaps most amusingly illustrated by the tongue-in-cheek work of Ben-Bassat et al. (2005). Inspired by the adage "never underestimate the bandwidth of a station wagon filled with tapes"[5] they showed that modern communication was literally slower than a snail. By constructing a harness by which a snail could pull a pair of DVD discs forward, they achieved a bandwidth comparable to modern typical home internet DSL/cable upload speeds, or 4G/LTE download speeds (see Figure 14.8).

This lighthearted demonstration has several serious lessons for SETI. The first is that the distance the material must travel is irrelevant for its bandwidth: the bandwidth of an ethernet cable is, to first order, independent of its length or, as Ben-Bassat et al. (2005) put it "data transfer is a continuous affair: once a communications channel is established between two points, packets of bits flow continuously from source to destination. In our case, it can be assumed that a new [snail] leaves the source every second with a payload of 9.4 gigabytes, yielding a pure delivery rate of 37,000 Kbps."

This is in strong contrast to photons, which lose information content as $1/r^2$ (each photon can carry around 1 bit of information, (Section 13.3), so fewer photons per

[5] Quoting Andrew Tanenbaum (1988) in his book *Computer Networks*.

Figure 14.8. Experimental design of Ben-Bassat et al. (2005), risibly but vividly illustrating how even extremely slow forms of transport of physical media can have extremely high data transmission speeds compared to photons in free space, or electrical signals in cables. In this case, the snail transports the data encoded on the DVDs faster than could be typically accomplished at the time through the Internet. By analogy, physical transportation of data with rockets can have a far higher throughput than data via radio or laser communication, at the cost of much higher latency. Courtesy of and © Improbable Research.

second necessarily means fewer bits per second). Inscribed matter thus preserves its content over arbitrary distances, and so it can easily overcome its disadvantage in speed compared to photons if it has an extremely high information density, such as provided in modern data storage.

The sacrifice is that there is a large *latency*: instead of having to wait, say, 10 years for a signal to arrive from a nearby star, one needs to wait, say, 100,000 years. The entire logic of the Fermi Paradox, though, rests on this not being a large amount of time on a Galactic timescale (10^8–10^{10} yr; Chapter 6), and the success of SETI depends on Drake's L being quite large anyway.

Bracewell imagined that the probes would be able to contact Earth via radio, but physical contact and inspection would also be possible, providing additional opportunities for high information content exchange.

14.3 Other Rationales for Solar System SETI

While Bracewell's suggestions regarding probes were quite specific, his general logic about probes in the solar system found purchase in the SETI community. Since then, many suggestions for the form and origin of potential alien technology in the solar system have been proposed.

14.3.1 Prior Indigenous Technological Species

If ancient technology is found, its most natural origin might not be at interstellar distances, but from what Wright (2018) dubbed a *prior indigenous technological species*, meaning a species originating in the solar system prior to humans. After all, we know that Earth is capable of producing technological species, and we have little evidence either way regarding whether we are the first such species to arise (Section 9.1.5).

We also understand that Mars likely had liquid surface water in its past (Carr & Head 2010), and Venus may have been habitable prior to its runaway greenhouse warming (Way et al. 2016). The question of whether they hosted prior life is thus open, and leaves open the possibility that they might have hosted technological life. More exotically, we might even consider whether indigenous life has already arisen beneath the ice of Europa or Enceladus (Kargel et al. 2000).

In any case, if prior technological life was spacefaring, then remnants of its activities might remain in the solar system, perhaps on or just below the surfaces of asteroids, the Moon, or Mars, or in orbit around a planet or the Sun.

14.3.2 Interstellar Objects

Technology might be in the solar system if it is just "passing through" and not bound to the Sun at all. For instance, we might imagine that alien spacecraft use the Sun's gravity to change their velocity through the Galaxy, akin to the way human spacecraft use the planets to accelerate via gravitational slingshot (Section 6.2.3).

Another possibility is that there are many inoperative spacecraft or fragments of spacecraft that fill nearby space—if their number density is high enough, some of these will naturally pass close to the Sun and Earth, where we can detect them.

Rationales for such technology are abundant, but in a SETI context have been made most forcefully by Loeb (2021), who argued that the interstellar object 1I/'Oumuamua might be an example of alien technology. Loeb's arguments centered largely on the object's anomalous acceleration (Section 14.5.2) and lack of any obvious coma that would be expected for accelerations that large due to outgassing, and on its extremely thin or elongated shape. In the case of 'Oumuamua, Loeb argued it could be a discarded fragment of a solar sail used to launch a spacecraft, that its anomalous acceleration could be due to radiation pressure, and that species that use such technology might often litter space with such debris.[6]

In order for us to detect such an object, they must be abundant enough that there is one large enough and close enough to the Sun and Earth. The details depend

[6] Loeb's arguments that an anomalous acceleration might be due to the object having low mass, leading to large nongravitational accelerations echo the earlier suggestion by Shklovskiĭ about Phobos we encountered in Section 14.1.1. However, Loeb's interpretation seems to be both inconsistent with the data and unnecessary to explain the properties of 'Oumuamua: it would need to be extremely thin to be subject to radiation pressure, which is inconsistent with the dimensions inferred from its light curve (Mashchenko 2019; Zhou et al. 2022; Taylor et al. 2023), and there exist multiple alternative models, such as a comet outgassing hydrogen or nitrogen (Jackson & Desch 2021; Desch & Jackson 2021; Bergner & Seligman 2023). See Wright, Raymond, and Desch (2023) for a longer discussion on Loeb's claims about 'Oumuamua.

strongly on the reflectivity and size of such objects, but roughly speaking we might choose 1 au as a distance from Earth within which we might plausibly detect a large (100 m) object. If the object is not bound to the Sun, then near the Earth it must have velocity of at least 40 km s^{-1} (Equation (6.4), where $M = M_{\odot}$), meaning it will only be visible to us for around 40 days or less.

For us to detect such an object, they must therefore come by quite often, since they do not linger. Roughly speaking, we can think of the Sun and typical objects moving in the Milky Way disk with mutual velocity of ~km s^{-1}. How many of these will encounter a sphere centered on Earth with radius 1 au?[7]

The answer is in the equation

$$n\sigma l = 1 \tag{14.1}$$

which relates the typical distance l an object with cross section σ can travel among targets randomly distributed with number density n. We can take a fiducial lower rate r of such objects entering our ken in order for us to find them of one per decade, in which time the Sun will have moved $l = vt$ with $t = 1/r$. Solving for n we have

$$n = \frac{1}{\sigma l} = \frac{1}{(\pi \text{ au}^2)(20 \text{ km s}^{-1})(10\text{yr})} \approx 10^{14} \text{ pc}^{-3} \tag{14.2}$$

or a number density of around 100 trillion such objects per cubic parsec (the more thorough calculation of Do et al. (2018) yields a number closer to 10^{15} pc^{-3}). Space is very large, and we need large numbers of objects for there to be a chance we'd encounter one. For us to find them, either vast numbers of spacecraft must exist, or they must visit the Sun specifically.

14.3.3 Exploration

Perhaps the most natural reason for there to be probes in the solar system is exploration: when we find interesting things in the solar system, we often send probes to visit them. If there are other species with a similar inclination, they might do the same, and Earth, with its life, might be an attractive target for such a visit.

A probe visiting the solar system might thus be positioned such that it could easily study targets of interest, such as the Earth or other planets or moons. A natural place to find such probes might then be in orbit around these bodies, or near associated Lagrange points (as suggested by Freitas (1983b), Section 5.7, Figure 5.7).

[7] We choose this number just for the purposes of this Fermi Problem (Box 6.1)—'Oumuamua was only detected because it was in a very favorable geometry, and we would not detect every such object within 1 au.

14.4 Nodes, Networks, and the SGL

14.4.1 The Network Hypothesis

Another way that artifacts in the solar system could address the problem of $1/r^2$ interstellar communication is if a species has already populated the Milky Way with settlements or probes. These probes could then form a network through which long-distance interstellar transmissions could flow. This was in fact one of the motivations for Bracewell (1960)'s original suggestion:

> We need not expect, however, that any community other than the nearest is trying to reach us, because the superior communities throughout the galaxy are probably already linked together into an existing galaxy-wide chain of communication.

On Earth, for instance, our long-distance transmissions of data and phone transmissions is generally not done point-to-point—mobile phones do not talk directly with each other but to nearby nodes of a vast planet-spanning data network.

Networked communication can have somewhat longer latency than point-to-point communication because it generally involves a longer, meandering path through nodes of a network, but it offers many advantages. A multiply connected network can send messages along multiple channels to make itself robust against node failure, and the proximity of the nodes reduces transmission losses with distance. Sending a signal 1000 pc is much easier if it is boosted every 1–10 pc by nodes that only need to get the signal 1–10 pc farther along on its journey.

It would make sense, then, for a Galaxy-spanning species to place nodes throughout the Galaxy and communicate only with some of their nearest nodes, with instructions to pass their messages on to the appropriate destination in the network.

There are many implications for such a network, two of which were highlighted by Bracewell: if this is the preferred method of all kinds of communication in the Galaxy, it might be worthwhile to focus efforts on long, sensitive monitoring of very nearby stars instead of shallow surveys of distant stars. Another implication is that the solar system itself might host a node in the network, and that we might be able to detect this node itself or its transmissions. This hypothesis has also been explored in series of papers by, for instance, Gertz (2018) and Hippke (2020).

14.4.2 The Solar Gravitational Lens

While it might seem that the engineering of an intra-Galactic network of communicative nodes has too many parameters for us to confidently predict anything about its structure or the locations of the nodes, there is an aspect of physics that offers a seemingly irresistible advantage to a certain architecture: stellar gravitational lenses.

Einstein famously predicted that large masses like the Sun would noticeably bend light rays passing them. This effect is exploited by astronomers using *microlensing*: the slight deflection and amplification of background light by a foreground star.

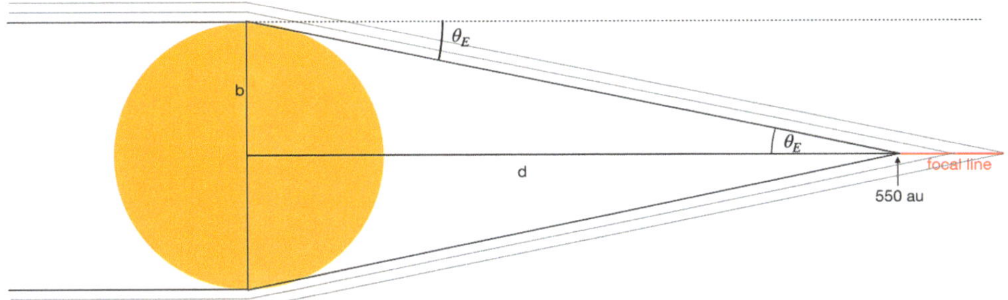

Figure 14.9. The geometry of the solar gravitational lens (not to scale). Rays from an infinitely distant source on the left pass just beyond the solar limb at distance $b = R_\odot$ (black lines). The Sun's gravity causes them to arc downward (here shown simply as a sharp corner) by angle θ_E, coming to a focus at distance $d = 550$ au. Rays passing farther from the limb (gray) feel weaker gravity and come to a focus farther from the Sun at some distance $d > 550$ au (gray lines). All rays passing a distance $b > R_\odot$ past the Sun thus form a focus at some point on the focal line (red).

We can estimate the amount of deflection light will have passing by the Sun by considering photons as objects moving at velocity c. The acceleration they feel from the Sun at closest approach b is $GM_\odot/b^2$, and they feel roughly this acceleration for a time b/c as they pass by. They thus pick up some velocity in the direction perpendicular to their travel $GM_\odot/bc$. The angle of deflection is then just the ratio of this velocity to the original velocity. A rigorous derivation reveals a factor of 4, giving us an angle of deflection $\theta_E = 4GM_\odot/bc^2$. The geometry is shown in Figure 14.9.

At some distance d, rays from an annulus radius b around the Sun will come to a focus, where $d = b/\theta_E = b^2c^2/GM_\odot$. Since rays cannot pass through the Sun, we have $b \geqslant R_\odot$, or $d \geqslant R_\odot^2 c^2/GM_\odot \approx 550$ au. This means that starting 550 au from the Sun, there exists a focal line where rays from background stars converge. If one looks back toward the Sun along this line, stars *precisely* behind the Sun will appear as a ring around the Sun with an angular radius given by this angle, called the *Einstein radius*. Solving in terms of d, we have

$$\theta_E \approx \sqrt{\frac{4GM_\odot}{c^2 d}} \tag{14.3}$$

(This is only approximate because we have made the assumption that the source is infinitely far away.) For rays just grazing the Sun's surface (i.e. $b = R_\odot$, and $d = 550$ au), this angle is $\theta_E \approx 1\overset{''}{.}7$.

Because this annulus acts as such a large lens, the gain and magnification is enormous. The actual value depends on many details of the wavelength (λ) and receiver used, but roughly speaking the gain can be approximated as

$$g_{\rm SGL} \sim \frac{8\pi GM}{c^2\lambda} \tag{14.4}$$

For light at 1 μ used through the Solar gravitational lens, this corresponds to gains of order $g_{\mathrm{SGL}} \approx 10^{11}$.

The insight into a potential Galactic communication network here is that stars are natural signal boosters which provide both power to node relays (via their starlight) and the gravity needed to tightly beam signals towards distant targets.

There are many details of such a network that complicate this picture—for instance, the solar corona interferes with radio waves, meaning communication through a stellar lens would likely be with frequencies above 100 GHz/wavelengths shorter than 3 mm.

It is always perilous to predict too much about what alien technology would be like, but it would seem strange if a spacefaring species undertaking interstellar communication did *not* use stellar lenses to boost their signals, given the huge gain they offer for "free."

The implications for SETI are several:

- Signals directed at Earth or the inner solar system generally might arrive from the very nearest stars, being the closest node in a network, and having been lensed by that nearby star.
- The Sun might host relays as part of interstellar communication schemes.
- We might detect communication to or from these relays when Earth passes through the beam.
- We might detect communication between such a relay and and probes in the inner solar system (since "phoning home" is most efficiently done via a relay).

Figure 14.10 schematically illustrates how a solar or stellar gravitational lens could be used as part of an interstellar communication scheme. In this highly stylized figure, in which nothing is to scale, the Sun is on the left, the Earth and a distant planet are blue, the distant planet's host star is red, and the relay probe in the solar system is magenta.

In the top scheme, a relay probe farther than 550 au from Earth can send or receive transmissions through the solar gravitational lens. These communications, shown as black lines, are focused by the Sun, greatly increasing their power when they arrive at their destination.[8] The relay probe can then communicate with probes in the inner solar system (shown as a magenta double arrow), and these transmissions might be detectable if Earth is in their path.

In this scheme it would be difficult to detect the transmissions from the distant planet to the relay probe, because they would arrive at Earth unmagnified. It might be possible to detect transmissions from the relay probe through the SGL back to the distant planet if and when Earth's orbit takes it between the probe and the limb of the Sun (which will only happen for stars on the ecliptic.)

In the bottom scheme, a probe in the solar system communicates with a relay in the distant star system at its gravitational focus. In this scheme, we would receive

[8] Note that outgoing transmissions toward the distant star will not be cleanly focused on the distant planet as implied here—even though such transmissions will be highly focused, the beam will still be somewhat divergent and be detectable everywhere in a distant star system.

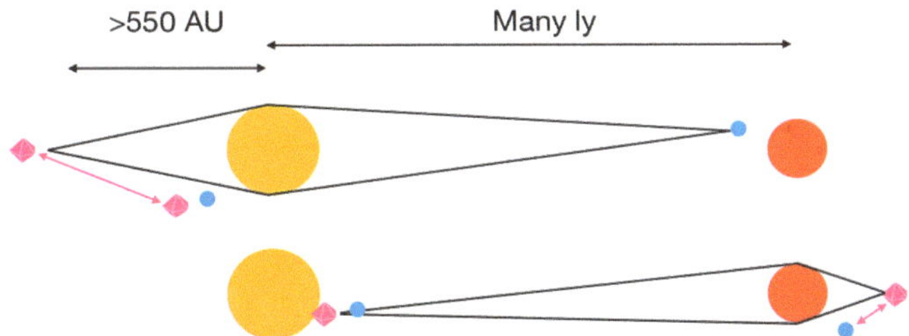

Figure 14.10. Highly schematic figure illustrating two transmission schemes via a stellar or solar gravitational lens. Nothing is to scale. *Top*: transmissions (black lines) between a distant planet (blue, right) and a relay probe in the solar system (magenta, far left), focused by the Sun (yellow). The relay must be at least 550 au away from the Sun or else there can be no focusing. Earth might be able to intercept outgoing transmissions from the probe if Earth's orbit fortuitously takes it across the black lines (which will generally happen only certain times of year, and only for communication with stars near the Earth's ecliptic). Earth might also be in the beam of any signals the relay sends to a probe near Earth (magenta arrows). Note that in these diagrams only the beams that reach their target are shown—as with any lens, any outgoing beam through a gravitational lens will diverge as $1/r^2$ and so the will generally have a footprint that encompass the entire planet-star system. *Bottom*: An alternative scheme in which a solar system probe communicates with a relay at the gravitational focus of a distant star. In this case we would be able to intercept the incoming transmissions from the relay, whose beam would encompass much of the solar system. Figure created by the author.

magnified signals from the distant relay probe just as well as the probe in the solar solar system does but would be unlikely to be in the beam of outgoing transmissions from the local probe.

Many details of the physics of the solar gravitational lenses for interstellar communication have been worked out in series of papers by Michael Hippke (e.g., Hippke 2020) and Slava Turyschev (e.g., Turyshev 2024).

The locations of any relay probes are well determined by the geometry of the lens. Each star has a corresponding focal line on the opposite part of the sky, so a receiving or transmitting probe using the Sun as a lens must lie on its corresponding line.[9] Figure 14.11 shows an example of the locations of the SGL focus of the α Centauri system.

14.4.3 Some Past Searches for Solar Gravitational Lens Signals

Tusay & Huston et al. (2022) performed a search for radio transmissions from probes at the focus of the α Centauri system, reasoning that if there were probes in the inner solar system receiving signals from a node at α Centauri, they might do so

[9] Because of the mutual motions of the Sun and other stars, this line is different for receivers (which are opposite the apparent position of the distant star) and transmitters (which are opposite where the true (light-travel-time corrected) position of the star will be when the signal arrives).

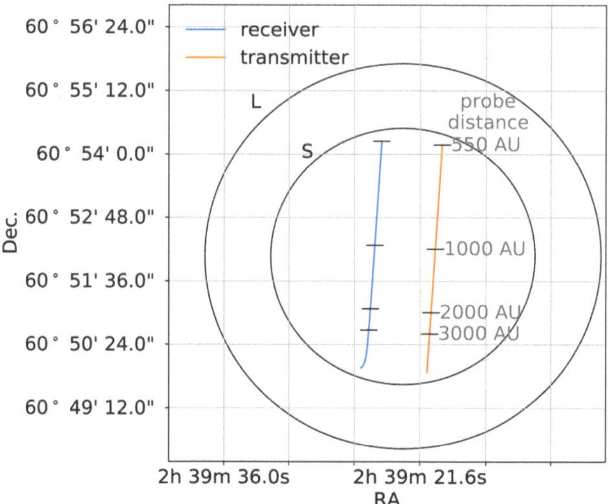

Figure 14.11. Figure 2 from Tusay et al. (2022) showing the location of the solar gravitational lens on a particular day for the focus for α Centauri for receivers (opposite the sky from the current *apparent* position of the stars, i.e., where they were one light-travel-time ago) and transmitters (opposite the sky from the true position of the star when the signal will arrive there, essentially correcting for the proper-motion over two light-travel-times). The lines point directly away from the Sun, and are visible as lines on the sky because of parallax from the position of the Earth, which changes over the course of a year. The circles refer to the beam sizes of the Green Bank 100 m telescope in L and S bands.

via radio communications in both directions with relay probes at the Sun's focal line.

One challenge in intercepting node communications is that communication *through* a stellar lens can generally only be done at infrared wavelengths or shorter, in which case the beams are likely rather tightly collimated and so might miss the Earth. Marcy et al. (2022) searched for optical emission from the α Centauri focal lines that, for whatever reason, might be detected at the Earth (for instance if it were communicating with probe visiting the Earth).

One exception to this difficulty is for stars on the ecliptic, for which the Earth appears to pass very close to the Sun twice per year, and so might pass through the tightly collimated beam from the relay on the focal line. Gillon et al. (2022) observed the focal line opposite the very nearby ecliptic star Wolf 359[10] during the time that that star would see Earth pass behind the Sun. This search would be sensitive to some optical transmissions sent toward Wolf 359 from a probe in the solar system if they were sent through the entire annulus at the Einstein radius.

[10] Many of the most nearby stars have names familiar to fans of science fiction. For instance, Wolf 359 was the site of a major battle between the Borg collective and the United Federation of Planets in *Star Trek: The Next Generation* (S4E1).

14.5 Orbital Material in Reflected Light

14.5.1 Identifying Anomalous Artifacts

The first searches for free-floating artifacts (such as *active* or inoperative spacecraft)[11] in the solar system was a program by Freitas and Valdes (Freitas & Valdes 1980; Freitas 1983a; Valdes & Freitas 1983), who searched for objects at the Earth's Lagrange points in reflected light, finding none. One challenge to such a search is that if one does find an object one is probably more likely to have discovered an asteroid or human-made probe than an alien spacecraft.

One aspect of an object's orbit that might indicate artifice is if it is in an orbit that has a short lifetime compared to the age of the solar system. Many possible orbits around the Sun will eventually bring an object close to a planet, which can dramatically change its orbit during a close encounter, or are affected by orbital resonances with the planets (regular perturbations which can cause large changes in an object's orbit over time). Over the course of the solar system's life, the planets (especially Jupiter) have thus cleared out large regions of orbital parameter space around the Sun of asteroids. Most of the asteroids that have survived to present day are present in a range of orbital distances that form the asteroid belt, and even in this range there are ranges of mean orbital distances, called Kirkwood gaps, in which few or no asteroids orbit because of resonances with Jupiter.

What objects we do find in orbits with short lifetimes must have achieved those orbits relatively recently, for instance because of a close encounter with a planet that altered its orbit.

But an artificial object that is actively thrusting could enter an orbit with a short lifetime and maintain it as long as it has the ability to thrust. Freitas and Valdes searched the Lagrange points because objects in orbits near these points require minimal amounts of thrust to stabilize themselves, so natural objects in such orbits should be rare, but they might form an attractive place for an artificial object to reside.

Once found, there are at least two ways beyond detecting communicative signals that technology orbiting the Sun or a planet might be readily distinguished from natural objects: nongravitational accelerations, and spectral reflectance (i.e., unusual colors or spectra).

14.5.2 Radiation Pressure and the Yarkovsky Effect

14.5.2.1 Radiation Pressure

Photons carry not just energy but also momentum, equal to $p = E/c = h\nu/c$ where E is the photon's energy and ν is its frequency. This means that photons that are absorbed will impart their momentum on the object that absorbs them; that photons that are reflected will impart up to *twice* their momentum; and that objects that radiate (for instance, as they cool) will recoil as their thermal photons escape.

[11] We use the term "active" simply to mean that an artificial object is not dormant or inoperative. In the context of comets, "activity" refers to signs of outgassing such as a coma, tail, or acceleration.

The acceleration of an object due to these photons can be complex, but a good first-order model (see Marsden et al. 1973; Farnocchia et al. 2015) is to parameterize generic *nongravitational accelerations* (NGAs) as:

$$\hat{a}_{\rm NG} = (A_1\hat{r} + A_2\hat{t} + A_3\hat{n})g(r) \tag{14.5}$$

Here, the A coefficients are accelerations in three orthogonal directions: radial (anti-sunward, $\hat{r}$), along-track ($\hat{t}$), and the out-of-plane ($\hat{n}$). The function $g(r)$ describes how these accelerations depend on distance from the Sun, and in its simplest form is $g(r) = (r/\text{au})^{-2}$.[12]

The dominant effect of photons striking an object is to push the object away from the Sun, providing some acceleration in the $\hat{r}$ direction.

The radial force $F_{\rm r}$ due to sunlight striking an object with cross sectional area A is given by the momentum it receives from that sunlight per unit time, equal to the energy deposited by the photons per unit time over c:

$$F_{\rm r} \rightarrow A\frac{\mathcal{F}}{c} = \frac{A}{4\pi d^2}\frac{L_\odot}{c} \tag{14.6}$$

where we have introduced the symbol $\mathcal{F}$ for the flux of photons to distinguish it from the force F. The radial acceleration due to this force is then

$$a_r = \frac{A}{M}\frac{L_\odot}{4\pi d^2 c} \tag{14.7}$$

where M is the mass of the object.

Reflected photons will generally scatter out in many directions, but on average will be pointed back toward the Sun, increasing the effect by some fraction.

Because the flux from the photons falls as $1/r^2$, just as solar gravity does, this effect is a constant fraction of the force of gravity, meaning we can model it as a weakening of gravity in a distance-independent way, as if the Sun effectively had less mass. This slightly changes the relationship between the orbital distance of an object and its period, following Newton's version of Kepler's third law (Equation (2.25)).

14.5.2.2 The Yarkovsky effect

Photons absorbed by an object in space will generally heat the surface, and their energy (and momentum) will then be reradiated via thermal photons. An object in space will generally be hotter on the side facing the Sun, and cooler on its nightside. The hotter side will cool more efficiently, its surface flux going roughly as a blackbody, or proportional to T^4. This creates an asymmetry in the recoil from photons, resulting in an outward (anti-sunward) force of order $A\sigma(T_{\rm day}^4 - T_{\rm night}^4)/c$. Since this is opposite the direction of the Sun's gravitational force, the first-order result is to enhance the effects of radiation pressure.

[12] That is, the effect of photons declines as the inverse square law. This may not be the correct assumption for other NGAs, for instance for thrust from sublimating ices on comets.

The surface takes time to change temperature, so as the object rotates the warm dayside will carry heat onto the evening side of the object, and the cooler nightside will remain cold as it rotates through the morning side. As a result, the re-emission of thermal photons will have a preferred direction, dictated by the thermal properties of the surface, such as the characteristic timescale over which the thermal reradiation occurs, and the rotation, which determines the direction of this radiation with respect to the Sun. This is called the *Yarkovsky effect*.[13]

For an object with zero obliquity,[14] the warmer evening side of the object will be pointed opposite the direction of its orbital motion, so it will experience a slight acceleration in the $\hat{t}$ direction. This adds to its orbital energy, causing its orbit to grow. Objects with obliquities greater than 90° (i.e., objects that spin retrograde) will experience the opposite effect. See Figure 14.12.

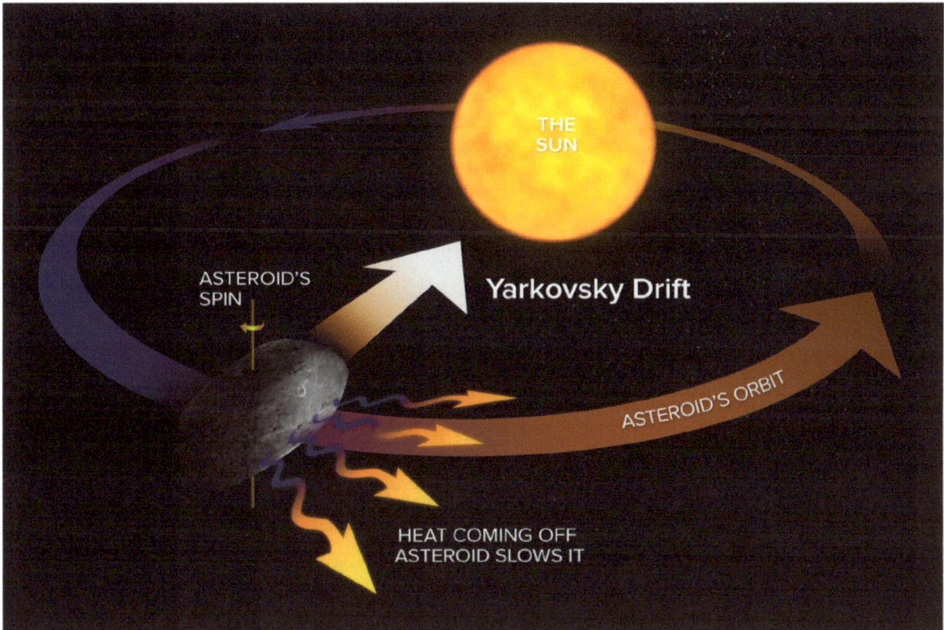

Figure 14.12. Schematic illustrating how photons on the warmer, evening side of an asteroid are preferentially emitted in the direction of the asteroid's orbit. In this example, the obliquity of the planet is 180°, so the asymmetric thrust counteracts its orbital motion, causing it to lose orbital energy. Over the course of an entire orbit, this will cause its orbit to shrink, moving it closer to the Sun. Credit: A. Angelich, NRAO/AUI/NSF.

[13] The naming of this effect is a testament to the desire to credit early publishers for their ideas. Ivan Yarkovsky was a Polish-Russian civil engineer who, two years before his death in 1902, published a 17 page pamphlet in Bryansk attempting to explain effects of gravity and light using concepts like the luminiferous ether, including the effect that now bears his name. The pamphlet was read by Ernst Öpik when he was in Estonia in 1909, who amazingly remembered it well enough in 1951 to credit Yarkovsky when describing the effect. The effect was also independently discovered by Viktor Radzievskii in his thesis in 1948. For more, see Beekman (2005).

[14] Obliquity is the angle between an object's rotation axis and its orbital axis, also called axial tilt. Earth's obliquity of 23°.4 is the primary reason for Earth's seasons. Zero obliquity means the object rotates in the same sense that it orbits, with its North Ecliptic Pole aligned with its North Celestial Pole.

Secondary effects include the seasonal Yarkovsky effect, affecting objects with nonzero obliquity. This effect is similar, except here one considers how the pole of the object changes orientation with respect to the Sun over the course of a year. The asymmetry arises because thermal inertia (the time it takes a surface to cool or heat when its illumination changes) will keep the autumn-time surface warmer than the springtime surface. Since the pole in autumn will always point in the direction of the object's motion, this effect always causes the orbit to shrink.

The Yarkovsky effect is modulated by many parameters, including the nonspherical shape of the object (which can result in an asymmetric radiation pattern that generates a thrust or torque on the object even if it is not spinning) and its spin rate.

Objects in orbit around planets are affected differently from objects orbiting the Sun because in this case the central force of gravity is not aligned with $\hat{r}$ when considering solar photons. Thermal and reflected photons from the planet have a similar effect as described above, but the acceleration from solar photons, to first order, averages out over an orbit, except when the object is in the planet's shadow. These variations of the Yarkovsky effect are generally thus much smaller than for objects orbiting the Sun.

The Yarkovsky effect depends strongly on the thermal timescale over which energy is reradiated from the object. If this timescale is very short compared to the orbital period (or, for the seasonal effect, the orbital period). The surface heats and cools almost immediately and there is little asymmetry between the evening and morning sides of the object. If the timescale is extremely long compared to the rotational (or orbital) period, then the entire surface will be of uniform temperature and again there will be no asymmetry.

14.5.2.3 The β Parameter for Technology and "Rocks"

The ratio β of acceleration due to radiation pressure a_r from absorbed sunlight versus that due to the gravity of the Sun $a_g = GM_\odot/d^2$ is called the *lightness number*. For a perfectly black object with cross section A and mass M, and is given by

$$\beta = \frac{a_r}{a_g} = \frac{L_\odot/c}{4\pi GM_\odot}\frac{A}{M} \approx 8 \times 10^{-5}\frac{A}{\text{cm}^2}\left(\frac{M}{\text{g}}\right)^{-1} = 8 \times 10^{-4}\frac{A}{\text{m}^2}\left(\frac{M}{\text{kg}}\right)^{-1} \quad (14.8)$$

independent of orbital distance. For scale, minor planets with density of 0.5 g cc^{-1} and radius 1 km, have an A/M ratio of order 10^{-5} cm^2 g^{-1}, so $\beta \sim 10^{-8}$. Photons contribute a correction of order a few parts per billion to objects of this size. For objects of different size, we expect $A \sim D^2$ and $M \sim D^3$, so this ratio $\beta \propto A/M \sim 1/D$, where D is the diameter of the object. Smaller objects are more easily affected by sunlight.

Human technology in interplanetary space generally has an extremely high surface area-to-mass ratio, for instance because it has extensive solar panels, antennae, and booms, or because it is largely hollow, having spent significant amounts of propellant. For instance, the second stage of a Falcon 9 rocket has maximal cross sectional area of $\sim$50 m^2 and an empty mass of 3900 kg, for a ratio of $A/M\sim$ 0.1 cm^2 g^{-1} and $\beta \sim 10^{-5}$, which is two orders of magnitude greater than an

Table 14.1. Some Radiation (β) Parameters for Some Artificial and Natural Objects

Object	Cross Section (m^2)	Mass (kg)	β
Falcon 9 (2nd stage)	50	3.9×10^3	1×10^{-5}
ISS	2500	4.5×10^5	4.3×10^{-6}
Cassini (end of mission)	23	2500	7.1×10^{-6}
Asteroid (r = 10 m)	310	2.1×10^6	1.5×10^{-7}
Asteroid (r = 100 m)	3.1×10^4	2.1×10^9	2.9×10^{-8}
Asteroid (r = 1 km)	3.1×10^6	2.1×10^{12}	2.9×10^{-9}
Asteroid (r = 10 km)	3.1×10^8	2.1×10^{15}	2.9×10^{-10}
Asteroid (r = 100 km)	3.1×10^{10}	2.1×10^{18}	2.9×10^{-11}

Note that human-made artificial objects all have similar β values despite having a large range of sizes. The asteroid values assume a spherical asteroid with density 2 g/cm^3.

asteroid of the same dimensions. Interstellar probes, such as New Horizons or Cassini, generally have low masses once they have expelled their propellant—both had $A/M \sim 0.15$ cm^2 g^{-1} at the ends of their missions—and the International Space Station weighs 450,000 kg with 2500 m^2 of solar panels—again yielding $A/M \sim 0.1$ cm^2 g^{-1} and $\beta \sim 10^{-5}$.

Table 14.1 compiles some characteristic β values for asteroids and artificial objects in space. The lightness numbers β are for a perfectly absorbing surface of that size, and are for approximate reference.

Human technology, it seems, has a rather consistent β value regardless of whether it is discarded material, a space station constructed in low-Earth orbit, or a deep space probe. Even if alien interplanetary probe favors a different engineering optimum, it may still be unlikely to have β values near that of natural objects.

14.5.2.4 Natural Confounders

Comets also exhibit significant NGAs, which can be distinguished from radiation effects based on the direction of the acceleration. Jets on comets are generally most active at the subsolar point, and so in some cases generate predominantly radial accelerations, however the complex spins and geometries of comets can complicate this picture. Yarkovsky effects are generally in-plane (i.e., either along or against the orbital motion) since asteroids generally have low obliquities.

It is now appreciated that the boundaries between asteroids and comets may not be clean. Traditionally, comets were characterized by large volatile content (ices that evaporate when comets near the Sun) and wide and eccentric orbits around the Sun. The discovery of Main Belt Comets (Hsieh & Jewitt 2006) revealed that the main belt of asteroids does include some objects with significant volatile content, and that orbital properties alone do not distinguish the populations. More recently, the discovery of "dark comets" (Seligman et al. 2023)—asteroids exhibiting very low but detectable NGA's characteristic of weak outgassing of volatiles—suggests that there may be a continuum of objects between true asteroids (which would suffer NGA's only from radiation pressure, which is tiny) and comets (which have substantial and easily detected NGAs). See Figure 14.13.

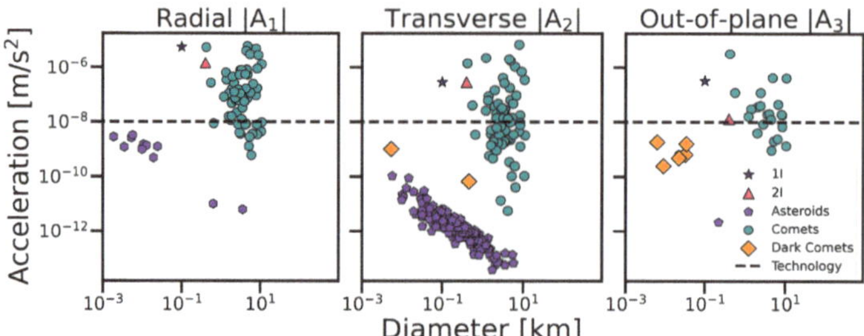

Figure 14.13. Measured nongravitational accelerations in the radial (A_1), along-track (A_2), and out-of-plane (A_3) direction for a collection of asteroids (purple), comets (green), and "dark comets" (yellow). The asteroids are affected primarily by radiation pressure, which is more effective on smaller objects with larger area-to-mass ratios. The horizontal line represents rough magnitude of the accelerations one might expect for artificial objects with area-to-mass ratios characteristic of human technology. Also shown are the measured accelerations of 1I/'Oumuamua and the second interstellar object known, 2I/Borosov. Figure courtesy of Darryl Seligman, adapted from Seligman & Moro-Martín (2022) with permission of the publisher Taylor & Francis Ltd.

Any detections of NGAs beyond that expected by radiation pressure on rocks thus reveals a scientifically interesting class of objects, or new, ordinary comets (which are usually easily identified via their comae and tails).

Figure 14.3 shows the NGAs expressed as components the components given in Equation 14.5 measured for a large number of asteroids, comets, and two interstellar objects. Also shown is the order of magnitude of acceleration expected for objects with β values typical of human technology, illustrating how they are typically much higher than expected for asteroids.

14.5.2.5 Thrust

It is possible that an active artificial object in the solar system might employ rockets to change their orbital parameters dramatically. Such actions would make the objects extremely hard to track, and indeed they might be detected multiple times and mistaken for a series of different asteroids with different orbits. Correctly identifying such an object as artificial would involve noticing that the object failed to be detected along its predicted orbit at some point before or after its discovery, a much more challenging task than merely measuring tiny deviations from a predictable orbit.

Being able to connect two objects discovered with anomalous spectra on different orbits that can be connected via a single thrust event would be strong evidence of artificiality.[15]

14.5.3 Colors and Spectra

The most suspicious aspect of an artificial object is that natural objects orbiting the Sun or planets have reflectance spectra characteristic of rocks and ices, while technology has no reason to share those particular features.

[15] For instance, Micheli et al. (2018) were able to show the anomalous orbit of 'Oumuamua was inconsistent with a single, impulsive acceleration.

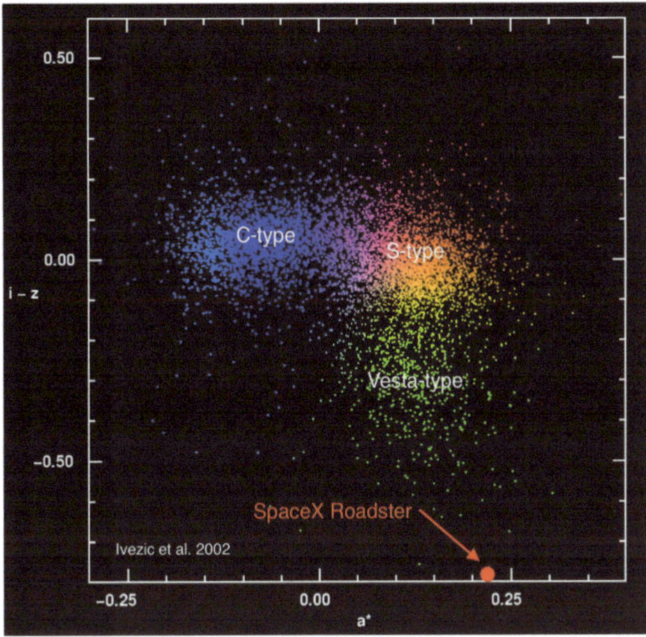

Figure 14.14. Color–color diagram showing the photometric properties of asteroids, using the Sloan Digital Sky Survey filters. a^* is a linear combination of the g, r, and i filters chosen to emphasize differences among the reflection properties of the asteroids as a function of wavelength, which in part reflects compositional differences (Ivezić et al. 2001). The printed colors of the points on this plot are just to distinguish the types, and are not meant to represent the literal appearance of the asteroids. The highlighted point is the upper stage of the Falcon Heavy rocket carrying a Tesla Roadster, measured shortly after its launch into interplanetary space when it was still close to Earth. This technosignature is clearly distinct from the vast majority of natural objects. Underlying figure from Ivezić et al. (2002), adapted by David Gerdes and first appearing in Karouzos (2018), reproduced here with permission from Springer Nature.

As an example, consider Figure 14.14, which shows a color–color plot of many asteroids. The axes represent flux ratios in various bands of optical light that are diagnostic of particular spectral features characteristic of different classes of asteroids.

The figure has been modified to include the colors of a rocket stage launched into interplanetary orbit.[16] The colors are likely that of the paint on the rocket, which may change with time in the future. The artificial object in this plot clearly stands out as anomalous.

This plot also shows many other objects far from the main clustering. Because it does not include error bars is it unclear how significant these outliers are, but a

[16] The rocket gained notoriety because it was a high-profile and successful test of the Falcon Heavy class of rocket by the SpaceX corporation, and because as a publicity stunt its payload was a red Tesla Roadster (both Tesla and SpaceX both being run by CEO Elon Musk).

search for artificial objects among them might first proceed by acquiring high quality photometry or reflectance spectra to confirm their anomalous nature in more detail.

14.6 Surface Artifacts

14.6.1 Technosignatures on the Moon

The Moon has a substantial number of technosignatures on its surface, put there by humans, mostly in the 1960s and 1970s. Figure 14.15 shows imagery from the Narrow Angle Camera of the Lunar Reconnaissance Orbiter, which can achieve roughly 50 cm resolution over a 5 km swath. This image clearly shows many unnatural features caused by the Apollo 17 astronaut's visit to the Moon, including leftover scientific and transportation equipment, and even their tracks in the lunar dirt. LRO also observed many other landing sites across the lunar surface at high resolution, and mapped the entire lunar surface at lower (around 100 m) resolution.

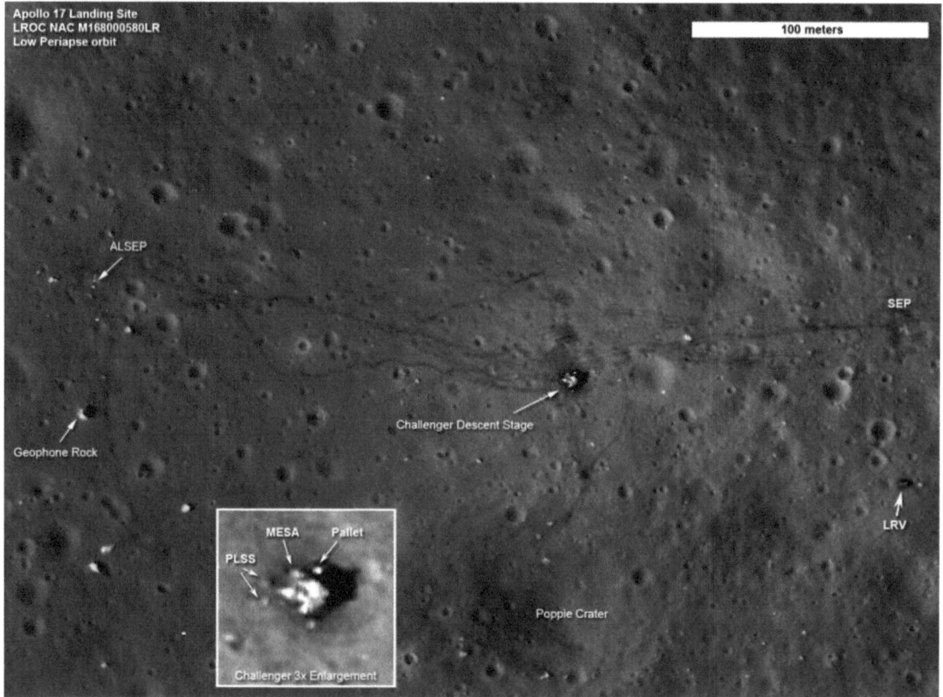

Figure 14.15. The Apollo 17 landing site as seen by the Lunar Reconnaissance Orbiter at high resolution. Visible are the parts of the Challenger lander left behind, the Lunar Roving Vehicle (LRV, the "Moon Buggy"), two science packages (ALSEP and SEP), and tracks and footpaths from the astronauts' explorations of the surface. Credit: NASA.

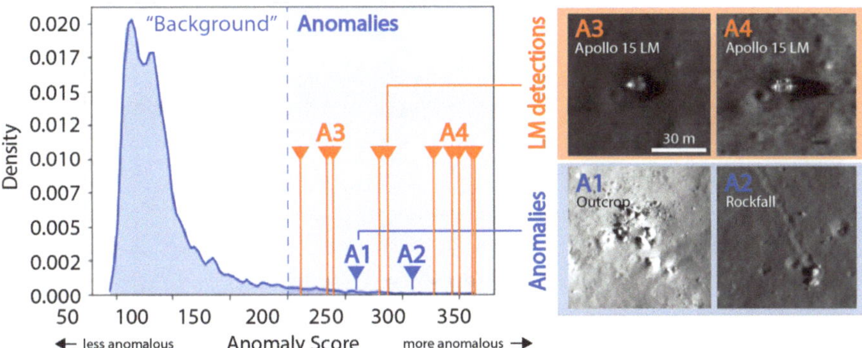

Figure 14.16. From Figure 10 of Lesnikowski et al. (2024). Using a neural network, they developed an anomaly detection algorithm that scrutinized lunar imagery for unusual features. Shown on the left is the distribution of anomaly scores, which have been divided into two categories: ordinary (or "background") scores less than a arbitrary threshold chosen for illustrative purposes, in which most image cutouts fall, and "anomalies" with a high score. Four examples of the algorithm successfully identifying rare and unusual features are shown, including two sets of images containing human technosignatures, the Apollo 15 landing modules. This paper was published under CC BY-NC-ND 4.0.

Having been thoroughly mapped, and having had genuine (human) technosignatures identified, one might naturally wonder if *non-human* technosignatures might be present on the Moon, and if so how one might find them in LRO or other lunar imagery.

One problem is that such imagery is quite extensive. Davies & Wagner (2013) suggested harnessing enthusiasm for citizen science on a platform like Zooniverse to have members of the public examine the imagery by eye for anomalies. Lesnikowski et al. (2024) explored the use of machine learning to identify anomalies in lunar imagery as being quicker than human review.

Lesnikowski et al. (2024) used machine learning methods to perform training on a large number of images of the lunar surface. The method is *unsupervised* meaning they did not label known anomalous features or otherwise "teach" the algorithm what to look for. Instead, the algorithm developed a model for what a "typical" image of the lunar surface looked like.

This method involves an autoencoder, which is a way to very efficiently represent typical images of the sort the algorithm was trained on, but which struggles to represent other kinds of images. By comparing the autoencoder's representation of an image it was not trained on with the actual image, one can get a sense of how anomalous or unlike the training set the image is, represented by an anomaly score.

Figure 14.16 shows the results of their validation, in which a large number of validation images were scored. Most images had low anomaly scores, but a small number were scored highly anomalous, out to a large tail. Among these anomalous images were some of the leftover material from the Apollo 15 landing, suggesting that these methods might be good ways to identify potential surface technosignatures.

Box 14.1: Jargon Review: Solar System SETI

- **Yarkovsky effect**: A gentle acceleration felt by asteroids an other objects due to asymmetric emission of thermal photons, for instance because the object is rotating.
- **Nongravitational acceleration (NGA)**: Any acceleration of an asteroid or other object not due to gravity, for instance from the Yarkovsky effect or from cometary outgassing.
- **Prior indigenous technological species**: Any (hypothetical) technological species that arose in the solar system prior to humanity, for instance on an early Venus biosphere.
- **Pareidolia**: The tendency for the human brain to identify familiar features in ambiguous or random images, such as seeing human-face-like features in surface imagery of other planets.
- **Microlensing**: The deflection of light by a massive object via gravity, such that a background object appears out of position or magnified.
- **Solar gravitational lens**: The focusing of light by the Sun's gravity, such that a focus occurs at distances beyond 550 au. Such a lens could be used as a large telescope or for interstellar communication.
- **Einstein radius**: The angular distance from an object at which background light from a particular object is most strongly focused. This depends on the distance to the lens and the source. When the Einstein radius appears smaller than the angular size of the object, microlensing can produce high magnification.

14.6.2 Lifetime of Surface Features

Planetary surfaces are subject to erosion and overturning, and so searches for artifacts upon them must necessarily be subject to some temporal horizon beyond which we cannot expect to find anything.

We encountered this in Section 9.1.5 when discussing to what degree we can rule out the presence of prior episodes of technology on Earth: the temporal horizon for a previous episode of global change like the modern-day "anthropocene" is surprisingly short—perhaps 10 million years. Certainly the vast majority of surface artifacts last a much shorter time than this thanks to the fast action of both natural erosion and the biosphere. While certain large stone structures in arid environments like the Great Pyramids might take tens or hundreds of thousands of years to erode away, the vast majority of unmaintained technology decays in months to years and likely disappears entirely in a matter of centuries.

The surfaces of the icy satellites turn over relatively quickly and are generally quite young, as evidenced by the relatively pristine and crater-free ice of Europa and the young surface of Miranda. Mars, the Moon, and the asteroids are all subject to impacts which will "garden" and erode surface features. Jupiter's moon Io is constantly resurfaced by volcanoes. On the Moon, the regolith grows very slowly but should eventually bury features over time, while Martian sandstorms should erode and bury features there.

The ice caps of Mars, and perhaps even ice deposits in caves and other permanently shadowed areas of the Moon, have grown and receded over millions of years or longer, and so may contain records of the atmospheres of these bodies, similar to how the Greenland ice sheets contain a record of Earth's atmospheric history. These records could conceivably allow for a great temporal horizon for atmospheric technosignatures.

The Venusian surface was thought to be quite stable, but has recently been observed to have volcanism, and may have had a resurfacing event around 500 Myr ago.

Overall, however, the temporal horizons for most surfaces have not been rigorously calculated for various plausible technosignatures, and will likely vary depending on the sort of technosignature being studied.

Exercises

1. Estimate the distance out to which a spacecraft is amenable to serendipitous discovery with a $D = 10$ m Earth-based optical telescope in reflected light. Some things to keep in mind:
   - The answer depends on geometry, but for now let's say the spacecraft is directly opposite the Sun in the sky, so appears in "full phase." This means its distance from Earth is $d_\oplus$ and its distance from the Sun is $d_\odot = d_\oplus + 1$ au. For distant objects, the extra 1 au is not a big effect and we can take $d_\odot \approx d_\oplus$, and for objects near the Earth we have $d_\oplus \ll 1$ au so $d_\odot \approx 1$ au.
   - The spacecraft will reflect only some fraction of the light that strikes it. This is called its albedo a, which we'll guess is 0.5.
   - Let's say the linear dimensions of the spacecraft are some length l, so its cross section to sunlight is l^2.
   - The telescope will need to collect a lot of photons to make a detection —let's say 100. This is enough that even with background photons and the light spread over many pixels, we can still get a $\sim 5 - \sigma$ result.
   - The telescope has some efficiency of detecting photons. Let's call this ε and guess it's 0.5, which accounts for 1%–10% losses at each the many optical surfaces between the top of the telescope and the detector.
   - The spacecraft will be moving. For rough purposes, let's assume its relative velocity with respect to Earth is about Earth's orbital velocity about the Sun, or $v = 30$ km s^{-1}. This means its light will be smeared into a streak if we expose too long, lowering its detectability.
   - The usual way to account for an object's motion when looking for it is to either track the telescope according to its known motion, or to "shift and stack" many short exposures along all of the potential directions the object could be moving. We are concerned here with serendipitous discovery where these techniques are not employed.
   - This means that to make a robust detection we must collect those 100 photons before it has moved by more than its apparent size in the

image, which is set by atmospheric blurring, usually around $\theta = 0.''5$. Recall that there are about 206,265″ in a radian.

So let's walk through it. For now, keep everything in terms of the variables listed above (D, a, ε, etc.). We'll plug in numbers at the end.

(a) Write down an expression for roughly how many optical photons from the Sun strike it every second in terms of $L_\odot$ and the variables listed above. You can assume half of the luminosity of the Sun is at optical wavelengths, and that the craft is 1 au from the Sun. You can approximate this number assuming all of the photons have, say, $\lambda = 0.5\mu$.

(b) Now assume it reflects about half the light that strikes it. How many photons per second are detected by our telescope?

(c) Next, let's calculate how fast the object is moving across the sky. Its angular speed is v/d in radians per time—convert this to arcseconds per unit time. Calculate the time it takes to move by $0.''5$ as a function of d and v.

(d) Next, put it all together, plug in numbers and calculate the number of photons we collect in the time it takes the object to move $0.''5$. Express as a dimensionless number times powers of (D/m), (l/m), $(d_\oplus/\mathrm{m})$, and $(d_\odot/\mathrm{m})$.

(e) Now, let's assume that we need to collect 100 photons to make a detection. How far out can we see the object? First, let's take the case of a large object, $l = 1$ km across. If we can see it far away then we can take $d_\oplus \approx d_\odot$, and solve for this d. Then plug in values for l and D. Check this value in terms of the smallest natural solar system objects known at this distance (for instance, planetary moons).

(f) Solve again for small objects with $l = 1$ m. In this case, we have $d_\odot = 1$ au and just solve for $d_\oplus$. Are there human-made objects at this distance from Earth? (And can you tell if they are monitored telescopically?)

2. The observed minimum rotational period for large asteroids is 2.2 hr. Compute the density this implies for these objects. Stepping through it:

(a) Assume the asteroid is spherical. What is its equatorial rotational velocity in terms of its radius R and orbital period P?

(b) What is the escape velocity of the object in terms of its radius R? (Hint: use Equation (6.4). Alternatively, you can calculate the acceleration due to gravity and compare this to the circular acceleration $a = v^2/R$.)

(c) Set these two parameters equal, and solve for the density.

References

Adams, W. S., & St. John, C. E. 1926, ApJ, 63, 133

Beekman, G. 2005, J. Br. Astron. Assoc., 115, 207

Ben-Bassat, A., Ben-David-Zaslow, R., Schocken, S., & Vardi, Y. 2005, AIR, 11, 4

Bergner, J. B., & Seligman, D. Z. 2023, Natur, 615, 610

Bracewell, R. N. 1960, Natur, 186, 670
Campbell, J. B. 2004, Bioastronomy 2002: Life Among the Stars, Proc. IAU Symp. 213, ed. R. Norris, & F. Stootman (San Francisco, CA: ASP) 505
Carlotto, M. J. 1997, Journal of Scientific Exploration, 11, 1
Carr, M. H., & Head, J. W. 2010, E&PSL, 294, 185
Cocconi, G., & Morrison, P. 1959, Nature, 184, 844
Davies, P. C. W., & Wagner, R. V. 2013, AcAau, 89, 261
Desch, S. J., & Jackson, A. P. 2021, JGRE, 126, e06807
Dipietro, V., & Molenaar, G. 1980, BAAS, 12, 450
DiPietro, V., & Molenaar, G. 1982, Unusual Mars Surface Features (Charlottesville, VA: Mars Research)
Do, A., Tucker, M. A., & Tonry, J. 2018, ApL, 855, L10
Farnocchia, D., Chesley, S. R., Milani, A., Gronchi, G. F., & Chodas, P. W. 2015, Asteroids IV, ed. P. Michel, F. E. DeMeo, & W. F. Bottke (Tucson, AZ: Univ. Arizona Press) 815
Freitas, R. A. 1983a, Icar, 55, 337
Freitas, R. A. 1983b, JBIS, 36, 501
Freitas, R. A., & Valdes, F. 1980, Icar, 42, 442
Gertz, J. 2018, JBIS, 71, 375
Gillon, M., Burdanov, A., & Wright, J. T. 2022, AJ, 164, 221
Hippke, M. 2020, AJ, 159, 85
Ilsich, H. H., & Jewitt, D. 2006, Sci, 312, 561
Ivezić, Ž., Tabachnik, S., Rafikov, R., et al. 2001, AJ, 122, 2749
Ivezić, Ž., Lupton, R. H., Jurić, M., et al. 2002, AJ, 124, 2943
Jackson, A. P., & Desch, S. J. 2021, JGRE, 126, e06706
Kargel, J. S., Kaye, J. Z., Head, J. W., et al. 2000, Icar, 148, 226
Karouzos, M. 2018, NatAs, 2, 286
Lesnikowski, A., Bickel, V. T., & Angerhausen, D. 2024, IJSTA, 17, 6589
Liu, J., Li, J., Feng, L., et al. 2014, Cortex, 53, 60
Loeb, A. 2021, Extraterrestrial: The First Sign of Intelligent Life Beyond Earth (Boston, MA: Houghton Mifflin Harcourt)
Marcy, G. W., Tellis, N. K., & Wishnow, E. H. 2022, MNRAS, 509, 3798
Marsden, B. G., Sekanina, Z., & Yeomans, D. K. 1973, AJ, 78, 211
Mashchenko, S. 2019, MNRAS, 489, 3003
Micheli, M., Farnocchia, D., Meech, K. J., et al. 2018, Natur, 559, 223
Murray, B. C. 1972, EnSci, 35, 10
Pettit, E., & Nicholson, S. B. 1924, PASP, 36, 269
Rose, C., & Wright, G. 2004, Natur, 431, 47
Sagan, C., & Lederberg, J. 1976, Icar, 28, 291
Seligman, D. Z., & Moro-Martín, A. 2022, ConPh, 63, 200
Seligman, D. Z., Farnocchia, D., Micheli, M., et al. 2023, PSJ, 4, 35
Shklovskiĭ, I. S., & Sagan, C. 1966, Intelligent Life in the Universe (Delta-Books) (San Francisco, CA: Holden-Day)
Sinton, W. M. 1959, Sci, 130, 1234
Slipher, E. C. 1962, The Photographic Story of Mars (Cambridge, MA: Sky Publishing)
Størmer, C. 1928, Natur, 122, 681
Tanenbaum, A. S. 1988, Computer Networks (Hoboken, NJ: Prentice-Hall)

Taylor, A. G., Seligman, D. Z., Hainaut, O. R., et al. 2023, PSJ, 4, 186
Turyshev, S. G. 2024, PhRvD, 110, 084010
Tusay, N., Huston, M. J., Dedrick, C. M., et al. 2022, AJ, 164, 116
Valdes, F., & Freitas, R. A. 1983, Icar, 53, 453
Ward, P., & Brownlee, D. 2000, Rare Earth: Why Complex Life Is Uncommon in the Universe (New York: Copernicus)
Way, M. J., Del Genio, A. D., Kiang, N. Y., et al. 2016, GeoRL, 43, 8376
Wright, J. T. 2018, IJAsB, 17, 96
Wright, J. T., Desch, S., & Raymond S., 2023, '*Oumuamua: Natural or Artificial?* Medium https://medium.com/@astrowright/oumuamua-natural-or-artificial-f744b70f40d5
Zhou, W.-H., Liu, S.-F., Zhang, Y., & Lin, N. C. 2022, A&A, 667, A108

Chapter 15

Infrared SETI

> I know I have the best of time and space, and was never measured and never will be measured.
>
> I tramp a perpetual journey, (come listen all!)
>
> ⋮
>
> This day before dawn I ascended a hill and look'd at the crowded heaven,
> And I said to my spirit *When we become the enfolders of those orbs, and the pleasure and knowledge of every thing in them, shall we be fill'd and satisfied then?*
> And my spirit said *No, we but level that lift to pass and continue beyond.*
>
> — Walt Whitman, *Song of Myself*, Section 46, lines 1–2 and 20–22

15.1 Origin of the Idea

The physics and observational consequences of Dyson spheres were first described by their eponym Dyson (1960a) at the dawn of modern SETI. Dyson (1960a)'s original suggestion was quite general, pointing out that the exponential growth of an extraterrestrial species' energy supply could lead it to occupy "an artificial biosphere which completely surrounds its parent star." He argued that the infrared heat signatures of the "energy metabolism" of this extraterrestrial technology would likely be detectable at 10 μ, and that a search of the sky at mid-infrared wavelengths for point sources of such emission would be worthwhile.

In response to letters to the editor reacting to his work, Dyson (1960b) clarified that (1) he had not envisioned a monolithic shell or ring—which he wrote would be "mechanically impossible"—but rather a "loose collection or swarm of objects traveling on independent orbits around the star"; (2) that his description of infrared emission was independent of the specifics of how or why such a swarm might be constructed; and that (3) detection of infrared excess around nearby stars "would not

doi:10.1088/2514-3433/adf5aech15

by itself imply that extraterrestrial intelligence had been found," but would be an important discovery regardless of its cause.

Box 15.1: Freeman Dyson (1923–2020)

The author obtained and cropped this image from the Wikimedia website where it is credited to and made available Monroem under a CC BY-SA 3.0 license.

Professor Dyson was a British-American polymath and physicist who contributed foundational ideas to physics, mathematics, and many other fields. Dyson was raised in the United Kingdom, and studied maths at Cambridge under G. H. Hardy. He interrupted his studies to serve as a civilian scientist during World War II advising the British Royal Air Force's Bomber Command.

He did postgraduate work at Cornell working with Hans Bethe and Richard Feynman, and participated in the "grand cleanup" of quantum electrodynamics, unifying Feynman's approach with Schwinger and Tomonaga's. In 1951 he became a professor at Cornell, and just a year later accepted a lifetime appointment to Princeton's Institute for Advanced Study.

Dyson was perhaps the most celebrated physicist never to win the Nobel Prize. He worked on problems involving nuclear power policy and weapons nonproliferation; random matrices, topology, and number theory; the origin of life; and biostatistics. He is the eponym of the Schwinger–Dyson Equation, the Dyson Series, and the Dyson Transform. He was disdainful of academic elitism and was proud that neither he nor any of his children held PhDs. He discovered a knack for popular writing at 56, after which he wrote several books and contributed regularly to the *New York Review of Books*.

Dyson was intellectually playful and described his attitude toward science and creativity as "subversive." He wrote many compelling and at times fanciful papers, including a proof that matter is stable only due to non-electromagnetic effects like electron degeneracy forces that infinite consciousness is possible in certain cosmologies, that the graviton cannot be detected, an example of a spacefaring von Neuman machine he called the "astrochicken," a summary of a novel proof of Maxwell's Equations he attributed to Feynman, using "pitlamping" to detect vegetation in the Kuiper Belt, and how trees could grow on comets. He gained some notoriety for having heterodox views on climate change models.

In popular culture, and in SETI, he is best known for his 1960 paper in *Science* describing an "artificial biosphere," a concept Kardashev would dub a "Dyson Sphere." He was quite amused when the idea was the premise of a *Star Trek: The Next Generation* episode.

For more on Freeman Dyson, see his memoir "Maker of Patterns: An Autobiography Through Letters" (Dyson 2018).

Dyson (1966) later elaborated on the engineering behind the idea, showing there is no physical barrier to building large rigid structures in space (where "large" was up to 10^6 km), that such structures can be very light, and that planets can (in principle) be deconstructed for their material in sufficient quantities to build a sphere of useful thickness (of order meters). He also outlined how such structures would be discovered: first by their 3–10 μ excess consistent with being a protostar, then diagnosed as artificial via "irregular light variations due to starlight shining through chinks in the curtain, and for stray electromagnetic fields and radio noise produced by large-scale electrical operations."

But unlike the radio program envisioned by Cocconi & Morrison (1959) and executed by Drake (1961) in 1960, searches for infrared radiation as a technosignature would not begin for decades after Dyson's suggestion, and only recently have the observational consequences of large amounts of circumstellar technology been computed. Bradbury (2001) presents a good analysis of the history of the idea, the confusion in the literature regarding the shape and purpose of the technology in Dyson (1960a)'s original suggestion, and an extensive catalog of the literature on the subject. In particular, he examines the assumptions that the sphere would necessarily constitute a habitat and be a true sphere, and concludes that energy harvesting alone would be a sufficient function to justify searches for such technology (Wright et al. 2014b).

Dyson (1960a)'s approach to SETI has been generalized to include searches for most other kinds of extraterrestrial technology exclusive of deliberate communicative transmissions, in what has been called "Dysonian SETI" by Bradbury et al. (2011) (and "artifact SETI" by others, see also Wright 2018; Wright et al. 2018).

The first use of the term "Dyson sphere" in the literature is apparently due to Kardashev (1964), but since then the term has been occasionally reserved for the idea of a complete or nearly complete monolithic shell, with other configurations getting other names (such as a "Dyson swarm"). Rather than fuss over correct and precise terms for what are, after all, entirely hypothetical objects, it is best to just use the well-known term "Dyson sphere" generically, to refer to any collection of artificial material around a star that produces significant amounts of infrared heat signatures, regardless of its specific geometry or purpose. This preserves the best known term for such material in a manner more or less consistent with Dyson (1960a)'s original suggestion. That said, in what follows we will often implicitly invoke spherical symmetry, a common orbital radius, and common physical properties of the swarm components (at least roughly). Nonetheless, much of the analysis will extend to other configurations, as well, and so should still be useful as a way to guide observational searches for them.

15.1.1 Development of the Idea

The concept of a Dyson sphere in popular culture actually predates Dyson (1960a): Dyson (1960a) credits the novel *Star Maker* (Stapledon 1937) with the original idea. Indeed, in science fiction the idea usually manifests not as a swarm of small objects but as a single object of planet size or larger (a *megastructure*). Perhaps the most famous appearance of a Dyson sphere in popular culture is in the *Star Trek: The Next Generation* television episode "Relics," which depicts a monolithic sphere with an entire biosphere on its inside surface (which makes sense only because artificial gravity is ubiquitous technology in the *Star Trek* universe). Megastructures of other configurations are also common in popular culture, for instance in *House of Suns* (Reynolds 2008) and perhaps most famously in the *Ringworld* series of novels (Niven 1970), where the megastructure is a single, gigantic, spinning ring of impossibly strong material centered on a star, with the biosphere held on the interior surface by centrifugal forces.

Kardashev defined his Type II "civilizations" as those capable of commanding the entire energy output of their star (Section 4.2.2). His approach was general, but many other analyses of the gravitational, radiative, and thermodynamic properties of Dyson spheres invoke specific geometries, purposes, energy generation schemes, or other activities for the Dyson spheres. Some examples are studies of the gravitational dynamics of monolithic rings around stars (McInnes 2003; Rippert 2014 and references therein), analyses of very cold Dyson spheres (Lacki 2016), and partial shells used for stellar propulsion or energy extraction (Badescu & Cathcart 2000, 2006).

Badescu & Cathcart (2000) describe the concept of a *stellar engine*, in which the industry around the star is collectively used for some single purpose. They define a Class A engine as a general case of a Shkadov thruster (Shkadov 1988), in which material around the star of very low surface density ($\beta = 1$, see Sections 14.5.2 and 15.2.1) is held in place by stellar radiation pressure in a hemisphere or other configuration of an incomplete sphere. The result is to redirect the stellar radiation in one direction, creating a photon rocket effect, providing a very gentle thrust that, over the course of billions of years, can slowly nudge a star on a slightly different orbit through the Galaxy.[1]

They define a Class B engine as one that extracts mechanical energy from starlight to do work by operating a heat engine between two concentric spheres around the star. We will work out the details of such engines below, in a more geometry-agnostic way. They define a Class C engine as an intermediate case of a Dyson sphere that performs both functions.

Interestingly, there is little or no discussion in the literature of the possibility of directly *imaging* a Dyson sphere. If such a sphere were sufficiently close to Earth, it might be easily resolved at microwave, infrared, or optical wavelengths: a 1 au sphere around the closest stars to the Sun would subtend nearly an arcsecond on the sky. Indeed, we now have many images of protoplanetary and debris disks in reflected and emitted light, and such methods should be sensitive to Dyson spheres, as well.

[1] Shkadov thrusters are a relatively inefficient way to move a star because photons have so little momentum. The Caplan thruster, which captures solar wind particles and photons and uses them to both push the star with a forward jet and thrust forward with a backward jet, is more efficient (Caplan 2019).

15.1.2 Why Dyson Spheres?

Why build a Dyson sphere? Dyson suggested it would be as a living surface, but a more general answer is in order to collect starlight as energy, and/or to reradiate that energy away.

Because Dyson spheres would be swarms of orbiting satellites and not a single construction project, they need not necessarily be planned out in advance. As an analogy, consider a large city like New York City. Over hundreds of years, the city and its environs have grown into a large "megastructure," a nearly continuous network of concrete and steel in roads, sewers, tunnels, and buildings many miles across. While certainly significant advance planning went into many components of the city, it is not a single engineering project, but a loosely coordinated set of innumerable interacting projects.

Now, consider our own technology in space; our satellites and interplanetary probes all have solar panels, shields, and radiators that block or absorb sunlight as part of their energy and thermal management. In some sense, these represent a very sparse Dyson sphere of our own. A detectable Dyson sphere might "grow" organically, as a species expands into its circumstellar environment, collecting more and more energy that otherwise would escape to space.

Any energy collected by these swarm elements used for some purpose—onboard computing, radio signaling—would eventually be re-radiated as "waste" heat, but these activities would constitute the sort of work envisioned by Badescu & Cathcart (2000) as part of a Class B stellar engine.

15.2 Engineering of Dyson Spheres

15.2.1 Stability of a Monolithic Sphere

Dyson spheres cannot be solid objects supported by their own gravity for several reasons. We cover two of them here.

15.2.1.1 Dynamical Non-stability

The first illustration of the non-stability of a monolithic Dyson sphere comes from Gauss's Law. Consider first the gravitational potential inside of a uniform, empty, massive, perfectly rigid sphere far from any other masses (i.e., we ignore the central star for the moment). By symmetry, all points on the sphere must be at the same gravitational potential, given by $-GM/R$ where M is the mass of the sphere and R is its radius.

Within the sphere, the potential must be constant because it is hollow, which means there are no gravitational forces within the sphere. To an object floating inside the sphere, it is as if the sphere is not there at all. Evidently, the gravitational force from all of the interior points on the sphere cancel each other out, resulting in no net forces.

By Newton's second law if the sphere is not exerting any net force on an interior object, the object must exert no net force on the sphere. This means that if we put a star at the center of the sphere, it and the sphere will be decoupled gravitationally: the sphere will not be "orbiting" the star, and so it could drift in any direction and crash into the star unless it is actively controlled.

This is illustrated heuristically in Figure 15.1. When the star sits at the center of the sphere, $\Delta R = 0$ and by symmetry there is no force pushing the star or sphere in any direction.

If the star becomes displaced by some amount in the x-axis, we can see by symmetry that any resulting force must be along that horizontal axis. We can consider a small, circular patch of the sphere at either end of the x-axis, subtending an angle $d\theta$ as seen from the star.

The surface area of that patch far from the star is $((R + \Delta R)d\theta)^2$, and the corresponding patch on the near side has surface area $((R - \Delta R)d\theta)^2$. These two patches experience gravitational attraction toward the star inversely proportional to their distance squared, so proportional only to $(d\theta)^2$ which is the same for both patches (i.e., the $(R + \Delta R)^2$ and $(R - \Delta R)^2$ terms cancel.) This calculation is performed more rigorously in Wright (2020).

One might wonder if photon pressure might stabilize a very thin sphere, but one can demonstrate it by considering the momentum of the ensemble of photons emitted by the star. By symmetry, the sum of their momenta is zero (or else their recoil would push the star in some direction as they left). Therefore whether the

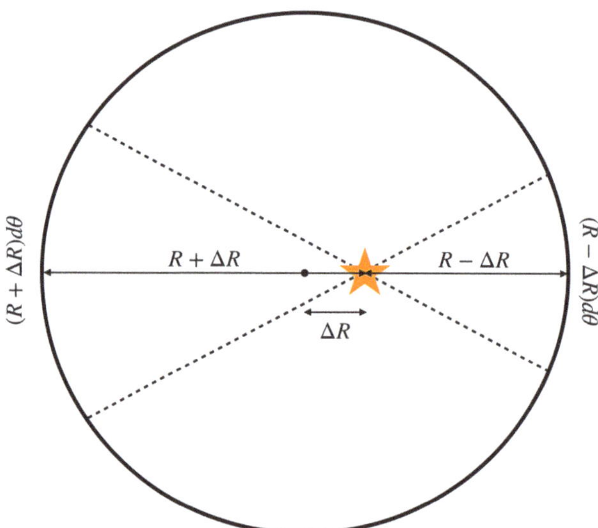

Figure 15.1. Cutaway of a monolithic Dyson sphere of radius R with a star displaced from the center by amount ΔR. A small arc subtending an angle $d\theta$ (greatly exaggerated here for clarity) is shown on either side of the star. Each arc spans the diameter of a circular patch on the sphere of area $((R \pm \Delta R)d\theta)^2$. The gravitational force from each on the star, which is inversely proportional to $(R \pm \Delta R)^2$, is the same, so there is no net force. Figure inspired by a slide by Samual de Jong.

sphere reflects or absorbs the photons, the momentum they impart on it must also be zero, regardless of the location of the star.

This situation is one of "neutral" stability. It implies that monolithic spheres would need some mode of thrust to maintain themselves against perturbations that would crash them into the star.

We can perform an order of magnitude calculation to see how long stability could be maintained via thrust. We will consider a monolithic Dyson sphere of arbitrary shape (so, perhaps a ring), mass M, of characteristic radius R from the star. The Dyson sphere suffers perturbative forces due to other masses m_p (such as a planet in the system) at closest approach distance r from the Dyson sphere (for instance, a distant ice giant planet in the system).

These accelerations will cause the Dyson sphere to distort and suffer differential acceleration with respect to the star in complex ways that depend on the geometry of the problem. To a rough order of magnitude, we can calculate the force to be balanced simply as

$$F_\mathrm{pert} = \frac{GMm_\mathrm{p}}{r^2} \tag{15.1}$$

with the caveat that the true force may be much smaller for many geometries, and r may vary greatly across the different parts of the Dyson sphere. The Dyson sphere can cancel these accelerations by using some of its mass as thrust, expelled at velocity v, producing a force on the Dyson sphere

$$F = \dot{M}v \tag{15.2}$$

The power P used to generate this thrust is

$$P = \frac{1}{2}\dot{M}v^2 \tag{15.3}$$

and so the mass loss rate of the Dyson sphere is

$$\dot{M} = \frac{F^2}{2P} \tag{15.4}$$

meaning that the Dyson sphere will have a lifetime of order

$$\tau_\mathrm{life} = \frac{M}{\dot{M}} = \frac{2PM}{F^2}. \tag{15.5}$$

Equating this to the perturbative forces of other masses in the system via $F = F_\mathrm{pert}$ yields

$$\tau_\mathrm{life} = \frac{2Pr^4}{MG^2m_p^2} = \left(\frac{P}{L_\odot}\right)\left(\frac{r}{\mathrm{au}}\right)^4\left(\frac{M}{M_{♃}}\right)^{-1}\left(\frac{m_p}{M_{♃}}\right)^{-2} 200\ \mathrm{yr} \tag{15.6}$$

where $M_{♃}$ is the mass of Jupiter and where we expect that P will be some (potentially very small) fraction of the total output of the star. Thus, for a monolithic sphere of mass $M_{♃}$ to be stable for timescales long enough for it to be discovered (say, $\tau_{\rm life} > 10^7\,{\rm yr}$) there can be no perturbers of more than roughly $10^{-2}M_{♃}$ within 1 au, and no gas giant planets within 10 au. It is worth reiterating that this estimate (and its scalings with r and R) is extremely sensitive to the assumed geometry of the problem.

If there are no planets in the system because they were used to construct the Dyson sphere, then it seems stability via starlight-powered thrust could be maintained with sufficient active control for cosmic timescales. Alternatively, the Dyson sphere might use mass from the star as fuel to stabilize itself.

15.2.1.2 Mechanical Instability

As asserted by Dyson (1960b) and shown quantitatively by Papagiannis (1985), the gravitational forces acting on a Dyson sphere are so extreme as to render the possibility of a real, rigid Dyson sphere supported by its own material strength dubious.

To see the difficulty, consider why an area element dA does not accelerate toward the star under the force of gravity. In the case of a free-floating object, one can invoke centrifugal force from an orbit, but this is not possible for a monolithic sphere containing the star (even for a rotating sphere, the poles will have zero centrifugal force, see Covington 1991).

Instead, all of the area elements feel the same radial force, which would naturally cause the sphere to shrink and therefore generate lateral, compressive forces between the area elements. Because of the curvature of the sphere, there is a small outward component of this compression force, which increases as the sphere shrinks until it is sufficiently large to counteract the star's gravity, and the sphere achieves equilibrium. This is also how arches work: the curve at the top of an arch provides a way for the keystone at the top to be supported vertically by the two stones on either side (and slightly lower) than it. A monolithic Dyson sphere is essentially an arch the size of a solar system. In this formal sense, monolithic spheres thus appear mechanically stable.

Real structures are not infinitely rigid, however, and will deform if the forces on them exceed their elastic strength. For perfect spheres subject to uniform external pressure, collapse can occur for two reasons: because the material yields, or because it buckles.

When a material experiences forces beyond its yield strength σ_y, it deforms. Imagine trying to use a stick of room-temperature butter stood on its end to support a dictionary: the yield strength of the butter is low, so it will squash beneath the book's weight.

For a sphere, the yield strength is exceeded if uniform pressure on it exceeds

$$P_{{\rm crit},y} = 2\sigma_y \frac{\Delta R}{R} \tag{15.7}$$

where ΔR is the thickness of the sphere (Pan & Cui 2010).

Buckling occurs when an object folds under weight. Consider trying to support a dictionary with a stick of dried spaghetti: it will not squash like the butter, but it will bow until it snaps. For a sphere, this failure mode occurs above a critical pressure, given to within a cofactor of order unity by

$$P_{\text{crit,b}} \approx E\left(\frac{\Delta R}{R}\right)^2 \tag{15.8}$$

where E is the elastic (Young) modulus (Papagiannis 1985; Pan & Cui 2010). For monolithic Dyson spheres, we (presumably) are in the thin regime so Equation (15.8) (the elastic buckling condition) controls.

The force on a Dyson sphere per unit area from the star's gravity is given by

$$P_{\text{grav}} = \frac{GM_*\rho\Delta R}{R^2} \tag{15.9}$$

where ρ is the density of the sphere material. This means that to avoid catastrophic buckling the sphere material must have an elastic modulus

$$E \gtrsim \frac{GM_*\rho}{\Delta R} \tag{15.10}$$

which, interestingly, is independent of the radius of the sphere. Plugging in some characteristic values yields

$$E \gtrsim \frac{GM_\odot(1\ \text{g/cm}^3)}{10\ \text{m}} \approx 10^{13}\ \text{GPa} \tag{15.11}$$

which exceeds the elastic modulus of the strongest known material, carbyne, by nine orders of magnitude. Indeed, since E is proportional to the strength of the atomic bonds of the material, and since the carbon–carbon bond is the strongest in nature[2] it appears impossible for a Dyson sphere of any size to support itself via elastic forces against gravity.

One tempting solution that fails is to thicken the sphere and lower its density, perhaps to 100 km and 0.1 g cc^{-1}, bringing the necessary strength of the material within a mere four orders of magnitude of the carbon–carbon bond. But for $R = 1$ au this would give the sphere a total mass of $\sim 1.4\text{M}_\odot$, and so the sphere's self-gravity would then become the dominant source of buckling and no further thickening would help. Adding struts or other internal bolsters is also no help, because the struts will suffer the same difficulties as the sphere itself.

[2] Invoking nuclear material as a building block only makes things worse: the density of such material is 10^{14} times higher, but its Young modulus is "only" of order 10^8 GPa (Scheuer 1981), and so it is *nineteen* orders of magnitude too weak.

A better solution is to make the sphere *thinner*, and so lessen the forces on the sphere via radiation pressure. For instance, the ratio of radiation forces to gravitational forces $\beta \sim 1$ (Section 14.5.2) when the surface mass density of a patch of material normal to the Sun is

$$\sigma \approx \frac{L_\odot}{4\pi G c M_\odot} \approx 0.8 \text{ g/m}^2 \tag{15.12}$$

which is comparable to the thinnest gold leaf. A monolithic Dyson sphere of this surface density is thus not mechanically unstable, although it is still dynamically neutral and it is not obvious what the utility of such a thin sphere would be (but see Sections 15.4.3 and 15.6). This surface density also serves as a lower limit on material bound by gravity to a star. A partial sphere in such a configuration would be a Class A stellar engine or Shkadov thruster (Section 15.1.1).

We thus see that monolithic spheres are essentially impossible to build, barring the existence of a new form of matter or some way to stabilize the sphere's material.[3]

Instead, we should consider, as Dyson did, a collection of independently orbiting structures, akin to the satellite constellations orbiting Earth but on a much larger scale.

The lessons of managing objects in Earth orbit might have important implications for detecting Dyson spheres: Lacki (2025) finds that unless the swarm elements are carefully controlled, it is susceptible to Kessler Syndrome, in which the fragments of collisions lead to a collisional cascade that eventually destroys the entire swarm. This suggests that Dyson spheres will not last much longer than they are maintained, and that any Dyson spheres discovered might be accompanied by other, active technosignatures.

15.2.2 Mass Constraints

Another important element to consider is whether enough mass exists in a planetary system to construct the sphere. Since we imagine the sphere should have some substance and not be merely paper thin, we might assign it a mass per unit area similar to a building or a space station. A building might weigh 50,000 kg across a couple hundred square meters; the Mir space station weighed around 150 tonnes and had dimensions of order 20–30 m. In both cases we end up with a surface mass density of around 20 g cm^{-2} or so.

[3] There are some more complicated ideas that get around these limits. For instance, one might be able to somehow actively stabilize the sphere to prevent it from buckling, which would mean Equation (15.7) would hold instead of Equation (15.8). This would improve the limits by a factor of $(R/\Delta R)$ or perhaps about a factor of 10^{11}, requiring yield strengths of "only" 1–100 GPa, which is plausibly within the outer limits of physically possible materials. Such ideas generally invoke rather complex and delicate engineering whose slightest failure would result in the immediate and catastrophic destruction of the sphere, without providing any obvious benefit in return. A swarm of material is a simpler and safer solution.

If a sphere's components were at the typical distance of 1 au, then a sphere covering all of the star's output would have an area of around 10^{27} cm^2, so would weigh around 10^{28} g, which is a few times the mass of Jupiter.

From this order of magnitude estimate, we see that there exists enough mass in planets to build a complete sphere around a star of about the mass density of typical buildings. Of course, there are a lot of assumptions that went into this calculation: that it is possible to deconstruct a planet the size of Jupiter, that the mass in Jupiter can be converted to a form useful for building things, and that any species would ever do such a thing.

But, again, Dyson's point was just to show such a thing is possible, not advisable or practical. Nonetheless, to answer some of those objections: Dyson (1966) outlined a method of deconstructing planets (it is not impossible); Jupiter is mostly hydrogen, but the non-hydrogen components could be sifted out, or fusion could be used to convert the hydrogen to other elements; the sphere might be built from the mass of the star itself instead of a planet.

15.3 Thermodynamics of Dyson Spheres

To understand the energetics of Dyson spheres and their infrared heat signatures, we need to think about the entropy of the radiation it emits. The concept of thermodynamic entropy can be tricky to grasp, but in the case of the radiation of Dyson spheres it is rather straightforward. Let's start by developing a heuristic understanding of entropy and exergy in the context of a machine doing work, like a computer or a car.

15.3.1 Energy Use

Consider some useful energy you have and want to put to work: perhaps it is stored in a battery, or it is sunlight about to strike your solar panels. Because energy is conserved, we can follow the energy from its origins and see where it ends up.

The computer will take that energy and use it to push electrons through semiconductors to perform its calculations, but only at finite efficiency. It can perform so many calculations per erg of electricity before the energy is "used up," as the electrons moving through those semiconductors lose their kinetic energy to their surroundings, heating them up. The computer thus converts virtually all of its electric energy to heat, which it uses fans and radiators to dispose of by blowing or radiating it away.

In a car, the energy from the battery used to propel it forward is slowly lost to friction with the road and air, the brake pads when you stop suddenly, and to mechanical and electrical inefficiencies, all of which generate heat.

Virtually all energy use on Earth follows this pattern, ultimately ending up as heat.

Note that here we distinguish energy *use* from energy *storage*, which might take many forms. If you use a crane to build a skyscraper, you are converting some of the crane's mechanical energy into gravitational potential energy of the building materials, which is a kind of energy storage, as long as the building stays up.

Only a very small fraction of all energy humans use on Earth goes to long-term storage of energy like this. In the large majority of other cases on Earth, energy eventually ends up as heat, which is ultimately radiated away to space. We call this heat *waste heat* because we can't do any more useful work with it.

15.3.2 Efficiency

Interestingly, if 100 ergs of sunlight strikes our solar panels, it is literally *impossible* to extract 100 ergs of useful work from it. The amount we can extract depends on the temperature of the machine performing the work. This is because the thermal motion of the components of the machine is itself a form of energy, and you can only extract energy from heat (like in an internal combustion engine, coal plant, or nuclear plant) or radiation (as with solar panels) when that heat or radiation has a higher temperature than the machine doing the work. The fraction of energy you can extract is called the *Carnot efficiency* of a machine, given by

$$\eta = 1 - T_{\text{waste}}/T_{\text{h}} \tag{15.13}$$

where T_{waste} is the temperature of the machine and its waste heat, and T_{h} is the hotter temperature associated with the energy before it is used. This means we can heuristically think of any collection of energy as having a temperature, even if it is not characterized by random thermal motions, based on how much of it is available to do work.

Let's say you are collecting sunlight for electricity, which has a temperature characteristic of the surface of the Sun, around 6000 K, and using that energy to power your computer. Your computer works at around 300 K, so the Carnot efficiency is around 95%: at best you can get 95 useful ergs of work for every 100 ergs of sunlight that strikes the solar panels.[4]

15.3.3 Exergy

In our example above using sunlight to power a computer at 300 K, the 95 ergs of useful work we could theoretically get from 100 ergs of sunlight is called the *exergy* available in the sunlight. If you were doing your calculations on Pluto, you would be able to get more out of the sunlight because your computer would be colder, so it would have more exergy.[5] Exergy is thus situationally dependent: it depends on the environment the energy is in.

[4] Note this is a maximum! This is also a heuristic argument—in reality there are many thermodynamic nuances and practical inefficiencies that mean you end up with much less than this in practice.

[5] Note that you can't get around this limit on Earth by cooling your computer to make it more efficient: your cooling system takes energy to operate, and when you factor that energy into your budget, at best you'll break even.

This is the way to think of energy use from sunlight: if it had hit the ground and warmed it, it would have done no useful work for us, and just radiated away back to space, its exergy all wasted. If instead we capture it via solar panels, it can do some work for us until its exergy is all used up, at which point it once again warms something and radiates away to space.

15.3.4 Entropy

But what is different about the energy as sunlight and that same energy as waste heat? What is it about 100 ergs of waste heat that it has 0 ergs of exergy, while 100 ergs of sunlight has 95 ergs? The difference is *entropy*.

Entropy has many different definitions in different situations. It can represent the amount of information in a system, it can be calculated in an absolute sense as the logarithm of the number of microstates in a system under certain constraints, and it can heuristically be thought of as the amount of "disorder" in a system. Thermodynamically, it is part of a differential relationship between heat and temperature, but for our purposes it is usefully phrased as the amount of energy that is unavailable to do work (per unit temperature). In this sense, it is the opposite or complement of exergy (although it has different units). All of these definitions turn out to be mathematically consistent in deep and subtle ways.

So, in our heuristic, as energy gets "used up" it goes down in temperature and gains entropy, until it has the same temperature as the machine it is operating in, at which point there is no exergy left and we can get no more use out of it (the Carnot efficiency becomes zero). At this point it is at maximum entropy given its surroundings.[6]

In our example of a computer, the energy of the electric fields manipulating the electrons, and the kinetic energy of those electrons' motions, is at low entropy (so has a high temperature in our heuristic). As the electrons interact with the finite-conductivity semiconductor they transfer their kinetic energy to the atomic lattice that makes up the circuit, where the energy gets dissipated as heat and warms up the lattice. That energy, now in the form of heat, then has high entropy, a lower temperature (basically the same temperature of the lattice) and the exergy of that energy has been lost.

In our example of a car, the high-temperature, low-entropy energy stored in the batteries gets transferred to the car's kinetic energy, with some finite efficiency. As the car brakes and moves on the road and through the air, portions of the energy are lost to inefficiencies and are converted to heat, gaining entropy and going down in temperature.

The second law of thermodynamics states that the entropy associated with some amount of energy inevitably goes up, as it did in these examples—it cannot go down unless the entropy of some other energy or material goes up to compensate.

[6] If the energy is a radiation field, it will then be characterized by blackbody radiation at that temperature.

15.3.5 Application to Dyson Spheres

Light that strikes a Dyson sphere might reflect away, or it might be absorbed by solar panels or just a passive dark surface. If reflected, the entropy will not change much. If absorbed by solar panels, it might be efficiently used for some purpose, such as electricity generation to power computers or machines, the energy being converted to some useful form. However, it must ultimately suffer the same fate as light that struck a passive dark surface, and be radiated away as waste heat.

But what temperature will the waste heat be? This depends on how large the radiating surfaces of the Dyson sphere are. It has to get rid of energy at the same rate it collects it, and the only way to do this with a small radiating surface is to have that surface be very hot, in which case the Carnot efficiency will be very low—if the builders of the Dyson sphere want to maximize efficiency, they'll want the biggest radiators they can get. If the radiators have the same surface area as the collectors (e.g., they are on opposite sides of shared panels) then its temperature will be about the same as anything else at that orbital distance from the star.

So regardless of what use a Dyson sphere puts its stellar energy to, we can, from a global energy and entropy budget perspective, treat it like any other material orbiting the star, like asteroids or dust: the energy will end up as infrared radiation from the warmth of the surfaces.

15.3.6 Kinds of Energy Use

If Dyson spheres are common enough to be observed by us, then they must be long-lived, so the nature of their work must be sustainable on cosmic timescales. The Sun generates enough energy to completely unbind[7] itself in $\sim$30 Myr. The gravitational binding energy of the planets is significantly smaller than this: the Sun generates enough energy to unbind every atom in Jupiter and eject it from the solar system in less than 1000 years. This provides a rough limit on how much energy a Dyson sphere—or anything near a star—could manage at once.

Only if the work of the Dyson sphere is to deconstruct the star should we expect to find it doing lasting work on its system. Whatever work a Dyson sphere does must therefore result in the energy leaving the system.[8]

Of course, it is possible that a species may be able to tap additional sources of energy using physics or technology not obvious to us. Speculation into the nature of such technology is perilous, but we can explore the limitations imposed on it by our

[7] "Unbind" here refers to disassembling the Sun and widely separating its components. This takes energy because the Sun has significant self-gravity, so it takes energy to lift mass from its surface.

[8] Or, in principle, being stored in some manner with significantly more capacity than heat or gravitational potential energy. The only obvious such mechanism would be to store the energy as mass. While converting starlight into, say, protons would seem inefficient (collecting mass from the stellar wind would be at least as effective), the production of antimatter is plausible activity.

current understanding of the laws of physics. The conversion of mass to energy through fusion or fission is generally inefficient (and, at any rate, stars already do this "for free"), but in theory the use of a rotating black hole (through the Penrose mechanism) can provide very high efficiencies. The star itself might provide the fuel for such an engine. If such a process operated at 40% efficiency and consumed the entire star over its natural lifetime, it would effectively increase the luminosity of a star system by a factor of $(0.4/0.001) = 4 \times 10^4$ (if the higher luminosity were to be sustained over 10 Gyr). More exotically, alien species might be able to tap zero-point energy or cosmic dark energy for their needs. It is unclear to what degree this is allowed or prohibited by the laws of physics.

Regardless of the energy source, however, one is bound by the laws of thermodynamics: one must radiate the energy away, which requires large surface areas. The ultimate necessity of a Dyson sphere is thus to provide surfaces to dispose of waste heat, which is an inevitable consequence of any energy use, regardless of whether the interiors of those surfaces are used to collect starlight.

The only exception is if the Dyson sphere stores energy or emits it away in a low-entropy form. As we have seen, if a Dyson sphere is to be long-lasting it can't store very much starlight—it will collect enough energy to completely melt and unbind itself gravitationally in just centuries. But it might radiate some of it away as, say, radio waves or laser communication, which would lower the amount of energy it radiated as waste heat. We will return to this idea later.

Box 15.2: Jargon Review: Dyson Spheres

- **Megastructure**: Any large structure, but specifically a planet-sized one that might be observable as a technosignature, especially in transit.
- **Shkadov thruster**: Also a "Class A" stellar engine, a large mirror held up against gravity by stellar radiation pressure that redirects starlight in one direction to provide gentle thrust that can slowly move the star.
- **Work**: A rich term in physics, here defined as some useful application of energy, such as a calculation in a computer.
- **Entropy**: A rich term in thermodynamics, heuristically associated with "degree of disorder" and formally defined in terms of amount of energy that is unavailable to do useful work, per unit temperature.
- **Exergy**: The amount of useful work that can be done with a certain amount of energy. Energy that has low entropy, such as a radiation field that is much hotter than its surroundings, has high exergy. A radiation field at the same temperature as its surroundings has zero exergy.
- **Efficiency**: A term with many meanings, here defined as what fraction of energy is available to do useful work. The **Carnot efficiency** is defined as $\eta = 1 - T_{\rm waste}/T_{\rm h}$, where $T_{\rm h}$ is the temperature of the radiation field from which work is extracted, and $T_{\rm waste}$ is the temperature of the machine doing the work.
- **Waste heat**: Energy leaving a system at high entropy, having been "used up." This energy cannot be used to do any more useful work in its current environment.

Box 15.3: Efficiency, "Waste," and the Jevons Paradox
The term "waste heat" is somewhat misleading because it suggests inefficiency or some sort of failing of a machine doing work. A common objection to infrared SETI is that a species energy hungry enough to collect most of a star's light would "waste" as little of that energy as possible, leaving nothing to detect. But one must keep in mind that energy is conserved, so any energy collected must be expelled after it has been "used up."

This is somewhat at odds with the usual sense of thermodynamic efficiency, in which waste heat is calculated as the input energy *not* doing useful work. The resolution of the apparent contradiction is that we are concerned only with the long-run, global fate of the energy, not the work it does between being absorbed and radiated. Consider cars and computers: both do very useful work, and potentially at high efficiency, but all of the useful work they do (i.e., increasing the car's velocity, pushing electrons through semiconductors) ultimately becomes waste heat (via wind resistance or electric resistance, for instance).

So infrared SETI can also be called "waste heat SETI," but one should not take this term to imply that only the inefficiencies of a species can be detected. Except for the energy being emitted at low entropy (via radio waves, for instance) *all* of the energy used by species is "waste heat."

Another common objection is that no species would build a Dyson sphere because any truly long-lived species would be so efficient at energy use they would not need that much energy. Perhaps so, but the lesson of human technology teaches the opposite case is also possible: as computers have become more energy efficient, we have devoted *more* of our energy to them.

In economics this effect is sometimes (misleadingly) called the Jevons Paradox, in which demand for a good is so sensitive to price (i.e., its price is highly *elastic*) that efficiencies that make the good cheaper to produce end up increasing its demand so much that *more* resources are spent producing it, overall. This is in contrast to a relatively inelastic good like food: per capita food consumption is more-or-less fixed, so agricultural innovations that have made food cheaper to produce have not caused much more of it to be bought, and the total amount of resources spent producing it per capita have gone down.

The Jevons Paradox is dramatically illustrated by computing: since 1940 the energy cost per calculation has dropped by over 15 orders of magnitude (Theis & Wong 2017), but the energy we spend on calculations has gone from essentially zero to, today, about 10% of humanity's entire energy supply. Computation is so versatile and useful for our society that it is easy to imagine its price elasticity might be a pervasive, and perhaps universal, phenomenon among technological species, and that they might all use much or most of their energy supply for it.

15.4 Optimal Sizes and Temperatures of Dyson Spheres

15.4.1 Optimizing Thermodynamic Efficiency

Some authors (e.g., Suffern 1977; Badescu 1995) have presumed that Dyson spheres would constitute living surfaces (either on the inside, outside, or in between) and/or that they would be engineered to maximize the work that could be done or the efficiency with which starlight could be processed.

Wright (2020) provides a detailed calculation of Dyson sphere efficiency as a function of size and temperature; here we present a simplified approach that captures the essence of the conclusions.

The effective temperature of a Dyson sphere collecting starlight is set by the fraction of starlight it collects and its distance from a star of some temperature T_*. Larger spheres will be at lower temperatures, and so have higher maximum efficiencies, but will also need more mass.

In general, because the Carnot efficiency increases with decreasing temperature, there is a tension between an efficient Dyson sphere (which is large) and the mass needed to build it (which favors smaller spheres).

In this work, we have generally remained agnostic with respect to the purpose or design of a Dyson sphere. Here, we will follow the spirit of Dyson (1966) and explore the maximum efficiency of a Dyson sphere allowed by physics as a function of its size, with minimal appeal to engineering and function.

For scale, we will refer to nominal numbers for a "canonical" Dyson sphere consisting of a complete shell of orbiting material at $R = 1$ au from a Sun-like star with $T_* = 5772$ and $R_* = R_\odot$. To be useful, the elements of the shell must have some substance, which requires a large amount of mass. If the density of the shell is of order 1 g cc^{-1}, then one Jupiter mass would provide for components a few meters thick.

We will make the assumption that the star is its only important source of energy (i.e., $\epsilon = 0$).

For a black solar panel of area A at distance d from a star of luminosity $L = 4\pi R_* \sigma T_*^4$ that collects all light that strikes it and radiates it away from both sides, we have from energy balance that the power absorbed is

$$P_{\text{in}} = L\frac{A}{4\pi d^2} \tag{15.14}$$

and the power it radiates away depends on its effective temperature T_{waste}:

$$P_{\text{out}} = 2A\sigma T_{\text{waste}}{}^4 \tag{15.15}$$

setting these two equal shows its temperature is

$$T_{\text{waste}} = \left(\frac{T_*^4}{8\sigma d^2}\right)^{\frac{1}{4}}. \tag{15.16}$$

If we have so many collectors that they start to form a sphere, then they cannot radiate back toward the star because there are lots of other collectors in that direction also radiating—to stay efficient they will only be able to radiate on their outward facing sides, in which case

$$T_{\text{waste}} = \left(\frac{T_*^4}{4\sigma d^2}\right)^{\frac{1}{4}}. \tag{15.17}$$

We'll use this latter case for concreteness. The amount of exergy per unit time available to them is given by the Carnot efficiency η times P_{in}:

$$\eta P_{\text{in}} = \eta P_{\text{out}} = \left(1 - \frac{T_{\text{waste}}}{T_*}\right) A_{\text{rad}} \sigma T_{\text{waste}}{}^4. \tag{15.18}$$

This is a fifth order polynomial, and for fixed area is maximized at $T_{\text{waste}} = \frac{4}{5}T_*$. The balance here is between being close to the star (which gives you more energy per unit collecting area) and being colder (which makes you more efficient).

This optimum is at a very high temperature: for the Sun it is around 4600 K! Solid materials do not exist at this temperature, so the conclusion is that the most efficient way to use a fixed amount of radiating area (which is presumably proportional to mass) is to operate at the highest possible temperature, with a Carnot efficiency of around 20%. This might argue for small, hot Dyson spheres. Yes, lots of energy gets wasted with such hot spheres, but you're always better off putting new mass close to the star than trying to capture the high-entropy waste heat of existing material around the star.

However, this is only optimal if mass is at a premium and one cannot build an entire sphere. Once that sphere is essentially complete—all of the energy is being captured—then an additional, larger Dyson sphere could in principle be constructed that captured the waste heat of the inner sphere and made use of some of the remaining exergy (being farther from the star, it would operate at a lower temperature).

The lesson is that if energy use is being optimized, that once the sphere is largely complete, its size will depend on the total mass available. This trend has strongly diminishing returns, however. The total energy flux into a sphere at distance R is just the luminosity of the star, given by $4\pi R_*^2\sigma T_*^4$. Similarly, the total energy flux out in waste heat is given by $4\pi R^2\sigma T_{\text{waste}}^4$. Setting these equal yields

$$A = \left(\frac{T_*}{T_{\text{waste}}}\right)^4 4\pi R_*^2 \tag{15.19}$$

$$=4\pi R_*^2(1 - \eta)^{-4}. \tag{15.20}$$

This shows how much area is needed to achieve a certain thermodynamic efficiency on the energy being used, and that it is a very steep function of the inefficiency of the system (it goes as the fourth power). For our nominal 1 au Dyson sphere, we have $T_e = 390$ K and efficiency $\eta = 93\%$, meaning only 7% of the exergy of the starlight is not being exploited. Achieving 99% efficiency with a larger Dyson sphere at the same surface mass density (i.e., going from 7% inefficiency to 1% inefficiency) would require 2400 times as much material—if we needed an entire Jupiter mass of material at 1 au, then we now need more like 2.4 solar masses worth, which is more mass than exists in the star itself.

For every additional "nine" of efficiency, the Dyson sphere requires four orders of magnitude more radiating material.[9]

[9] If the energy is being put to use in a low-entropy way, for instance for powering communication beacons, then the scaling goes as $A \sim (1 - \eta)^3$, which is still quite steep and does not significantly change our argument.

15.4.2 Efficiency of Computation

In the case of purely dissipative work, one might follow Sandberg et al. (2016) and consider computations the nominal kind of work performed since the amount of such work that can be done scales with exergy and does not result in any stored or emitted energy—an ideal function for a Dyson sphere. Indeed, Sandberg (1999) provides an extensive discussion of the limits of many aspects of Dyson sphere computation including memory storage density, computational speed, and energy use.

Briefly, Landauer's principle sets the minimum amount of heat generated by a single binary logical operation such as AND or NOR at $kT\ \ln(2)$. This has the nice consequence that colder computers can do more operations with a given amount of energy. This means that large Dyson spheres win twice when their work is computational: once because they can extract more exergy from starlight, and again because they need less energy per computation.

The rate of classical computation r at a distance R from a star at power level L at T_e, and is

$$r = \frac{\eta L}{kT_e \ln(2)} \tag{15.21}$$

which for our nominal Dyson sphere is approximately 10^{47} logical operations per second. This rate is maximized for very large Dyson spheres with $T_e \to T_{min}$, which for the Sun would yield 10^{49} logical operations per second.

We can then define a *computational* efficiency η_{comp} calculated as a Dyson sphere's computational rate compared to the maximum possible rate at $T_e \to T_{min}$.

$$\eta_{comp} = \frac{r}{r_{max}} = \frac{\eta}{\eta_{max}} \frac{T_{min}}{T_e}$$

which we have seen for our nominal Dyson sphere is around 1%. It would seem, then, that when optimizing for computational rate instead of work, there might be great profit in investing in larger Dyson spheres.[10]

For Dyson spheres far from their limits (i.e., neglecting outside radiation and with $\eta \sim 1$), we can write $L \approx A\sigma T_e^4$, and so the scaling of computational rate with Dyson sphere area is

$$A = \frac{(k\ \ln(2))^4}{L^3 \sigma} r^4 \tag{15.22}$$

[10] Or in waiting long enough that T_{min} falls. Sandberg et al. (2016) suggest that the reason we do not see Dyson spheres is that the would-be builders are "aestivating," i.e., in a state of suspended animation until the Universe cools enough for their computers to run more efficiently. See, however, Bennett et al. (2019) for a critique of this idea.

The cost function here is not as steep as with thermodynamic efficiency until $T_e \rightarrow T_{min}$, but it still requires a factor of 16 in material to double computing power. Dyson spheres used for computing might thus have larger optimal sizes and lower optimal temperatures than other Dyson spheres, but there are still strongly diminishing returns.

We use the qualifier "classical" to describe this computing because it remains unclear if the entropy produced by logic gates represents a fundamental limit of computing, or whether reversible computing techniques can surpass those limits (see, e.g., Frank 2002). If so, then it is not immediately clear which aspects of Dyson spheres would optimize such computation, although Sandberg (1999) provides some possibilities.

For instance, all physical systems are subject to noise and error, and so the entropy generated (and energy disposed of) during error correction is likely a universal feature of all computers. A fully reversible computer might do many more logical operations with a given amount of energy than its classical counterpart, but it still needs a source of low entropy power to clear its computer memory for error correction. In this case, the rate of memory clearings a Dyson sphere can perform for a given exergy is equal to its classical computation rate r, and this might drive a similar design optimization.

15.4.3 Optimal Temperatures for Dyson Spheres

All of these calculations presume that a Dyson sphere is being engineered in some optimal way for energy use. The fact that multiple optima exist depending on the constraint we assume shows that we cannot really predict the sizes or temperatures or covering fractions of spheres, because they depend on how and why the spheres were built. After all, one would hardly look at humanity's megastructures—cities and megalopolises—and conclude they were optimizing something as simple as energy use or population density. A Dyson sphere that grew organically might be a compromise among these optima, or even far from any of them.

However, given the extremely steep efficiency dependence and the fact that solid material melts at high temperatures, and that the optima in mass-constrained calculations are for small, hot spheres, suggest that all but the largest and most ambitious projects may involve spheres at temperatures of hundreds of Kelvins—as hot as possible without melting important materials, essentially.

From Wien's law (Section 2.2.3) we can calculate we thus expect the majority of the infrared heat signatures of Dyson spheres at temperatures of 100–1000 K to be emitted between 3 and 30 μ. This is in the near and mid-infrared regions of the EM spectrum.

Nonetheless, it is possible that a species might find ways to make the coefficient so small and the benefit function so large that very cold Dyson spheres would be common. For instance, if the purpose of the Dyson sphere was only to shift starlight to the lowest possible frequency, and not to do any useful work, then many of the above considerations are not relevant. In this case the minimum surface density allowed by physics is quite low, because its only function is to absorb starlight.

Recently developed opaque metallic metamaterials are only tens of nanometers thick (Hägglund et al. 2013), which approaches theoretical limits (Hägglund et al. 2010). At these surface densities (of order 10 g $cm^{-3} \times 10$ nm $=10^{-5}$ g cm^{-2}) one Jupiter mass would provide for a shell with $R = 9000$ au. Lacki (2016) describes a scheme with dipole antennae that would use even less material for similarly sized Dyson spheres. If such a Dyson sphere could be engineered with significant microwave emissivity (and if the issue of radiation pressure discussed above), its temperature would be close to $T_{\min}$.

This does not mean that the star would be completely hidden. In order to "blend in" with the cosmic microwave background at $T = 2.7$ K, the Dyson sphere would have to be cooler than $T_{\min}$, which would require both actively cooling the Dyson sphere and disposing of the excess stellar luminosity nonthermally. It would also be an incomplete "cloak" because the microwave background and foreground are variable across the Dyson sphere's sky, meaning that the Dyson sphere would need to present a different thermal profile to observers in different directions.

15.5 Observability of Dyson Spheres

Dyson noted two methods by which Dyson spheres might be identified: their inevitable infrared heat signatures, and flickering of light that passes through the swarm of satellites. Both methods offer strong opportunities for discovering or putting upper limits on their existence.

15.5.1 The AGENT Formalism

Wright et al. (2014a) described a general formalism for parameterizing the effects of a Dyson sphere's infrared heat signatures on the spectrum of a star or galaxy.

We divide energy supply (that is, generation and collection) into two components: starlight collection and other sources, which we specify with the symbols α (which is meant, for mnemonic purposes, to remind the reader of an absorption coefficient) and ϵ (meant to remind the reader of the energy generation per unit volume in stellar structure equations).

We divide emission of energy into two categories: waste heat in the form of radiation (denoted γ, meant to evoke a common symbol for the photon) and all other forms (denoted ν, meant to suggest neutrino radiation, an example of a source of non-thermal or non-photon energy losses, as in, for example, supernovae). Finally, where appropriate, we assume that infrared heat signatures have some characteristic temperature T_{waste}.

While these parameters could be expressed in physical units, it is more convenient to normalize them by the energy available to the sphere in starlight. That is, we express them as fractions of the total luminosity of the star. This restricts α to lie between zero and one, but the other parameters can take any positive value.

This is the $AGENT$ parameterization, after the parameters $\alpha\gamma\epsilon\nu T_{\text{waste}}$, shown in Table 15.1.

Table 15.1. Definition of Parameters in the AGENT Formalism

Parameter	Meaning (All Powers are Normalized by Available Starlight)
α	Power of intercepted starlight
ϵ	Power of non-starlight energy supply (e.g., fossil fuel use, zero-point energy, nuclear energy)
γ	Power of waste heat in the form of thermal radiation of photons
ν	Power of other waste disposal (e.g., neutrino radiation, non-thermal emission, kinetic energy, energy-to-mass conversion)
T_{waste}	Characteristic color temperature of thermal waste photons

Conservation of energy and steady-state together require that the energy a civilization collects or generates must equal the energy it emits or disposes of, most naturally as waste heat. This yields:

$$\alpha + \epsilon = \gamma + \nu. \tag{15.23}$$

The assumption that no energy is sent away in a low entropy form before it can do work ($\nu \sim 0$) and the further assumption that all sources of power other than starlight are negligible ($\epsilon \sim 0$) yields the useful simplification of $\alpha = \gamma$.

To illustrate the AGENT parameters, consider humanity's presence on Earth as if it were a small fraction of a Dyson sphere (which it is, in some sense). Humanity has an energy supply of approximately 1.4×10^5 TWh yr^{-1}, or approximately 1.7×10^7 MW, mostly from fossil fuels. Instead of normalizing by all sunlight, for illustrative purposes we will normalize by the total amount of sunlight incident on the Earth, which is 1.7×10^{11} MW. Then we can parameterize humanity as having values given Table 15.2.

We have made several simplifying assumptions for this calculation. We have ignored sunlight used for passive heating, agriculture, and the many other ways it makes Earth inhabitable for us, because we are here isolating our civilization's contribution to the global energy budget. We have also assumed steady state, and so ignored forms of energy storage such as the potential energy in architecture, the chemical energy in stored food and batteries, and similar considerations, because they are negligibly small and do not affect these numbers much. Fossil fuels are stored solar energy, but the time over which this storage took place is so long that their use does not badly violate our steady-state assumption.

Put this way, we see that humanity warms the Earth with its energy generation by 0.01% above the effects of sunlight alone, and this effect has elevated the Earth's infrared heat signature by that amount above its pre-industrial levels. In terms of global temperature change, this effect (direct heating) is currently small compared to greenhouse warming (which insulates the Earth's surface but does not change the global energy balance) and albedo effects (such as loss of reflective sea ice or changes in cloud cover).

Table 15.2. Approximate Values for Humanity in the AGENT Formalis

Parameter	Value for Humanity	Notes
α	$\sim 10^{-7}$	Counting only photovoltaic generation (ignoring passive heating, agriculture, etc.)
ϵ	$\sim 10^{-4}$	Mostly fossil fuel, some nuclear
γ	$\sim 10^{-4}$	$\sim \epsilon$
ν	negligible	Radio transmission and radar, neutrino losses in nuclear reactors, kinetic and potential energy in spacecraft
T_{waste}	$\sim$285 K	Typical operating/ambient temperature

The energy budget of life on Earth is surprisingly poorly known, but on average photosynthetic life on Earth generates $\sim 2 \times 10^8$ MW from sunlight with approximately 2% efficiency (Kleidon 2010), implying that it ultimately intercepts $\sim$5% of sunlight that strikes the top of the atmosphere. Since photosynthetic energy production dominates Earth's biological work budget, we would write that for all life on Earth $\alpha = 0.05$ and, since all of this energy eventually cascades into waste heat, $\gamma = 0.05$.

15.5.2 Illustrative Examples

We now illustrate our formalism by translating some speculative forms of advanced civilizations from the scientific and popular literature into the AGENT parameterization.

First, we consider civilizations with energy supplies comparable to their star's total luminosity. A "classical" Dyson sphere that generated power solely from its completely enfolded solar-luminosity star and with a radius of 1 au would have $\alpha = \gamma = 1$; $T_{\text{waste}} \sim 250$ K; and $\nu = \epsilon = 0$. A "ring world" with radius $r = 1$ au and with a width of $w = 1.6 \times 10^6$ km would intercept $\frac{1}{2}w/r$ of its star's light, and so have $\alpha = 5 \times 10^{-4}$.

Radio or optical communication SETI can be thought of as searches for the deliberate or leaked power parameterized by ν. Indeed, radiation used for communicative purposes is inevitably emitted at low entropy in order to distinguish it from natural sources of radiation, which tend to be high entropy (Section 11.1.3).

In Kardashev's original formulation of types of civilizations, he made an order of magnitude estimate of the power available for (low entropy) radio emission from civilizations given that it must be some fraction of the energy supply. In AGENT terms, Kardashev estimated $\nu \lesssim (\alpha + \epsilon) \approx 1$.

Viewed this way, the principal advantage to searches for electromagnetically telecommunicating civilizations (i.e., searches for high-ν civilizations; Blair et al. 1992) is that the power emitted might be in a narrow bandwidth (and so have high flux density), might be obviously of alien origin, and might thus be easily distinguished from background or natural sources of radiation. The principal disadvantage is that we must guess at the form of ν and/or hope it is large enough for us to detect with our technology and from our vantage. In contrast, the principal advantage of searches for infrared heat signatures is that we expect $\gamma \gg \nu$, and we expect the waste heat to have a characteristic (thermal) spectrum; their principal disadvantages are that they are only sensitive to large values of γ, and that there are many confounding sources that will mimic this signature. Thus, the two forms of SETI are most sensitive to different kinds of activities of alien civilizations and face different challenges.

15.5.3 Spectral Implications

We can use the AGENT parameters to roughly calculate the infrared spectrum we would expect for large civilizations.

Consider a simple model in which a star at distance d has a stellar luminosity L_{host} and a flux density received at Earth F_λ. We describe the dimensionless spectrum of the host star, f_{host}, as:

$$\lambda F_\lambda = f_{\text{host}} \frac{L_{\text{host}}}{4\pi d^2}. \tag{15.24}$$

This star hosts a Dyson sphere with typical waste heat temperature T_{waste} blocking a fraction α of the stellar luminosity and generating other emission, such as communicative radio waves, with normalized SED f_{nt}.

The presence of the civilization will, to first order, alter the spectrum received at Earth to be:

$$\lambda F_\lambda = \left[(1 - \alpha) f_{\text{host}} + \gamma \frac{\lambda \pi B_\lambda(T_{\text{waste}})}{\sigma T_{\text{waste}}{}^4} + \nu f_{\text{nt}}\right] \frac{L_{\text{host}}}{4\pi d^2} \tag{15.25}$$

where $B_\lambda(T_{\text{waste}})$ is the specific intensity of the Planck function at $T = T_{\text{waste}}$.

If we crudely approximate a stellar spectrum as a blackbody with temperature T_* and ignore ν, then we can rewrite the expected spectrum as

$$F_\lambda \approx \left[(1 - \alpha) B_\lambda(T_*) + \gamma B_\lambda(T_{\text{waste}}) \left(\frac{T_*}{T_{\text{waste}}}\right)^4\right] \frac{\pi}{\sigma T_*^4} \frac{L_{\text{host}}}{4\pi d^2}. \tag{15.26}$$

If, more realistically, we restrict this last approximation to the Rayleigh–Jeans limit (i.e., to infrared radiation), we have

$$F_\lambda(\lambda \gg hc/(kT_*)) \approx \left[1 - \alpha + \gamma \left(\frac{T_*}{T_{\text{waste}}}\right)^3\right] \frac{ck}{2\lambda^4 \sigma T_*^3} \frac{L_{\text{host}}}{d^2} \tag{15.27}$$

$$\approx \left[1 - \alpha + \gamma \left(\frac{T_*}{T_{\text{waste}}}\right)^3\right] F_\lambda(\alpha = \gamma = 0) \tag{15.28}$$

This form reveals the power of working in the infrared when searching for Dyson spheres. We see that the infrared flux will be strongly enhanced by the large $(T_*/T_{\rm waste})^3$ term, which will be on the order of 10^3 *even for modest values of* γ. Indeed, for $\gamma = 0.15$, $T_* = 4000$ K, and $T_{\rm waste} = 255$ K, this amounts to an increase in infrared flux by a factor of nearly 1000. By contrast, the dimming at optical wavelengths is linear in α, and so is a much smaller effect, only observationally distinguishable for values of α near 1 (i.e., nearly complete Dyson spheres).

The lesson here is that to have detectable infrared heat signatures, a Dyson sphere need not completely obfuscate its star. The infrared excess scales, to first order, as $\gamma(T_*/T_{\rm waste})^3$, which is large enough that the primary obstacle to detecting infrared heat signatures from Dyson spheres will not necessarily be their brightness, but confusion with other astrophysical sources with similar broadband colors or spectra.

15.6 More Detailed Spectral Modeling

Wright (2020) provided a more detailed calculation for predicting the spectra of Dyson spheres, extending the AGENT formalism to Dyson spheres with different interior and exterior temperatures, and accounting for stellar feedback onto the star.

The essence of the calculation is that a Dyson sphere that was either very small or very complete would have difficulty radiating energy into space. Some photons would be reflected or re-emitted toward the interior of the sphere, where they would strike the star or Dyson sphere again, warming them. For very small or very complete spheres, some of the stellar photons return to the star itself, warming its surface.[11] This feedback complicates the calculation of the AGENT parameters.

In this formalism, a Dyson sphere allows some fraction of all starlight to escape freely into space, absorbs some fraction, and reflects the rest. The reflection is somewhere between specular (like off of a mirror) and diffuse (like off of a white piece of paper). The sphere has some characteristic radius from the star, and has different properties on the interior surfaces that starlight strikes, and exterior surfaces pointing away from the star. The sphere further manages its heat such that these two surfaces can be at different temperatures.

Figure 15.2 shows an example of the spectrum that can be calculated for a Dyson sphere with radius $R = 1$ au, orbiting a central, Sun-like star, as observed at a distance of 100 pc, under an assumed set of properties described in the caption.

15.7 Dyson Spheres in Transit

Dyson (1960) presumed that a Dyson sphere would completely block all optical light from escaping, but in his followup lecture in Dyson (1966) he suggested how a suspiciously bright infrared source could be identified as artificial: "One would look in particular for irregular light variations due to starlight shining through chinks in

[11] And, potentially, its interior. Huston & Wright (2022) explored this possibility, showing that for fully convective stars a Dyson sphere returning more than a few percent of the star's light could dramatically alter the star's structure and extend its lifetime.

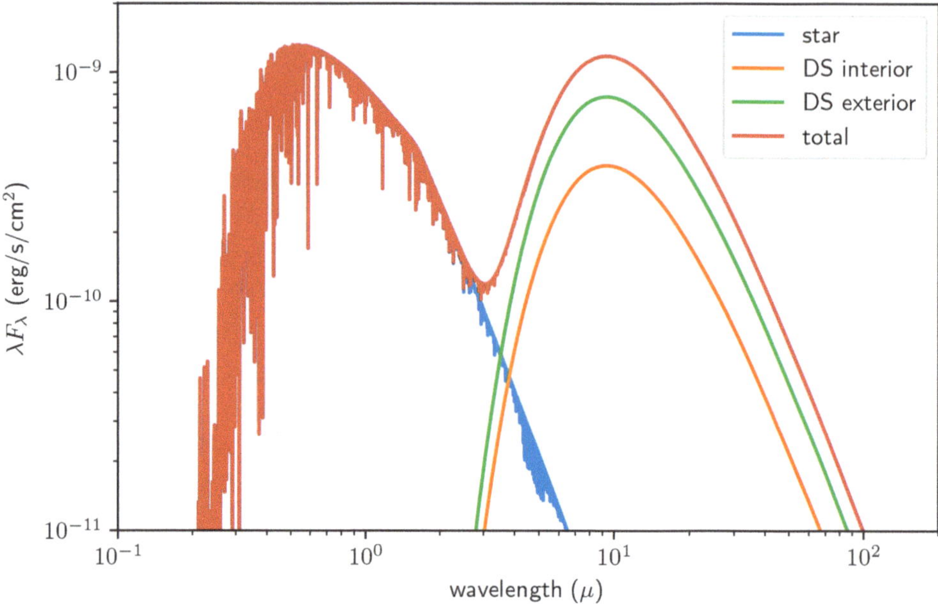

Figure 15.2. Spectrum of a Sun-like star surrounded by a 1 au Dyson sphere, as observed from a distance of 100 pc. The sphere transmits 1/3, absorbs 1/3, and reflects 1/3 of the photons that reach it, and radiates waste heat evenly between its interior and exterior surfaces, which have the same effective temperatures. The interior surface is partially obscured by the sphere itself and so appears dimmer. The stellar spectrum blueward of 24 μ is a $T_{\rm eff} = 5800$ K dwarf star spectrum from the NextGen stellar atmosphere grid (Allard et al. 1997; Hauschildt et al. 1999). Reprinted with permission from Wright (2020) © 2020 the author. CC BY 4.0.

the curtain, and for stray electromagnetic fields and radio noise produced by large-scale electrical operations (not necessarily carrying any...message)."

Of course, if a star was surrounded by a less-than-completely obscuring swarm of material, some starlight would emerge. The effect this would have on the brightness of starlight observed from Earth depends on the details of the sizes, numbers, and orbits of the swarm components.

15.7.1 Single Objects in Transit

In general, the amount a star's brightness appears to drop due to an opaque object transiting[12] it is given by the fraction of the star's area blocked. Since stars and the material orbiting them are approximately the same distance from Earth, this fraction is simply the ratio of the cross-sectional areas of the star and intervening object.

The typical amount of dimming during the transits of giant exoplanets like Jupiter (roughly ten times the linear size of Earth) in front of Sun-like stars (roughly 100

[12] When two astronomical objects align and one blocks our line of sight to the other, we call this an *eclipse*. In the special case where the foreground object is much smaller than the background object, we refer to its passage between us and the background object as a *transit*. Mercury and Venus are both seen to transit the Sun, and astronomers have detected thousands of exoplanets by their transits in front of their host stars.

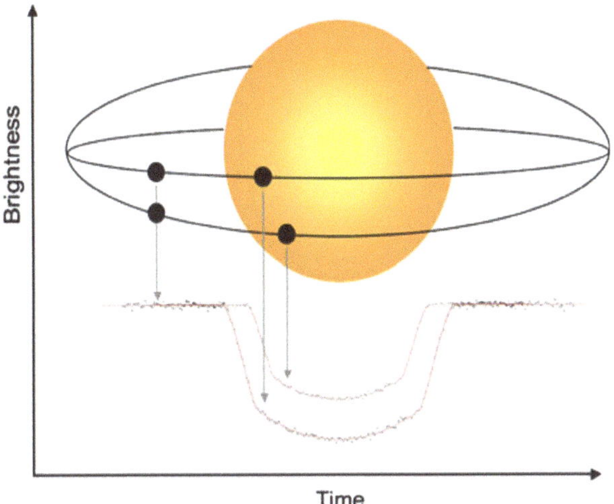

Figure 15.3. Schematic showing the approximate transit shape expected from a planet crossing a stellar surface. The bottom red lines represent the light curve of the star during a transit, that is the brightness of the star as a function of time. The black points are simulated data of the star's brightness. The figure above illustrates the geometry that gives rise to these light curves, for different orientations of the orbits of a planet with a circular aspect. Because of limb darkening (shown here somewhat exaggerated) a transit that passes nearer the center of the star will be longer, somewhat deeper, and have steeper ingress and egress. Reproduced from Rouan (2015) with permission from Springer Nature.

Earth radii) are of order $(R_p/R_*)^2 \sim 1\%$. Earth-sized planets, whose transits are only barely detectable with space observatories, have transit depths 100 times smaller, or 100 parts per million.

The pattern of dimming of a planet transiting a star has a characteristic shape that depends on the particular chord the planet takes across the stellar disk, and the amount and pattern of *limb darkening* on the star.[13]

Figure 15.3 shows this characteristic transit shape. The brief period between when a planet begins to block part of the stellar disk and is completely in front of the edge of the star is called *ingress*, and the corresponding time on the way out is called *egress*.

A single, large artificial object orbiting a star would be seen in transit once per orbital period, and if it were sufficiently large might be detectable. Such a large object is often referred to as a *megastructure*.

Arnold (2005) noted that NASA's *Kepler* mission, which observed 100,000 stars in search of their transiting planets, might be able to distinguish between transiting planets and megastructures. Figure 15.4 shows four illustrations of objects of different shapes transiting a background star.[14] Arnold showed that if these

[13] The outer portions of the disk of the Sun and other stars is dimmer than the center of the disk because our view penetrates to deeper, hotter layers of the solar atmosphere at the disk center. The amount and pattern of limb darkening depends on the kind of star and what wavelength one is observing at.

[14] These illustrations are what the stars would looks like up close—astronomers detecting transits work entirely with the total light arrive at Earth, and do not generally image the surfaces of other stars like this.

Figure 15.4. Compilation of figures showing four examples of objects in silhouette in front of portions of their star as proposed by Arnold (2005). Each object has the same overall cross section, but each produces a subtly different light curve, especially during ingress and egress. © 2005 the American Astronomical Society, all rights reserved.

Jupiter-sized objects were orbiting stars observed by *Kepler*, they would produce similar total transit depths (because they have the same cross-sectional area) but very slightly different transit shapes, at a level that in would be in principle detectable by *Kepler*. Most of the differences in the light curves of these objects occur during the brief ingress and egress portions of the light curve.

Similar results can be derived for exoplanets with rings or moons, meaning transit light curve modeling can in principle be used to detect these orbiting other planets, although no such detection has yet been made.

Arnold (2005) suggested that megastructures might be intentionally constructed and placed in short-period orbits around stars as beacons. Advantages to such a beacon over a radio or other communicative beacon are that it would be passive (not requiring constant power to operate), long lived, and extremely noticeable—its detectability would be set by the brightness of the star and the fraction of the starlight it blocked. If its shape were sufficiently non-circular, its artificial nature might be readily apparent, and it could even employ louvres or other moving parts to alter its shape and convey small amounts of information.

Arnold (2005) suggested that beacons like this might include co-orbital structures that generated transits whose spacings counted out the prime numbers, or whose sizes were different such that the transit depths traced out a multiple of prime numbers. Planetary transits, in general, are constant in depth and period.

Sandford & Kipping (2019) and Luger et al. (2022) explored how one might invert light curves to derive the shape of the transiting object. They showed that considerable information about the object's shape is not recoverable due to degeneracies in the problem (that is, very different shapes can generate identical light curves), however one can usually establish whether the underlying object has a circular aspect or not.

Except for the complications of moons or rings, exoplanets are expected to be almost perfectly circular (with only small amounts of oblateness due to rotation), whereas there is no reason that megastructures should have circular aspects. Transit surveys such as those performed by *Kepler* are thus in principle sensitive to megastructures orbiting stars.

15.7.2 Swarms and Boyajian's Star

It is perhaps more likely that a Dyson swarm would be constructed without any communicative purpose, in which case we might not observe a single repeating megastructure transit but many.

If the swarm elements were much smaller than large planets, then no individual element would produce a detectable transit, but the starlight might be seen to "flicker" due to the collective transits of many swarm elements, or to have slow variations in brightness as the total number of elements in front of the star varies.

If some swarm elements are very large, or if many smaller swarm elements fly in formation such that they effectively block light as a single large object, then a star might display many dimmings nonperiodically as the different swarm elements transit.

Arnold's description of a star being occulted by multiple, large, non-circular objects in less-than-strictly-periodic patterns found a fascinating case study in KIC 8462852, a star observed by NASA's *Kepler* mission to have an extremely strange light curve.

First identified by citizen scientists of the Zooniverse Planet Hunters program, the star came to widespread attention when Boyajian et al. (2015) cataloged its strange properties. It exhibited large, aperiodic dimming events hours to weeks long, some deeper than 20%, indicating it was being obscured by multiple, large (stellar-sized) collections of material. Boyajian et al. (2015) hypothesized this was the result of a family of extremely large comets orbiting the star.

Additional work by Schaefer (2016) and Montet & Simon (2016) showed that the star also exhibited long-term dimming over the past century and during the *Kepler* mission. This sort of behavior, like the dips, is all but unprecedented for this kind of star.

Followup observations by Bodman et al. (2018) showed that the dimming events redden the star in a manner characteristic of dust. This strongly implies that the material is not a collection of solid, opaque objects, but more likely large collections of dust, consistent with the comet hypothesis of Boyajian et al. (2015).

We will discuss the media's reaction to the suggestion that Boyajian's Star is orbited by megastructures in Section 20.7.2.

15.8 Extragalactic Dyson Spheres

The expectations that guide the logic of the Fermi Paradox (Chapter 6) (to wit, that at least one spacefaring species will settle an entire galaxy if technological life is common) argue that even if humanity is alone in the Milky Way, we should be able to search the many *other* galaxies in the Universe for evidence of such expansion. This is also the logic that leads to expectations of Type III Kardashev species (Section 4.2.2).

A detectable Dyson sphere would seem to require a much larger ichnoscale value (Section 4.2.1) than traveling to nearby stars (i.e., we need extend our robotic probe range by "only" a factor of 10,000 or so to reach nearby stars, but we must multiply our hardware in interplanetary space by a factor of 10^{18} to build a Dyson sphere.) Thus, if a species builds Dyson spheres, it is perhaps likely to already be interstellar, and thus might build them everywhere (indeed, building a Dyson sphere in one's own stellar system might be gravely detrimental to one's home planet). We can thus probe the assumptions of the Fermi Paradox by looking for distant galaxies endemic with Dyson spheres.

In this case, the spectral modeling is more complex because galactic emission comes from several components, including its stars, gas, dust, and nuclear region (powered by supermassive black holes). Presumably, only the stars (and perhaps the nuclear region) are the sites of Dyson spheres, and so only these components will be optically dimmed and reradiate in the infrared.

What's more, most of the starlight from galaxies comes from the brightest stars[15] In old, elliptical galaxies these are typically red giants—stars of approximately the mass of the Sun in a relatively brief stage at the ends of their lives when they become extremely luminous as they burn helium in their cores and/or hydrogen in a shell around their cores. In younger, spiral and irregular galaxies where star formation is still underway, the most luminous stars (especially at shorter wavelengths) tend to be young, hot, very massive stars with short lifetimes. These stars are found near their birthplaces in the spiral arms of galaxies, giving spirals their distinctive appearance.

At first blush, one might expect that these bright stars might be poor sites for Dyson spheres because these stars are rather volatile and short lived (the most massive stars will explode as supernovae, and even the giants have highly variable luminosities over their brief giant-hood.) On the other hand, the timescales involved are still measured in millions of years, and these are the most abundant sources of free energy in a galaxy (aside from the central black hole), and so perhaps still attractive targets for Dyson spheres.

At any rate, a species that builds Dyson spheres throughout a galaxy will block the overall optical emission and reradiate it in the infrared much like for a single star but instead for the entire galaxy. See Section 17.5 for more on this topic.

15.9 Observations and Upper Limits

15.9.1 Confounders

Consistent with his First Law of SETI Investigations (Section 11.4.2), Dyson (1966) wrote of his eponymous spheres:

---

[15] This is not obvious: it could have been the case that very faint stars, which are far more numerous dominate the light of the very rare luminous stars. It just doesn't work out that way.

> ...we can look for infrared sources in the 3–10 micron band using ground-based telescopes. The astronomers are of course professionally interested in looking for natural objects in this band. I am happy if they look for natural objects; this preserves their professional purity and serves my purpose just as well as an avowed search for artificial objects. The sort of natural object one expects to see is a protostar, a cloud of gas and dust in the process of condensing into a star and not yet hot enough to radiate in the visible. To observe such natural objects is important for the study of star formation, which is one of the central problems of astronomy. When such objects are found, I hope nobody will rush to the newspapers with claims that we have found something artificial. I say only that if highly expanded technologies exist then they are to be found among such objects...
>
> Certainly one will not claim any object to be artificial unless it appears pathological in at least two independent modes of observation.

Infrared astronomy has indeed found that stars and galaxies can be copious emitters of infrared radiation for a variety of natural reasons, making technological infrared heat signatures a highly ambiguous technosignature.

The most common confounder for stars is circumstellar material, which, just as circumstellar technology would, absorbs starlight, heats up, and reradiates the energy at infrared wavelengths. Such material is generally somewhat short lived, coalescing into planets or being cleared out by radiation pressure on timescales of 10^6–10^7 yr. We thus in general do not expect to find much circumstellar material or infrared excess around most stars, which are significantly older than this.

Young stars can also typically be identified by their environment in star-forming regions or stellar clusters, or by their strong emission lines, for instance in the H-α line, characteristic of accretion (as the star forms) and magnetic activity (because young stars tend to rotate quickly and so have strong magnetic dynamos). This can help serve as a filter to find astrophysically anomalous infrared emission.

The second major stellar confounder is very cool giant stars. Some stars, especially evolved, massive ones that have exhausted their central helium supply, generate extremely high luminosities, which makes them expand to the point that their outer atmospheres become cool enough to form molecules, and even in some cases dust particles.[16] These stars can have extremely high infrared luminosities compared to their optical radiation, and (as we will see) were the first false positives for Dyson spheres. Fortunately, today we have good distance measurements or lower limits for stars in the Milky Way, which allows these stars to be easily identified and filtered out by their extremely high intrinsic luminosities.

[16] These have the strange name *asymptotic giant branch stars*, and they fuse helium in a shell around their central cores, now composed of carbon and oxygen. Only the most massive stars burning shell helium generate enough luminosity to have such cool atmospheres.

In spiral and irregular galaxies, most of the infrared emission will come either from the central black hole[17] or its star-forming regions, which are generally clustered, for instance along spiral arms.[18] For Type III species, the spatial distribution of Dyson spheres will likely follow the distribution of stars in the galaxy, which is generally smooth, giving it an overall "glow" in the infrared. For spiral and irregular galaxies, this should prove to be a good discriminant regarding the source of the infrared radiation.

Elliptical galaxies generally have very little gas and dust or star formation, so most of their infrared radiation will be from giant stars and the dust they generate. This is a rather low level of infrared radiation, meaning that almost any detected level of it is anomalous.

15.9.2 Early Searches

Though the technical feasibility of searches for Dyson spheres was noted early on (Dyson 1960a; Sagan & Walker 1966), thorough surveys could not begin in earnest until the launch of *IRAS*, which provided the first mid-infrared survey of the sky with the aereal coverage and sensitivity necessary for such work. Slysh (1985) describes the first interpretations of the data set, which confirmed Dyson (1960a)'s prediction: there would be many bright infrared Galactic sources, and that the primary difficulty would not be detecting such sources but distinguishing them from natural sources.

Analysis of this data set proceeded slowly over the next 25 years. Timofeev et al. (2000) conducted a cursory search identifying a few objects with blackbody-like SEDs among the *IRAS* point sources. Jugaku & Nishimura (2004) used the results of *IRAS* and near-infrared measurements to search for Dyson spheres, finding no good candidates among 384 nearby solar-type stars. Carrigan (2009) used the spectroscopic information from *IRAS* to examine sources across the sky for potential Dyson spheres, concluding that all four good candidates were likely distant red giants.

Teodorani (2014) outlined a more modern approach to the problem, using *Spitzer* as a more sensitive probe for infrared excesses and, following the suggestion of Arnold (2005), *Kepler* to look for transiting megastructures. Wright et al. (2014a) describe a program using the *WISE* all sky survey to search for Dyson spheres (and also other galaxies endemic with them), including a parameterization of their properties in terms of observables (the "AGENT" formalism after its five parameters, see Section 15.5.1). Wright et al. (2016) outlined a general set of photometric

[17] This region is various referred to as the *nuclear region*, the *active galactic nucleus (AGN)*, or the *central engine*.

[18] The characteristic arms of spiral galaxies are actually just slight overdensities in the underlying stellar and dark matter population. Like ripples in a pond, they are not composed of a consistent set of stars, but are a spiral-patterned disturbance that propagates around the galaxy. The compression of gas and dust as it "falls into" these mass overdensities due to their slightly higher gravitational pull triggers star formation, the brightest products of which are blue stars that live and die in roughly the time it takes the overdensity wave to pass by. These stars thus illuminate and trace the spiral arms, making them obvious.

anomalies in the light curves of stars that would be indicative of transiting megastructures, and identified one particular case study illustrative of the concept ("Boyajian's Star," Boyajian et al. 2016; Wright & Sigurdsson 2016, references therein, and Sections 15.7.2 & 20.7.2).

Wright et al. (2014a) pointed out that unless virtually no starlight escapes a Dyson sphere, the flux decrement from blocked starlight is many times smaller than the infrared excess from the Dyson sphere's infrared heat signatures. Nonetheless, Zackrisson et al. (2015) recommended searching first not for infrared excesses, but for optically underluminous stars (i.e., stars with disparate trigonometric and spectroscopic parallaxes), and then following up with sensitive infrared measurements to search for the "missing" luminosity. Both Wright et al. (2014a) and Zackrisson et al. (2018) emphasized the value of *Gaia* as a way to greatly improve search capabilities for Dyson spheres, since they would help distinguish Galactic sources (i.e., stars) from the far more ubiquitous extragalactic infrared sources (such as AGN); and since they would allow for underluminous stars to be more precisely identified.

The first observational search for galaxy-wide populations of Dyson spheres was that of Annis (1999), who examined a sample of over 100 galaxies of similar distance to search for any that were optically underluminous, consistent with a significant amount of starlight blocked by Dyson spheres. Griffith et al. (2015) used the *WISE* all-sky survey to search for resolved sources of extended infrared emission to search for populations of Dyson spheres by their infrared heat signatures, and Lacki (2016) used *Planck* data to put upper limits on galaxies with all of their starlight being reprocessed to very low temperatures. Zackrisson et al. (2015) combined these strategies, searching for optically underluminous galaxies and following up with infrared measurements to identify whether the "missing" luminosity was being emitted there. None of these searches resulted in particularly good candidates for Type III species.

15.9.3 Modern Upper Limits

Searches for Dyson spheres in the Milky Way were greatly aided by the ESA *Gaia* spacecraft, which measured parallaxes and space motions of over a billion stars near the Earth. This allows one to filter out all very luminous confounders—like distant galaxies and giant stars—and study a rather clean sample of nearby stars to determine their level of infrared radiation.

The most comprehensive such study is by Suazo et al. (2022), who cross-matched the *Gaia* catalog with that of NASA's *WISE* infrared all-sky survey and other photometric catalogs to compute upper limits within the AGENT formalism. They find that Dyson spheres with more than $\sim 15\%$ coverage are very rare near the Sun: no more than 1% of stars within 200 pc have such spheres, and out to 5000 pc no more than 1% of stars have more than 30% coverage (see Figure 15.5).

Suazo et al. (2024) followed up this study by examining the stars consistent with Dyson spheres to find candidates. This study took the opposite approach to that of Suazo et al. (2022): instead of counting the fraction of stars definitively *not* consistent

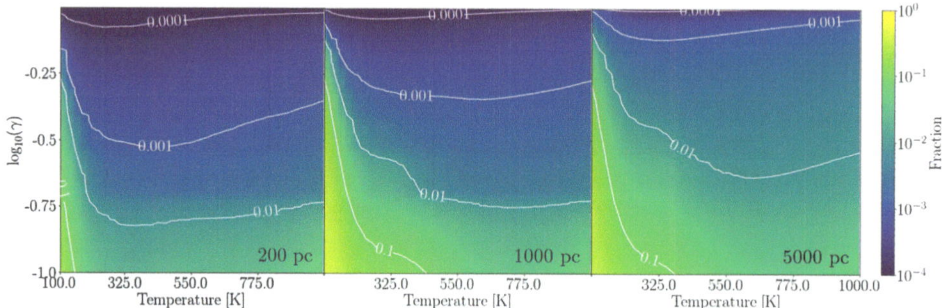

Figure 15.5. Limits from Suazo et al. (2022) on Dyson spheres in Earth's part of the Milky Way. The y-axis represents the (logarithm of the) γ component of the AGENT formalism (Section 15.5.1, i.e., the fraction of starlight re-radiated into the infrared, and presumed equal to α, the fraction of starlight blocked by a Dyson sphere). Complete spheres are represented at the top of the figure, and those with 10% coverage at the bottom. The x-axis of each figure represents the characteristic temperature of the material from 100 to 1000 K. The contours represent the upper limits set by the search: no more than 1% of stars have Dyson spheres characterized by the region above the line labeled 0.01. The three panels represent the results for different distance thresholds.

with Dyson spheres of certain AGENT parameterization, they examined stars with infrared excesses that were definitively not consistent with the most common confounders of stellar Dyson spheres (primarily young stars and chance alignments of stars with unrelated infrared sources). They found seven candidates worthy of follow up, although some of these have proven to be rather insidious chance alignments with background galaxies (Ren et al. 2024).

The best upper limits on extragalactic Dyson spheres are from the Ĝ survey (Wright et al. 2014a, 2014b), which similarly used the NASA *WISE* mission's all-sky survey but focused on most the 100,000 nearby galaxies with the largest apparent sizes in the sky. Griffith et al. (2015) found a loose upper limit of zero galaxies with more than 85% of their starlight reprocessed in the infrared, and $\sim 0.05\%$ of galaxies with more than 50% reprocessing. The primary confounders in this search were starburst galaxies, and this upper limit can presumably be greatly improved by removing these confounders.

Exercises

1. **Matrioshka brains:** A specific suggestion in the gray literature by Robert Bradbury for a maximally computationally efficient Dyson sphere is a "Matrioshka Brain"[19]: a nested set of Dyson spheres, with each sphere capturing and exploiting the waste heat energy of the sphere interior to it.

[19] The name refers to "Matrioshka dolls" or "nesting dolls" (also known as "babushka dolls" in English), a folk craft from Russia in which a set of hollow, wooden figures painted as a women are nested inside each other. Bradbury's suggestion can be found at https://web.archive.org/web/20080918090527/http://www.aeiveos.com:8080/~bradbury/MatrioshkaBrains/MatrioshkaBrainsPaper.html.

A typical objection to this sort of scheme in general points out that capturing waste heat produces feedback on the system that generated the heat that makes it less efficient, negating your gains. You can't run a computer server farm on the exhaust of the cooling system of a computer server farm!

For Dyson spheres, If the inner sphere cannot radiate freely into dark space, but "sees" the inside of an outer sphere, the incoming radiation from the outer sphere will heat it up, lowering its efficiency.

It turns out, there's a bigger problem with Matrioshka Brains, at least when it comes to optimizing computation rates. Let's explore this with the simplest Matrioshka Brain, with just two layers, where the outer layer is very far from the inner one, so is provides minimal thermal feedback.

(a) Let's start with Dyson spheres that extract useful work from starlight and remove it from the system, perhaps as coherent radio emission. Consider a shell at some large distance from the star doing this—how much work can it do at the Carnot limit? Express this work P in terms of the stellar luminosity L, shell equilibrium temperature T_l, and stellar radiation temperature T_*.

(b) Write down the same expression for a smaller shell closer to the star at temperature T_h.

(c) The leftover energy from the smaller shell that does not escape the system in part (b) is emitted as waste heat. If that energy is completely captured by an outer shell at temperature T_l, how much useful work can the outer shell extract?

(d) Look at the sum of the work done by the two shells in parts (b) and (c). How does it compare to the single shell in part (a)?

(e) Now let's look at the case of dissipative work (i.e., work that does not result in energy leaving the system in any form other than waste heat, like computation). Repeat the calculations in parts (a)–(d), but instead of calculating the amount of work done, calculate number of binary operations per time that can be performed at the Landauer limit of $\ln(2)kT$ in waste heat per operation. *Hint: remember in this case that the outer shell will see the full luminosity of the star with or without the inner shell!*

(f) Now, repeat the calculation for calculations above the Landauer limit, that is where the cost per computation is fixed at $c > \ln 2kT_h$.

(g) Interpret your results. Are Matrioshka Brains a good idea? Why, other than the reasons we've considered here, might you want to do computations closer to the star than the optimal distance?

References

Allard, F., Hauschildt, P. H., Alexander, D. R., & Starrfield, S. 1997, ARA&A, 35, 137

Annis, J. 1999, JBIS, 52, 33

Arnold, L. F. A. 2005, ApJ, 627, 534

Badescu, V. 1995, AcAau, 36, 135
Badescu, V., & Cathcart, R. B. 2000, JBIS, 53, 297
Badescu, V., & Cathcart, R. B. 2006, AcAau, 58, 119
Bennett, C. H., Hanson, R., & Riedel, C. J. 2019, FoPh, 49, 820
Blair, D. G., Norris, R. P., Troup, E. R., et al. 1992, MNRAS, 257, 105
Bodman, E., Wright, J., Boyajian, T., & Ellis, T. 2018, arXiv:1806.08842
Boyajian, T. S., LaCourse, D. M., Rappaport, S. A., et al. 2015, MNRAS, 457, 3988
Boyajian, T. S., LaCourse, D. M., Rappaport, S. A., et al. 2016, MNRAS, 457, 3988
Bradbury, R. J., Cirkovic, M. M., & Dvorsky, G. 2011, JBIS, 64, 156
Bradbury, R. J. 2001, Proc. SPIE, 4273, 56
Caplan, M. E. 2019, AcAau, 165, 96
Carrigan, R. A. 2009, ApJ, 698, 2075
Cocconi, G., & Morrison, P. 1959, Natur, 184, 844
Covington, T. 1991, Resources of Near-Earth Space, Vol. 36, ed. J. S. Lewis, M. Shapley Matthews, & M. L. Guerrieri (Tucson, AZ: Univ. Arizona Press)
Drake, F. D. 1961, PhT, 14, 40
Dyson, F. 2018, Maker of Patterns: An Autobiography Through Letters (New York: WW Norton)
Dyson, F. J. 1960a, Sci., 131, 1667
Dyson, F. J. 1960b, Sci., 132, 250
Dyson, F. J. 1960c, Sci., 131, 1667
Dyson, F. J. 1966, Perspectives in Modern Physics: Essays in Honor of Hans A. Bethe on the Occasion of His 60th Birthday, July, 1966, ed. R. E. Marshak, J. W. Blaker, & H. A. Bethe (New York: Wiley) 641
Frank, M. P. 2002, CSE, 4, 16
Griffith, R. L., Wright, J. T., Maldonado, J., et al. 2015, ApJS, 217, 25
Hägglund, C., Apell, S. P., & Kasemo, B. 2010, NanoL, 10, 3135
Hägglund, C., Zeltzer, G., Ruiz, R., et al. 2013, NanoL, 13, 3352
Hauschildt, P. H., Allard, F., & Baron, E. 1999, ApJ, 512, 377
Huston, M., & Wright, J. 2022, ApJ, 924, 78
Jugaku, J., & Nishimura, S. 2004, Bioastronomy 2002: Life Among the Stars; Proc. IAU Symp. 213 ed. R. Norris, & F. Stootman (San Francisco, CA) 437
Kardashev, N. S. 1964, SvA, 8, 217
Kleidon, A. 2010, PhLRv, 7, 424
Lacki, B. C. 2016, arXiv:1604.07844
Lacki, B. C. 2025, ApJ, 985, 191
Luger, R., Agol, E., Bartolić, F., et al. 2022, AJ, 164, 4
McInnes, C. R. 2003, JBIS, 56, 308
Montet, B. T., & Simon, J. D. 2016, ApJL, 830, L39
Niven, L. 1970, Ringworld (New York: Ballantine)
Pan, B., & Cui, W. 2010, MaStr, 23, 227
Papagiannis, M. D. 1985, IAU Symp. 112, The Search for Extraterrestrial Life: Recent Developments ed. M. D. Papagiannis (Cambridge: Cambridge Univ. Press) 543
Ren, T., Garrett, M. A., & Siemion, A. P. V. 2024, RNAAS, 8, 145
Reynolds, A. 2008, House of Suns (London: Gollancz)
Rippert, E. D. 2014, arXiv:1412.1881

Rouan, D. 2015, Encyclopedia of Astrobiology eds. M. Gargaud et al. (Berlin: Springer)
Sagan, C., & Walker, R. G. 1966, ApJ, 144, 1216
Sandberg, A. 1999, J. Evolution and Technology, 5, 1
Sandberg, A., Armstrong, S., & Cirkovic, M. M. 2016, JBIS, 69, 406
Sandford, E., & Kipping, D. 2019, AJ, 157, 42
Schaefer, B. E. 2016, ApJL, 822, L34
Scheuer, P. A. G. 1981, JApA, 2, 165
Shkadov, L. M. 1988, SoSyR, 22, 210
Slysh, V. I. 1985, IAU Symp. 112 The Search for Extraterrestrial Life: Recent Developments, ed. M. D. Papagiannis (Dordrecht: Reidel Publishing) 315
Stapledon, O. 1937, Star Maker (Malton, UK: Methuen)
Suazo, M., Zackrisson, E., Wright, J. T., Korn, A. J., & Huston, M. 2022, MNRAS, 512, 2988
Suazo, M., Zackrisson, E., Mahto, P. K., et al. 2024, MNRAS, 531, 695
Suffern, K. G. 1977, PASA, 3, 177
Teodorani, M. 2014, AcAau, 105, 512
Theis, T. N., & Wong, H. S. P. 2017, CSE, 19, 41
Timofeev, M. Y., Kardashev, N. S., & Promyslov, V. G. 2000, AcAau, 46, 655
Wright, J. T. 2018, arXiv:1803.06972
Wright, J. T., & Sigurdsson, S. 2016, ApL, 829, L3
Wright, J. T., Griffith, R. L., Sigurdsson, S., Povich, M. S., & Mullan, B. 2014a, ApJ, 792, 27
Wright, J. T., Mullan, B., Sigurdsson, S., & Povich, M. S. 2014b, ApJ, 792, 26
Wright, J. T., Cartier, K. M. S., Zhao, M., Jontof-Hutter, D., & Ford, E. B. 2016, ApJ, 816, 17
Wright, J. T. 2020, SerAJ, 200, 1
Wright, J. T., Sheikh, S., Almár, I., et al. 2018, arXiv:1809.06857
Zackrisson, E., Calissendorff, P., Asadi, S., & Nyholm, A. 2015, ApJ, 810, 23
Zackrisson, E., Korn, A. J., Wehrhahn, A., & Reiter, J. 2018, ApJ, 862, 21

AAS | IOP Astronomy

Chapter 16

Exoplanetary SETI

> If we found a totally bizarre situation—such as the presence of traces of fluorocarbons in our own atmosphere—we might be inclined to invoke the action of intelligent life, thereby assigning a value to f_i. With our own civilization as an example, we would be very surprised if beings that had developed fluorocarbons were not also capable of and interested in interstellar communication, so we would strongly lobby for an intense effort by radio astronomers to search for signals and give us a number for f_c. A new minimum for L would follow once we had made contact.
>
> — Owen (1980)

> If the chemistry and biochemistry of Earth-like planets...can be detected and analyzed...then in theory we could detect some signs of possible technologies (pollution, "greenhouse" effects, infrared excess, variable ozone levels, nuclear accidents or wars, etc.)...
>
> From archaeological and SETT* points of view the development of instruments and software for the direct imaging of exoplanets should allow for detection and analysis of chemical, biochemical, and radioactive traces of technological activity.
>
> — Campbell (2006)
>
> *Search for Extraterrestrial Technology, see Section 1.1.1.

doi:10.1088/2514-3433/adf5aech16

16.1 Origins

16.1.1 Owen and Campbell's Vision

For over 20 years there were four "pillars" of SETI—radio (Chapter 12), optical (Chapter 13), solar system (Chapter 14), and infrared (Chapter 15)—anchored by a series of papers published in the span of two years following the Catalytic Insight of Modern SETI (Section 4.1).

A fifth pillar was erected by American Astronomer Tobias Owen and Australian archaeologist John B. Campbell (Campbell et al. 2006) in a pair of proceedings of meetings of the International Astronomical Union. Owen made his suggestion (above) in 1980 in an almost offhanded way when discussing the possibility of the spectroscopic detection of life on exoplanets via direct imaging generally.

Since exoplanets would not even be discovered for another 13 years, the topic lay largely dormant, and was apparently revived by Campbell in 2006 in a session on the direct imaging of exoplanets. Writing of previous work to detect passive signs of technology such as Dyson spheres, he complained "[n]one of this work was carried out by archaeologists, even though by their very nature archaeologists are experts in the detection of ancient technologies."

Campbell connected the new work being done by NASA and ESA to plan space missions that might characterize the atmospheres of exoplanets and search for biosignatures to the possibility that they might also detect technosignatures (see the quote above).

As exoplanetary astrophysics has matured (Section 2.14), astronomers eventually caught up with Owen and Campbell's vision, and have begun to develop a library of possible exoplanetary technosignatures and techniques to discover them. In retrospect, it is perhaps somewhat surprising that it took so long for the idea of looking for technosignatures on exoplanets to enter the literature, but it does make sense that the idea is only flourishing now that the possibility of studying exoplanetary spectra has moved from a distant dream to concrete reality.

16.1.2 Exoplanetary Considerations Before 1995

Many searches for technosignatures in the first 35 years of SETI focused on slowly rotating Sun-like stars, on the reasoning that they were most likely to host Earth-like planets.[1] Two of the terms in the Drake Equation illustrated the importance of planet formation to SETI because they helped determine the number of sites where life could form: f_p and n_e, the fraction of stars that form planets, and the typical number of Earth-like planets around each of those stars. It is not a stretch to write that the primary goal of the *Kepler* mission (Section 2.14.1) was to determine the values of these quantities (expressed, roughly but not precisely, as their product "eta sub Earth" $\eta_\oplus \approx f_p n_e$, see Section 8.1).

[1] This is because slow rotation was thought, at the time, to be a sign that the process of planet formation had created a stellar rotational angular momentum deficit. We now understand that all cool stars slow down naturally with time, via a process unrelated to planet formation.

The discovery of exoplanets created new opportunities to target systems known to host an exoplanet. Although most of the first exoplanets discovered prior to *Kepler* were gas or ice giants within 1 au of their host star, these systems at least offered a target list enriched in stars that had formed some planets at all. The *Kepler* mission provided thousands of new targets hosting smaller, presumably terrestrial planets, although virtually none of these were in the Habitable Zones of their stars. The few potentially habitable, apparently rocky planets discovered via transit have been routinely searched for signs of radio transmissions.

More importantly, the *Kepler* mission demonstrated that planet formation was an extremely common outcome of star formation; that nearly all stars host exoplanets of some sort; and that $\eta_\oplus \sim 0.5$ with large uncertainties (e.g., Bryson et al. 2021). This, somewhat ironically, validated SETI researchers' original "look everywhere" strategy from before any exoplanets were known.

16.1.3 Radio Technosignatures of Earth as an Exoplanet

Sullivan et al. (1978) performed an early calculation of Earth's leaked radio technosignatures, concluding that (at the time) Earth's most obvious and unambiguously technological radio signatures were that of powerful radars used as early warning systems to detect intercontinental ballistic missiles, and ordinary television broadcasts.

They found that long-term monitoring of the Earth would not be able to necessarily disentangle or decode any of the signals from Earth, but the time variation of their aggregate power levels would reveal a wealth of information about humanity. For instance, the Earth's sidereal rotation period could be seen in the daily modulation as major broadcast centers like the East Coast of the US came in and out of view. Over the course of Earth's year, distant observers might also be able to deduce that some broadcasters have broadcast powers that vary with the solar day.

Powerful receivers that could distinguish among individual broadcasters could also use the Doppler shift due to Earth's rotation and orbit to deduce things like the inclination of Earth's rotation and orbit, and the location of transmitters across its surface.

They found that the distance at which Earth could be detectable with an Arecibo-style telescope (Box 12.1) was about 2 lt-yr, less than the distance to the nearest star, but that radar might reach farther, and Earth's transmission powers might grow in the future.

Variations on this calculation continue to help guide the field, essentially asking at what distance we could detect a planet whose technosignatures all have an ichnoscale value (Section 4.2.1) of $\iota = 1$ (see Sheikh et al. 2025 for a good example). In these calculations, high gain radio transmissions such as interplanetary radar are by far our most obvious technosignature.

16.1.4 Galileo's Observations of Earth as an Exoplanet

In 1990, the *Galileo* spacecraft, as part of its circuitous route to Jupiter, performed a gravitational slingshot maneuver with Earth, coming within 1000 km of the surface at high velocity. Carl Sagan (Box 3.5) organized a life-detection "control experiment" whereby the probe observed Earth as an alien planet, looking for signs of biology and technology.

Sagan et al. (1993) reported positive detections of life on four fronts (Figure 16.1).

In reflected light, the probe's spectrographs detected absorption features due to H_2O, CO_2, CH_4, N_2O, O_3, and O_2. They argued that the atmosphere appears to be in chemical disequilibrium, in the sense that these four gases cannot coexist for long in a planetary atmosphere, so at least some of them must have a source that replenishes them.

First, they argued first that the high abundance of O_2, which is highly reactive, requires a strong continuous source, and that most natural sources that could be responsible on Earth are insufficient. This implied a biological source (which is, in fact, the case: photosynthesis is the primary source of O_2 on Earth).

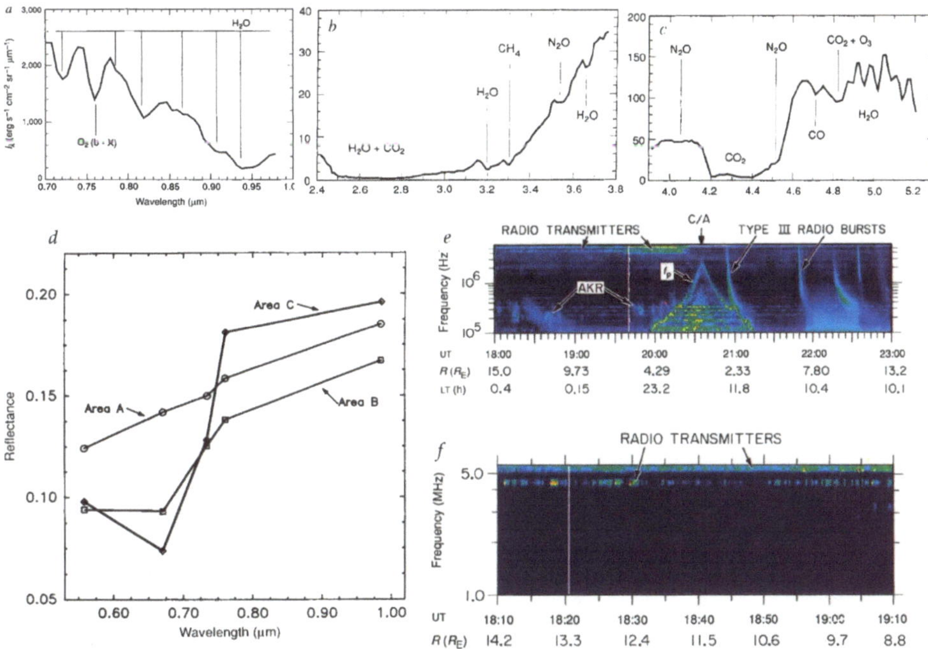

Figure 16.1. Collection of Figures 1, 3, and 4 from Sagan et al. (1993) showing the detection of technosignatures and biosignatures on Earth. *Top:* Reflectance above a cloud-free region of the Pacific Ocean from 0.7–5.2 μ, showing the presence of many biosignature gases, including O_2, CH_4, and N_2O, which is produced by both life and technology. *Bottom left:* Reflectance of Earth in three regions, showing a spatially varying "red edge" at 0.7μ due to chlorophyll. *Bottom right:* Figure *e* (labeled *a* in the original) shows a dynamic spectrum including power from Earth's magnetosphere, but also radio transmitters near 4 MHz. Figure *f* (labeled *b* in the original) is a detail showing the modulation of the signal more clearly. [Figure lettering changed from the original for consistency with this caption. With permission from Springer Nature.]

Second, carbon appears in both fully oxidized (CO_2) and reduced (CH_4) forms, and in the presence of O_2 the CH_4 should eventually become fully oxidized. Its presence thus suggests a strong source, which on Earth we know is due to both ordinary biological processes such as methanogens (methane-producing bacteria), and industrial agriculture, for instance rice cultivation and the flatulence of herds of cattle.

N_2O, also, is subject to photodissociation and will eventually reduce to N_2. On Earth, its source is nitrogen-fixing[2] bacteria and fertilizer produced by humans for industrial agriculture.

Third, they noted a “red edge” in reflected light, that is, a jump in reflectance at around 750 nm on *some* portions of the surface. They argued this was not consistent with any known or plausible soil or rock types, and so evidence for photosynthetic pigments. We know that this edge is due to the reflectance of chlorophyll: plants use this pigment to absorb light very well, but only blueward of 750 nm.

They noted that all three of these detected biosignatures (some of which, we argue here, are technosignatures) were strong but ambiguous.

Galileo’s scientific instruments[3] also made a fourth detection of life, however: its plasma wave radio detector saw persistent, narrowband 4 MHz radio emission on Earth’s nightside, which was not observable on the dayside. This emission could be localized to the surface of the Earth by its strength during the encounter (getting stronger as *Galileo* approached) and by the fact that the Earth’s ionosphere blocks such transmissions during the day, but becomes more transparent to them at night.[4] Sagan et al. (1993) argued that this was the only unambiguous detection of life on Earth made during the encounter.

This experiment gives us a guide to fulfilling Owen and Campbell’s vision of technosignature searches on exoplanets. It also reinforces the possibility that technosignatures might be a stronger and less ambiguous path to life detection than biosignatures (Chapter 9).

[2] To *fix* nitrogen means to change it from its molecular state N_2 to a more oxidized or reduced state, such as N_2O or NH_3. Molecular nitrogen is extremely tightly bound and generally not available for life to use despite being the most abundant component of Earth’s atmosphere. The natural nitrogen cycle relies on a few species of bacteria that fix nitrogen to more available forms. A major innovation for industrial agriculture is the Haber–Bosch process which performs nitrogen fixation by converting atmospheric nitrogen to ammonia for fertilizer.

[3] Sagan et al. (1993) did not “cheat” and take into account data from *Galileo*’s communication antenna to “discover” radio transmissions from Earth, for instance those from NASA controlling it; they restricted their analysis to the instruments designed to make scientific measurements.

[4] The ionosphere is a layer of plasma that surrounds the Earth, generated by ionization of the upper atmosphere by high-energy radiation from the Sun. Plasmas are conductive (and, so, opaque) below a frequency called the plasma frequency, and so this layer effectively traps low frequency radio waves on the Earth. At night, the atoms recombine, which causes its ionosphere’s plasma frequency to drop, makes the ionosphere less opaque, and raises the altitude of its base. This last point has many implications for radio transmissions: long-range radio transmissions on Earth use this lower surface as a mirror, allowing their signals to reach over the horizon, and this is why AM radio stations and amateur radio broadcasters, for instance, can have much longer range in the early hours of the morning, before sunrise reionizes the upper atmosphere.

16.2 Abundance, Planetary Ichnoscales, and Service Worlds

Most work on exoplanetary biosignatures (Section 1.1.3) focuses on terrestrial planets in the Habitable Zones of their stars, because these have the best chance of having life like that on Earth, and of us being able to detect signs of that life in reflected or transmitted light (Section 2.14). Following the logic of the Drake Equation (Chapter 8), some fraction of these planets will have liquid surface water and atmospheres like Earth's, some of these will be host to life, and on some of these will that life's biosignatures be detectable. The Galileo example suggests that on some fraction of the planets with life, that life will produce detectable surface or atmospheric technosignatures, providing a second opportunity to detect such life.

The case for exoplanetary technosignatures is stronger than this, however. The reason is that, as we discussed in Chapter 9, technological life can expand beyond its home planet, and we saw Mars was an example of a planet outside of its host star's Habitable Zone with a growing "technosphere" devoid of obvious biosignatures. Our focus on Habitable Zone terrestrial planets, likely appropriate for biosignature searches, might be too narrow for life-detection experiments in general.

Many signs of technology on Earth, from our atmospheric pollution to our agriculture, have some maximum ichnoscale value consistent with the maintenance of Earth's biosphere. For instance, there are negative ecological effects of putting too much CO_2 or chlorofluorocarbons (CFCs) in our atmosphere, and we could hardly cover the surface of the planet in solar panels without major repercussions for life on Earth.

We might then expect that the ichnoscale of certain technosignatures on a species' home planet or on one they have terraformed to be limited to small values, perhaps out of a sense of environmentalism or simply the survival necessity of having a healthy biosphere to live in. But as we saw in Section 9.1.2 (and as the example of Mars illustrates) this concern only applies to planets with biospheres, and the (many?) worlds onto which technology spreads could have very large ichnoscale values: there are no ecological consequences to, say, covering Mercury with solar panels or producing so much rocket exhaust on the Moon that it grows an atmosphere. Such service worlds might be the most abundant and salient signs of life in the Universe.

16.3 Atmospheric Technosignatures

The two primary methods of exploring the atmospheres of exoplanets are via transmission spectroscopy (Section 2.14.1) and direct imaging (Section 2.14.2). Both offer opportunities to detect artificial gases, and only recently has a library of potential technosignature gases started being compiled.

On Earth, the most obvious technosignature gases are CO_2, produced by fossil fuel use, various nitrogen compounds produced during combustion and agriculture, and CFCs, produced for many industrial and commercial purposes.

Haqq-Misra et al. (2022) explored the detectability of CFCs, which despite their very low concentration in Earth's atmosphere are so highly optically active[5] that

[5] This means they have a high absorption cross section to certain wavelengths of light per molecule.

they are a potent greenhouse gases and produce noticeable absorption features in Earth's atmosphere. They found that detecting such gases in transmission on exoplanets would be challenging even for optimistic scenarios involving M dwarf planets and JWST, but at high concentrations or when observed with future missions might be within reach.

Haqq-Misra et al. (2022) explored changes in Earth's atmospheric nitrogen gas abundances, including CH_4 and N_2O, and found that their simultaneous presence, especially in an O_2-, H_2O-, and CO_2-rich atmosphere, is a strong indicator of industrial-scale nitrogen fixation, which is required for widespread agriculture. They found that agricultural ichnoscales of 4–12 could be detectable on exoplanets, in principle.

Kopparapu et al. (2021) explored the possibility of using NO_2, a byproduct of combustion in a nitrogen-rich atmosphere, as a technosignature. They found that future space missions could detect ichnoscales of 3 or more in reflected light in optimistic conditions.

Seager et al. (2023) argued that fully fluorinated gases such as NF_3 and SF_6 make outstanding technosignature gases because they fulfill several criteria. They are unambiguous in the sense that they have no significant natural sources on Earth, and Petkowski et al. (2024) provided a strong argument that the reasons Earth life does not produce them might be universal. They also have a long residence time in Earth's atmosphere in part due to their low water solubility. They also have numerous industrial uses, providing a plausible reason why other species might produce them. Importantly, they tend to be very optically active, meaning they could plausibly be detected in exoplanetary atmospheres even at relatively low concentrations.

Because such gases tend to be transparent at visible wavelengths but opaque in the infrared, they are strong greenhouse gases and have the potential even at relatively low concentrations to drive climate change (indeed, this is one of the problems with CFC pollution on Earth). Schwieterman et al. (2024) studied their detectability on the theory that large amounts of them might be part of efforts to "terraform" planets—i.e., to adjust their climates and atmospheres to make them habitable, as some have proposed humans may someday do for Mars. They found that under optimistic conditions such gases could be detectable in transmission with near-future observatories if they had concentrations of many parts per million (i.e., thousands to millions of times higher than on Earth today).

More generally, the atmospheres of exoplanets that lack biospheres or technospheres are generally going to be dominated by a small set of common gases such as H_2, N_2, CO_2, H_2O, or CH_4, which have distinct spectral signatures. So, while the universe of technosignature gases used by other species might be so large that it is not possible to correctly guess which would be found first, the universe of natural gases might be small enough that anomalous exoplanetary spectra could be identified as due to artificial gases by exclusion, even if the specific gas involved is not obvious.

Catling et al. (2025) argued that low deuterium abundances on a planet might be a good technosignature because deuterium is such a useful element for energy generation via fusion. He argued that, although deuterium is not a very abundant isotope of hydrogen, there are sufficient quantities of it in seawater (for instance) that it could provide nearly all of a species' planetary energy needs for hundreds of millions or even billions of years, over which time the oceanic and atmospheric concentrations of deuterium would noticeably diminish.

Catling et al. (2025) argued that the deuterium abundance on a planet (expressed relative to hydrogen as a D/H ratio) is generally set by the abundance in the interstellar medium, which is rarely much different from the initial abundance set during the Big Bang. While this abundance can be greatly enhanced by mechanisms such as fractionation during atmospheric escape of hydrogen, it would be very strange to see a planetary D/H ratio far below the cosmic value, and that its consumption by an energy-hungry species would be a likely explanation.

16.4 Surface Technosignatures

The dawn of direct imaging of exoplanets raises the possibility that we might be able to detect surface technosignatures in reflected or even emitted light.

Lingam & Loeb (2017) explored the possibility that photovoltaic cells across a planetary surface might be a distinct and detectable technosignature. They noted that just as the "red edge" of chlorophyll is a feature of plants' ability to harvest free energy from sunlight, silicon similarly has a spectroscopic "blue edge." They argued that silicon is sufficiently abundant, easy to use, and efficient as a method of energy harvesting that it is a plausible material even for alien solar panels.

They hypothesized that a service world might have a substantial fraction of its surface covered in such solar panels, and so present such an anomalous spectral edge in reflected light. Kopparapu et al. (2024) explored the feasibility of detecting such an edge and found that even with upcoming space observatories and optimistic assumptions it would be extremely challenging.

Beatty (2022) explored the possibility of detecting emission from city lights, and found that optimistically, an ecumenopolis (a planet-wide city, akin to the planet-cities of Coruscant or Trantor from science fiction) might be detectable coming from the night side of a terrestrial planet in the Habitable Zone of a nearby star with upcoming space observatories.

Beech (2008) suggested that one might find signs of planetary terraforming, as some have suggested humans might perform on Mars in the distant future to make it hospitable for life. They argued that finding planets interior or exterior to the traditional Habitable Zone (Section 2.14.3) with clement surface conditions because of very high albedos or strong strong greenhouse gases would be indicative of technological climate manipulation. Jill Tarter (Section 1.2) has argued that finding multiple planets in a system with similar atmospheric and surface properties, including temperature, would also be a sign of terraforming.

An ambitious possibility is that planetary surfaces might be imaged by very large telescopes through rotational modulation. The idea is that as a planet rotates its brightness changes due to the differing reflectance or emission properties across its surface, and it might be possible to reconstruct the distribution of these features from a high SNR light curve taken throughout the planetary day and year. Berdyugina & Kuhn (2019) explored this possibility, suggesting that widespread photosynthesis or solar panels could be detected with a very large telescope (Figure 16.2).

Even more ambitiously, a telescope placed at the Solar Gravitational Lens (Section 14.4.2) could in principle resolve the surface features of a planet orbiting

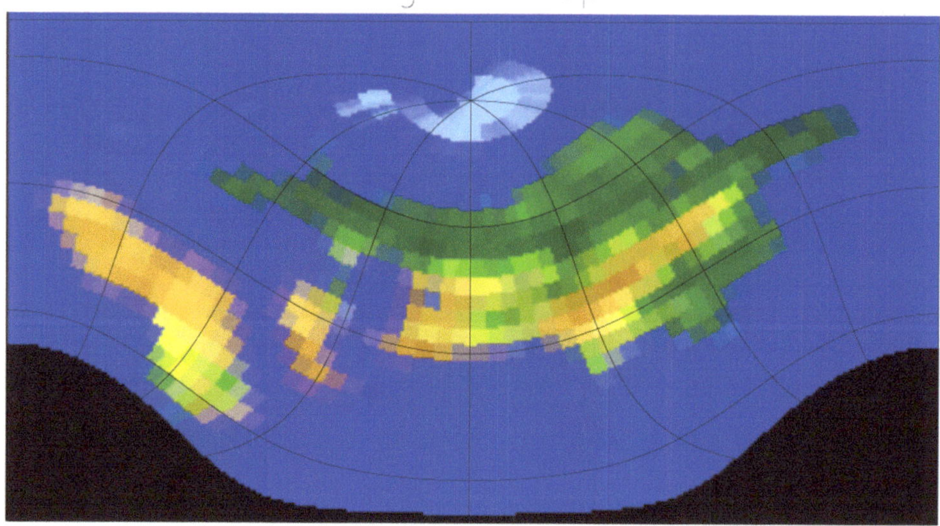

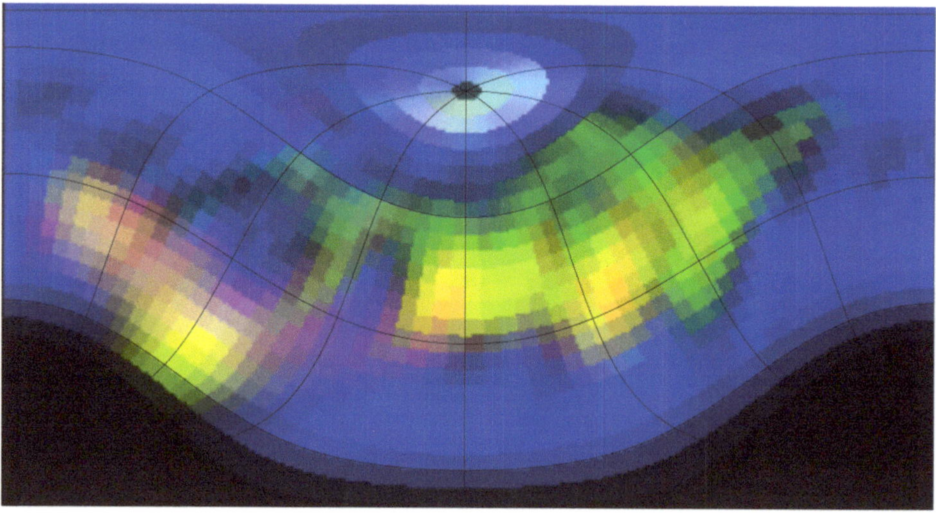

Figure 16.2. Figure 17 from Berdyugina & Kuhn (2019), showing the input and output of an algorithm that inverts light curves at different wavelengths to reconstruct surface features, in this case of a cloudless version of Earth. Widespread surface features like deserts, forests, and perhaps widespread technosignatures might be localized in this way. © 2019 the American Astronomical Society, all rights reserved.

a nearby star at scales of 10s of km, raising the possibility of detecting cities and other large surface features indicative of technology. This possibility has been explored in a series of papers by Russian-American physicist Slava Turyshev and others (e.g., Turyshev & Toth 2022; Toth & Turyshev 2023).

Box 16.1: Jargon Review: Exoplanetary SETI

- **Ecliptic (plane)**: The plane in space in which the Earth (and, very nearly, the other solar system planets) orbits the Sun. By extension, a similar plane in other planetary systems. Seen from Earth, the ecliptic appears as a great circle in the sky.
- **Eavesdropping**: The interception of communicative signals, for instance via spillover radiation from signals between planets in another system. Despite the connotation, eavesdropping need not (and likely would not) imply the signal could be interpreted by us.

16.5 Planet-to-Planet Eavesdropping

Some of the most powerful radio transmissions from Earth—those from interplanetary radar and those to our interplanetary probes— are not directed randomly but are generally sent out to low ecliptic[6] latitudes, and often directed at other planets (Ashtari 2023; Derrick & Isaacson 2023).

This behavior—radar to monitor potentially dangerous asteroids and planet-to-planet communication—are both plausibly universal for spacefaring species, and might grow in strength and frequency as a species spreads through its planetary system. If this is the case, then it might make sense to choose the direction and time of radio (or optical) observations of *other* planetary systems such that we might intercept such signals.

To detect ecliptic radar generally, this would mean focusing observations on planetary systems for whom we are in their ecliptic plane, which is conveniently also the vantage from which we can detect planets in that system transit (because planetary systems are generally "flat" like the solar system, seeing just a single, usually innermost planet transit is sufficient to establish that we are in or near that system's ecliptic plane). This has the additional advantage that such observations can also take advantage of the temporal Schelling point when planets transit (Section 5.5), and that species might be more likely to intentionally transmit beacons out toward stars that can detect their planets in transit.

Siemion et al. (2013) suggested that in cases where exoplanetary orbits are sufficiently well known, we might time our observations during planet–planet conjunctions, i.e., when we are in the path of such transmissions, behind the

[6] The *ecliptic* is the plane in which the Earth orbits the Sun, which, from our vantage, appears as a line across the sky. Because the orbits of the other planets are inclined by, at most, a few degrees from Earth's orbital plane, from Earth they appear to travel very nearly along the ecliptic as they (and we) orbit the Sun. Ecliptic latitude measures the angular distance between a point on the sky and the ecliptic, which for the other planets is never more than 5°.

intended receiver. Tusay et al. (2024) found that the TRAPPIST-1 system of planets had such a sufficiently precise ephemeris, and performed a long stare using the Allen Telescope Array to search for such transmissions. This is sometimes called planetary *eavesdropping*, although it should be emphasized that the chances of decoding such a transmission would seem to be minuscule, both because a transmission not intended for us is unlikely to be sufficiently overpowered that it would still have all of its information content when it arrived at Earth,[7] and because it is presumably very difficult or impossible to deduce an alien encoding scheme.

At any rate, as of this writing few systems were known with sufficiently precise ephemerides for such work (TRAPPIST-1 may be the only one, in fact), and the planets involved in such measurements are generally very close to their stars (so unlikely to be the home planet of any species, though they could certainly be service worlds or otherwise occupied by technology). It is also possible that optical or higher frequencies would be employed for such transmissions (as our own space agencies have plans to do), in which case the "spillover" would be much smaller than for radio, and we might be much less likely to be in the background of such a beam.

Exercises

1. Consider an Earth-like planet orbiting a Sun-like star 10 pc away. Assume half of the luminous energy of the star is in optical photons, with $\lambda \approx 0.5\ \mu$, and that we want to observe it with a $D = 10$ m telescope.
   (a) What is the rate of photons r from the star that are collected at Earth in a $D = 10$ m telescope every second?
   (b) First, let's think about transits. What fraction of the star's surface is blocked by the planet (that is, how "deep" is the transit)?
   (c) The Earth's atmospheric scale height is about 8.5 km. Taking this to be the height of the atmosphere, what fraction of the star's light passes through the atmosphere of this distant planet as seen from Earth?
   (d) How many photons $r_{\rm atm}$ arrive at our 10 m telescope on Earth every second that have passed through the planet's atmosphere?
   (e) Next, let's think about direct imaging in reflected light. What fraction of the star's photons strike the planet's surface?
   (f) Assume half of these photons are reflected from the planet, and we image the planet when we see half of its surface illuminated (quarter phase). How many of these reflected photons arrive at our 10 m telescope on Earth per second, $r_{\rm DI}$?
   (g) In both cases, assume that you want to collect a spectrum from the planet with a signal-to-noise ratio of 10 in each of 100 spectral channels. Let's assume we can collect these photons with 10% efficiency.

[7] As we explored in Section 13.4, the information content of a message decreases with signal strength, so if a signal has a power level that is well chosen for interplanetary communication, it is presumably far too weak to retain its full *information content* over interstellar distances, even if the signal's *existence* is obvious for many light years.

For the transit case, the signal-to-noise ratio is the number of photons from the atmosphere $r_{\text{atm}}t$ divided by the number from the star, $\sqrt{rt}$. How much time will this take?

(h) How does this compare to the time it takes the planet to transit its star? Assume an orbital velocity of 30 km s^{-1}.

(i) Now do the direct imaging case. Here, assume that the coronagraph can completely separate the photons from the star, so the noise is just from the $r_{\text{DI}}t$ photons one receives.

2. Ideally, one would be able to image the surface of distant exoplanets.

(a) What is the linear extent of an optical telescope (either single dish or interferometer) with sufficient angular resolution to image an Earth-like exoplanet 10 pc away with 50 km resolution?

(b) The solar gravitational lens provides focusing from a ring the size of the entire solar diameter, effectively allowing one to use the Sun as a primary mirror, with a similar resolution as if it were a disk of the same size. What resolution can a telescope employing the solar gravitational lens achieve on a planet 10 pc away?

(In practice many other effects limit the resolution of an SGL telescope).

3. **Planet–Planet eavesdropping**:

NASA's deep space network sends signals to Mars using 34 m dishes at 7.2 GHz with a power of up to 20 kW with a duty cycle of around 80%.

Mars has a physical radius of 3400 km, an orbital radius of around 1.5 au, and an orbital velocity of around 24 km s^{-1}. Earth's orbital velocity is around 30 km s^{-1}. Both have essentially circular orbits.

In the problems below you may look numbers up (for instance in an ephemeris) to check your answers but do the calculations given only the numbers above.

(a) How large, in angular size, is the outgoing DSN signal beam? How does this compare to the angular size of Mars? Roughly what fraction of the energy of the beam "misses" Mars?

(b) Approximately how fast does the outgoing beam sweep across the sky as the DSN tracks Mars? Remember that both Mars and the Earth are moving.

(c) For how much time will a distant observer "behind" Mars from Earth's perspective be in the path of outgoing DSN transmissions?

(d) Over the course of a synodic period (e.g., the time it takes Earth and Mars to go from conjunction (lined up with the Sun) to conjunction) what fraction of the sky receives these DSN transmissions? (*Hint*: either look up the area of a sphere in square arcseconds or (easier) calculate the requested solid angle in square radians which, for small sections of a sphere, are equal to steradians (sr). There are 4π sr in a sphere.)

(e) Because it has a different orbital inclination from Earth, over the course of many decades Mars's ecliptic latitude spans roughly $\pm 6^{\circ}$.

If the DSN network continues to transmit far into the future, what fraction of the sky will eventually receive one of its transmissions to Mars?

(f) Assume that half of the power is in a 1 Hz wide carrier wave. Out to what distance can these transmissions be detected by a 100 m receiver with SEFD of 15 Jy (comparable to the Green Bank Telescope) and an efficiency $\epsilon = 0.5$ at SNR 10? Assume that they integrate and build up signal for as long as the signal lasts (from part c).

(g) There are around 0.1 stars per cubic parsec near the Sun. About how many stars are within range of the DSN's transmissions to Mars, given your answer above?

References

Ashtari, R. 2023, ApJ, 957, 15

Beatty, T. G. 2022, MNRAS, 513, 2652

Beech, M. 2008, JBIS, 61, 43

Berdyugina, S. V., & Kuhn, J. R. 2019, AJ, 158, 246

Bryson, S., Kunimoto, M., Kopparapu, R. K., et al. 2021, AJ, 161, 36

Campbell, J. B. 2006, IAU Colloq. 200: Direct Imaging of Exoplanets: Science & Techniques ed. C. Aime, & F. Vakili (Cambridge: Cambridge Univ. Press) 247

Catling, D. C., Krissansen-Totton, J., & Robinson, T. D. 2025, ApJ, 979, 137

Derrick, R., & Isaacson, H. 2023, PASP, 135, 034201

Haqq-Misra, J., Fauchez, T. J., Schwieterman, E. W., & Kopparapu, R. 2022a, ApJL, 929, L28

Haqq-Misra, J., Kopparapu, R., Fauchez, T. J., et al. 2022b, PSJ, 3, 60

Kopparapu, R., Arney, G., Haqq-Misra, J., Lustig-Yaeger, J., & Villanueva, G. 2021, ApJ, 908, 164

Kopparapu, R., Kofman, V., Haqq-Misra, J., Kopparapu, V., & Lingam, M. 2024, ApJ, 967, 119

Lingam, M., & Loeb, A. 2017, MNRAS, 470, L82

Owen, T. 1980, Strategies for the Search for Life in the Universe, Astrophysics and Space Science Library, Vol. 83, M.D. Papagiannis (Dordrecht: Reidel Publishing Co.) 177

Petkowski, J. J., Seager, S., & Bains, W. 2024, NatSR, 14, 15575

Sagan, C., Thompson, W. R., Carlson, R., Gurnett, D., & Hord, C. 1993, Natur, 365, 715

Schwieterman, E. W., Fauchez, T. J., Haqq-Misra, J., et al. 2024, ApJ, 969, 20

Seager, S., Petkowski, J. J., Huang, J., et al. 2023, NatSR, 13, 13576

Sheikh, S. Z., Huston, M. J., Fan, P., et al. 2025, AJ, 169, 118

Siemion, A. P. V., Demorest, P., Korpela, E., et al. 2013, ApJ, 767, 94

Sullivan, W. T., Brown, S., & Wetherill, C. 1978, Sci, 199, 377

Toth, V. T., & Turyshev, S. G. 2023, MNRAS, 525, 5846

Turyshev, S. G., & Toth, V. T. 2022, MNRAS, 515, 6122

Tusay, N., Sheikh, S. Z., Sneed, E. L., et al. 2024, AJ, 168, 283

AAS | IOP Astronomy

The Search for Extraterrestrial Intelligence
Theory and practice
Jason T Wright

Chapter 17

Other Technosignatures

> The main thing about aliens is that they are alien. They feel no responsibility for fulfilling any of your expectations.
>
> — Silverberg (2015)

17.1 Miscellany

While most SETI efforts can be at least roughly sorted into the five pillars described in the previous five chapters (radio, optical, infrared, solar system, and exoplanetary SETI), there have been many suggestions for technosignatures that escape these categories. The panoply of other suggested technosignatures is too vast and varied to exhaustively enumerate or categorize, so here we will instead focus on a few of the better explored suggestions.

The most obvious is perhaps to look for communicative or other signals beyond the electromagnetic spectrum, in carriers (*messengers* in astronomy jargon) of the other three fundamental forces: neutrinos, gravitational waves, and cosmic rays (Section 2.6). Might not another species, perhaps with greater mastery over physics and matter, have found a way to use these other forces to communicate?

There have also been many suggestions for alien activities that might be detectable via ordinary electromagnetic radiation, including gamma-rays or radio waves from spacecraft propulsion byproducts, searching the Earth for technosignatures, and looking for signs of grand engineering projects such as artificial exoplanetary systems or anomalous stellar populations in other galaxies.

doi:10.1088/2514-3433/adf5aech17

17.2 Gravitational Waves

> In any search for evidences of technologically advanced societies in the universe, an investigation of anomalously intense sources of gravitational radiation ought to be included.
>
> — Dyson (1963)

Gravitational waves have some properties that make them an intriguing possibility for a technosignature. Roughly speaking, when masses accelerate they launch waves in spacetime, similar to how accelerating charges launch waves in the electromagnetic field, and in doing so generate a signal that is in principle detectable at great distances.

17.2.1 Gravitational Waves from Binaries and Dyson's Suggestion

The first to mention such waves in a SETI context was Freeman Dyson, in a prescient article in 1963 (Dyson 1963).[1] In it, Dyson explores binary stars as a source of energy, in the form of "gravitational machines."

Dyson's insight is that as masses approach each other, their mutual binding energy becomes *more negative*, and so by conservation of energy they release energy that can be extracted. He imagined that by manipulating masses into trajectories bringing them into close encounters with one of two orbiting bodies, the mass can leave with more energy than it had when sent in, having tapped some of the orbital energy of the binary. This is the idea of gravitational slingshots or assists that we encountered in Section 6.2.3 in orbital dynamics, in which our interplanetary spacecraft use close encounters with planets to gain or lose energy beyond which they could generate from their rockets. In such an encounter, the craft gains (or loses) some energy at the expense of (or to) the planet.

The maximum amount of energy available to be tapped in such a system is

$$E = \frac{GM_1M_2}{2(R_1 + R_2)} \tag{17.1}$$

where M and R are the masses and radii of the two objects, and $R_1 + R_2$ represents the closest the two objects may approach before colliding. This is roughly the amount of energy released bringing the objects from very far apart until they are almost touching in a circular orbit.

For two stars like the Sun, this energy is 10^{48} erg, which is about 0.1% of the energy of a supernova and equivalent to the output of the Sun over 60 million years.

[1] The article was actually Dyson's entry to a competition held by the Gravity Research Foundation in 1962 (it won fourth place). It appeared in Alastair Cameron's influential compilation of SETI-related articles *Interstellar Communication* in 1963, but was not widely available until put on the arXiv with the permission of all involved by David Derbes in 2023 at https://arxiv.org/abs/2305.10470.

Dyson imagined that it would be more profitable in the long run to tap such stars' luminous energy for their lifetime (over 10 billion years), but noted that the situation is "entirely reversed" for white dwarfs.

Such stars have about the mass of the Sun but are only about 1% of its radius (similar to the Earth), so they can get much closer and more energy is available for extraction. In this case the energy available[2] is 10^{50} ergs, which is much more than the white dwarfs themselves will emit as they cool.

Dyson noted that ships using such systems for gravitational assists would be accelerated from rest to 2000 km s^{-1} (or nearly 1% the speed of light) without sensing any acceleration, making such a system an outstanding method of initiating (or completing) interstellar travel.

Not content to stop there, Dyson speculated that neutron stars[3] might form in binaries. Being only 10 km across, but more massive than the Sun, their total binding energy is 10^{53} erg.

Dyson noted (in what may be the first published appearance of the relevant equation) that these machines would not be able to be tapped quite down to their full potential, because once the objects became close they would begin to emit gravitational waves and radiate away their orbital energy at a rate of

$$W = \frac{128}{5}\frac{V^{10}}{Gc^5} = \frac{4}{5}\frac{G^4M^5}{c^5d^5} \tag{17.2}$$

where V is their orbital velocity and d is their separation. He noted that the extremely steep dependence on d (or V) meant that there was a sharp boundary beyond which the system would spontaneously begin to shrink and eventually coalesce, giving off intense gravitational radiation in the process.

He imagined that a technological species would stop extracting energy from its machine before reaching the point that it began to radiate significantly, but noted that if tight binary neutron stars ever did form they would generate a short burst of intense gravitational radiation that could be detectable on Earth.

He concluded that the coalescence of neutron stars might be a technosignature, representing the death throes of a once useful gravitational machine.

17.2.2 Detections of Gravitational Waves

As discussed in Section 2.6.2, gravitational waves from binary neutron stars have been observed. In 1974, Hulse and Taylor inferred the existence of such waves being emitted from a binary pulsar by their slow orbital decay. Then, in 2015, the LIGO experiment detected the waves of merging black holes directly, and since then many black-hole and neutron-star binary mergers have been detected (both discoveries garnered the Nobel Prize).

---

[2] Dyson used slightly different numbers in his paper than I do here, but the point is the same.

[3] At the time neutron stars were purely hypothetical! They would not be discovered for another five years (Box 3.1).

The characteristic luminosity of a gravitational wave source is c^5/G where G is Newton's constant (this works out to 3×10^{59} erg s^{-1}, which is more than the light put out by all of the stars in the observable Universe). Roughly speaking, gravitational waves' power are determined by the third time derivative of the moment of inertia of a system, $\dddot{I}$. This has units of luminosity, but the emitted luminosity in gravitational waves is much smaller than this for most sources, scaling as $\dddot{I}^2/(c^5/G)$. When $\dddot{I} \gtrsim c^5/G$ gravitational waves become so efficient at removing energy from the system that it decays on an orbital timescale; this occurs when the objects are separated by roughly their own event horizon, which only happens for black holes and neutron stars.

Another way of expressing this is that gravitational waves are very weakly coupled to matter, so matter in their path does not lose much energy to them, and in turn they do not move matter much. In order for matter to be strongly coupled to spacetime, it must be extremely dense—dense enough to be nearly a black hole. Thus, only black holes and neutron stars can lose much energy to gravitational waves, and even then only when moving near the speed of light.

Because our are so weakly coupled to gravitational waves, we can only detect such waves when they contain an enormous amount of energy. For scale, LIGO's first detection was of a black hole merger that converted three times the mass-energy of the Sun ($3\mathrm{M}_\odot c^2 = 6 \times 10^{54}$ erg) worth of black hole rest mass into gravitational waves, and yet it was still just barely detectable here on Earth. The effect is given by the dimensionless *strain* of spacetime, which describes how much space is distorted by the wave (Section 2.6.2).

Roughly speaking, the strain of gravitational waves given by neutron stars or black holes merging is given by

$$h \approx \frac{R_\mathrm{s}}{r} = \frac{GM}{c^2 r} \tag{17.3}$$

where R_s is the *Schwarzschild radius* of the black hole[4] and r is our distance to it. For scale, LIGO's first detection was of the merger of $30\mathrm{M}_\odot$ black holes, whose radii are around 300 km. Since the merger occurred at a distance of about 1.3 billion lt-yr, corresponding to a strain of around 10^{-21}, this means that as the gravitational waves passed by, the Earth was distorted by about $10^{-21}\mathrm{R}_\oplus \approx \times 10^{-14}$ m, or the size of a large atomic nucleus.

17.2.3 Energy Based Gravitational Wave Detection

Figure 17.1 shows the spectrograms (Section 11.1.4, compare Figure 11.3) of the detection of gravitational wave event GW170817, the result of the merger of two neutron stars, as described by Dyson (1963). The signal is strongly *chirped*, with the

[4] The Schwarzschild radius is effectively the "size" of a black hole, within which no information or material can escape. Neutron stars are only barely bigger than their Schwarzschild radii, so this is close enough for an order of magnitude estimate.

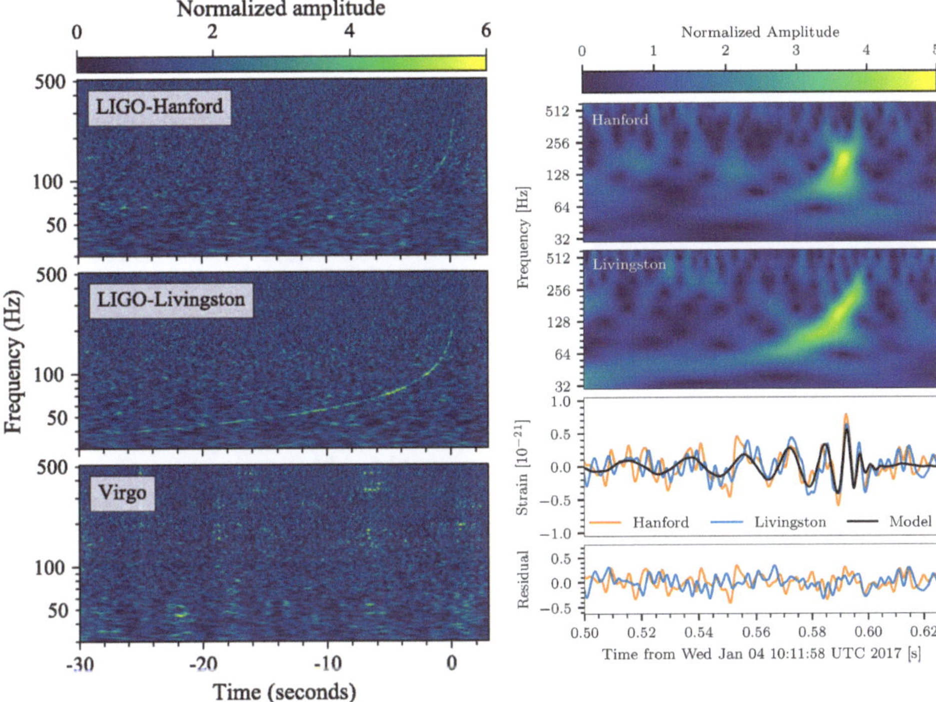

Figure 17.1. Spectrograms of the detection of gravitational wave event GW170817, a neutron star-neutron star merger. These are essentially the same as dynamic spectra used in radio astronomy (Section 12.3). *Left:* The 30 s before the merger. The two LIGO detectors see power at around 50 Hz, as the neutron stars orbit each other 25 times per second. *Right:* Zoom in on the final 0.1 s of the merger event. The bottom panel shows the actual waveform of the strain measured by LIGO. Right figure from Abbott et al. (2017a). Left figure from Abbott et al. (2017b) CC BY 4.0.

frequency of the signal (equal to twice the orbital frequency of the binary system) increasing as the neutron stars inspiral and eventually merge. Note how quickly the system accelerates, as the power lost to gravitational waves increases according to Equation (17.2).

Because the waveform of inspiraling binaries is well known and modeled with few free parameters, searches in LIGO are generally done using *matched filters* on the strain data for exactly these waveforms. Two sites are used both to provide coincidence detection to reject false positives from spurious vibrations at one site, and to help localize the source on the sky.

An interesting factoid about gravitational wave detectors is that like radio receivers they are antennae that measure the *amplitude* of waves, not the *energy* of waves, as a photon-collecting detector would. Because the energy carried by a wave is the square of its amplitude, this means that the strength of the signal of a precisely known waveform decays only as $1/r$, not as the usual $1/r^2$—this is very good for detecting extremely distant events, but means the gains from having a nearby source are smaller.

Radio receivers also have this property, but it typically cannot be exploited, and so radio telescopes are effectively $1/r^2$ detectors.[5] This is because the precise waveform of radio signals from astronomical sources cannot be known in advance, and so one must do energy-based searches for them, rather than the waveform template-matching done at LIGO.

Gravitational wave detectors looking for novel waveforms, perhaps as a form of communication, will generally need to be energy-based, and so be $1/r^2$, and performed on the spectrograms, which can be made at the Gabor limit (set by the uncertainty principle, Section 11.10).

17.2.4 Gravitational Wave Technosignatures

17.2.4.1 Communicative Gravitational Waves

Gravitational waves have some qualities that would seem to make them very good for communication: they move at the speed of light, and they are not extinguished by intervening material (so travel unimpeded through anything). Also, there is essentially no background at frequencies above 1 Hz, so the signals are very clean.

The major disadvantage is that they are extremely hard to generate or detect, although in principle this might be considered a benefit to a species which has found a way to use them and wishes to keep its signals hidden from species that have not figured out the trick.

It is not clear how to build a useful gravitational wave transmitter, although it is probably not impossible. Like electromagnetic waves, gravitational waves can be focused using a phased array (Section 2.3.3), somewhat reducing the power needed to generate detectable strain at large distances. Rana Adhikari has calculated that an array of a billion 100 kg dumbbells 3 m in length spinning at 20 kHz could be phased up to generate a strain detectable by an instrument like LIGO at a distance of 50 km (NASA Technosignatures Workshop Participants 2018).

To extend this possibility to interstellar distances, Jackson & Benford (2018) suggest other species might build transmitters out of small black holes instead of ordinary materials.

Another option for generating gravitational wave signals is to adopt the philosophy of the heliotrope (Section 11.1.6) and modulate the signal of a natural transmitter as a power source: that is, to deliberately manipulate the orbits of stars near a supermassive black hole that are generating gravitational waves anyway.

Abramowicz et al. (2020) describe a "messenger" in the form of a Jupiter-mass black hole placed in a tight orbit around the supermassive black hole at the center of the Milky Way, manipulated in a manner similar to Dyson's idea for gravitational machines such that its orbit does not decay (that is, energy is *added* to its orbit to prevent it from merging and ensure a steady production of gravitational waves.) The orbits of the black hole would be unambiguously artificial because they would not decay as expected via gravitational waves, and such a signal might be detectable with future gravitational wave detectors such as LISA.

---

[5] The $1/r$ benefit of gravitational wave astronomy over other forms of astronomy is also often overstated: the significance of all detections made by LIGO so far are only a few times stronger than they would be as energy-based detections.

17.2.4.2 Other Uses

It is also possible that gravitational waves could be associated with very fast travel. While no realistic solution to Einstein's field equations is known that admits superluminal "warp drives," there remains the logical possibility that they exist and are exploited by technology elsewhere. If ships or other matter is accelerated via spacetime warping, it is possible that the acceleration of the mass itself or the creation or destruction of the warp itself could generate a detectable signal (Sellers et al. 2022; Lentz & Felton 2024).

And while it is not a gravitational wave technosignature itself, Dyson's original idea of identifying neutron star binaries used as gravitational machines is still valid. The Hulse and Taylor binary pulsar was the subject of a Nobel Prize because its orbital decay rate matches the expectations of general relativity essentially exactly, suggesting no other important sources of energy loss from the system exist. A binary pulsar whose decay rate exceeded the GR expectation might therefore be explained by an alien species tapping its orbital energy (although other, natural explanations might easily be found, as well, for instance involving a third body in the system).

17.3 Neutrinos

Electromagnetic and gravitational waves are only two of the ways that communicative signals could, in principle, be sent. Might an old technological species use something else?

Hippke (2018) explored the properties of various carriers, and, consistent with many previous studies of the subject concluded that there are good reasons to favor photons. There are indeed many particles in the subatomic menagerie, but most of them suffer from the problems of either having short lifetimes or being charged, or both. Being charged is a problem because over interstellar distances a charged particle's path will be deflected by galactic magnetic fields, causing the particle to lose energy and veer from its intended target.

The only exceptions are the photon (which we already know is a good carrier), gravitational waves (which we discussed above) and neutrinos (Section 2.6.1).[6] Neutrinos offer an intriguing possibility because, like gravitational waves, they travel without scattering or extinction through virtually any ordinary matter, and because they are challenging, but not impossible, to detect (so might make a good covert channel). And, unlike gravitational waves, a neutrino beam can be relatively easily created and collimated in a particle accelerator.

The primary challenge of neutrino communication is that, unlike photons, which can be collected with near-perfect efficiency over a wide telescope aperture, neutrinos hardly ever interact with ordinary matter. A neutrino detector generally

[6] There are also hypothetical particles, like axions, that could satisfy these criteria. The fact that we do not easily detect such particles (or we would have verified them experimentally by now) means they will share many of the challenges of neutrinos for communication.

consists of a vast amount of transparent material—water ice in the case of the Ice Cube neutrino detector at the South Pole—within which one searches for the occasional byproducts of a high-energy neutrino interacting with an atomic nucleus and producing a shower of subatomic particles, some of which are charged and produce Cherenkov radiation. Neutrino communication would thus be rather expensive in the sense that most of the neutrinos produced would be "wasted" because they would pass through the recipient's detector.

Nonetheless, the problem has been considered by a number of authors. Saenz et al. (1977) offered an early proposal for neutrino communication through the Earth using neutrinos from particle accelerators, and Pasachoff & Kutner (1979) and Subotowicz (1979) offered early suggestions that they might be used for interstellar communication, the former noting that a single detector would be omnidirectional, greatly easing the problem of knowing when and where to look for a signal. Learned et al. (2009) explored some practical issues around the construction of neutrino beams and information encoding.

Learned et al. (1994) explored the utility of neutrino pulses to provide a galactic timing standard akin to the Global Positioning System, arguing that neutrino pulses could serve the purpose well because they could be extremely narrow and universally accessible.

Interestingly, neutrino beams are common and perhaps inevitable byproducts of any particle accelerator, including the various ones used by humanity to explore subatomic physics. This implies that any extraterrestrial particle accelerator could be detected on Earth if it is pointed in our direction and has sufficient strength. For instance, Silagadze (2008) explored the signal of an extraterrestrial muon collider, and Lacki (2015) explored the consequences of a species that had constructed a particle accelerator that could explore the extraordinarily high energies associated with physics at the Planck scale.

One advantage of SETI with neutrinos is that it can be performed commensally with ordinary high-energy neutrino physics. A challenge will be identifying a reason to suspect any detected neutrinos are artificial and not natural.

17.4 Monumental Architecture

Some of Earth's most obvious and enduring technosignatures are not the sorts of things aliens might expect to find: our monumental architecture. The Great Pyramid of Giza and the Great Wall of China, while perhaps not quite visible from space as a popular misconception would have it, are some of the largest structures ever built by humans, and they persist long after their original purposes have faded away.

Might an alien species similarly construct enormous and persistent "monuments" for reasons we cannot guess and perhaps even they have forgotten, that might similarly be their most ostentatious technosignatures? As Dyson (1966) argued (Section 4.2):

> …we are not interested in guessing what an average technological society might look like. We have to think instead of what the most conspicuous out of a million technologies might look like. The technology which we have a chance to detect is by definition one which has grown to the greatest possible extent…
>
> I do not need to discuss questions of motivation, who would want to do these things or why. Why does the human species explode hydrogen bombs or send rockets to the Moon? It is difficult to say exactly why. My rule is, there is nothing so big nor so crazy that one out of a million technological societies may not feel itself driven to do, provided it is physically possible.

What sort of monumental architecture might be detectable? We discussed megastructures in transit in Section 15.7, which is certainly one possibility.

Clement et al. (2022) and Raymond et al. (2023) explored the architecture of planetary systems themselves as potential technosignatures. Our expectations from planet formation theory and observations of exoplanets reveal that tightly packed terrestrial planetary systems are common—we often see many coplanar terrestrial planets in short period orbits around a single star, often in *mean-motion resonances*, meaning that these planets' orbital periods are related through ratios of small integers. In our own solar system, for instance, Neptune and Pluto are in a 3:2 resonance, and the innermost Galilean satellites of Jupiter are in a 1:2:4 resonance.

Some resonances are highly stable, and the mechanics of planet formation and migration can lead to planets becoming trapped in these configurations, explaining why we see them. Other configurations are particularly unstable, or would not be an natural outcome of planet formation. Clement et al. (2022) explored unnatural configurations, such as a prime number sequence of resonances (periods in the ratio 2:3:5:7:11), the Fibonacci sequence and the "lazy caterer" sequence. They found, somewhat surprisingly, that these sequences are stable, and so if found might represent deliberate sculpting by an intelligent architect. Raymond et al. (2023) found a perhaps even more surprising and artificial stable family of configurations in co-orbital terrestrial planets, in which many planets like Earth share a common orbit. He found if they were carefully placed, they would stably execute so-called horseshoe orbits.

17.5 Galaxy-scale Engineering

A mature space-faring species that has settled its entire galaxy might radically alter its environment such that the galaxy was hardly recognizable as natural, much like city centers on Earth could hardly be mistaken for unspoiled wilderness. Kardashev imagined they would surround their stars with Dyson spheres to capture all of the starlight, but what other signs of their presence might we look for?

Carrigan (2012) suggested that a rapidly expanding species surrounding stars with Dyson spheres might, at an intermediate stage of growth, result in what appear to be "bubbles" in their galaxy within which all stars had Dyson spheres. Voros (2013) and Wright et al. (2021) pointed out that such a phase would be short-lived if it existed at all, because a galaxy's differential rotation and random stellar motions would smear out such a bubble, first roughly into a ring then progressing toward the center of the galaxy, where stellar densities are higher (and, so one would presume interstellar travel is easier). Infrared bright but optically dark bubbles, rings, or nuclear regions in galaxies might thus be technosignatures. Lacki (2019) explored the observational consequences of galaxies in which certain stars are prefentially cloaked in Dyson spheres.

Voros (2013) specifically pointed to Hoag's object, a galaxy with a highly unusual ring-like morphology as an example (this galaxy's appearance is widely believed to be due to a galactic collision, which can result in such morphologies, although this object in particular has not been rigorously explained, see Figure 17.2).

Figure 17.2. Hoag's Object, a galaxy of highly unusual morphology, perhaps due to a galaxy collision. The central region is not extremely infrared bright, so cannot be explained by a central concentration of Dyson spheres. Credit: NASA and the Hubble Heritage Team (STScI/AURA); Acknowledgment: Ray A. Lucas (STScI/AURA).

Griffith et al. (2015) and others have suggested that another galactic engineering project might involve "weeding" a galaxy of dangerous phenomena like supernovae or gamma-ray bursts that might destroy settlements or inhabited planets in some region of the galaxy. For instance, both phenomena can be the result of the final stage of evolution for high mass stars, which are rare and relatively short-lived (Section 2.11). A species might therefore somehow disrupt the process of such stars' formation, which could result in a relative dearth of bright blue stars in some region of the galaxy, or even of any gas that could form stars (indeed, Hoag's object would seem to conform to this description in its inner region). Some gamma-ray bursts are understood to be due to neutron-star mergers, so perhaps, contrary to Section 17.2.1, a *lack* of neutron star binaries in a galaxy would be a technosignature.

17.6 Black Holes and Neutron Stars

17.6.1 Black Holes and Neutron Stars as Power Sources

An energy-hungry species of the sort that we discussed in Section 4.2.2 and Chapter 15 might, either through the course of its settlement of the Galaxy or by design, find some of its members near a black hole or neutron star, or perhaps even somehow be able to manufacture one. In this case the species would have an opportunity to use the object as an engine to generate large amounts of useful energy that could be extracted.

One option would be along the lines of Dyson's gravitational machines above, but another would be to extract the energy associated with its accretion. Black holes (Section 2.12) are perhaps the most efficient energy-generation engines in the Universe. Over its lifetime of billions of years, a star might convert a few tenths of a percent of its rest mass into helium via fusion. A black hole has no intrinsic source of luminosity, but when it accretes matter that matter ends up radiating energy up to a few percent of the material's rest mass. This would also work for neutron stars, though at lower efficiencies.

Many black holes naturally accrete material around them—stellar mass black holes can do this via mass from a binary companion, and supermassive black holes at the centers of galaxies from the surrounding stars and gas.

A species near such a black hole might then capture this naturally occurring source of energy. They might also deliberately "feed" a black hole in a way that throttles its output, or even, if they are galactic engineers, manage the natural flow of mass into a supermassive black hole to control its output, much as water flow is throttled to control power generation in hydroelectric dams and water levels in rivers and lakes.

Hsiao et al. (2021) explored some of the physics of energy collection from black holes and its observational consequences. As with Dyson spheres around ordinary stars, a primary observational signature would be an infrared excess that depends on the temperature of the technology's radiators, which might be challenging to distinguish from natural infrared radiation from the accretion disk. A highly efficient Dyson sphere might capture most or all of the high-energy radiation, but this is also how natural active galactic nuclei appear to us when viewed edge-on.

Hsiao et al. (2021) also noted that in many systems the majority of the energy released from accretion is not in luminosity near the black hole, but in kinetic energy of material ejected at relativistic speeds in polar jets. This energy is ultimately dissipated into luminous energy that makes, for instance, quasars and blazars so bright, but efficiently harnessing this energy would seem to require something beyond a Dyson sphere. Inoue & Yokoo (2011) explored a scheme by which the energy from a galactic central engine could be collected by a hub and beamed out for use elsewhere.

17.6.2 Energy of Accretion and the Eddington Luminosity

Material falling onto the surface of an object accelerates via gravity. The total energy gained by the material when arriving at the surface is given by the gravitational potential of the surface

$$E = \frac{GMm}{R} \tag{17.4}$$

where M is the mass of the object and m is the mass of the material falling. For a black hole, the relevant "surface" is near the Schwarzchild radius $R_S = 2GM/c^2$ (Section 2.12), so we have

$$E_{\mathrm{BH}} = \frac{1}{2}mc^2 \tag{17.5}$$

which is half of the rest mass of the material falling. If this occurs in free fall then the object will have achieved nearly the speed of light as it arrives at the event horizon. As the material forms an accretion disk, a fraction of energy is converted to luminosity of the disk and kinetic energy in polar jets. Expressed as a fraction of the rest mass of the infalling material, this is called the efficiency, usually denoted by η, and for black holes it can have a value of a few percent or higher. The luminosity of a black hole accretion disk is thus

$$L = \eta \dot{M} c^2 \tag{17.6}$$

where $\dot{M}$ is the rate of accretion (in units of mass per time) (Section 2.12).

An approximate upper limit on the luminosity that can be achieved in this manner is given by the *Eddington luminosity*, set by the effects of radiation pressure near very luminous sources, which will ionize and blow away material, preventing further accretion.

Consider a very thin shell of hydrogen plasma with radius r and mass m around a source with luminosity L and mass M. The pressure produced by the radiation on the shell is given by $P_{\mathrm{rad}} = F/c$, where F is the total energy flux at r, given by $F = L/(4\pi r^2)$.

The plasma has an opacity κ due to electron scattering, which has a value of $\sigma_{\mathrm{T}}/m_{\mathrm{p}}$ where σ_T is the Thomson cross section of the electron and m_{p} is the mass of a proton. This quantity, with units of $\mathrm{cm}^2\ \mathrm{g}^{-1}$, represents the effective area a unit mass of hydrogen plasma presents to photons, and is independent of photon frequency. This means that the total cross section of the shell to photons is κm, so the fraction of light encountered by the shell that is scattered and contributes pressure is $\kappa m/(4\pi r^2)$. So the total radiation pressure on the shell is

$$P_{\rm rad} = \frac{L}{4\pi r^2 c}\frac{\kappa m}{4\pi r^2} = \frac{L\sigma_{\rm T} m}{cm_{\rm p}(4\pi r^2)^2} \tag{17.7}$$

The force of gravity on this shell per unit area is

$$P_{\rm grav} = \frac{GMm}{r^2}\frac{1}{4\pi r^2} \tag{17.8}$$

and setting these equal to one another determines the limiting luminosity beyond which radiation is stronger than gravity and so will stop accretion. This is called the Eddington luminosity and is given by

$$L_{\rm Edd} = \frac{4\pi GMm_{\rm p}c}{\sigma_{\rm T}} = 3.2 \times 10^4\left(\frac{M}{M_\odot}\right){\rm L}_\odot \tag{17.9}$$

For a stellar mass black hole this is around 10^5 solar luminosities, implying a Dyson sphere radius of around 300 au if it has temperatures around 300 K, requiring perhaps 100 solar masses of material following the assumptions in Section 15.4.1 (so, outweighing the black hole itself). The engineering of such a black hole Dyson sphere might be very different from that around a star. For a supermassive black hole with $4 \times 10^6 {\rm M}_\odot$ (similar to Sgr A*) material at 300 K would need to be at a distance of around 3×10^5 au, or about 1 pc, and have a mass of many millions of solar masses.

17.6.3 Other Black Hole and Neutron Star Technosignatures

Black holes and neutron stars might be used in other ways as well. For instance, their extreme gravities make them outstanding gravitational lenses (Section 14.4.2), which could make them nodes in a galactic or intergalactic communication network (an idea explored in Cohen 1989, Maccone 2012, Jackson 2015, and Jackson & Benford 2020). Opatrný et al. (2017) explored the possibility that black holes might be used as an entropy *sink* (instead of a low-entropy energy *source*) to run a sort of Dyson sphere in reverse, exploiting energy falling onto the outer surface of the sphere, which would radiate its waste heat *inwards* (and so be virtually undetectable!).

Vidal (2011) explored the many extraordinary characteristics of black holes that might make them "attractors of intelligence," and argued that the control of binary systems containing them would in some sense represent the ultimate limit of technology. Black holes also exhibit many properties that might make them valuable places to perform calculations, a concept explored by Lloyd (2000) and Smart (2012).

17.7 Almost Natural Technosignatures

> Any sufficiently advanced technology is indistinguishable from nature.
> — Attributed to Karl Schroeder

17.7.1 Manipulating Natural Signals

In Section 11.2.7 we encountered "generalized SETI" as the search for astronomical outliers as potential signs of technology. This in some sense is a catchall category that has the advantage that we need not hypothesize any particular alien motivations or engineering, merely that their technology is detectable as an obviously unnatural phenomenon.

Such a search would be most effective if a given technosignature were strongly distinct from any natural phenomenon, such as an extremely narrowband radio transmission. But it is also possible that such an anomaly could be *very nearly* natural. One reason for such a thing could be that a species trying to attract attention might make use of interesting natural phenomena as a way to draw attention to their beacon.

For instance, we saw in Section 5.5 how one might send signals from a planet in the direction opposite its host star such that they would arrive at distant stars at the same time that the planet appeared to transit its star. Kipping & Teachey (2016) explored how a species might use lasers to either cloak its own transit signature, or to sculpt it in ways that would be obviously artificial. In this way, such a species could inject a message into data being taken for ordinary scientific study of transiting exoplanets.

Learned et al. (2008) proposed that Cepheid variable stars could have their brightness variations, which are normally quite regular, be deliberately modulated by a technological species to broadcast messages. This would have the advantage that the power for the transmitter would be that of a Cepheid variable—one of the brightest classes of stars in the Universe—thus sparing the transmitting species from producing that much power themselves. We saw a similar scheme proposed for a gravitational wave transmitter in Section 17.2.4.

In addition to drawing the attention of those studying natural phenomena, such schemes can be energy efficient in the sense that the power of their transmitter is a naturally bright source, and they need spend only the energy necessary to modulate the signal or otherwise render it obviously artificial.

17.7.2 Stellar Markers and Przybylski's Star

Shklovskiĭ & Sagan (1966) write that Drake and Shklovskiĭ independently suggested that an alien species might deliberately *mark* or *salt* a star with unlikely abundances of elements to make them stand out in studies of stars. They specifically suggested using radioactive elements with short enough half-lives that they should not appear naturally, and Whitmire & Wright (1980) suggested that such salting might be done not as part of a beacon, but simply because a species was using its star as a repository for its fissile waste.

Intriguingly, there is a class of star with highly peculiar abundance patters: Ap stars. These are highly magnetic hot stars that rotate extremely slowly, and whose atmospheres appear to be highly stratified, with high concentrations of certain elements in layers that make them extremely visible in stellar spectroscopy. The result is abundance patterns that land very far from the typical pattern, and the appearance of many spectral lines from species rarely or never seen in most stellar spectra.

The most extreme example of the phenomenon is Przybylski's[7] star (Przybylski 1961; Cowley et al. 2000), whose spectrum is so complex and rich in overlapping lines that making any line identifications is very difficult. Struggling to identify the lines, Cowley et al. (2004), Bidelman (2005), Quinet et al. (2007), Fivet et al. (2007), and Gopka et al. (2008) reported that the best fit for many lines were short-lived isotopes, including actinium and promethium. This should not be naturally possible: such isotopes have can half-lives of *decades*, so all such isotopes, whatever their source, should have decayed away.

The most likely explanation for these "detections" is line misidentifications (Andrievsky et al. 2023) since the atmospheres of these stars are challenging to model, and spectroscopic information for elements heavier than barium is lacking (Shulyak et al. 2010).

Box 17.1: Jargon Review: Other Technosignatures

- **Messengers**: Any of the different carriers of information from space astronomers can use to learn about the Universe, usually comprising photons, neutrinos, cosmic rays, and gravitational waves.
- **Mean-motion resonance**: A planetary configuration that occurs when the orbital periods of planets are related by small integer ratios, as with Neptune and Pluto.
- **Stellar marking** or **Salting**: To deliberately pollute a star's atmosphere with short-lived or other unusual elements to produce a bright technosignature.
- **Schwarzschild radius**: The physical radius of a black hole within which no matter, energy, or information can escape.
- **Eddington luminosity**: The maximum luminosity of an object under certain assumptions, beyond which radiation pressure will overpower the object's gravity.

17.7.3 Mistaking Technology for Nature

In Section 18.5 we will explore the assumptions that often underlie what we expect from "advanced" technology or alien life, but one of the lessons is that it can be easy to misidentify technology as a natural phenomenon if one is not familiar with it, as European colonists did with the products of Native American agricultural practices.

[7] For Antoni Przybylski, whose name is pronounced roughly "(p)shi-BILL-skee," where the opening plosive is roughly that in "pshaw."

One might imagine that there is already some puzzling astrophysical phenomenon that defies easy explanation by Earth's astrophysicists today because it is, in fact, a technosignature and does not occur naturally. Indeed, technology is in some sense the manipulation of nature to achieve some end, and so technology with a very large ichnoscale might appear as some generic astrophysical phenomenon operating in an unusual manner.

Science fiction is filled with such examples, for instance astrophysicist Fred Hoyle's novel *The Black Cloud* (Hoyle 1957) in which an interstellar cloud turns out to be a living, sentient being.

An example in the scientific literature for such phenomenon is from Beech (1990), who suggested that "blue stragglers" might be stars that have had their lives extended artificially (blue stragglers are stars that appear to be far too old to have not yet exhausted their hydrogen supply, and are generally understood to be the product of stellar mergers). Clément Vidal has suggested that black hole or neutron star binaries might actually be technology, or even themselves living creatures (Vidal 2011, 2016).

Exercises

1.
   (a) Assume a species harnessed a stellar mass black hole of $5M_\odot$, and fed it mass at the Eddington limit. What would its luminosity be in solar units?
   (b) If the species wanted to harness this energy with a Dyson sphere and have the swarm components be at a temperature of roughly 300 K, what distance would they need to be around the black hole in au?
   (c) If the species had 1 Jupiter mass of fuel for the black hole, how long would it last? Assume a mass-to-energy efficiency η of 5%.
   (d) What if it were in a binary star system and the species had 1 solar mass to feed it?
   (e) Now assume they operate at a supermassive black hole in the center of a Galaxy with 10^6 solar masses. What is the Eddington luminosity, and what is the mass accretion rate in solar masses per year?
   (f) How far from the black hole would Dyson sphere swarm components need to be to be at 300 K in pc?
2. The Nine Axes of Merit

   You have now seen a survey of most of the kinds of technosignatures, and the considerations that go into looking for them.

   Apply Sheikh's nine axes of merit to specific examples of:
   - narrowband radio technosignatures;
   - pulsed laser technosignatures;
   - Dyson spheres;
   - hunts for free-floating solar system objects;
   - searches for astrophysical anomalies.

   That is, in each case articulate a specific example of such a search (you might use ADS to find such a search) and consider, qualitatively, where you feel it ranks on the nine axes. Use the code provided in Sheikh (2020) to create graphics for each case. Justify the axes in each case with a few sentences.

References

Abbott, B. P., Abbott, R., Abbott, T. D., et al. 2017a, PhRvL, 118, 221101
Abbott, B. P., Abbott, R., Abbott, T. D., et al. 2017b, PhRvL, 119, 161101
Abramowicz, M., Bejger, M., Gourgoulhon, E., & Straub, O. 2020, NatSR, 10, 7054
Andrievsky, S. M., Korotin, S. A., Werner, K., & Kovtyukh, V. V. 2023, AN, 344, e20230056
Beech, M. 1990, EM&P, 49, 177
Bidelman, W. P. 2005, ASP Conf. Ser. 336, Cosmic Abundances as Records of Stellar Evolution and Nucleosynthesis ed. T. G. Barnes III, & F. N. Bash (San Francisco, CA: ASP)
Carrigan, R. A. 2012, AcAau, 78, 121
Clement, M. S., Raymond, S. N., Veras, D., & Kipping, D. 2022, MNRAS, 513, 4945
Cohen, N. 1989, IAF 40th International Astronautical Congress (Paris: IAF)
Cowley, C. R., Ryabchikova, T., Kupka, F., et al. 2000, MNRAS, 317, 299
Cowley, C. R., Bidelman, W. P., Hubrig, S., Mathys, G., & Bord, D. J. 2004, A&A, 419, 1087
Dyson, F. 1963, Interstellar Communication: A Collection of Reprints and Original Contributions, ed. A. G. W. Cameron (New York: Benjamin)
Dyson, F. J. 1966, Perspectives in Modern Physics: Essays in Honor of Hans A. Bethe on the Occasion of His 60th Birthday, July, 1966, ed. R. E. Marshak, J. W. Blaker, & H. A. Bethe (New York: Wiley) 641
Fivet, V., Quinet, P., Biémont, E., et al. 2007, MNRAS, 380, 771
Gopka, V. F., Yushchenko, A. V., Yushchenko, V. A., Panov, I. V., & Kim, Ch 2008, KPCB, 24, 89
Griffith, R. L., Wright, J. T., Maldonado, J., et al. 2015, ApJS, 217, 25
Hippke, M. 2018, AcAau, 151, 53
Hoyle, F. 1957, The Black Cloud (New Zealand: Pearson)
Hsiao, T. Y.-Y., Goto, T., Hashimoto, T., et al. 2021, MNRAS, 506, 1723
Inoue, M., & Yokoo, H. 2011, arXiv:1112.5519
Jackson, A. A. 2015, JBIS, 68, 342
Jackson, A. A., & Benford, G. 2018, arXiv:1806.02334
Jackson, A. A., & Benford, G. 2020, arXiv:2009.08873
Kipping, D. M., & Teachey, A. 2016, MNRAS, 459, 1233
Lacki, B. C. 2015, arXiv:1503.01509
Lacki, B. C. 2019, PASP, 131, 024102
Learned, J. G., Pakvasa, S., Simmons, W. A., & Tata, X. 1994, QJRAS, 35, 321
Learned, J. G., Kudritzki, R. P., Pakvasa, S., & Zee, A. 2008, arXiv:0809.0339
Learned, J. G., Pakvasa, S., & Zee, A. 2009, PhLB, 671, 15
Lentz, E. W., & Felton, R. C. 2024, arXiv:2405.19381
Lloyd, S. 2000, Natur, 406, 1047
Maccone, C. 2012, AcAau, 78, 109
NASA Technosignatures Workshop Participants 2018, arXiv:1812.08681
Opatrný, T., Richterek, L., & Bakala, P. 2017, AmJPh, 85, 14
Pasachoff, J. M., & Kutner, M. L. 1979, CosSe, 1, 2
Przybylski, A. 1961, Natur, 189, 739

Quinet, P., Argante, C., Fivet, V., et al. 2007, A&A, 474, 307
Raymond, S. N., Veras, D., Clement, M. S., et al. 2023, MNRAS, 521, 2002
Saenz, A. W., Uberall, H., Kelly, F. J., Padgett, D. W., & Seeman., N. 1977, Sci, 198, 295
Sellers, L., Bobrick, A., Martire, G., Andrews, M., & Paulini, M. 2022, arXiv:2212.02065
Sheikh, S. Z. 2020, IJAsB, 19, 237
Shklovskiĭ, I. S., & Sagan, C. 1966, Intelligent Life in the Universe (Delta-Books) (San Francisco, CA: Holden-Day)
Shulyak, D., Ryabchikova, T., Kildiyarova, R., & Kochukhov, O. 2010, A&A, 520, A88
Silagadze, Z. K. 2008, AcPPB, 39, 2943
Silverberg, R. 2015, Dark City Lights: New York Stories, Have a NYC Series, ed. L. Block (New York, NY: Three Rooms Press)
Smart, J. M. 2012, AcAau, 78, 55
Subotowicz, M. 1979, AcAau, 6, 213
Vidal, C. 2011, arXiv:1104.4362
Vidal, C. 2016, AcAau, 128, 251
Voros, J. 2013, arXiv:1412.4011
Whitmire, D. P., & Wright, D. P. 1980, Icar, 42, 149
Wright, J. T., Carroll-Nellenback, J., Frank, A., & Scharf, C. 2021, RNAAS, 5, 141

Part IV

SETI and Society

AAS | IOP Astronomy

Chapter 18

Popular Culture

Michael Rennie was ill
The Day the Earth Stood Still,
But he told us where we stand.

And Flash Gordon was there
In silver underwear;
Claude Raines was the Invisible Man.

Then something went wrong
For Fay Wray and King Kong:
They got caught in a celluloid jam.

Then at a deadly pace
It came from outer space,
And this is how the message ran:

Science fiction double feature!
Doctor X will build a creature!
See androids fighting Brad and Janet!
Anne Francis stars in Forbidden Planet!
At the late night double-feature picture show.

— Richard O'Brien, Science Fiction/Double Feature,
The Rocky Horror Show (1973) and *The Rocky Horror Picture Show* (1975)

doi:10.1088/2514-3433/adf5aech18

18.1 The Big Question: "Are We Alone?"

> Are We Alone?
>
> Since the dawn of human civilization, humankind has looked to the stars and wondered if life, whether it be similar to our own or not, exists beyond our planet.
>
> — NASA Exoplanet Exploration Website—The Big Questions—Are We Alone? Retrieved 2024 March 13.

It's common, practically to the point of cliché, for astronomers and astrobiologists exploring exoplanets and the possibility of extraterrestrial life to begin a talk, proposal, or research paper with a sentiment along the lines of the quote above. And indeed, the question of humanity's place in the Universe is one of the most profound and debated in all of philosophy and theology.

But we should be careful when making sweeping statements like this about the history of human thought. Many traditions have not wondered about the existence of life in the heavens at all, because they have long taught that the question has a positive answer. As we saw in Section 3.1.1, many pre-Copernican traditions did not really distinguish between the divine beings and those in the sky, i.e., aliens and angels: both were "heavenly." The concept of literal alien species on planets like (or unlike) the Earth certainly existed to some degree and in some cultures, but curiosity at whether such creatures really existed probably did not compose a large part of humanity's wonder at the night sky. As Canadian anthropologist and archaeologist Kathryn Denning[1] puts it "Asking the proverbial question 'Are we alone?' presupposes that we don't already know the answer, and that may actually be a recent phenomenon born of modern astronomy which effectively 'emptied out' the previously very busy imagined cosmos by showing that, for example, the Moon and Mars actually didn't have obvious signs of life and were kinda boring."

While planetary scientists would hardly call the Moon and Mars "boring," those bodies' barren surfaces are certainly less interesting to most of the lay public than the fantasies we explore in Section 18.2. It is thus natural and proper for scientists concerned with their work's place in society and popular consciousness to tap into the (perhaps quite modern and even parochial) fascination many members of the public have in the question of alien life to educate and inspire them with stories of modern astrobiological research. Even beyond that, such fascination surely and appropriately drives many to careers in the many fields of astrobiology, and SETI specifically, and there is nothing wrong with sharing and celebrating that.

So the point here is not that the question "Are we alone?" is not actually profound or inspiring (it is!) or that the public doesn't really care (they do!) or that the discovery of technological alien life would not mark an epoch in human thought and understanding of the Universe (it would!); the point is just that we should be mindful not to overstate the case.

[1] In an as-yet unpublished contribution to the "Communicating Discoveries in the Search for Life in the Universe" NASA Astrobiology workshop.

The questions of how a big a deal the public would make of the discovery of various kinds of alien life, and of how much they have wondered about it across culture and history are complex, and not ones where we have much data or good answers. In Chapter 20 we will explore what little we do know about current beliefs and attitudes toward the topic are, especially how large sections of humanity believe that aliens have already been discovered, either because they are only vaguely familiar with the progress of astrobiology and think we have already found such life, or because they subscribe to popular beliefs about UFOs or alien visitations.

In short, to a large extent most people today and throughout history probably do not or did not spend much time actually wondering "are we alone?" To the degree they care or cared at all, they probably presumed the answer is "yes."

18.2 From Monsters to Aliens

Cultures around the world have told stories of the strange things that lie beyond the lands they know, inhabited by alien creatures. These stories play many roles, from entertainment to religion, and are a mixture of fabulation and nuggets of truth gleaned from travelers' stories.

Some examples from Western literature include, obviously, Greek mythological figures, such as gods and titans that lived in the sky and upon Mount Olympus, where people generally could not visit. They also include stories of monsters—hybrid creatures that crossed and violated natural categories of animals, like minotaurs and harpies—and how humans interacted with them.

Whether these stories were taken as true or not probably depends on the storyteller and the listener, but not all such stories were so obviously mythological. The *Histories* of Herodotus include more-or-less accurate, if embellished, descriptions of many historical events and are sprinkled with fables or (generously) misunderstood stories, such as giant ants that mine for gold in Persia.

The Syrian author Lucian of Samosata in the second century CE wrote *True History* ('Αληθῆ διηγήματα) as a satire of mythology passing as history, describing a journey around the world and to the Moon and Sun, filled with outrageous monsters like fish people and giant spiders (see Figure 18.1). It is sometimes called the first science fiction story.[2]

Lucian's stories take the form of a literary tradition called travel literature or travelogues. Some travelogues are intended to be as truthful as possible, but the genre always appeals to the desire to learn about strange beasts, people, and lands far away.

When Marco Polo returned from the East he shared a prison cell with Rustichello da Pisa, an author previously most famous for writing the first version of the King Arthur myth in Italian, *Roman de Roi Artus*. Rustichello's telling of The Travels of Marco Polo are a strange mix of boring trade details of cities on the road to the Khan's palace (which were presumably of great interest to Polo), and fantastic tales

[2] Another contender for that title goes to Johannes Kepler's posthumous novel *Somnium* (Dream), in which he describes the environment and life on the Moon, including its intelligent inhabitants, based on his telescopic observations of the lunar surface and the new appreciation, as part of the Copernican Revolution, that the planets were likely to be similar in character to the Earth.

Figure 18.1. Illustration by William Strang depicting a scene from True History obtained by the author from the Wikimedia website, where it is stated to have been released into the public domain.

of the magic of the Khan's court, clothes that one cleans by burning them, and a heroic rescue of a princess on the high seas. The book was said to contain "a million lies" but was nonetheless read and appreciated worldwide.

Perhaps the most famous satirical travelogue in English is Jonathan Swift's *Gulliver's Travels*, describing travels to distant lands like Lilliput (inhabited by tiny people) and Brobdingnag (inhabited by giants). Soon thereafter Voltaire wrote a satirical story *Micromégas* ("Small Large") about a traveler's exploration of creatures both incredibly small and incredibly large, including giants from Saturn.

Around this time (the 18th century) most of the world was being explored by Europeans, and so there were few uncharted places left for fantastic beasts to be found in these tales. Jules Verne wrote travelogues like *20,000 Leagues Under the Sea* in the still largely unexplored depths of the ocean and *Journey to the Center of the Earth* below the Earth's surface. H. G. Wells put his fantastic people and societies in the distant future in *The Time Machine*, and from Mars in *War of the Worlds*. Arthur Conan Doyle imagined dinosaurs existing in jungles on a distant plateau in South America in *The Lost World*.

In the early 20th century, the most famous fictional travelogues in English are perhaps the *Oz* books by Frank L. Baum, describing the travels of Dorothy and other protagonists through the various lands of Oz. These books were meant primarily to entertain children but were not short on political dimensions—except for a brief stint when it is ruled by the Wizard and the Scarecrow, the fairy kingdoms of Oz and Ev are traditionally governed by women like Princess Ozma and the witches.

The stories are almost all travelogues, as the reader learns of the fantastic people and beasts that inhabit the various fairy lands, including a thinking robot (Tik Tok, the first such character in literature), talking animals, and animated objects. See Figure 18.2.

By the mid 20th century the excitement of the Space Race and thorough exploration of the planet moved a substantial portion of the genre firmly to space.

Figure 18.2. Some of the various "alien" characters that inhabit Oz and other fairy lands in the Baum series of books. From left to right: Jack Pumpkinhead (top), Pip (middle), Nick Chopper (the Tin Woodman, top), the Scarecrow (bottom), Professor Wogglebug, the Nome King, Quox the Dragon, the Nome King again, Dorothy Gale, and Princess Ozma (after whom Drake's inaugural SETI search was named). Art by John R. Neill from the books in the public domain.

The best example is the original *Star Trek* series from the 1960s, a travelogue of a ship moving among the stars and encountering alien creatures from different planets each episode.

American pioneers of modern SETI would have been raised as children on the Oz books, and the second generation grew up with *Star Trek*. These literary traditions —stories of fantastic creatures living beyond where we can reach now—would have informed their views and expectations of what we can expect to find beyond the Earth. Indeed, Drake named his first search for radio signals from space "Project Ozma" in honor of Baum's princess, explaining she ruled "a land far away, difficult to reach, and populated by strange and exotic beings."[3]

Box 18.1: *Contact*

When federal funding for the NASA SETI program was cut in 1978 (Section 3.3.4), Carl Sagan and others began a campaign to gather popular support and private funding for the endeavor, especially in Hollywood. He and Ann Druyan began a science-fiction film treatment dramatizing the political and financial struggles of SETI, and the benefits of a successful detection.

After production stalled, in 1981 Sagan secured a record-breaking $2 million advance from Simon and Schuster to convert the treatment into a novel. The book was extremely successful, eventually selling millions of copies. This generated studio interest in the film, and after many false starts Warner Bros. in 1996 secured director Robert Zemeckis to direct the film, starring Jodie Foster and Matthew McConaughey.

The film was a box office success. Iconic elements include the opening sequence, which tracks the leakage of human transmissions into space; the protagonist "listening" to signals from space with headphones at the VLA; and an eerily prescient Howard-Hughes-like billionaire philanthropist reminiscent of some present-day Silicon Valley personalities.

The film is noteworthy for its extremely accurate portrayal of many technical aspects of radio SETI and the political struggles of the field. The film's protagonist, Ellie Arroway (Foster), is a composite character but is at least partly based on the biography of Jill Tarter (Box 1.2), including her father's support of her scientific career and death when she was young. The character Kent Clark, a blind radio astronomer, is based on the real SETI researcher Kent Cullers. The film dramatizes the benefits to humanity of contact, and explores the possible geopolitical fallout.

The romantic interest in the film, the lapsed priest Palmer Joss (McConaughey), provides a vehicle to highlight the human, emotional, and religious aspects of SETI. The film explores the ways in which SETI is in many ways a search for ourselves and meaning in the Universe, and the role of faith in science. It also echoes the historically fuzzy line in the pre-history of SETI between spiritual and literal belief in and communion with extraterrestrial beings (Section 3.1.1).

[3] Also relevant is Baum's conceit that he is recounting actual events, communicated to him by Dorothy via—what else—radio: "I am obliged to talk these matters over with Dorothy by means of the 'wireless," for that is the only way I can communicate with the Land of Oz" (Baum 1914).

18.3 Popular Portrayals of Aliens in Science Fiction

Science fiction has provided some of the most pervasive and important guides to thought on what aliens and their technology would be like. In some cases, authors have made an attempt to imagine actual alien contact; in most, the story takes precedence over such concerns. Certainly many important ideas in SETI have come from or been amplified by science fiction, and nearly all SETI practitioners will bring ideas from science fiction into their work.

Science fiction is a wide and varied genre. "Hard" science fiction is often characterized by an exploration of the possible that is informed by science, or at least scientific plausibility. "Soft" science fiction often uses the tropes or settings of hard science fiction but is more concerned with other aspects of storytelling than exploring possible futures or realistic portrayals of aliens or spaceflight. Science fiction has a fuzzy boundary with fantasy and other genres unconcerned with realism and plausibility. Many works along this boundary are set in space and employ tropes of hard science fiction, but also freely employ fantastical elements, for instance in the *Star Wars* franchise.

Hard science fiction has explored and inspired many aspects of human space exploration and possible contact with extraterrestrial species, and has likely inspired many young readers to pursue careers in science, the space industries, and astronomy, including SETI. Certainly science fiction, especially the vast trove of hard science fiction in print, has had a largely positive and important role in shaping and inspiring scientists and science.

The most popular forms of science fiction often conform to certain expectations of the genre in order to attract the broadest audience. Especially because it shapes the broader cultural expectations of contact and work in SETI, we will take a moment to try to develop a clear-eyed view of how the most popular forms of science fiction often portray the alien, and why.

In many cases in television and film, practical matters such as budget and the limitations of special effects directly impact the appearance of aliens. In all cases, the tropes of the genre and the constraints of the kinds of stories that are popular drive particular portrayals of aliens' form, language, technology, and behavior. For instance, standard plot structures require some sort of conflict, so conflict with aliens is a popular theme. Science fiction also often contains social commentary, in which case aliens can serve as stand-ins for certain kinds of people, which requires them to share many human traits.

For instance, take the original series *Star Trek* episode "Let That Be Your Last Battlefield," a paper-thin allegory for racial or ethnic strife on Earth. The aliens in this episode appear entirely human, except for the coloration of their skin: half of their faces are bright white, and the opposite half is dark black. The punchline (spoiler alert) is that the crew of the *Enterprise* completely overlooks what the aliens perceive to be the obvious and essential difference among themselves: *which* side of their face is white. Much of the power and point of the episode would be lost if the aliens were not so obviously relatable to humans.

The most popular modern narrative structures are also generally quite streamlined, with few digressions that would distract from the driving action. This can make it difficult to describe and portray an alien species that differs from standard tropes of the genre in more than a handful of ways, except in extended series that have the space to do extensive world building.

18.3.1 Physical Form and Language

Actual aliens' body plans have no particular reason to follow humans', but in television and film makeup and costume design gets more expensive the further from an actor's actual appearance one gets. Many or most aliens are typically depicted as "bumpy forehead" humans, i.e., actors wearing a lot of facial makeup or a mask. Even more elaborate costumes typically have aliens with generally animal-like[4] body plans, including bilateral symmetry, a head with eyes, a mouth from which to speak, and a brain, and so on.

Although some have argued that such a body plan is so useful and has elements so conducive to tool development that there might be elements of it that other species converge on, it is worth noting that even on Earth we have examples of intelligent, tool-using species that do not follow this plans (such as octopi). Indeed, in Wells-Jenson (2023) and other publications Sheri Wells-Jenson has argued that we should not even assume that aliens can see light with eyes at all, and explored what the consequences of such an assumption have on our conclusions about alien cognition and language.

In order to share a scene with a human, aliens also are typically portrayed as being comfortable in a typical human indoor environment—one with the temperature, gravity, gas pressure, atmospheric composition, light and radiation levels similar to our everyday lives (or a movie set). A more accurate portrayal—one in which aliens must be separated from human environments by some sort of hermetic suit or wall, or in which they are never in the same room as humans at all—generally requires narrative inconvenience or exposition that might detract from a well-designed story.

When universal translators are not invoked, sci-fi aliens generally have such similar modes of communication and cognition to humans that we can simply read subtitles of their speech, or understand that we are hearing a translation. Alternatively, the aliens have already learned enough about human language and cognition that they can communicate with us in our own terms.

In truth, we might expect their communication modes to be even more alien to us than, say, whale song, and have cognitions that render our typical forms of communicative interactions unintelligible. Science fiction rarely deals with a perhaps more probable outcome of first contact, namely that we will simply not understand what they are saying, or that such understanding will take a very long time to develop.

---

[4] Or, more specifically, bilaterian, see Section 7.5.3.

18.3.2 Technology

Warp drive is a staple of science fiction because without faster-than-light travel interstellar stories would have to take place over millennia, and any human embarking on such a journey would not be able to return in a human lifetime. Similarly, faster-than-light communication allows characters to have ordinary conversations across interstellar distances as if communicating by telephone. These contrivances thus allow for familiar and relatable story lines to be transposed to an interstellar setting.

In television and film, practical matters drive the existence of other technology. For instance, the writers of *Star Trek* invoked teleportation technology ("transporters") because the television production team could not afford the special effects necessary to show the *USS Enterprise* landing on alien planets. Artificial gravity is a sci-fi staple because it is challenging and expensive to convincingly depict a zero-g environment—with artificial gravity, the inside of a spaceship is much like the inside of any building (or movie set) on Earth.

But all of these technologies are, as far as we know, probably *impossible*. We discussed in Section 4.4 how SETI must deal with the possibility that the technology we find might be of a form we cannot understand or that would seem as magic to us, following Clarke's "Third Law" (see also Section 18.5 below). But while it is *conceivable* that impossible-seeming technology is out there, it's important to keep in mind *why* the above technologies appear so often in science fiction (i.e., for narrative convenience), and not mistake their ubiquity in fiction for a genuine expectation by those who have considered the matter that such technology is possible and exists.

18.3.3 Behavior

Perhaps the first widespread portrayal of alien life in popular culture is the Martians in H. G. Wells' *War of the Worlds*. In this novel, a malevolent, colonizing armada from Mars attacks the Earth with overwhelming force, in the end only defeated by their lack of resistance to Earthly diseases.

Since then aliens have appeared throughout science fiction in a variety of forms, from similarly violent armies, to refugees (*Alien Nation*, *District 9*), to equals in the Galaxy (*Star Trek*, *Star Wars*), to benevolent overlords (*Childhood's End*).

The forms and behaviors of these aliens in film are often primarily dictated by the genre, not educated expectations of contact. Two diametric examples might be the titular aliens seen in the *Alien* franchise, who are hideous, animalistic, vicious predators with a parasitic life cycle, befitting the horror and action movie genres; and those in *E.T.: The Extra-Terrestrial*, who are kind to human children and life generally, physically vulnerable, highly intelligent, and technologically sophisticated, befitting feel-good family fare.

Unless the story has space and time to develop the details of an alien society, alien species are generally presented as much more homogeneous than humanity. The entire species might be characterized by a particular trait—for instance, on the original *Star Trek* Vulcans were logical and ritualistic, Klingons were fierce and evil caricatures of Communists, Romulans were militaristic and xenophobic, and so on.

When they have social organizations like humans, they are generally presented as having a single unified government or military—a single ambassador might reliably speak for the entire species (similarly, humanity is sometimes portrayed as having been unified under a single society or government).

Such homogeneity is a useful device it lends itself to metaphor about humans and to narrative simplicity. In truth, even if, unlike humans, an interstellar species somehow had a single, uniform culture, it would be challenged to remain homogeneous because large light-travel times would make rapid exchange of information impossible. We might expect social species whose presence in the Galaxy spans more than one light-lifetime to suffer strong cultural drift.

18.4 European Colonialism in SETI

Some of the earliest serious contemplation of the existence of alien lifeforms occurred right at the time that European powers were extending their colonial ambitions around the globe.

Bishop Wilkins' account of future travels to the Moon (Section 3.1.2) used exploration of the New World as an analogy, imagining that just as European sailing had rapidly advanced to the point that Europeans could discover new lands and peoples, so might they eventually explore the Moon. Kepler explicitly argued that once a nation developed flying machines, it would be able to colonize the Moon (Crowe 1986).

Perhaps the best articulation of this attitude came from Bernard le Bovier de Fontelle, whose *Conversations on the Plurality of Worlds* was a widely popular discussion of the possibility of lunar life, and what it would be like. Written in 1686, it explicitly considers whether the inhabitants of the Moon will have inferior technology to Europeans in an analogy with European's views of the native peoples of the New World, Australia, and Polynesia. It also considers whether the Moon people might have *superior* technology to Europeans, and might make the trip to Earth to visit us.

Fontelle makes extrapolations from his understandings of the variations on peoples on Earth to aliens: that physical differences among people increase with geographic distance, and with climate, which he ascribes to distances from the Sun. He uses this to hypothesize how creatures on Mercury and the Moon might look. Kepler, in his *Somnium*, similarly projected his understanding of the natural order of human culture to the Lunarians:

> Once a comparison is instituted between the populations of the Moon and the Earth, the judgement about similar things is the same. Since we see that the Moon's spotted parts are civilized, we shall assign to the surrounding rough and mountainous regions wild and savage bands of thieves. Let these be the enemies of the more civilized people, who build their fortifications against them. But in this connection recall the observation [in a previously referenced work], which can be explained in no other way than as a fortification against hostile attack. (translation by E. Rosen in Kepler & Rosen 2003)

When speculation on intelligent life shifted to Mars, H. G. Wells' *War of the Worlds* used the Martians' violent attempt to conquer Earth as a metaphor of colonization, which he made explicit in the opening chapter:

> And we men, the creatures who inhabit this Earth, must be to them at least as alien and lowly as are the monkeys and lemurs to us. The intellectual side of man already admits that life is an incessant struggle for existence, and it would seem that this too is the belief of the minds upon Mars. Their world is far gone in its cooling and this world is still crowded with life, but crowded only with what they regard as inferior animals. To carry warfare sunward is, indeed, their only escape from the destruction that, generation after generation, creeps upon them.
>
> And before we judge of them too harshly we must remember what ruthless and utter destruction our own species has wrought, not only upon animals, such as the vanished bison and the dodo, but upon its inferior races. The Tasmanians, in spite of their human likeness, were entirely swept out of existence in a war of extermination waged by European immigrants, in the space of 50 years. Are we such apostles of mercy as to complain if the Martians warred in the same spirit?

This analogy that mapped European exploration of the globe to human exploration of outer space, and of contact between "inferior" and "superior" communities, has never waned. Arthur C. Clarke, a scientist and science-fiction author of great poplar influence, was one of the primary mythmakers of the future of space exploration. Writing in the heydey of the early Space Age, he explicitly used the language of colonialism to describe humanity's future in space and our interactions with aliens:

> The age of mass colonization has gone forever. Space has room for many things, but not for "your tired, your poor, your huddled masses yearning to breathe free." Any statue of liberty on Martian soil will have inscribed upon its base "Give me your nuclear physicists, your chemical engineers, your biologists and mathematicians." The immigrants of the twenty-first century will have much more in common with those of the seventeenth century than the nineteenth. For the *Mayflower*, it is worth remembering, was loaded to the scuppers with eggheads. (Clarke 1962)

and

> Even if we never reach the stars by our own efforts, in the millions of years that lie ahead it is almost certain that the stars will come to us… And when the first contact with the outer universe is made, one would like to think that Mankind played an active and not merely a passive role —that we were the discoverers, not the discovered. (Clarke 1951)

NASA embraced this comparison, launching its SETI program on 1992 October 12, the 500th anniversary of Columbus's landing in the New World. At the inauguration event, Carl Sagan spoke:

> This inauguration is occurring on the five hundredth anniversary of Columbus arriving in this part of the world, hardly the discovery of America because there were hundreds of thousands of people here already, and as you know there is substantial controversy about the wisdom of celebrating this event. There's no question that in the annals of exploration, the course of bringing together the planet…Columbus's voyage was of historical significance…
>
> On the other hand, there are many other aspects of that voyage which we could properly have reasons to regret. But I want to stress that this project, beginning today, has all the positive and none of the negative aspects of that historic journey. We are not seeking slaves as Columbus and his funding agency was, we are not out to forcibly convert the heathen, we are not out to steal spices and gold.
>
> We are seeking knowledge only, exploration and discovery. And that being the case, it is really remarkable that a government is willing to fund, although sometimes with second thoughts, (audience laughter) so far seeing a project devoted so purely to exploration and discovery, and I think it speaks extremely well of our civilization and our nation, it is a sign of a hopeful and confident society.[5]

British physicist Stephen Hawking painted a vivid and widely reported picture of the possible consequences of contact, saying "If aliens visit us, the outcome would be much as when Columbus landed in America, which didn't turn out well for the Native Americans…We only have to look at ourselves to see how intelligent life might develop into something we wouldn't want to meet." He supposed alien technology "would be only limited by how much power they could harness and control, and that could be far more than we might first imagine." He imagined resource scarcity might drive them to the stars, and that "such advanced aliens would perhaps become nomads, looking to conquer and colonize whatever planets they can reach."[6]

The legacy of this association of colonialism with alien contact lives throughout SETI. Wright & Oman-Reagan (2018) explored many of these influences, and recommend that practitioners be careful with terms like "alien race," or "colonize" and its cognates, understanding how they entered the field. Shorter & TallBear (2021a) and Atalay et al. (2021) describe a negative interaction between indigenous scholars and SETI practitioners on the nature of contact that highlights the role colonialism continues to play in SETI.

---

[5] Transcribed by Don Davis from a personal video recording of the event (typo corrected). https://web.archive.org/web/20230501182722/http://www.donaldedavis.com/2012%20new/HRMSSPEECH.html.

[6] As quoted from the documentary "Into the Universe with Stephen Hawking" by ABC News online, https://abcnews.go.com/Technology/Space/stephen-hawking-alien-contact-risky/story?id=10478157 retrieved 2024 March 18.

18.5 "Advanced" Aliens

> And when will we have [a space elevator]? I wouldn't like to hazard a guess, so I'll adapt the reply that Arthur Kantrowitz gave, when someone asked a similar question about his laser propulsion system:
>
> The Space Elevator will be built about 50 years after everyone stops laughing.
>
> — Arthur C. Clarke, Address to the XXXth International Astronautical Congress, Munich, 1979 September 20 (Clarke 1981)

We briefly touched on the mistaken idea that the technology SETI seeks necessarily is more "advanced" than ours in Section 4.2.3. We develop this idea further here.

Related to the idea of colonialism are ideas held over from the philosophy of *progressivist social evolutionism*, which holds that human culture has a natural, inevitable movement toward an improved and better state. It has many forms, but perhaps most relevant to SETI is the belief that technology progresses in a particular way toward certain more sophisticated goals.

Progressivism has been largely or totally rejected by the social sciences, but still maintains a strong hold on popular understandings of the arc of human history and alien technology. In this way it has similarities to fallacious understandings of the evolution of life that imagines the process to be directed or to inevitably produce certain sorts of life (similar to ideas we explored in Section 7.5).

With respect to concepts like morality or social structures like government, the idea of a natural direction of social evolution is old. The ancient Greek author Hesiod taught of the degradation of society through a Golden Age, Silver Age, Bronze Age, and finally to his present-day Iron Age. The rapid development of technology and social innovations after the Renaissance in Europe led to progressivism, which argued things would move in the opposite direction.

Applying social progressivism to alien species also leads to expectations regarding their behavior during contact. For instance, Jill Tarter (Box 1.2) was confident that contact would be with peaceful aliens:

> The idea of a civilization, which has managed to survive far longer than we have…and the fact that that technology remains an aggressive one, to me, doesn't make sense…The pressure of long-term survival—of limiting population…I think requires that the evolutionary trends that ratcheted up our intelligence…continues to evolve into something that's cooperative and [so we can] take on global scale problems.[7]

[7] Interview with Jessica Orwig in *Business Insider*, 2016 January 20, "A world leading scientist on the search for extraterrestrials pointed out a flaw in Stephen Hawking's fear of finding intelligent aliens" http://www.businessinsider.com/jill-tarter-says-stephen-hawking-is-wrong-about-aliens-2016-1 (retrieved 2024 June 20).

With respect to technology, the issue is more nuanced since technology and science are largely *cumulative*, meaning as long as methods are not forgotten they can be built upon and improved generation after generation. From this it is easy to incorrectly conclude that there is an inevitable direction of the kinds of technology that get developed or what they can accomplish.[8] It is easy to imagine that if history were replayed with different people (or on another planet) a similar progression of technology development would occur.

Extrapolations of this idea into the future would imply that is is all but inevitable that certain technologies will eventually be developed. Science fiction is filled with such technologies, like rockets, space stations, powerful weapons, remote sensors, elimination of cancers and infectious diseases, and mastery of genetics.

More exotic, less plausible technologies such as warp drives, teleporters, or faster-than-light communication, are often invoked primarily for plot purposes but justified via Arthur C. Clarke's "Third Law." Clarke's three laws (Clarke 1962) reflect a heavy influence of progressivism on his futurism and science fiction:

1. When a distinguished but elderly scientist states that something is possible, he is almost certainly right. When he states that something is impossible, he is very probably wrong.
2. The only way of discovering the limits of the possible is to venture a little way past them into the impossible.
3. Any sufficiently advanced technology is indistinguishable from magic.

The analogy that justifies these "laws" is how much of modern technology today might seem as magic to people of, say, 1000 years ago, such as computers, airplanes, rockets, cars, lasers, televisions, telephones, and so on. These technologies are so sophisticated it would be in some cases essentially impossible to describe how they work (how does one explain a wireless phone to someone who has no concept of radio waves?)

Clarke's third law, in particular, is widely cited, and implies that we should not assume we can put *any* physical constraints on what alien technology might accomplish (see Section 4.4 for how we confront this possibility in SETI). Coupled with the idea that alien technology will certainly be *older* than human technology (Section 4.2.3), and we arrive at the conclusion that any *spacefaring* species will necessarily have technologies of the sort we find in science fiction, and that it will include the ability to do things that we would consider impossible today. We see this in papers that refer to "superior" or "advanced" technologies possessed by aliens (such as Bracewell 1960, entitled "Communication From Superior Galactic Communities").

[8] This idea is a core mechanic in the popular video game series *Civilization*, in which a player guides a human civilization from Earth's actual history through thousands of years of development in part by selecting which technology they will develop next, selecting from a "tech tree" that describes which technologies are needed to develop others. As certain technologies are developed, the planet enters different "ages" of history, such as "the industrial age." While players can in principle take somewhat different routes through the tree, by the end of the game all players' civilizations will have developed nearly the same set of technologies.

The question here is similar to the contingency versus convergence in evolution we encountered in Section 7.5, and arises from a similar fallacy that imagines evolution to be directed, and that complex, multicellular life will follow an evolutionary path similar to the development of animals on Earth. Just as we should not necessarily expect aliens to evolve body plans and cognitions similar to ours, we should not expect that their technological developments have tracked ours, even if some of them converge on some similar technologies.

One way we can see this is in how our definition of technology is sometimes quite parochial, as in the example of the first European colonists to various parts of North America. Anderson (2005) describes how these settlers often marveled at the bounty of "unspoiled" nature in their new home, comparing it to gardens, parks, or orchards. What they did not realize is that they were looking at what might be described as "advanced" agricultural technology, that is, lands which had been tended and curated to be in that state. They had encountered the lands of a civilization that was perhaps just as old or older than theirs, and which had pursued a different direction for their agricultural technological sophistication. Nonetheless, because their social and technological development was in that different direction, the Native Americans who tended these lands were and continue to be portrayed as "primitive" for not having put into widespread use many technologies Europeans consider foundational, such as the wheel.

But does one really need the wheel to be "advanced?" The Inca built a civilization and managed an empire without making much use of the wheel, not because they did not invent it, but because they found other traditions of engineering that were better suited to their needs in the mountainous terrain of the Andes. And in *The War of the Worlds*, H. G. Wells imagines that the Martians, whose command of the technology of spaceflight and war is far superior to humans', make no use of wheels of any kind, large or small, in their machines.

In other words, one *can* define technology as it has developed in Europe or even the entire planet as part of a cumulative "technology tree," but it is probably inaccurate to gauge the *overall* technological sophistication of *other* civilizations or species based on how far they progressed along humans' perhaps idiosyncratic paths, or how far they are along the path we encounter in science fiction and which we imagine lies before us.

As we discussed in Section 4.2.3, it is better to focus only on the detectability of the technology we seek (as with the ichnoscale Section 4.2.1), and be prepared to discover that while a species may be quite old, their technology in some dimensions might strike us as rather primitive, if we even recognize it as technology at all.

18.6 "Ancient Aliens" and Psuedoscience

A significant challenge in public communication about SETI and the scientific search for alien life is the large industry that profits promoting psuedoscientific views of history and science across popular media not as fiction, but as truth.

Perhaps the two most famous examples of the form are "Chariots of the Gods?" by Erich von Däniken, and the *Ancient Aliens* television program.

In "Chariots of the Gods?" (von Däniken 1968) and other books (whose titles would later drop the fig-leaf question mark) von Däniken concocted an elaborate story of how aliens either built or taught humans to build many of the most famous ancient artifacts such as the pyramids in Egypt and Mesoamerica, the Nazca lines in Peru, Stonehenge, and so on. He also claims descriptions of divine beings in ancient religious texts (similar those in Section 3.1.1) in fact describe encounters with technologically advanced aliens.

Von Däniken's "evidence" largely consisted of dubious interpretations of cherry-picked imagery and iconography which he claimed depicted aliens and their spacecraft, and arguments that the construction of these structures required engineering techniques far beyond what their host civilizations could have possessed.

This theme has been shamelessly picked up by the American cable television series *Ancient Aliens*. Each episode of this faux-documentary credulously examines "evidence" of this sort to explore how aliens have influenced human history. *Ancient Aliens* began its run in 2009 and has become a staple of pop culture, in large part due to its longtime host and executive producer Giorgio A. Tsoukalos.

Tsoukalos's flamboyant hair and style of purporting to simply present evidence without judgment while strongly implying alien origins to human creations led to the creation of a widely used meme, "I'm not saying it's aliens...but it's aliens."

Von Däniken's and *Ancient Aliens*' claims are widely derided by anthropologists, but have found a wide and eager audience around the world, who now sees these human monuments as evidence of alien visitation. This presents a major and ongoing challenge to public outreach and education efforts in archaeology and history.

The ancient aliens claim is often couched in racist and progressivist assumptions, that "primitive" (and, largely, non-European) peoples could not have possibly constructed their technology, and must have needed help from an "advanced" civilization. In fact, the Egyptians (for instance) left behind a detailed description of how they constructed they pyramids using ancient techniques. *Ancient Aliens* thus does not just present a false narrative about aliens and SETI, it promulgates racist and progressivist ideas about history and anthropology, and dismisses and erases the accomplishments of many ancient peoples around the world.

For whatever reason, alien life is a common element of psuedoscientific and supernatural beliefs, and through these and many other venues this remains a challenge for science communication.

These sorts of unscientific associations with studies of aliens also contribute to the *giggle factor* that provides some of the challenges SETI has being taken seriously in Congress (Section 3.3.4) and in scientific circles.[9] The Tsoukalos meme also provides an unfortunate template for SETI researchers, who must distinguish their work carefully describing candidate signals from the fact-free hype machine that is *Ancient Aliens*.

[9] Wright & Oman-Reagan (2018) have argued that the association of aliens with campy science fiction contributes to this as well.

18.7 UFOs

> On this particular Thursday, something was moving quietly through the ionosphere many miles above the surface of the planet; several somethings in fact, several dozen huge yellow chunky slablike somethings, huge as office blocks, silent as birds. They soared with ease, basking in electromagnetic rays from the star Sol, biding their time, grouping, preparing.
>
> The planet beneath them was almost perfectly oblivious of their presence, which was just how they wanted it for the moment. The huge yellow something went unnoticed at Goonhilly, they passed over Cape Canaveral without a blip, Woomera and Jodrell Bank looked straight through them, which was a pity because it was exactly the sort of thing they'd been looking for all these years...
>
> Miles above the surface of the planet the huge yellow somethings began to fan out. At Jodrell Bank, someone decided it was time for a nice relaxing cup of tea.
>
> — Douglas Adams, *The Hitchhiker's Guide to the Galaxy*, describing the moments before the destruction of Earth by a Vogon Constructor Fleet.

In the popular consciousness, UFOs are closely related to SETI, for obvious reasons. The idea that unidentified objects in the sky might be alien spacecraft is a staple of pop culture, and there exists a vast trove of psuedoscientific lore about alien visitations pushed or credulously reported across a wide media ecosystem of fake television documentaries, books, websites, conferences, and beyond.

In this text, we consider the topics of SETI and UFO[10] studies wholly distinct, and have not addressed UFOs at all so far. But no study of SETI is complete without discussing its interactions with and demarcations from studies of UFOs, if only because practitioners will need to be able to articulate the similarities and differences.

Why, after all, should the sober, objective exploration of the origins of UFOs not be part of SETI? If SETI can look for alien spacecraft in the solar system, why should such investigations stop at the top of the Earth's atmosphere?

18.7.1 UFOs Should be Studied without Taboo

Studies of UFOs and SETI both have a history of not being taken seriously in scientific circles, in part for the common reason that discussion of technological alien life invites a strong "giggle factor." It does not help UFO studies that there is a large industry of grifters and fact-free sensationalism around the UFO phenomenon, some of which is connected with unscientific promotion of the supernatural.

This is unfortunate, because the topic of UFOs clearly deserves study. In addition to the intense popular interest in the topic, the possibility that there are unknown

[10] UFOs are also now sometimes referred to as Unidentified Aerial Phenomena (UAPs), a term that is meant to be more general than UFOs, and generally used by those taking a more scientific approach to studying the phenomenon, especially in a manner that is agnostic to their nature.

forms of aircraft in they sky is of understandable interest to military and governments around the world. Even if all UFOs turn out to not be any kind of aircraft (or spacecraft) at all, understanding why military sensors and personnel *seem* to detect such things is important.

18.7.2 UFO Studies and SETI Have a Tenuous Connection at Best

The primary connection between UFO studies and SETI is the common hypothesis that extraterrestrial technology exists and can be or has been detected. Beyond this, the connection is quite tenuous, and the methods typically used in the two efforts are in many ways antithetical.

First of all, to say UFOs are objects of SETI interest is to assume the consequence and conclusion of UFO studies. Including the alien hypothesis is a reasonable thing to do when studying UFOs, but this by itself does not make it SETI. After all, studies of anomalous astronomical objects do not become SETI simply because the alien hypothesis is kept in mind when studying the objects. This is important because there is actually very little reason to hypothesize that UFOs are due to aliens beyond the intense popular association with them thanks to urban legends and science fiction.

SETI proceeds via the systematic study of the sky searching for some phenomenon, by collecting data with astronomical instruments whose properties are understood. Surveys ideally end with upper limits placed on the phenomenon. In contrast, much of UFO studies seem to consist of interpreting or debunking claims of sightings of mysterious objects, from eyewitness testimony or military sensors. Attempting to explain data and anecdotes of uncertain provenance and reliability is in some cases a worthy endeavor, but bears little resemblance to SETI.

In addition, the alien hypothesis for (some) UFOs is extremely narrow: it is that alien spacecraft are at least occasionally visible in Earth's atmosphere but for some reason there is no easily confirmable or corroborating evidence for them, despite their appearances over many decades. The UFOs-as-alien-spacecraft hypothesis thus actually requires some additional hypotheses to justify why we only ever detect them in ambiguous ways: for instance that they are deliberately but imperfectly hiding, or that they are somehow manipulating humans' perceptions of them.

18.7.3 Why UFO Studies and SETI Are Best Considered Separate

UFO studies can, in principle, be done scientifically and systematically by collecting new data to detect and study the phenomenon in a reproducible way. Even in this case, however, there are good reasons to consider the efforts distinct.

One is that it is confusing to the public to conflate the two, and reinforces unscientific perceptions that UFOs have a strong association with alien life. If UFO studies proceed truly agnostically about the nature, frequency, and manifestations of the phenomenon, then there is no reason to brand it as studies of alien spacecraft. Keeping the studies distinct thus helps the public understand both efforts' purposes and methods better.

Another is that the expertise and methods used in the two fields have hardly any overlap. SETI is primarily an effort of astronomers and planetary scientists.

Depending on its methods, UFO studies need to incorporate atmospheric science, human perception and psychology, military sensor equipment operations, aeronautics, and so on. They are distinct fields.

18.7.4 The Origins of the UFOs-Are-Alien-Spacecraft Belief

The idea that UFOs might be alien spacecraft has a long history, but a key mythmaking element was the Roswell Incident.

In 1947, debris landed in a field in Roswell, New Mexico. The debris was from a balloon launched and lost by Project Mogul, a military intelligence-gathering operation to launch microphones high into the atmosphere to monitor for the sounds of above-ground nuclear testing around the world.

This occurred amidst a "flying saucer" craze based on a UFO report the previous month in Washington State. The rancher who collected the debris and the Roswell Air Force personnel who were contacted about it thought this might be connected, and an initial statement by the Roswell Army Air Field claimed they had recovered a "flying disc."

The government quickly changed its story to falsely claim it was debris from a weather balloon: Project Mogul was, after all, at the time a Top Secret program, unknown even to the Army Air Force personnel at Roswell. Not much was made of the incident at the time, but in the 1970s UFOlogists plucked it from obscurity and developed an elaborate conspiracy theory about an alien spacecraft that had landed and was being kept in secret by the US government.

It bears repeating that there is no basis for believing the Roswell wreckage was of extraterrestrial origin; the connection came about purely because of temporal proximity to the earlier "flying saucer" incident. Declassified documents tell us what happened. One cannot, of course, disprove a conspiracy theory—any such proof is countered by the claims of conspiracy—but to demand such proof is to get the burden of proof backwards.

18.7.5 Earth Skies Are Very Well Studied

Unlike the far reaches of space studied with modern telescopes across and beyond the electromagnetic spectrum, humans have been studying the skies for tens of thousands of years. If alien spacecraft were common visitors to Earth's atmosphere in ships, it would be a well-documented, global phenomenon. What's more, as our awareness of things in the sky has improved since the 20th century with radar and imagery to track planes and other aircraft, our ability to detect alien spacecraft has increased. There are millions of phones in people's pockets all over the world, further increasing our ability to document UFOs, but evidence that they are alien is as ambiguous and elusive as ever.

To be sure, there are many UFOs. To say they are probably not alien is not to say they probably do not *exist*. Reports of UFOs come from many sources, including misidentification of planets and other heavenly bodies, misidentification of ordinary and secret aircraft, malfunctioning radar and other equipment, and misuse of cameras and radar in a way that produces anomalous readings. Eyewitness accounts are especially unreliable, running the gamut from hoaxes to hallucinations to honest mistakes to correct identification of ordinary phenomena and aircraft to sightings of classified aircraft. It is worth noting that UFO reports are heavily concentrated in the United States, where people are culturally primed to believe UFOs might be alien.

At any rate, the skies are today filled with flying technology, much of which is unfamiliar to the lay public, and some of which, like the Project Mogul balloons, is quite secret. It is unsurprising that these things are routinely seen and unidentified or unidentifiable. None of this implicates aliens.

18.7.6 "Impossible" Maneuvering

The best argument to make that UFOs might be alien is that some are claimed to exhibit "impossible" flight characteristics, such as making right-angle turns or quickly accelerating to very high speeds. The implication is that only alien technology could do such things.

But this is quite the leap of logic. Confronted with an apparently impossible flight path, there are many explanations that should presumably come before aliens. The flight path could be mismeasured. Two objects quite far apart might be misidentified as a single object, moving impossibly quickly between two locations. And so on.

Indeed, there are dueling cottage industries of people who analyze videos, photographs, and eyewitness accounts to establish UFOs must be performing feats of aeronautics beyond human technology, and people who "solve" these videos by showing they are almost certainly commercial aircraft, balloons, birds, or other ordinary phenomena. Two common resolutions are that the distance to an object is misgauged, or the motions of the video camera were not taken into account, meaning the UFO was not actually moving in the manner claimed.

Further, one must ask whether the explanation that alien technology is at work to explain "impossible" maneuvering makes physical sense. Turning sharply or accelerating requires momentum conservation, meaning there must be some substantial exhaust, and moving hypersonically creates substantial friction with the atmosphere. Both effects might have observable effects, such as sonic booms and atmospheric heating, not typically observed with UFOs.

18.7.7 UFOs as Part of SETI

What would it take for studies of UFOs to be part of SETI? Certainly if extraterrestrial beings made themselves known as the pilots of craft, that would do it! But what short of that?

Let us, as a concrete example, make the strongest possible analogy between studies of UFOs and legitimate SETI research. In solar system SETI (Chapter 14), we might seek to put an upper limit on the existence of alien spacecraft in the solar system by cataloging all solar system objects (mostly asteroids) and checking if any have anomalous orbits indicative of being hollow or having thrust. These are presumably inherently interesting from a planetary science perspective (are they "dark comets"?) and worthy of followup. If we simply extend that to the Earth's atmosphere, cataloging all airborne objects and checking if any have anomalous trajectories indicative of propulsion no animal or human-made technology could produce, why is that not SETI?

There is a case to be made that it should be in a limited way. If one undertakes rigorous data collection procedures instead of relying on reports using data of uncertain quality and provenance, as much of UFOlogy does, then one is doing a real scientific experiment. The upper limits set might be quite interesting, and indeed the Galileo Project led by Avi Loeb purports to do just this.

The question of what comes next is challenging, and where the experiment and SETI might part ways: if one has a detection of an object in Earth's atmosphere that "should not be there" how does one follow it up? Presumably, most of those cleanly detected will end up being either lost or secret aircraft (stray weather or spy balloons, or experimental jets, for instance). In the case of the latter, one will struggle with the question of how to distinguish alien technology (which, again, we have no reason to think it could be) and secret human technology. The governments behind such technology (which we *know* exists) will presumably not be very helpful in resolving the matter, and indeed might both resist such investigations and strategically amplify suggestions it is "not theirs." One wonders to what degree the US government appreciated the "alien" interpretation of the Roswell Incident as a way to deflect attention from Project Mogul.

This is somewhat different from the situation in other forms of SETI, where a signal from space is either dispositive as being both alien and technological (as in the case of communication SETI) or results in astrophysically interesting outliers to be studied with conventional astronomical methods. The study of such false-positive outliers in the Earth's atmosphere is the study of human technology, a very different sort of task.

Efforts like the Galileo Project thus operate at the edge of SETI: when scientifically conducted, they are interesting and have the potential to provide data that, interpreted in a SETI context, might provide both upper limits on alien technology and potential candidates for it. This is similar to, in our analogy above, data from surveys of asteroid orbits: not a SETI project *per se*, but a way to collect data SETI could potentially use.

When candidates come up, one must then formulate a concrete hypothesis about them consistent with other data: how many of these could there be that we had not seen them before? What about them makes us think they cannot be natural or human-made? What additional studies could be performed to show they are both alien and artificial? One must then formulate a way to study those candidates in a way that either rules them out (as understood human technology) or proves they are alien.

In summary, UFOs are primarily not part of SETI because there is no reason to think they have any link to alien technology. While searches for alien technology in Earth's atmosphere could *in principle* be conducted as a serious SETI project, the methods for detecting such technology and ruling out confounders are far from astronomy, where the rest of SETI operates, and will involve many artificial phenomena that may be subject to government secrecy, which complicates their study.

But perhaps most importantly, calling UFO studies part of SETI in some sense assumes the conclusion, and favors a particular, perhaps narrow and highly contrived, hypothesis over many others in what should be an objective study.

18.8 Further Reading

History of the question of extraterrestrials: Crowe (1986), Basalla (2006).
Popular imaginings of aliens: Guthke (1990).
Visions of human futures and SETI: Wright & Oman-Reagan (2018).
Colonialism and gender in SETI: Charbonneau (2021).
SETI, CETI, and the Cold War: Charbonneau (2024).
Indigenous perspectives on SETI: Shorter & TallBear (2021b).
A history of the UFO phenomenon: Eghigian (2024).

References

Anderson, K. 2005, Tending the Wild: Native American Knowledge and the Management of Californiaas Natural Resources (Berkeley, CA: Univ. California Press)
Atalay, S., Lempert, W., Shorter, D. D., & TallBear, K. 2021, AICRJ, 45, 9
Basalla, G. 2006, Civilized Life in the Universe: Scientists on Intelligent Extraterrestrials (Oxford: Oxford Univ. Press)
Baum, L. F. 1914, Tik-Tok of Oz, The Oz Series (New York, NY: Open Road Media)
Bracewell, R. N. 1960, Natur, 186, 670
Charbonneau, R. 2024, Mixed Signals: Alien Communication Across the Iron Curtain (Cambridge, UK: Polity Press)
Charbonneau, R. 2021, Astronomy and the Search for Extraterrestrial American Indian Culture and Research Journal, 45, 71
Clarke, A. C. 1951, The Exploration of Space (PA, USA: Temple Press)
Clarke, A. C. 1962, Profiles of the Future: An Inquiry Into the Limits of the Possible, Bantam Science and Mathematics (New York: Harper and Row)
Clarke, A. C. 1981, Advances in Earth Oriented Applications of Space Technology, 1, 39
Crowe, M. J. 1986, The Extraterrestrial Life Debate, 1750-1900, Dover Books on Science (New York: Dover)
Eghigian, G. 2024, After the Flying Saucers Came: A Global History of the UFO Phenomenon (Oxford: Oxford Univ. Press)
Guthke, K. S. 1990, The Last Frontier: Imagining Other Worlds, from the Copernican Revolution to Modern Science Fiction (Ithaca, NY: Cornell Univ. Press)
Kepler, J., & Rosen, E. 2003, Kepler's Somnium: The Dream, Or Posthumous Work on Lunar Astronomy (New York: Dover)
Shorter, D. D., & TallBear, K. 2021a, AICRJ, 45, 1
Shorter, D. D., & TallBear, K. 2021b, AIRCJ, 45, 1
Von Daniken, E. 1968, Chariots of the Gods? Unsolved Mysteries of the Past (New York: Book Club)
Wells-Jenson, S. 2023, Xenolinguistics: Towards a Science of Extraterrestrial Language, ed. D. A. Vakoch, & J. Punske (London: Taylor and Francis)
Wright, J. T., & Oman-Reagan, M. P. 2018, IJAsB, 17, 177

Chapter 19

METI

> This Voyager spacecraft was constructed by the United States of America. We are a community of 240 million human beings among the more than 4 billion who inhabit the planet Earth. We human beings are still divided into nation states, but these states are rapidly becoming a single global civilization.
>
> We cast this message into the cosmos. It is likely to survive a billion years into our future, when our civilization is profoundly altered and the surface of the Earth may be vastly changed. Of the 200 billion stars in the Milky Way galaxy, some—perhaps many—may have inhabited planets and spacefaring civilizations. If one such civilization intercepts Voyager and can understand these recorded contents, here is our message:
>
> This is a present from a small distant world, a token of our sounds, our science, our images, our music, our thoughts, and our feelings. We are attempting to survive our time so we may live into yours. We hope someday, having solved the problems we face, to join a community of galactic civilizations. This record represents our hope and our determination, and our good will in a vast and awesome universe.
>
> — US President Jimmy Carter's message on the Voyager Golden Records

19.1 Messaging Extraterrestrial Intelligence

Much of this text has been concerned with the *receiving* of signals, but in the early days of SETI, especially in the Soviet Union, the problems of sending and receiving were considered together, and indeed the entire field was originally called CETI: *Communicating with ETI* (Section 3.2.5).[1]

[1] See Charbonneau (2024). The term was coined by first Chairman of the CETI committee of the International Academy of Astronautics, Czeck scientist Rudolph Pešek, as the name for a conference on the topic. The conference was canceled when superseded by the first Byurakan conference, but the name stuck: Sagan and others liked that *ceti* is the Latin genitive for "whale" (being an example of a nonhuman terrestrial intelligence) and coincidentally referenced Project Ozma's focus on the star τ Ceti. It was at the Second Byurakan Conference (Section 3.3.1) that Sagan, Shklovskiĭ, and others decided to focus efforts away from communication and toward searching.

doi:10.1088/2514-3433/adf5aech19

The name of this chapter comes from Alexander Zaitsev (2006a), who defined *Messeging Extraterrestrial Intelligence* or METI as the act of sending out a beacon to let other species understand that they are not alone in the Galaxy. In this text we will broadly refer to any attempt, superficial or otherwise, to contact ETIs or construct messages they could understand as "METI," but it is useful to also honor important distinctions among three broad classes of such activities: superficial METI, effective METI, and active SETI.

19.1.1 Flavors of METI

In arguments and discussions about METI there is often a lot of confusion about what sorts of activities are being discussed, and what constitutes "real" METI. We thus need to think precisely about the relevant properties that distinguish different METI activities.

The vast majority of METI signals or messages that have been sent, including arguably all of the most famous examples described below, are largely superficial efforts, performed primarily with humans as their audience. These messages have virtually no chance of being received; their purpose is primarily to inspire humans to think about the possibility. We might call this *ostensible* or *superficial METI.*[2]

We can distinguish these superficial efforts from activities that are both *intentional*[3] and *have a reasonable chance of being detectable*. We will refer to such METI activities as *effective*.

For instance, to be effective, radio transmissions should be:

- strong enough,
- of sufficient duration or sufficiently well timed, and
- sufficiently well targeted that they are actually detected by ETIs, and
- likely to be interpreted as technological or inspire further investigation as technological and not misinterpreted and ignored as natural emission.

That is, they should be *conspicuous* and *obviously technological. Conspicuousness* here roughly means the degree to which the signal is significantly more salient than, say, our ordinary leaked transmissions.

Of course, we do not know the capabilities of ETIs, so it is unclear how strong a signal needs to be to be effective by this definition. People thus may disagree, depending on their expectations for the abundance and capability of ETIs, about whether a METI program is effective.

We also need to consider that ETIs may already know we are here, rendering intentional signaling perhaps superfluous. We can then consider whether the message is *revealing*, because it contains or otherwise reveals to the alien species something they did not already know, such as

- the existence of Earth, or
- the existence of life on Earth, or
- the existence of technology on Earth, or

[2] Although the word "superficial" often comes with negative connotations, I mean none here. Zaitsev (2012), however, was not so charitable about radio signals sent without regard for their detectability, calling some such efforts "pseudo-METI," "stunts," and "idle METI profanations."

[3] They need to be intentional or else all of Earth's technosignatures would constitute METI.

- our interest in communicating with them, or
- some other property of humans.

Again, we cannot know whether a signal is revealing in an absolute sense because we do not know what, if anything, ETIs already know about us. Certainly, though, some METI signals are much more revealing than others.

19.1.2 Active SETI

One important kind of revealing signal is one that is *invitational* because it requests and generates or otherwise provokes a response from ETIs, thus revealing their existence to us. Because the purpose in this case is to make searches for alien life succeed, this is often called *active SETI*.[4]

We can thus characterize an active SETI program and one where:

- the message is designed and transmitted with the goal of triggering a response, and
- includes a program to detect that response.

In other words, an active SETI program would ideally send signals with sufficiently intelligible information that ETIs understand that we are requesting a response; the signal should be sent to a sufficiently nearby target that the signal will be strong when it arrives and there is a chance for a timely response; and involve a follow-up program to listen for a response for a reasonable amount of time beginning two light-travel-times after the transmissions.

Box 19.1: Jargon Review: Flavors of METI

- **METI**: Messaging Extraterrestrial Intelligence. The construction or sending of any message to contact ETIs, superficial or otherwise.
- **Superficial**: Describing METI primarily aimed at humans, with a negligible or irrelevant chance of actually being received by ETIs. Such signals are only **ostensibly** intended for ETIs. The Arecibo Message, Pioneer Plaques, and Voyager Golden Records are examples of superficial METI.
- **Effective METI**: Non-superficial METI. To be effective, a METI transmission must be truly intended for ETIs and conspicuous.
- **Conspicuous(ness)**: Describing the degree to which a METI signal is significantly more likely than our ordinary leaked transmissions to be detected.
- **Revealing**: The degree to which a METI signal communicates something ETIs did not already know. An information-free beacon might be minimally revealing, indicating only the existence of a technological species. An information-rich signal might not be particularly revealing if the ETIs are already aware we are here and are monitoring us unless it conveys a specific message to them, such as an invitation to communicate.
- **Invitational**: The property of a METI signal that specifically requests contact.
- **Active SETI**: A program to use invitational METI signals to produce a response from ETIs whose receipt by us would allow SETI to succeed.

[4] Although this term can be awkward because of the syntactic ambiguity it generates ("they are an active SETI scientist" could mean either they send signals or are currently working in the field).

19.1.3 Arguments for METI and the SETI Paradox

Zaitsev (2006b) made the perhaps obvious but important point that in order for communicative SETI to succeed, someone needs to be transmitting. He phrased this in terms of what he dubbed the *SETI Paradox*:[5] "Searching is meaningless if no one feels the need to transmit." Zaitsev (2011) put it slightly differently: "SETI makes sense only in a Universe that creates an Intellect which realizes the need, not only to search for another Intellect, but also for transmitting intelligent signals to it."

Zaitsev felt passionately that it was important not just for humans (and aliens) to search for ETI in the Galaxy, but that we produce the sorts of signals that communication SETI searches for. "Unless this process [of mutual contact] is triggered and is ongoing," Zaitsev (2011) wrote, "intelligent life in different parts of the Universe will remain lonely, isolated, and inclined to extinction."

He thus argued for highly effective and revealing signals that were not necessarily invitational—the motive here could be entirely altruistic, to let others know they are not alone.

de Magalhães (2016) proposed sending METI signals that were strongly invitational but not conspicuous by embedding requests for contact in existing television and radio broadcasts. The idea was to test a form of the Zoo Hypothesis (Section 6.6.3) that ETIs are aware of our existence but are hiding (or simply not making themselves known) because we have not asked them to reveal themselves.

19.1.4 Effective METI Transmissions

Zaitsev (2011) studied which METI transmissions were strong enough, sufficiently targeted, and sufficiently intelligible to qualify as what he considered to be a "valid" transmission. He dismissed publicity stunts by corporations (such as sending a Doritos ad into space[6]) both on principle and because they made no attempt to craft a message that would be universally intelligible as Sagan and Drake had done in their messages. His definition of "valid" roughly corresponds to our definitions of *effective* and *revealing*.

Zaitsev felt that only five messages were valid examples of METI or Active SETI: The Arecibo message (3 minutes long), and four transmissions in which Zaitsev participated: "Cosmic Call 1999" (960 minutes), "Teen Age Message" (366 minutes), "Cosmic Call 2023" (900 minutes), and "A Message from Earth" (240 minutes). The latter four were transmitted using the *Yevpatoriia* 70 m telescope using its planetary radar equipment.

[5] Chelsea Haramia has pointed out that this is not a logical paradox, and is actually formally "self-defeating": if the logic behind the paradox is sound and other species have the same reasoning we do, then they will transmit, defeating the need for us to go first.

[6] https://web.archive.org/web/https://www.sciencedaily.com/releases/2008/06/080612122817.htm. Similar stunts of simply pointing a transmission dish at a particular star and sending an ordinary human signal include a Netflix ad (https://web.archive.org/web/https://www.deeplocal.com/the-world-is-on-netflix).

The last four of these messages might seem to be quite *effective* because they were quite powerful and directed at nearby stars, but it is actually unclear how *conspicuous* they really are. Together they compose about 41 hr of total signaling, targeting nearby stars, typically 20–70 lt-yr distant, with no star getting more than a few hours of transmissions. In order for these signals to actually be detected requires that ETI at one of these targets be listening for such a signal at the right time on the right frequency. We must imagine that such monitoring is continuous for so fleeting a signal to be received, which is perhaps unlikely.

19.2 Some Famous METI

19.2.1 The Morse Message

In 1962, the Soviet Union sent a powerful radio signal to Venus including the words мир (mir, meaning “peace” or “world”) and Ленин (“Lenin”), and the initialism CCCP (“USSR”). Charbonneau (2024) argues persuasively that there was no actual expectation that Venusians would detect the signal, or that it was intended for them at all (after all, the signal used Morse Code and was in Russian, referencing contemporary politics).

The real purpose of the message was to generate press along the lines of the newspaper Krasnaia Zvezda (“Red Star”):

> The Institute of Radio Engineering and Electrical Engineering of the Academy of Sciences of the USSR informed [us] of a new victory for Soviet science and technology… For the first time in the history of mankind, Soviet scientists succeeded in carrying out radio communications through to the planet Venus.[7]

So although The Morse Message is listed in many places as the “first METI signal” it is not actually an example of the practice.

19.2.2 The Pioneer Plaques

The *Pioneer 10* and *Pioneer 11* probes were launched by NASA in 1972 and 1973, respectively. They were the first probes sent to the outer solar system, to explore Jupiter; *Pioneer 11* would continue on to visit Saturn. Their extremely high velocities meant that they would only be able to perform flybys of their targets, after which they would continue on with escape velocity from the solar system, becoming humanity’s first interstellar objects.

[7] “In the Words of the Cosmos: Lenin, USSR, Peace,” 1962 December 30, quoted in Charbonneau (2024).

Eric Burgess was a freelance journalist covering the *Pioneer* program at NASA and is widely credited with the idea that they should carry some sort of message or commemoration of Earth. He approached Carl Sagan, who worked with Frank Drake to design a plaque that would carry the message. Sagan's wife at the time Linda Salzman Sagan created the illustrations, which were engraved onto gold-plated aluminum plaques.

Figures 19.1 and 19.2 show their design, which contains five elements intended to provide a universally understandable message conveying information on the spacecraft's origins.

The upper left depicts a schematic of what is intended to be a single hydrogen atom undergoing the hyperfine transition corresponding to the 1420/21 cm line (Section 5.2.1). The circle depicts the 1s orbital of an electron in the ground state,

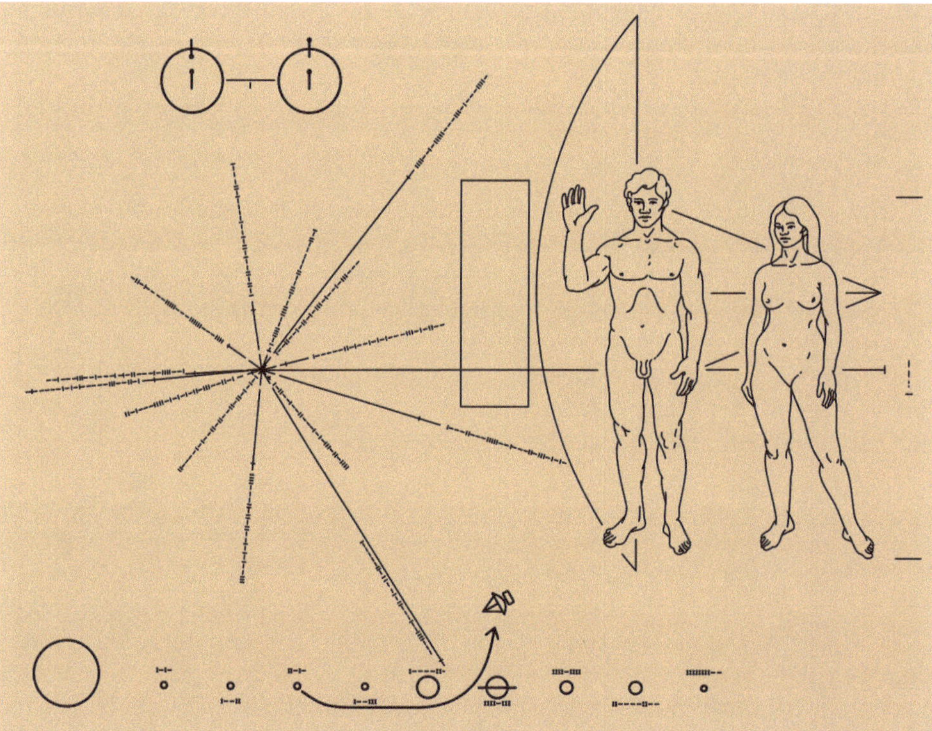

Figure 19.1. The Pioneer Plaque. The design elements from top-to-bottom, left-to-right are the hydrogen atom before and after a spin-flip that generates the 21 cm/1420 MHz line; a map of pulsars around Earth and the location of the Galactic Center with their periods given in binary in units of 1/1420 MHz; a schematic of the *Pioneer* probe; two human figures to scale, the height of the human on the right as a binary number in units of 21 cm; a schematic of the solar system with orbital distances given in binary (in units of 1/10 of Mercury's orbit), and the approximate path of *Pioneer* out of the solar system. Credit: NASA Ames.

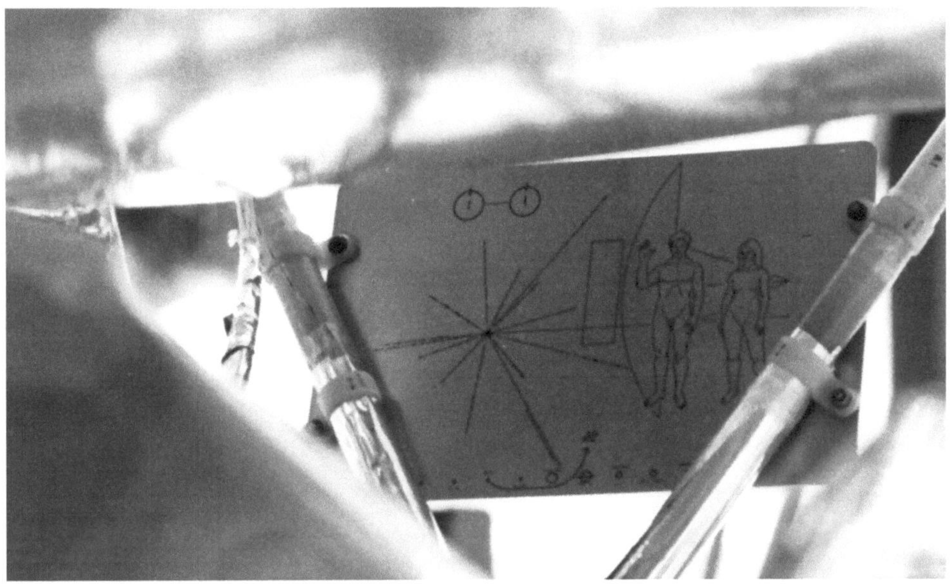

Figure 19.2. One of the plaques in context, attached to *Pioneer 10*. Credit: NASA.

and the lines with a ball at the end are intended to depict the angular momentum vector of a particle. The line in the center represents the proton, which is anti-aligned with that of the electron in the left-hand atom. The right hand atom shows them aligned, indicating a change has taken place. The line between them is labeled with a "1" indicating that the transition is one unit of measurement (either 21 cm in length or 1/1420 MHz = 0.7 ns in time).

The star pattern represents a "map" to the solar system. Each line contains a binary number represented as ones and zeros (with zeros represented by dashes). These numbers represent the spin periods of various pulsars around Earth, with their positions in space shown in a 2D projection. The long line to the right going behind the human figures represents the direction of the Galactic Center, to scale with the pulsar lines.

The idea here is that if one is aware of at least some of these pulsars and their spin periods (and if one can correct for their well-behaved spin-downs and motions through the Galaxy) then one could back out the position in the Galaxy from which the probe came.

The map at the bottom shows a schematic of the solar system. Nothing is to scale here, but the relative orbital distances of the planets are given in binary in units of 1/10 the Mercury–Sun distance. An arrow shows the rough path from Earth to beyond Jupiter and out of the solar system the probe took, with a schematic of the probe.

To the right are nude male and female human figures with a bar to the woman's right to indicate her height (in units of 21 cm in binary: "|- - -" means 1000, or eight, meaning she is 169 cm tall, or about 5′6″.) Sagan recounted that the figures were drawn in a manner intended to be "panracial" to represent all of humanity. The man's hand is raised in a "hello" gesture.

Behind the couple is a schematic of the *Pioneer* probe for scale, as a second way of conveying the units involved.

The *Pioneer* probes are extremely unlikely to ever encounter an alien species by chance. They are not headed toward any particular star (their trajectories were determined by their science missions at Jupiter and Saturn) and they no are longer functioning. As attempts to contact ETI they are thus largely ineffective, but unlike the Morse Message, the plaques were designed to be universally intelligible and genuinely addressed to an alien species that might encounter them, making them a legitimate example of METI.

19.2.3 The Arecibo Message

Around a year after the launch of *Pioneer 11* in 1974, Frank Drake, who was then serving as the director of NAIC, the organization operating Arecibo (Section 19.2.3), decided to use the occasion of the commissioning of the new radar at Arecibo to transmit another message to ETI.

He worked with Richard Isaacman, Linday May, and James C. G. Walker (and with input from Carl Sagan and others) to compose a 1679 bit message to be transmitted at the globular cluster M13. Staff at the National Astronomy and Ionosphere Center (1975) explained:

> As the first use of this remarkable new facility, the Arecibo staff thought it was highly appropriate to send a simple and brief signal to the fringes of the Galaxy telling of the existence and nature of human life…to transmit a message for possible reception by other intelligent creatures.

The signal was transmitted at 2.380 GHz with an EIRP (Equation (12.21)) of 3×10^{13} W, being the strongest radio signal ever produced at the time. The entire transmission lasted about 170 seconds.

The idea behind the message length was that the bits were intended to be arranged in a rectangular grid to be interpreted pictorially. Since $1679 = 23 \times 73$, and 23 and 73 are both prime numbers, there are only two ways to make this arrangement. One results in a more or less random arrangement of bits, but the other produces the long image shown in Figure 19.3.

The message is meant to be read in order from top to bottom in the figure. The first four rows represent a primer on counting, depicting a binary representation of the first 10 digits. The bit in the fourth row represents a "number label" showing the column containing the least significant bit. The third row above these labels has place value 1, the next row up has place value 2, and so on. The numbers 8, 9, and

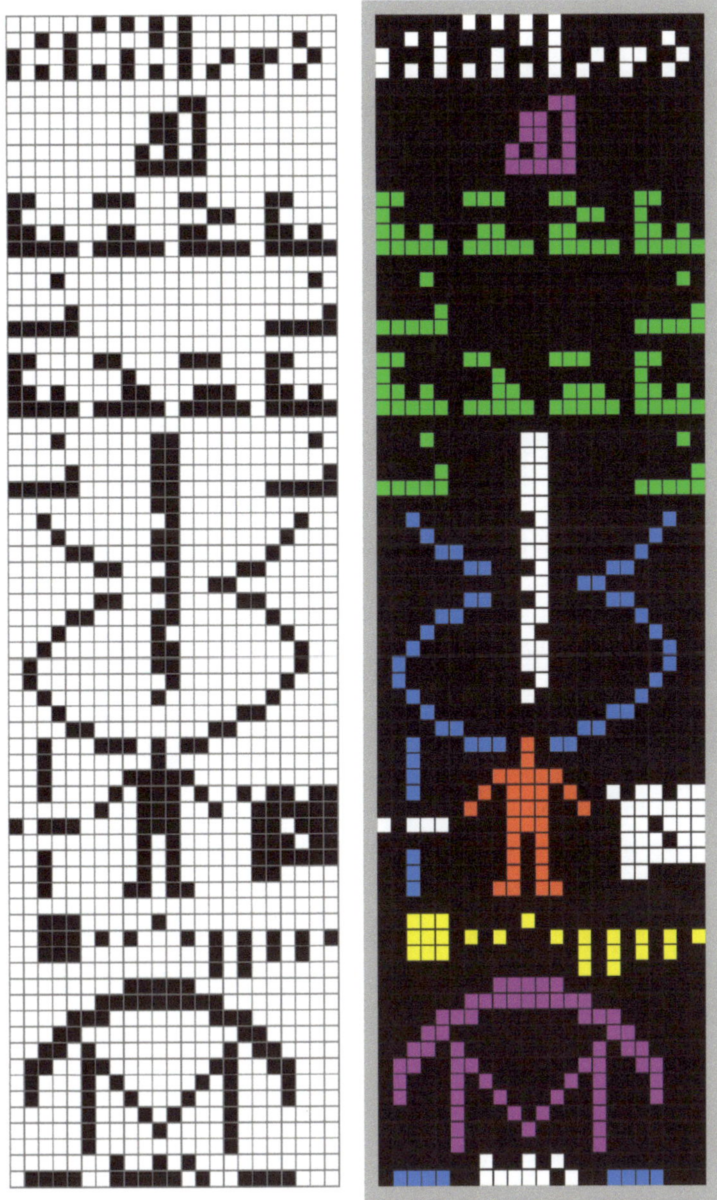

Figure 19.3. *Left:* The Arecibo message, rendered as an image with black and white corresponding to the bits in the message. *Right*: A version of the same color-coded to aid with interpretation. See text for an explanation of the features. The author obtained the left and right images from the Wikimedia website, where they are attributed to Arne Nordmann, Pengo and AnonMoos and made available under a CC BY-SA 3.0 license.

10 need more rows than are available, and so claim the following column to represent the place values 8, 16, and so on.

After a blank row, the next five rows represent five numbers, 1, 6, 7, 8, and 15, intended to represent HCNOP, the five most common elements in Earth life. The intention is that when five numbers appear below, they will represent the numbers of these elements in a chemical formula.

After another blank row, Drake schematically represents DNA. Along the left and right sides are four formulae stacked vertically, alternating H_7C_5O (deoxyribose) and O_4P (phosphate). These units alternate to form the backbone of DNA. Between these in the center of the message, connecting the deoxyribose from left to right, then top to bottom, are $H_4C_5N_5$ (adenine), $H_5C_5N_2O_2$ (thymine), $H_4C_4N_3O$ (cytosine) and $H_4C_5N_5O$ (guanine), the four bases that make up the genetic code.[8]

From these chemical formulae, it is presumed that a competent chemist could deduce the specific isomers involved, and that they could form a chain. The curved lines (rendered in blue in the figure) represent the double-helix formed by the molecule. The tall row of bits up the middle (white in the figure) represent a number near 4.3 billion, at the time believed to be the approximate number of base pairs in DNA (and presumably a measure of our biological complexity).

Continuing the "zoom out" from base pairs to the double helix, below the DNA is a schematic representation of a person. As with the Pioneer plaque, to the left of the person is a height indicator, with a (horizontal to the right) number 14, representing the number of transmission frequency wavelengths tall the person is (the transmission was at 12.6 cm, so the person is a bit over 5′9″).

To the right of the person is a (horizontal to the right and down) representation of the population of Earth, again about 4.3 billion (the two numbers are not identical in the message, and both are rendered to far greater precision than they were known at the time).

In another callback to the Pioneer plaques, below the person is a schematic representation of the solar system, with the Sun warranting 9 bits, the gas giant planets 3, the ice giants 2, and the terrestrial planets and Pluto 1 each. Earth is displaced vertically, directly under the human.

The bottom of the message depicts the parabolic dish of Arecibo, along with the light ray path emerging from the focus, reflecting from the surface, and being sent in a collimated beam away from Earth and the human. A width indicator shows the

[8] The formulae are usually rendered in proper chemical order as $C_xH_xN_xO_xP_x$; I order them as they appear in the message for ease of comparison. Astute readers will note the bases lack one hydrogen from their usual formulae, and that deoxyribose is usually rendered $C_5H_{10}O_4$. The message contains these substances' properties as they exist within DNA, as opposed to their free forms. Deoxyribose forms bonds with the phosphates above and below it and a base, all three of which replace hydroxyl groups, thus accounting for the missing $(OH)_3$. Similarly, the bases give up one hydrogen in the bonds.

diameter of the telescope (horizontal to the *left*) to be 2430 units long, or 306 m. This allows the message recipient to deduce the power of the transmission as a function of the distance to Earth.

Like the *Pioneer* plaques, the Arecibo Message was primarily aimed at humans as Staff at the National Astronomy and Ionosphere Center (1975) explained:

> As the choice of frequency, duration of message, and distance of the target clearly shows, the Arecibo message is very unlikely to produce interstellar discourse in the foreseeable future. Rather it was intended as a concrete demonstration that terrestrial radio astronomy has now reached a level of advance entirely adequate for interstellar radio communication over immense distances.

In fact, according to Drake (Sagan 1978), the signal was not actually sent to the core of M13 because their pointing failed to account for the proper motion of the cluster.[9]

19.2.4 The Golden Records

The success of the *Pioneer* probes led NASA to launch another pair of probes, *Voyagers* 1 and 2, to explore the outer solar system. Both performed flybys of Jupiter and Saturn (and Voyager 2 would go on visit Uranus and Neptune) before leaving the solar system.

The *Voyager* project manager, John Casani, asked Carl Sagan to lead a committee to produce a more sophisticated version of the Pioneer Plaques with many parts and much more information. Frank Drake had the idea that the form of the message should be an LP phonograph record, because of their robustness and high information content density (and because it was the hundredth anniversary of their invention by Edison) (see Figure 19.4).

To ensure its longevity in interstellar space, the record was a copper positive master (as opposed to the usual vinyl copy of commercial records) plated in gold. Each record was then placed in a protective aluminum case, also plated in gold, and a cartridge and stylus were bracketed to the spider support on which the cover is mounted (Figure 19.5).

[9] The M13 cluster moves across the sky at around 4 mas/yr, and is about 22,000 lt-yr away, so will move by around 90″ in one light travel time, which is around the beamwidth of Arecibo at the wavelength of the message. Because we see the cluster where it *was* one light travel time ago, and need to send the signal to where it *will be in reality* (not where it will *appear* to be), one needs to apply this correction to the apparent position of the cluster *twice*, so the beam was mispointed by 180″. The half-light radius of M13 (corresponding to roughly half of its stars) is 90″, so the signal will miss most of the stars in the cluster. However, the cluster has stars extending to over 500″ from its center, so the signal will pass by at least some of its members.

Figure 19.4. The Voyager Golden Record itself. The author obtained this image from the Wikimedia website, where it is stated to have been released into the public domain.

The content of the record comprised four parts: the first 4 minutes contain audio of greetings in 55 languages, the last of which is English (spoken by Sagan's son Nick, on behalf of the Earth's children).

The second part is "sounds of Earth" containing a mix of music, nature sounds, and the sounds of technology of humans. The human sounds include Sagan laughing, and a sped-up sonification of an electroencephalogram (an EEG, which measures brain waves) of one of the message designers, Ann Druyan. This is followed by a 90-minute selection of music from around the world.

The next section of the record is a collection of digital images. Following the spirit of the Arecibo Message, this section begins with a primer on counting and units so that scale bars on subsequent images can be interpreted; DNA is explained in some detail; and Arecibo is shown. Some of the images in this section are shown in Figure 19.6.

The case (Figure 19.5, bottom) incorporated design elements from the Pioneer plaques, including the pulsar map (also repeated in the images section) and the hydrogen atom, which sets the timescale for two design elements that explain the

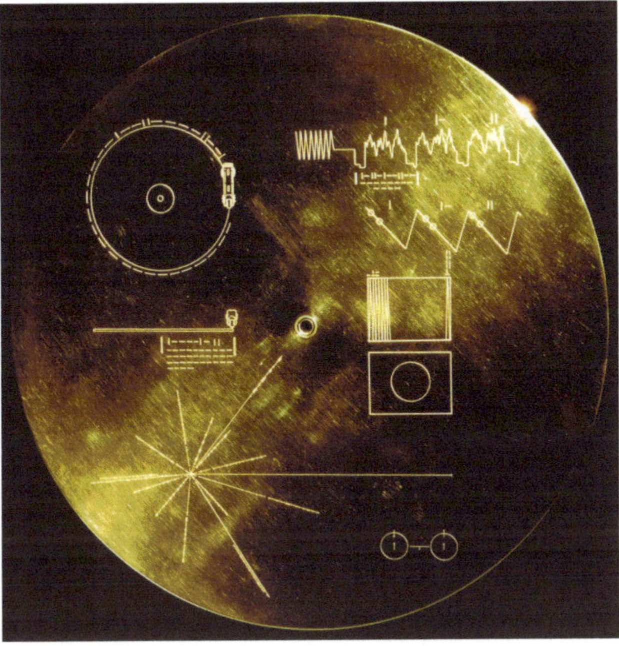

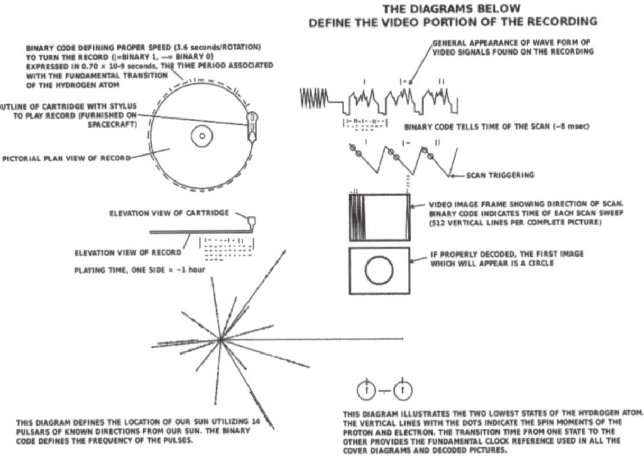

Figure 19.5. *Top*: The Voyager Golden Record cover. *Bottom*: Explanatory diagram of the cover components. The author obtained this image from the Wikimedia website, where it is stated to have been released into the public domain.

speed at which the record is to be played. The cover also shows schematically how to use the included cartridge and needle to play the record, and how to interpret the digital imagery section of the record, including a representation of the first image (a circle) to ensure the aspect ratio is correct.

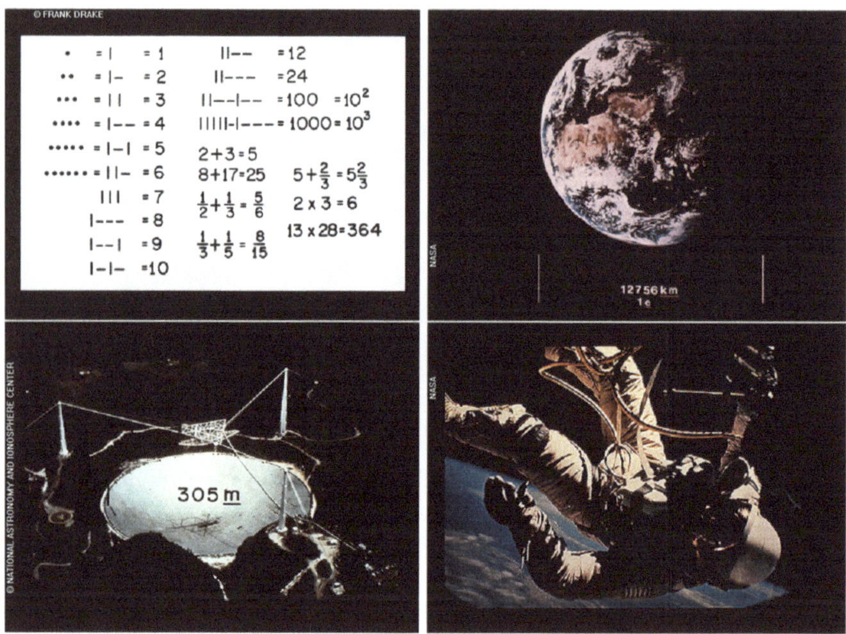

Figure 19.6. Four images on the images portion of the Voyager Golden Records. The first is one of several images explaining Arabic numerals and our units of measurement. Credit: NASA, except upper left with credit to Frank Drake.

Finally, a 2 cm diameter region on the case was electroplated with a layer of ^{238}U, which has a half life of around 4.5 billion years. The idea is that this material would slowly decay into the stable isotope of ^{206}Pb over billions of years, meaning one could determine the time since the plating took place from the ratio of these two elements.

As with the Pioneer plaques, there was little expectation that these records would ever be noticed by extraterrestrials. As Sagan wrote in Sagan (1978), describing the process of compiling the album contents:

> Many of the consultants emphasized that receipt of the message by an extraterrestrial civilization was chancy at best, while its receipt by the inhabitants of Earth was guaranteed: the public would eventually have access to the message contents, as is in fact accomplished by this book. As [consultant Barney] Oliver put it, "There is only an infinitesimal chance that the [message] will ever be seen by a single extraterrestrial, but it will certainly be seen by billions of terrestrials. Its real function, therefore, is to appeal to and expand the human spirit, and to make contact with extraterrestrials intelligence a welcome expectation of mankind."

In this the records were quite successful.

Box 19.2: Voyager Golden Records

Credit: NASA

The story of the making of the *Voyager* Golden Records (Section 19.2.4) is documented thoroughly in *Murmurs from Earth: The Voyager Interstellar Record* (Sagan 1978) by Carl Sagan and some of the other team members, Frank Drake, Ann Druyan, Timothy Ferris, Jon Lomberg, and Linda Salzman Sagan.

They report that the entire process felt very rushed and many compromises and last-minute changes had to be made. The record was actually designed to be played at 16 2/3 revolutions per minute (half the speed of typical LPs) so its audio portion could be twice as long (at the expense of some audio fidelity).

Sagan quickly regretted contacting the US State Department and United Nations to collect greetings in many languages at the United Nations—the exercise quickly devolved into a mess of both domestic and international diplomacy.

It was Ann Druyan who had the idea for including sonification of brain waves in the audio portion: "Would it be possible, I wondered, for a highly advanced technology of several million years from now to actually decipher human thoughts?" Two days before the recording, Sagan called Druyan and proposed marriage (they had not even dated at this point; Druyan and Sagan were married from 1981 until his death).

Druyan later said "My feelings as a 27 year old woman, madly fallen in love, they're on that record…It's forever. It'll be true 100 million years from now. For me *Voyager* is a kind of joy so powerful, it robs you of your fear of death."

In the music portion, the contributions from the United States are songs by Chuck Berry, Louis Armstrong, Blind Willie Johnson, and a Navajo Night Chant.

Sagan (1978) wrote that after word of the production process reached the press, "Very large numbers of people wrote to ask if a commercially available version of the record might be in the offing. We are still hopeful that such a nonprofit two-record album may be made…" In 2017, a group of enthusiasts under the label Ozma Records produced a re-creation of the records as a three vinyl album set available for purchase.

NASA's Jet Propulsion Laboratory maintains a website with many details and examples of the Golden Records' contents at https://voyager.jpl.nasa.gov/golden-record/.

19.3 Critiques of the Messages

19.3.1 Interpretibility

The interpretability of the above messages to an alien species is still largely unclear. For instance, they all presume that ETIs will scrutinize symbols meant to be interpreted in a 2D plane the way humans interpret words or pictures on a page (or cave wall). Would this be obvious, or even intelligible, to an alien species?

There is a widespread expectation that mathematics, and possibly some principles of engineering, are universal, an assumption we explored in Chapter 5. Quite aside from philosophical questions about whether mathematics is purely a human invention or "really" exists, certainly our approach to calculation and algebra is entirely specific to (certain cultures of) humans. Might ETIs have entirely different abstractions and heuristics for doing radio engineering calculations?

Since the purpose of the messages described in this chapter was largely to inspire further work on the topic and begin the conversation and study of what a truly universal message could or would look like, their universal interpretibility is not essential to their function, except to the degree it inspires more work on the topic.

19.3.2 Accuracy and Choice of Message

Sagan and Drake reported that some members of the public objected to the nudity depicted on the Pioneer plaques, and were "denouncing NASA for using taxpayers" money to send "smut" into space." (Sagan 1978) NASA informed Sagan that the pictures on the Voyager Golden Records could not contain anything that could be considered unnecessarily salacious, and an image the team chose depicting a nude couple was not included. "'There was no way that NASA was going to launch full frontal nudity to the stars." (Sagan 1978)

On the other side, the Pioneer plaques were critiqued for not being sufficiently explicit, in particular that the woman on the plaque is shown without any genitalia.[10] Others objected that the woman seems to be in a subservient position to the man, who gets the privilege of a greeting and stands straight and tall, while she stands contrapposto beside him. Linda Salzman Sagan, the artist, explained that the choices of pose and detail were inspired by classical Greek statuary, which made similar choices.

Carl Sagan also reported that the hair and features of the figures were criticized for being unrepresentative of humanity, despite their attempts to make them appear "panracial."

[10] The lack of proper biological representation for the woman inspired artist Joe Davis to produce *Poetica Vaginal*, a conceptual art piece in which he constructed a message consisting of analog audio, digital audio, and "English phonetic maps" of sounds recorded using a "vaginal detector," which was self-inserted by female volunteers so it could record their voices, heartbeats, and vaginal contractions. Davis claims to have sent 20 minutes of the message to τ Ceti and ϵ Eridani (targets of Project Ozma) before the Air Force, which operated the Millstone Hill Radar used for transmission, shut it down upon learning what he was doing (but see Zaitsev 2012, who is skeptical that any meaningful transmission took place.)

19.3.3 Propriety: "Who Speaks for Earth?"

> There are other species on Earth, doctor. Only human arrogance would assume the signal must be meant for mankind.
>
> — Spock, *Star Trek IV: The Voyage Home*

One of the most important debates in METI is "Who Speaks For Earth?" and the propriety of the information content of any revealing message.

Many have objected that Drake, Sagan, and others presume to speak for all humans and represent us without permission and without being truly representative. Sagan's efforts to engage the United Nations in the Voyager Golden Records were mostly limited to collecting audio "hellos" in many different languages, and did not constitute a genuine effort to have global input on the message design.

The idea that messages to ETI should be composed by some sort of inter-governmental or otherwise globally-representative body was articulated by Drake and others with respect to the Arecibo message:

> More extensive attempts at the transmission of radio messages from the Earth to extraterrestrial civilizations should be made only after international scientific consultations as recommended by the first Soviet–American conference on communication with extraterrestrial intelligence. (Staff at the National Astronomy and Ionosphere Center 1975)

The Declaration of Principles regarding post-detection protocols (which we will discuss in more detail in Section 20.3) state that any *response* to an alien signal should be made only with "guidance and consent of a globally representative body." In 2010 the Principles were revised and streamlined, and as part of this process it was proposed that this principle be expanded to include pre-detection (METI) activities as well. This change of scope of the principles was not incorporated into the final 2010 version, and as a result several members of the IAA permanent committee on SETI resigned in protest, including John Billingham (Sections 3.3.2 and 20.4).

There are a small number of vocal proponents for conspicuous METI. Most notable is METI International, an organization founded in 2015 by American psychologist and astrobiologist Douglas Vakoch to foster and conduct interdisciplinary research in METI and SETI internationally, and to do public outreach on the topics and the long-term future of humanity. METI International holds workshops, funds optical SETI programs, and occasionally sends signals to nearby stars (such as transmissions in 2017 to Luyten's Star). As the most visible organization promoting METI, it is a frequent target of METI critics (Section 19.4.1).

Hatfield & Trueblood (2021) analyzed the options for how a representative message might be constructed, conducting a poll of 2000 people in the United Kingdom about whether they thought a signal should be responded to, and whether they thought the response should be crafted by scientists, politicians, a random assembly, or global referendum (democratically). They found a majority thought we should send a response, and a plurality preference for scientists composing it.

At any rate, it seems unlikely that, in the absence of a confirmed detection of ETI to send a message to, the issue of an appropriate message will rise to the level of importance that the United Nations or other globally representative body would take the problem up in a rigorous way. Nonetheless, the problem is worthy of study and consideration, if only so the groundwork for a solution is laid in the event a target of such a message is found.

Many of these issues presume that the content of the message is all that the ETIs can learn about Earth. The fact that Earth and its biosignatures have been around for billions of years, and that we are *also* broadcasting our radio and television signals containing a great deal of information about humans and Earth, means that it is possible that ETIs *already* know a great deal about us, and that the messages we send will be in some ways superfluous.

19.4 The Potential Danger of METI

19.4.1 Danger

The strongest and perhaps most famous objections to these messages, and indeed METI in general, is that they are dangerous because an alien species might be hostile and harm humanity, and that METI transmissions might trigger such a response. These objections are primarily to *conspicuous and revealing* signals that reveal humanity's existence to a species that did not previously know about it.

Such concerns are as old as the first messages to space. Drake recounted that there was backlash after the sending of the Arecibo message that "we had exposed humanity as a whole to extraterrestrials of possibly evil intent, who might now prey on us" and that people were worried that "for all we know, any creatures out there were malevolent or hungry, and once they knew of us, they might come to attack or eat us."[11]

Such scenarios are common, of course, in science fiction, and this danger forms a central plot point of the Liu Cixin novel *The Three Body Problem*. We explored some rationales for why an alien species might be hostile in Section 6.6.3. Gertz (2016a) called METI "unwise, unscientific, potentially catastrophic, and unethical." We have also seen how these concerns are amplified by popular historical accounts of human history informed by colonialism (Section 18.4) and technological progressivism (Section 18.5), and contradicted by attitudes stemming from social progressivism (Section 18.5).

In 2015, 26 prominent scientists and others, including leaders of the Breakthrough Listen project (Section 3.5.1) and board members of the SETI Institute, published a statement arguing against METI efforts on several grounds, including the presumption to speak for Earth, the potential danger of making ourselves known, and the incidental damage to passive SETI efforts from vocal opposition to METI efforts (Box Box 19.3). They argued for passive SETI efforts, writing "it is prudent to listen before we shout."[12]

[11] These concerns, echoing the fate of humans at the hands of the Martians from *The War of the Worlds*, were supposedly expressed by British radio astronomer Martin Ryle. However Charbonneau (2024) has shown these are mischaracterizations of Ryle's concerns, and so are likely either exaggerations or misattributions.
[12] Original at https://setiathome.berkeley.edu/meti_statement_0.html.

Box 19.3: Statement Regarding METI/Active SETI

Messaging to Extraterrestrial Intelligence (METI), also called Active Searches for Extraterrestrial Intelligence (Active SETI), refers to the act of using high power communications equipment on Earth to transmit various messages to unknown extraterrestrial intelligences.

METI has been conducted sporadically in the past, but recently a surge of individuals and organizations have initiated or suggested new METI programs, both academic and commercial in nature. METI programs carry unknown and potentially enormous implications and consequences. We feel the decision whether or not to transmit must be based upon a worldwide consensus, and not a decision based upon the wishes of a few individuals with access to powerful communications equipment. We strongly encourage vigorous international debate by a broadly representative body prior to engaging further in this activity. We also note the following:

- ETI's reaction to a message from Earth cannot presently be known. We know nothing of ETI's intentions and capabilities, and it is impossible to predict whether ETI will be benign or hostile.
- Because we have just recently (in cosmic terms) attained an interstellar communications capability, it is likely that other communicative civilizations we encounter will be millions of years more advanced than us.
- The Search for Extraterrestrial Intelligence, the scientific effort to determine whether or not other advanced life exists in the Universe, is still in its infancy. Even though SETI experiments have been ongoing for more than 50 years, scientists are still decades away from completing a comprehensive search for radio signals similar to those produced by our own technology. As a newly emerging technological species, it is prudent to listen before we shout.
- Although a nearby advanced ETI may have already picked up Earth's omni-directional radio leakage, e.g., early television transmissions, or the presence of industrial wastes in the atmosphere of the Earth, such detections are far more difficult than detecting a focused radio or optical signal sent from a large telescope.
- It is not necessary to actually transmit powerful electromagnetic signals in order to study interstellar communication from the perspective of the transmitter, or to develop transmission techniques that might one day be used to respond to a message received from an ETI.
- Opponents of METI would vocally condemn METI transmissions, confusing the public about, and imperiling funding for, bona fide scientific endeavors related to extraterrestrial life.

Intentionally signaling other civilizations in the Milky Way Galaxy raises concerns from all the people of Earth, about both the message and the consequences of contact. A worldwide scientific, political and humanitarian discussion must occur before any message is sent.

[See Section 19.4.1. Original here.]

19.4.2 Danger in the Pioneer Plaques and the Voyager Messages

Traphagan (2021) argued that the Golden Records were "naive, biased, and misleading" and a "lie" because they featured, "no images of war or even weapons of war…poverty, suffering, starvation, overpopulation, shanty towns, industrial pollution…colonization and exploitation of indigenous peoples. Although there is an image of a gymnast and of runners, there are no images of…any sport that involves direct physical conflict."

One of Traphagan's concerns was that if aliens were to arrive and see that Earth is not as depicted on the records, they would conclude we were lying and being deceptive, and then "they might respond quite negatively and vaporize the planet." Traphagan also objected to the Golden Record the pulsar maps on the Voyager Golden Record covers and Pioneer Plaques: "If ET turns out to be belligerent or imperialistic, then the contents of the Voyager records may well send a very clear message: *there is easy prey on that planet situated here in relation to these 14 pulsars.*"

Sagan actually explained their reasoning for this in Sagan (1978):

> A few writers also criticized us for presenting only the favorable circumstances of mankind and chided us for not including scenes of famine, devastation, ravaged cities, and nuclear weapons explosions. This is an issue we debated long and hard during our deliberations on repertoire. There is no question that destruction is a characteristic aspect of what we are pleased to call human civilization. But such a message might be misinterpreted. Might a photo of a thermonuclear explosion be considered by an extraterrestrial civilization as a pathetically feeble but still palpably nasty attempt at threat?...Why not a hopeful rather than a despairing view of humanity and its possible future?

19.4.3 Pascal's Wager

Many of the arguments for and against conspicuous METI take the form of Pascal's Wager, a famous but fallacious argument for having faith in God.

French philosopher and mathematician Blaise Pascal (of Pascal's triangle fame) argued that it was irrational not to believe in God because there was infinite upside to such a belief (eternal life in heaven) but no downside: if God does not exist, then death would be the same state of nonexistence whether or not one believed in him. Nonbelief, by contrast, had no upside at all, and risked the infinite downside of eternal damnation.

This argument is often summarized in a matrix like this:

	Belief	Disbelief
God does not exist	nothing	nothing
God exists	∞ good	∞ bad

Pascal's Wager is fallacious for many reasons. One is the false choice it presents, which becomes clearer when we clarify that Pascal was Catholic and was advocating for belief in the Christian god. It is more precise then to express the wager this way:

	Belief in Christian god	Disbelief in Christian god
God does not exist	nothing	nothing
Christian god exists	∞ good	∞ bad

We now realize that we are missing possibilities in our grid. For instance, one might receive eternal damnation not just for atheism, but for belief in the *wrong* god (or even the wrong Christian denomination!). We might then expand our grid to accommodate more possibilities[13]:

	Belief in Christian god	Belief in God #2	Belief in God #3	Disbelief in god
God does not exist	nothing	nothing	nothing	nothing
Christian god exists	∞ good	∞ bad	∞ bad	∞ bad
God #2 exists	∞ bad	∞ good	∞ bad	∞ bad
God #3 exists	∞ bad	∞ bad	∞ good	∞ bad
. . .	?	?	?	?

If the choice is as stark as Pascal claimed, then it is not just important to have faith, but to correctly guess which faith to hold. The bottom row here represents the many other possibilities, including a god that provides eternal bliss or damnation or other afterlife experiences in ways not connected to one's faith.

Pascal's wager then, at best, argues that one should have *some* kind of faith in hopes that it is the kind of faith that, at a minimum, avoids damnation.

[13] This expanded grid is by no means comprehensive, of course. For instance, there could be eternal damnation or bliss based on something other than belief in God, or even in the absence of God.

The conspicuous METI question can similarly be phrased in terms of a Pascal's-Wager-like grid like this:

	Transmit	Do not transmit
Aliens are hostile	Very bad	Nothing
Aliens are not hostile	Contact	Nothing

Here, the choice is not even as stark as it is for Pascal's argument for faith. The case to transmit offers two possibilities: cataclysm, and some other form of contact, whereas not transmitting has no consequence at all (or, to be more precise, no consequence *that would not happen anyway*).

As with Pascal's Wager, we need to acknowledge that there are other possibilities. We might expand them like this:

	Transmit	Do not transmit
Aliens notice and are hostile	Very bad	Nothing
Aliens notice and are neutral	Contact	Nothing
Aliens notice and are benevolent	Very good	Nothing
Aliens do not notice	Nothing	Nothing
Aliens do not exist	Nothing	Nothing

In particular, we need to consider the possibility that there might be extremely high upside to contact from benevolent aliens. Of course, there are many more boxes we could fill in to make the choice even less clear (e.g., there might be high downside even to benevolent aliens, or downside to transmissions because of reactions of other Earthlings completely independently of whether aliens exist). Regardless, expressed this way, it's clear that the METI question is not even as clear-cut as Pascal's (fallacious) question: we have no objective way to weigh the potential upsides and downsides of contact.

For instance, cataclysmic outcomes probably require physical contact with the alien species, which requires them to physically travel to Earth. This might be so impractical as to be impossible or simply not worth the effort for even the most hostile alien species, depending on the distance and amount of mass involved. High upside, on the other hand, might simply come from the information we receive from the other species, which might come as soon as the message is received.

19.4.4 Risk and the Precautionary Principle

Another argument used to discount concerns that METI is dangerous is that the chances of harm are so remote as to be not worth worrying about. The rebuttal to this argument is the Precautionary Principle, which is an approach to risk that says that in the face of high uncertainty about the harms of an action, one should be extremely cautious.

In many contexts, risk is defined as likelihood times severity (or "impact"; see Section 20.6). The argument is that even very unlikely outcomes are worth worrying about if the consequence is severe enough; the classic example is asteroid impacts in the context of the Torino Scale (see Figure 20.3). Figure 19.7 shows a schematic of a *risk matrix*, a method of qualitatively assessing relative risks.

The colors here represent how hard one should work to reduce a given risk. One cannot eliminate *all* risk in any activity, but it is obviously more important to prevent severe or likely things than unimportant or very unlikely events. Risk matrices play many different roles in risk management, including helping determine the allocation of risk mitigation resources and identifying which activities one should participate in.

The exact placement of the black lines that demarcate different levels of caution is subjective and depends on how the risk matrix is being used. For instance, the degree to which one finds a risk acceptable may depend on a cost-benefit analysis or involve looking at opportunity costs involved in *not* taking a particular risk.

The color coding of risk matrices is standard but the labels here are not. Essentially, green means that the risk is low enough to not require substantial resources to address. Orange means the risk should be mitigated if possible, and red means the risk is so high as to be unacceptable, and so requires mitigation or retirement.

It is important to note that the lower right corner is color-coded green, not red. This is because there is some minimum probability below which one must be willing to tolerate risk or one could not function, and some even lower risk level where the

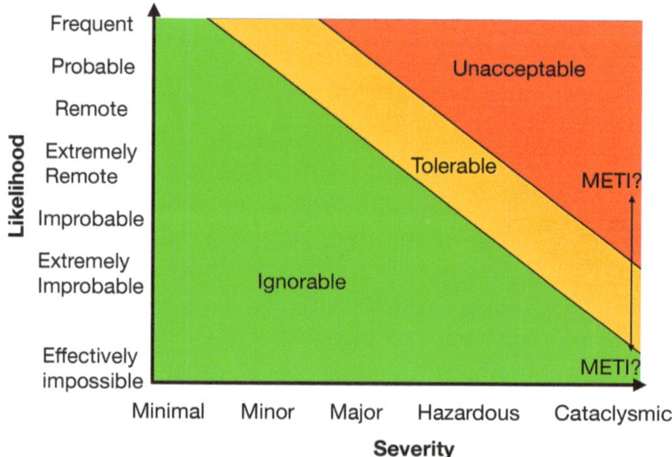

Figure 19.7. A schematic risk matrix. Risk matrices help qualitatively identify which risks to address, and are often discrete with 2–5 rows and columns. The Torino Scale (Figure 20.3) is an example of a risk matrix. The diagonal lines roughly track regions of equal risk, defined (qualitatively) as likelihood times severity. Concerns that METI will lead to humanity's destruction occupies some point on the right axis; part of the METI debate regards how for down the y-axis it belongs; another part of the debate is where the black lines should lie.

probability is so remote it can be ignored. For instance, driving a car, taking a plane flight, crossing the street, or even just eating some solid food, all come with some small amount of mortal danger, but are not considered unacceptable risks by most people.

If the lower right corner here were red, one would be so cautious that one could not do *anything* if one could imagine *any* reason it might be mortally dangerous. This would lead to a situation where anyone could propose any reason an action could be disastrous—no matter how improbable—and put an end to the activity via the Precautionary Principle (a "worrier's veto").

We also move the black lines up and down depending on how inconvenient avoiding a risk would be. If it costs nothing to avoid a risk, then even a small amount of the risk might be unacceptable. On the other hand, if the risk is very expensive or very inconvenient or impossible to avoid, then even high risks might be tolerable. For instance, much solid food comes with some risk of choking, but this does not prevent most people from eating it; driving cars comes with risk but driving is so useful that most people find the risk to be worth it.

The Precautionary Principle, in this context, effectively says that if one does not know the likelihood or severity of a bad outcome from an activity, then one cannot tell if the risk is ignorable, tolerable, or unacceptable, and so one should avoid the activity out of an abundance of caution. Of course, application of the principle cannot be strict and universal because one can never know with certainty what the severity and likelihood of a bad outcome is, so one can only apply the principle in a subjective way.

Because the downside of a hostile alien invasion is in principle maximally bad—leading to the extinction of all of humanity—it follows that even a very small probability of its outcome must be minimized. Since there is very little cost to not transmitting METI signals (except the potential opportunity cost from not getting a beneficial response), one might move the black lines in the figure quite low.

In Figure 19.7, the dangers of METI would occupy a spot on the far right of the grid—much of the debate over the existential dangers of METI is effectively about where it lies, and about where the line between tolerable and unacceptable lies.

19.4.5 Would METI Really Trigger an Alien Attack?

> "It's just men and ants. There's the ants builds their cities, live their lives, have wars, revolutions, until the men want them out of the way, and then they go out of the way. That's what we are now—just ants. Only—"
> "Yes," I said.
> "We're eatable ants."
>
> — H. G. Wells, *War of the Worlds*

The fear that our transmissions might attract the attention of malevolent aliens, or that aliens would wage war on humans, has obvious origins, some of which we explored in Section 18.3. One of the earliest portrayals of aliens in popular culture, Wells's Martians from *War of the Worlds*, came with malevolent intent after observing Earth. In that work, Martian civilization is vastly older and more advanced than ours, and alien war machines come to Earth because it is more hospitable than their dying world. Although humans have some small victories against the superior Martian military technology, they are ultimately completely defeated, and the Martians begin collecting people for their own sustenance. Wells's twist is that it is ultimately Earth microbes that defeat the invaders.

In the final chapter, after the Martians are defeated, Wells's protagonist warns "We have learned now that we cannot regard this planet as being fenced in and a secure abiding place for Man; we can never anticipate the unseen good or evil that may come upon us suddenly out of space."

Since then, alien invasion stories have remained a staple of widely popular science fiction, including *Independence Day* (in which it is a *computer* virus that allows human militaries to defeat the otherwise indestructible alien war machines) and Liu Cixin's *Remembrance of Earth's Past* series including *The Three Body Problem*. In the latter, a SETI scientist sends out a signal to contact aliens, only to find that the recipients are hostile (in the book, the scientist actually *invites* the aliens to invade Earth). This work also widely popularized the Dark Forest Hypothesis (Section 6.6.3), that the Galaxy is an extremely dangerous place in which any "loud" species are inevitably attacked by their neighbors.

Of course, an idea does not lack merit simply because it is a staple of science fiction (after all, science fiction has gotten a lot of things right about humanity's future!). It bears remembering in the METI debates, though, that the standards for plausibility in fiction are much lower than in reality, and that science fiction narratives are fundamentally about making exciting and appealing stories about humans, and not portraying the likely outcomes of actual alien encounters (Section 18.3).

To be specific about the dangers of METI and make them debatable, we need to identify what is dangerous about them, specifically what the signal reveals (e.g., our planet's existence, our species' existence, etc.) that would cause them to react in a way that harms us. That is, to be dangerous, not only must the METI signals succeed in signaling something new about our existence that Earth's biosignatures and radio leakage have not already revealed, and not only must the aliens happen to notice and correctly interpret the signal, but they must react in a way that harms us, for instance by physically coming to Earth with military force.

Much of the reasoning for *why* they might attack follows from perceived tendencies from human history. We explored projections of colonialist mindsets onto SETI in Section 18.4, including the idea that alien species would seek to colonize Earth and subjugate Earthlings just as Europeans did to the peoples of the New World. Other perspectives are that Earth would have resources needed by aliens which they would take by force (although it's unclear what resources that would be worth interstellar travel and warfare would be revealed by a METI signal.)

It is also often presumed that if an alien invasion force arrived it would likely or necessarily be so militarily superior to humanity that humans would have no chance to defend themselves. For instance, Gertz (2016a) writes "We are only two thousand years more advanced than Rome, yet the best Roman legion could be annihilated by a modern army in mere minutes. Never mind two thousand years ago, Napoleon's armies of barely two hundred years ago would face the same fate."

As we discussed in Sections 4.2.3 and 18.5, while it is certainly *possible* that if aliens were to arrive that we would be powerless against their military technology, we should be careful about presuming so. Certainly their spaceflight technology would need to be quite powerful, but it does not follow that they would have sufficient numbers or ammunition to subjugate or destroy humanity.

Another concern is that the presence of even neutral or benign ETIs could harm us in unintentional ways, much like smallpox and other diseases ravaged the New World population after the arrival of Europeans. While actual interstellar diseases might be unlikely, the presence of ETIs on Earth might cause major disruptions culturally or economically, depending on the nature of their interactions. Traphagan (2021) argued that even the perception that signal might be detected could be enough to start major geopolitical troubles—actual contact could be much more disruptive.

Finally, Carrigan (2006) and Hippke & Learned (2018) have suggested that physical contact might not even be necessary for an ETI to harm Earth, and that the danger might arrive via an ordinary SETI message containing some sort of dangerous content. For instance, if the message described instructions for creating a computer program that could do important calculations, or to build a machine that could accomplish a certain task, that program or machine might prove to be dangerous. (In the film *Contact* (Box 18.1) the alien message (spoiler alert) includes instructions on how to build an enormous machine of uncertain purpose that turns out to have significant (nonmalicious) destructive power).

19.4.6 Common Counter-arguments to METI Regulation

There are several counter-arguments commonly deployed to those who argue for regulation of radio METI activities.

One was well articulated by Drake (Sagan 1978) regarding objections to the Arecibo Message:

> The fact is, for better or for worse, we have already announced our presence and location to the universe, and continue to do so every day. There is a sphere of radio transmissions about thirty light-years thick expanding outward at the speed of light, announcing to every star it envelops that the Earth is full of people. Our television programs flood space with signals detectable at enormous distances by instruments not much greater than our own. It is a sobering thought that the first news of us may the be the outcome of the Super Bowl.
>
> Our radar transmissions similarly spread the signs of human activity to the far corners of the galaxy. Whether or not [critics of the Arecibo Message are] justified in [their] anxieties about revealing the location of our civilization is of course a debatable subject. Even so, it is too late to worry about it, so we might as well try to be friendly.

There are two essential arguments here. One is that signals like the Arecibo message are not likely to let aliens know we are here because if they can detect our signals, then they already know, so no harm can come from such signals. This is sometimes called the *Barn Door Argument*.[14]

There are several rebuttals to the Barn Door Argument. The first is that the entire point of conspicuous signals is to do something that makes the Earth more conspicuous than it otherwise is, and so one can either claim active SETI can be effective and conspicuous or make the Barn Door Argument, but one cannot have this cake and eat it too. If the Barn Door Argument is correct, then conspicuous METI is impossible. This rebuttal only works for those advocating for conspicuous signals, however. Many proponents of METI are more concerned with the information content of their signals than their conspicuousness.

A second rebuttal is that even if METI signals are not significantly more obvious than our leaked emission, they nonetheless contribute in aggregate to our global radio broadcasting. It is irresponsible, the rebuttal goes, to deliberately increase the chances of a catastrophic outcome, akin to purchasing some crude oil and lighting it on fire solely to (very slightly) hurry along catastrophic climate change.

But Drake is not just making the Barn Door Argument (he was not, after all, a proponent of effective METI). He is also, at the end, arguing that since we are going to be detected soon via our radio waves anyway, we should "try to be friendly" and be thoughtful about what messages we are sending. That is, the point of the

[14] After idioms along the lines of "locking the door after the horse has left the barn" or "shutting the stable door after the horse has bolted."

messaging is not to get attention, it is to establish friendly relations and begin diplomacy (or rather, to start a conversation *on Earth* about what that should look like). In that vein, the Arecibo Message, Pioneer Plaques, and Voyager Record are essentially ineffective practice runs to get the Earthbound conversation going in preparation for actual contact.

This is a strong argument for METI: that the Arecibo Message, Pioneer Plaques, and Voyager Golden Records have already succeeded in their purpose of illustrating issues surrounding and the importance of message composition and preparing for contact. All three messages have become quite famous, have begun a vigorous discussion in the literature, and have become staples of education and outreach on the topic of SETI. This would not have been true if the messages were merely composed and not actually sent. The sending of the messages is thus important because it increases their salience to *humans*!

In fact the possibility that ETIs have *already* detected us and are waiting only for an overture inspired the proposal of de Magalhães (2016) to embed invitations for contact into existing radio leakage. This nicely avoids concerns about conspicuous METI alcrting ETIs to our presence.

Another strong argument for active SETI—that is, for sending ostentatious, invitational signals to species that might not know we are here or know that we wish to initiate contact—is that it might initiate contact that could be extremely beneficial for humanity. Knowing and understanding that life exists elsewhere in the Galaxy is a major goal of astrobiology, and the discovery of technological life would be a profound one. To go further and actually make contact with such life would be even more so. The dangers are probably extremely remote, and the potential benefits are quite large.

19.4.7 Conspicuousness and the San Marino Scale

Almár & Shuch (2007) described a scale to rank the danger (really the conspicuousness) of METI transmissions. The scale is logarithmic in the strength of a transmission from Earth, normalized to the background radio emission of the solar system (for instance because of Jupiter or the Sun, both of which are broadband radio sources). A signal that has the same strength as the background is given a score of $I = 0$, and the scale increases by one for every power of ten in strength above this a transmission has.

To attempt to quantify the additional boost a signal gets from being transmitted for a long time or is targeted at a particular nearby star, they suggested an additional score C be added according to this schema:

Character of Transmission	Value of C
A beacon without any message in an arbitrary direction (e.g., interplanetary radar)	1
Informational message targeting an arbitrary direction for minutes or hours	2
Intentional, informational signal targeting a specific system	3
Continuous, omnidirectional, informational broadband transmission	4
Reply to a signal from an ETI unaware of our existence	5

The final score $S + I = C$. Almár & Shuch (2007) suggested that scores would rank "potential hazard," where a score of 1 would rank "insignificant," 5 would rank "intermediate," and 10 would rank "extraordinary."

In 2007, the IAA Permanent Committee on SETI adopted the San Marino scale, with the intention of continuing to improve it.

The San Marino scale has not found widespread use, and it is unclear how useful it is in its current incarnation for assessing the risk of transmissions.

19.4.8 What Should Be Regulated?

For the reasons above, Gertz (2016b) and others have argued that METI should be regulated, and effectively banned. Possible regulations come in a range of varieties, and would require the machinery of international law to manage.

An important question that would be confronted in any attempt at genuine METI regulation is exactly what kinds of signals are restricted.

If the purpose of the regulation is to avoid "shouting" into a potentially hostile Universe, then the intention of the signals is not really relevant, only their conspicuousness, so a regulation scheme focused entirely on METI transmissions may not be logically coherent. In this case, the appropriate restrictions are likely on *any* potential technosignature from Earth, but especially strong radio emissions. Zaitsev (2008) argued that interplanetary radar is probably more conspicuous than any METI transmission has been, and so would likely need to be severely curtailed.

Something like the San Marino scale would be needed to assess the risk associated with any transmissions of a given duration, EIRP, bandwidth, frequency, and pointing, but we do not know the locations or sensitivity of alien transmitters, or how often they might be pointed at the Earth. Regardless of its place on the San Marino scale, the persistent omnidirectional transmissions of the innumerable transmitters across the Earth might be much more conspicuous to hostile aliens than highly targeted and intermittent METI transmissions. So if one is following the Precautionary Principle one cannot assume that the San Marino scale accurately measures hazard, and then it is not at all clear which transmissions should fall below the threshold for regulation.

If the purpose of the regulation is to control the contents of messages to ETIs to ensure a few do not inappropriately speak for the entire planet, then a different set of regulations would need to be considered. In this case only effective and revealing messages would seem to be implicated (presumably superficial METI with practically zero chance of being detected would not be regulated). The boundary here is somewhat unclear: our television and radio broadcasts are information-rich and convey a great deal of information about humanity, but they are not very conspicuous and are encoded in a way that may not be decipherable by ETIs.

In this case, one would need to define some threshold of conspicuousness, intelligibility, and information-richness above which a transmission would need to be regulated or banned.

19.5 Universal Message Composition

The problem of effectively communicating with ETIs is a thorny one. The easiest solution would be a scenario in which they have already learned human languages and have solved the language barrier problem for us. METI presents the more challenging problem of how to craft a generic message that any technological species could decode and understand.

Below we discuss a few of the many attempts at how to construct such a message.

19.5.1 Attempts at Universal Human Communication

One proxy for technological ETIs might be future humans, and there has been some substantial work on how to communicate with them under the assumption that they would not understand any modern language.

19.5.1.1 LAGEOS

The Laser Geodynamic Satellites[15] are a pair of heavy (400 kg) 60 cm spheres launched in 1976 into medium Earth orbit. They are covered with highly reflective retroreflectors, which send optical or infrared light back to its origin, allowing for extremely precise distance measurements.

Their purpose is twofold. Because they are in extremely stable orbits, and they are heavy and not subject to significant radiation pressure, they can be used as position references on Earth. That is, because we can calculate their orbits precisely, one can use lasers launched from a position on Earth to determine one's position on the Earth very precisely (to within inches, in principle). These positions could then be used to track continental drift. A similar principle, using timing of radio signals from atomic clocks, underlies GPS positioning today.

The second purpose is that one can use ground stations of precisely known positions to track their orbits and understand the subtle effects of radiation pressure and gravitational perturbations on objects in Earth orbit. The LAGEOS satellites' orbits were chosen to be superlatively stable, and their expected lifetimes are around 8.4 million years.

Carl Sagan was asked by NASA to design a "time capsule" for the LAGEOS satellites, that might be recovered when they eventually crash into the Earth or when future humans discover the satellites in orbit (Figure 19.8). Based on the principles of the Arecibo Message and the Pioneer Plaques, it contains a primer on binary counting and a schematic to illustrate that units are expressed in Earth years.

Sagan decided to illustrate the date of the satellites' launch in two ways, using a numerical representation of the presumed time until the satellites' orbits decay, but also using continental drift as a clock. The plaque shows the organization of the Earth's landmasses around 270 million (2^{28}) years in the past, today, and 8.4 million (2^{23}) years in the future.

[15] While the satellite is certainly called LAGEOS, what these letters represent is inconsistent. It is variously known as the Laser Geodynamics Satellite, the Laser Geodynamic Satellite (on the plaque), and the Laser Geometric Environmental Observation Survey.

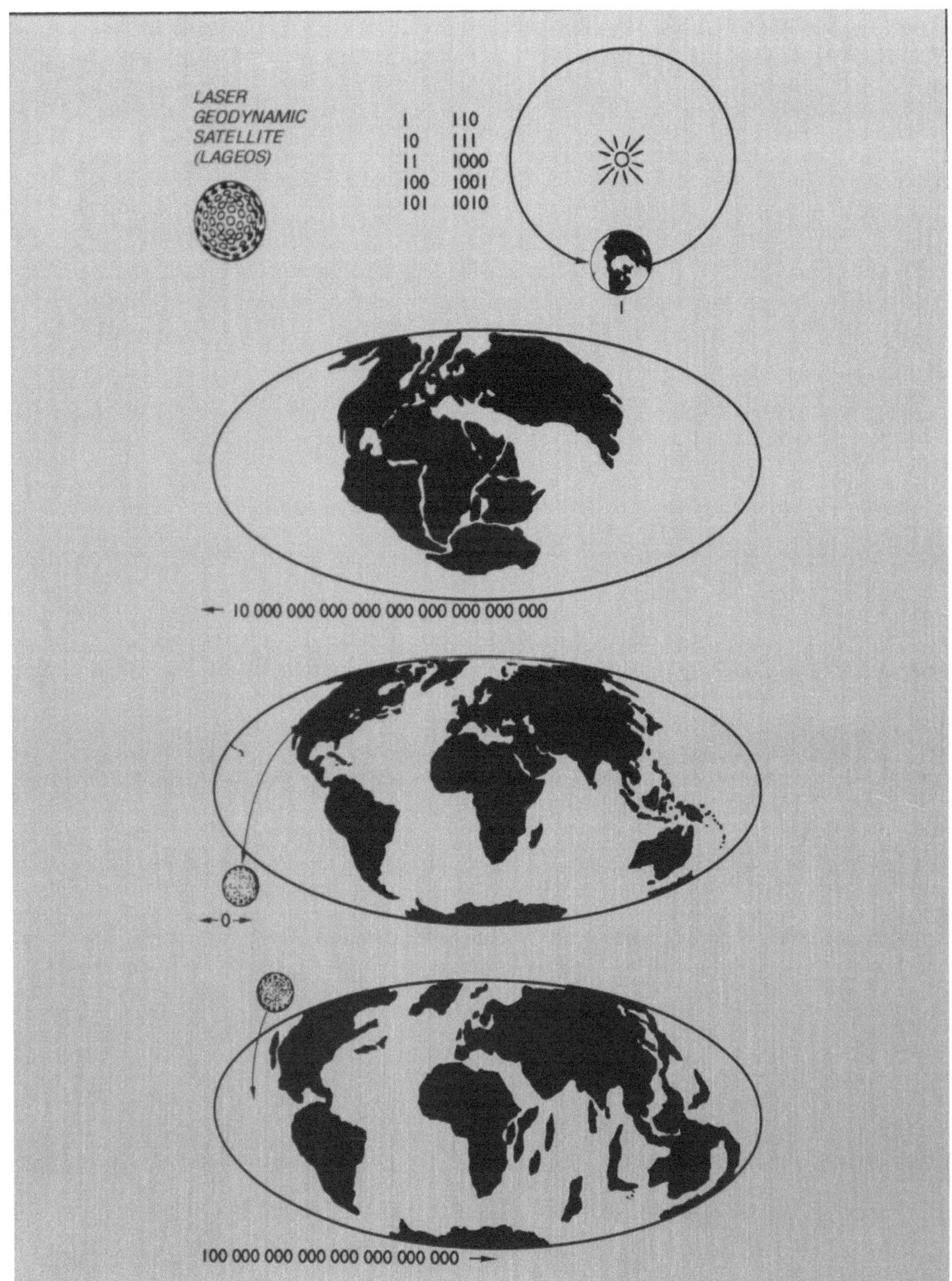

Figure 19.8. Plaque on the LAGEOS satellite, designed by Carl Sagan. Credit: NASA/Goddard.

19.5.1.2 Long-Term Nuclear Waste Warning Messages

The United States has several long-term nuclear waste storage facilities designed to safely contain nuclear waste underground over thousands of years. One concern is that future humans might come across these facilities and unearth the waste, perhaps thinking something valuable is being stored in them. The concern is not unfounded: looters and archaeologists have thoroughly explored, for instance, the tombs and temples of ancient Egypt in hopes of finding valuable treasure and information.

Linguists, engineers, and behavioral scientists have worked over the decades on the field of *nuclear semiotics* to design signs and signals that would outlast the English language and convey the danger associated with the sites. A 1993 report (Trauth et al. 1993) summarized the messages such signage should convey, including the now somewhat famous phrases

> This place is not a place of honor…no highly esteemed deed is commemorated here…nothing valued is here.
>
> What is here was dangerous and repulsive to us. This message is a warning about danger.

How to convey this information effectively has generated many suggestions. The message must be perceived as genuine and not a ruse to protect something valuable, and also must be unambiguous. Some suggestions have included "hostile architecture" such as fields of spikes that discourage coming near the site, and pictograms. One proffered sketch of such pictograms conveying the specific dangers is in Figure 19.9.

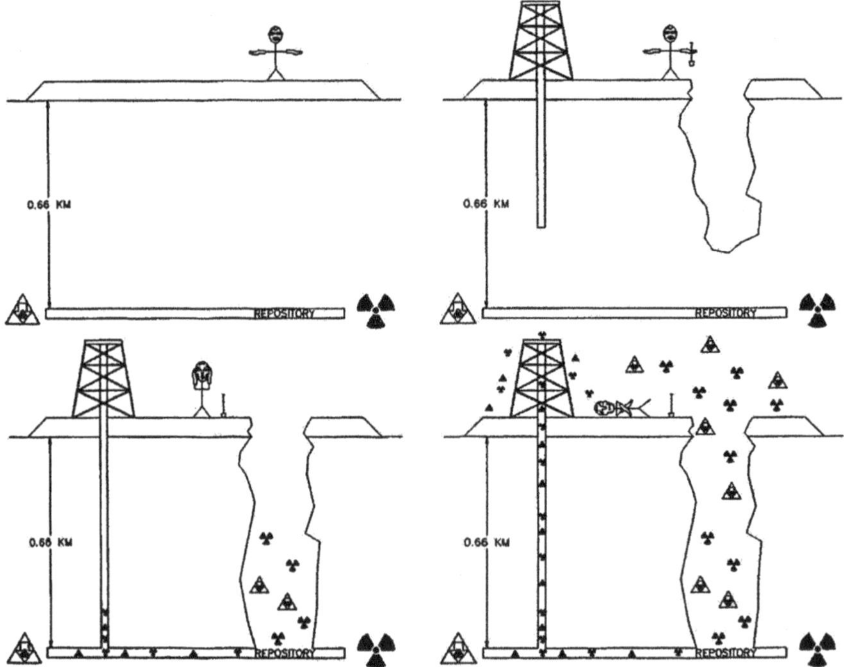

Figure 19.9. Sketch of a pictogram suggested to convey the dangers of digging at a long-term nuclear waste disposal site. The author obtained this image from the Wikimedia website, where it is stated to have been released into the public domain.

Sagan did not contribute to the effort due to scheduling conflicts, but offered this advice:

> ...we want a symbol that will be understandable not just to the most educated and scientifically literate members of the population, but to anyone who might come upon this repository. There is one such symbol. It is tried and true. It has been used transculturally for thousands of years, with unmistakable meaning. It is the symbol used on the lintels of cannibal dwellings, the flags of pirates, the insignia of SS divisions and motorcycle gangs, the labels of bottles of poisons—the skull and crossbones. (G-89 Trauth et al. 1993)

Sagan's logic might seem sound, but as Trauth et al. (1993) noted

> Graphics are likely to be culturally restricted in meaning. There are no conventional signs, such as the skull and crossbones, for example, that convey the same meaning across cultures.

with the footnote "In Mexico, the bones are the repository of the life force, and thus the skull and crossbones would have a very different meaning." Indeed, many might associate the skulls with el Día de Muertos and its attendant festivity and humor, conveying entirely the wrong message.

The report goes on to point out the problem has been confronted before:

> Representations of human faces and human and animal figures tend to be recognized for what they are, however, across cultural boundaries and millennia. For example, we have no trouble recognizing such figures in the Paleolithic cave paintings of Europe and in prehistoric rock carvings and rock shelter paintings in Africa, Australia, and the Americas. We can even recognize many of the activities in which the human figures in these paintings seem to be engaged. But why these representations were put there and what the beholders should infer from them are obscure and the subject of conflicting interpretations. Cross-cultural ambiguity of this kind is especially likely with the use of cartoons.

These efforts at communication to future humans convey some of the challenges of generic communication, even among members of the same species with identical forms of cognition and a common biology.

19.5.2 Terrestrial Interspecies Communication

Interspecies communication is a field that explores how messages are conveyed across terrestrial species, especially in a mutually beneficial way, and examples of such communication abound across the animal kingdom. Human-animal communication is an example of this, and such communication can take many auditory and gesture-based forms (as any pet owner can attest).

METI is generally more concerned with a higher level of communication that conveys specific and arbitrary messages, for instance as could be used to establish the exchange of scientific ideas or diplomacy. Such communication generally presumes the other side possesses human-like cognition that can abstract and understand the Universe as humans do, in a way that we presume many or most animals cannot.

The Order of the Dolphin (Section 3.2.4) was named based on the idea that cetaceans like dolphins might possess such a cognition, and have a language that humans could decode and participate in. Despite decades of work on the topic, however, our understanding of dolphin and whale languages is still highly incomplete, and we are not able to communicate with them the way that we hope to communicate with ETIs. One might imagine that the problem of communicating with ETIs, with whom we share no evolutionary descent at all, would be even more challenging.

There are two principles that can give us some hope for communication with ETIs, however. The first is that if they are receiving our transmissions (we will use radio as an example) then they must be able to build radio telescopes. This immediately gives us something to work with: they must understand, in some way, the principles of electromagnetic radiation and signal processing. From this, we can hope that there are sufficiently universal principles of mathematics and engineering, and that we could use those to identify common areas of understanding from which to build an intelligible language.

That said, it may also be that ETI cognition is sufficiently different from ours that their reactions to us and our messaging will not be what we expect or intend, even if we successfully communicate our message. For instance, Santana (2021) argued that ETIs might not be a society at all, but a "eusocial hive" or "superindividual," and that the best stance for humanity to take would be one of "defensive belligerence" to dissuade them of any hostile intentions.

At any rate, the lessons of our interactions with animals teach us that it is likely impossible to deduce how an alien species will react to us. For instance there is no universally proper response to hostility, even with terrestrial animals. Depending on the animal, the best action could be to play dead (for instance with a brown bear), run away (bees), stay perfectly still (some venomous snakes), or make oneself large and loud and prepare to fight back (black bear). Knowing the best message to send requires familiarity with the species, which we will not have in advance with an alien.

19.5.3 Lincos

In 1960, Dutch mathematician Hans Freudenthal, then a visiting professor at Yale, wrote *Lincos: Design of a Language for Cosmic Intercourse, Part 1*,[16] a book describing how to compose a message in an artificial language called *lincos*, a portmanteau of *Lingua Cosmica*, latin for "cosmic language."

[16] An anticipated Part 2 never materialized. Hans Freudenthal is well known for many contributions to mathematics, math education, and popular puzzles.

The message, which could be spoken, written, or transmitted via radio, is entirely artificial and formal. Freudenthal wrote that it his purpose was "to design a language that can be understood by a person not acquainted with any of our natural languages or even their syntactic structures." Freudenthal imagined the recipients would be "human or at least humanlike as to his mental state and experiences." "I should not know how to communicate with an individual who does not fulfil these requirements," he continued, "yet I shall not suppose that the receivers of my messages must be humans or humanlike in the sense of anatomy and physiology."

Freudenthal built on the ideas of the formal language of logicism, developed by Bertrand Russell and Alfred North Whitehead, which is used to formalize mathematics axiomatically. Freudenthal's language is the application of these ideas to arbitrary communication.

Freudenthal leaned on an analogy of how babies learn language by examples, so the book proceeds as a series of lessons that will allow a receiver to learn the language as they experience it. The first lessons introduce the language's symbols, then the natural numbers and relational operators, then the notion of binary representation of numbers. Later messages produce the tools of symbolic logic, then physics. Eventually, the language becomes a series of conversations among actors who discuss behavior, manners, ethics, and other social qualities.

Lincos proved very influential. Drake cites the book in Sagan (1978) as an inspiration for how to build a mathematical primer into a message to allow it to be interpreted, and would adopt the counting and binary methods Freudenthal described into the messages he composed. Lincos was also used as the basis of the messages of some METI transmissions called *Cosmic Call.*

19.5.4 Drake's Pictorial Message Composition

As the examples of METI in the previous section show, a common theme of message composition is that the message might contain a pictorial primer to teach the recipient symbols for abstractions such as binary numbers that can be used to convey more complex information. The Voyager Records, in particular, contain extensive instructional material to establish the meanings of units and numbers that annotate later photographs (see Figure 19.6).

Drake was particularly interested in the idea that the message should be pictorial. Freudenthal rejected the idea of sending images initially because he felt they would not be obvious how to decode, but Drake disagreed, writing

> About six months after the Green Bank meeting, it occurred to me that a powerful way to send unambiguous interstellar messages was to transmit pictures resembling ordinary television pictures, a procedure very similar to the procedure by which human infants learn to speak. Infants are shown objects and told what the names of these objects are. Would it not make sense in the interstellar context to send images of things and attach to them some linguistic representation, which could then be used to construct sophisticated texts? With a little experimenting, I found that quite a good picture could be made by simply drawing the picture with black and white

> spots…The sequence of black and white spots that make up the picture could then be sent as a sequence of two characters (one for black and one for white), two tones, or perhaps dots and dashes. It should be quite trivial for any intelligent civilization to decode. (writing in Sagan 1978)

Drake composed such a message using 551 characters. His reasoning was that 551 was the product of two prime numbers, 29 and 19, and that given the signal one only had two ways to arrange the characters into a rectangular grid. One produced gibberish, and the other produced the image shown in Figure 19.10.

Figure 19.10. Drake's original pictorial message, sent to the Order of the Dolphin for decoding. This includes many forms that would appear in later METI messages, including a planetary system (down the left side), an image of a person and their height (bottom), binary numbers, and the population of humans (to the right of the smaller planets); as well as images of atomic structure (upper right). Compare Figure 19.3. Figure generated by the author using the bits from the message.

This image bore many of the elements that would appear in later examples of METI including a human form, a graphic depicting a planetary system, and binary numbers. Drake sent the image to the Order of the Dolphin for decoding, but was disappointed that not only could none of them decode it, only one, Barney Oliver, deduced it was supposed to be an image (Oliver responded by inviting Drake to drinks via a similar message that rendered an image of a martini glass with an olive).

Drake and Oliver continued to work on easier to understand images building up to the Arecibo message, a draft of which Drake gave to Carl Sagan. Drake reported that Sagan eventually decoded the entire thing, giving him confidence that the message was appropriately universal.

Further Reading

Beyond Contact: McConnell (2001).
The Alien Communication Handbook: McConnell (2021).
The Construction of the Voyager Golden Records: Sagan (1978).
Unpacking the Great Transmission Debate: Denning (2010).
"METI or REGRETTI: Ethics, Risk, and Alien Contact" Smith (2020).

References

Almár, I., & Shuch, H. P. 2007, AcAau, 60, 57
Carrigan, R. A. 2006, AcAau, 58, 112
Charbonneau, R. 2024, Mixed Signals: Alien Communication Across the Iron Curtain (Cambridge, UK: Polity Press)
de Magalhães, J. 2016, SpPol, 38, 22
Denning, K. 2010, AcAau, 67, 1399
Gertz, J. 2016a, JBIS, 69, 263
Gertz, J. 2016b, JBIS, 69, 31
Hatfield, P., & Trueblood, L. 2021, AcAau, 180, 596
Hippke, M., & Learned, J. G. 2018, arXiv:1802.02180
McConnell, B. 2001, Beyond Contact: A Guide to SETI and Communicating with Alien Civilizations, O'Reilly Series (Sebastopol, CA: O'Reilly)
McConnell, B. S. 2021, The Alien Communication Handbook: So We Received a Signal–Now What? Astronomers' Universe (Berlin: Springer)
Sagan, C. 1978, Murmurs of Earth: The Voyager Interstellar Record (New York: Random House)
Santana, C. 2021, SpPol, 57, 101430
Smith, K. C. 2020, Social and Conceptual Issues in Astrobiology ed. K. C. Smith, & C. Mariscal (Oxford: Oxford Univ. Press) 209
Staff at the National Astronomy and Ionosphere Center 1975, Icar, 26, 462
Traphagan, J. W. 2021, SpPol, 57, 101440
Trauth, K. M., Hora, S. C., & Guzowski, R. V. 1993, Expert Judgment on Markers to Deter Inadvertent Human Intrusion into the Waste Isolation Pilot Plant Technical Report, *Sandia National Lab (SNL-NM)*

Zaitsev, A. 2006a, arXiv:astro-ph/0610031
Zaitsev, A. 2006b, arXiv:astro-ph/0611283
Zaitsev, A. 2012, AcAau, 78, 16
Zaitsev, A. L. 2008, arXiv:0804.2754
Zaitsev, A. L. 2011, Searching for Extraterrestrial Intelligence, ed. H. Paul (Berlin: Springer) 399

AAS | IOP Astronomy

Chapter 20

Post-detection Matters

Ellie Arroway: If we go public with this, and we're wrong, that's it; it's over, we're cooked. God, I wish Kent was here.
Willie: Whatever the signal is, we better do something soon. Vega's gonna set.

Cut to Ellie on a video call with Ian Broderick in Australia. The pulses continue.

Broderick: That position is confirmed. We've got 4.4623 GHz. Confirmed, we've got 112 Jy.
Ellie: All right, do you have a source location yet?
Broderick: We put it right, smack in the middle. Vega.
Ellie: OK, thanks Ian. Just keep tracking and we'll get back to you.
Broderick: Yup, righto.

Long pause while they reflect as the pulses continue, then stop. Fisher checks the monitor.

Fisher: OK, a hundred and one. The pulse sequenced through every prime number between two and a hundred and one.
Willie: Who we gonna call now?

Ellie: Everybody.

— *Contact* (Box 18.1)

20.1 Post-detection Concerns

The implications and aftermath of contact with alien life has been the subject of speculation and science fiction for over 100 years (Chapter 18), but the imminence of such a possibility pointed out by Cocconi & Morrison (1959) eventually gave rise to the realization that real plans needed to be in place for success, which we call post-detection matters.

doi:10.1088/2514-3433/adf5aech20

Some such matters are:

- How should searches be communicated to the public, scientists, and governments?
- How should false alarms and hoaxes be handled?
- How should candidate signals be classified and monitored?
- How should signals be confirmed?
- How should the discovery of communicative alien life be communicated to the public, scientists, and governments?
- How can we ensure that the discovery of alien life is not kept secret?
- How will humanity react to the news?
- When and how should we respond to a signal?
- If a signal is communicative, how can we go about understanding it?
- Who should speak for Earth?
- How should who speaks for Earth be determined?
- What should we say to aliens?

Early considerations focused on the discovery of a quickly confirmed radio signal in a pre-Internet world. It is now more widely recognized that the discovery may be a process, not a moment or event, and that the modern media landscape means that scientists may have very little control over the narrative. As such, many additional complications to this picture have emerged as components of the post-detection landscape, such as:

- How should leaked news of candidate signals be handled?
- How can the definitive detection of an ambiguous technosignature be responsibly communicated?
- How should widespread mis- and disinformation[1] be handled?
- How can the safety of the scientists involved in discovery be maintained?

Since the beginning of modern SETI, there has been a strong sense among SETI practitioners that it was essential to cooperate internationally on SETI, and to incorporate expertise beyond astronomers in preparing for success. This need for cooperation led American and Soviet scientists to meet at the Second Byurakan conference (Section 3.3.1), and to work through organizations that spanned the Iron Curtain such as the International Academy of Astronautics (IAA) to establish fora for such cooperation. In 1965 the IAA formed a committee that began planning coordination on these matters, and in 1974 the IAA established what today is known as the Permanent Committee on SETI.

Since then, there have been numerous workshops and conferences devoted to the need to do more work on these matters, and to thoroughly integrate the expertise of social scientists, legal experts, diplomats, and scientists so that humanity will have prepared as best it can for the day contact is made. Below we will look at the output

[1] Misinformation is false. Disinformation is misinformation promulgated with the intention to mislead.

of some of these efforts, noting that the overall consensus is that while the need for such work is widely and often noted and appreciated, the most of the actual hard work still lies in our future.

20.2 The Different Forms of Contact

When considering how to prepare for success, it's important to clearly consider what success looks like, and to recognize it may arrive in a wide variety of forms. For ambiguous technosignatures, the discovery of alien technology may be a long process where scientific and popular consensus that a discovery has been made shifts as evidence grows and alternative interpretations fall out of favor. For a dispositive detection rich with information and opportunities for followup, SETI may evolve very so rapidly that the 'S' in its name will have to change from "Search for" to "Study of."

A thorough post-detection analysis needs to prepare for a variety of potential contact scenarios; should draw on lessons from across astrobiology; and can likely be integrated with post-detection matters for the discovery of biosignatures (Race 2008).

Below are some examples of very different (though not necessarily mutually exclusive) categories of potential SETI success that might warrant different post-detection preparations and protocols.

20.2.1 Bare Contact

Bare Contact is a term coined by Ćirković (2016) and refers to contact scenarios in which only the existence of a particular technology is inferred, with little or no additional information.

An example might be the weak detection of radio "chatter" in a distant star cluster or galaxy. We might understand only that there must be artificial radio transmitters because we detect fleeting instances of narrowband radio carrier wave emission that favor particular channels, but not be able to detect any modulation that would convey information. We might know only the general region the signals come from, and not be able to associate the transmissions with any particular star or planet.

This scenario has perhaps the least social and scientific impact of any definitive detection scenario, and would in some sense be the most frustrating. SETI would have succeeded, but building on that success would require finding another, more study-able example of the phenomenon.

20.2.2 Ambiguous Technosignatures

Many searches for technosignatures, such as for the infrared heat signatures characteristic of Dyson spheres, search for definitive detections of ambiguous technosignatures. Unlike the detection of a narrowband radio signal, such technosignatures might always have some, however contrived, natural explanation, making the road to the acceptance of the discovery of alien life long and largely based on circumstantial evidence.

Many discoveries in science proceed this way, including the discovery of black holes and the acceptance of timelines of ancient human history. There is no "moment" of discovery, and a consensus story only slowly emerges as more evidence is gathered and scientists propose, champion, and abandon various interpretations of it.

This scenario may be the hardest to cleanly communicate, since the issue of whether SETI has succeeded will be subjective and subject to significant nuance, and discovery will happen over an extended period of time, perhaps decades.

20.2.3 Unambiguous Communicative Technosignatures

The holy grail of SETI is a persistent, confirmable, unambiguous technosignature, such as an always-on narrowband radio signal. In this case, there are no natural confounders (Section 11.2.4) and the only possible source of confusion is interference from human transmitters. Such a discovery could be made and confirmed in a matter of days.

In this case, one can expect significant attention from worldwide mainstream media and the public. The possibility of decoding the message would likely lead to a large effort to do so, and success might have profound impact, depending on what the message says.

20.2.4 Nearby Technosignatures

If technological life is detected within, say 100 lt-yr, and especially if it is communicative, it will offer the possibility that we might send a message that would be received and answered in the foreseeable future. This would immediately raise concrete questions of who speaks for Earth; whether, and how to respond; and how to coordinate or control the many actors inclined to do so.

20.2.5 Solar System Artifacts

Most SETI projects do not implicate the possibility of physical contact with aliens or alien technology, but solar system SETI (Chapter 14) is the exception. If a probe were found in the solar system it would be amenable to close inspection via interplanetary probes from Earth. Indeed, it is not hard to imagine some space agency or private company being unable to resist the opportunity to get a close look at alien technology. The apparition of 'Oumuamua excited planetary scientists and opened up the possibility of an intercept mission to a future interstellar visitor, which would certainly be a high priority if a future visitor showed signs of being artificial.

Such a discovery has an extremely high potential for followup and escalation to physical contact with the artifact. It would also potentially involve the world's militaries because of the inevitable concerns that the artifact could be dangerous.

Figure 20.1. Still from the 1951 film *The Day the Earth Stood Still*, depicting a flying saucer landing on The Ellipse, a short distance from the White House (so not literally the White House lawn, but still likely the origination of the trope.) Image courtesy 20th Century Fox.

20.2.6 Close Encounters

The most dramatic contact scenario involves what in UFOlogy is known as a *close encounter*, meaning that aliens or their technology are in close proximity (500 ft) or in physical contact with Earth or humans.[2]

This situation comes with the highest impact and consequence, but also is least contemplated in SETI because it is not generally something that can be triggered by mere searching. It is, of course, a common trope in science fiction (the proverbial "Flying saucer on the White House lawn," see Figure 20.1).

20.3 Declaration of Principles Concerning Activities Following the Detection of Extraterrestrial Intelligence

The International Academy of Astronautics has since 1974 hosted a committee on SETI, whose focus has largely been on the development of uniform procedures for the verification and announcement of the discovery of contact with an alien species.

[2] The term originates with American astronomer J. Allen Hynek's scale to categorize UFO reports in an attempt to study them rigorously (Hynek 1972), where three kinds of alleged "close" encounters occupy the highest positions on a 6 point scale. Close encounters of the first, second, and third kind refer to seeing an object closer than 500 ft, experiencing or witnessing some physical effect from such an object, and encounters with alien beings themselves, respectively. The scale was made popular in the film *Close Encounters of the Third Kind*, in which encounters with alien spacecraft climb the scale as the plot progresses, until the climax in which (spoiler alert) several people are brought aboard a ship crewed by apparently benevolent aliens.

In 1989, the SETI committee adopted the *Declaration of Principles Concerning Activities Following the Detection of Extraterrestrial Intelligence*,[3] and it strives to follow the principles of *scientific validity* and *transparency*, to alleviate concern that the search is frivolous, or that researchers might make a discovery without informing the public or researchers in other countries. (The current version of the declaration, described here, is from a revision in 2010.)

The eight principles it lays out are (paraphrased here, with the degree to which signatories are bound by the principles emphasized):

1. *Searching*: All researchers are *free* to discuss their activities in public and professional fora, and *will* be responsive to public communications media.
2. *Handling Candidate Evidence*: Researchers *will* make "all efforts" to confirm a suspected detection at other facilities and with other organizations. Discoverers *need not* disclose these efforts publicly, and there *should be* no "premature disclosures pending verification." Nonetheless, inquiries from the media "*should be* responded to promptly and honestly." These results *should be* treated like provisional results in any other scientific context. The Rio Scale or its equivalent *should be* used when speaking to the media.
3. *Confirmed Detections*: Once all investigators involved in verification agree that a real discovery has been made, the discoverer *shall* make a full, complete, and open report to "the public, the scientific community, and the Secretary General of the UN," and the International Astronomical Union, which *will* include the "basic data," the process, the verification process, and any information in the signal.
4. All data *should be* made available to the scientific community.
5. The discovery *should be* monitored and all data *should be* recorded, stored permanently, and shared.
6. If the discovery is in the radio, observers *should* seek international agreement to protect the frequency.
7. *Post Detection*: A post-detection group within the IAA can assist with post-detection issues.
8. *Response to signals*: Signatories *will not* respond "without first seeking guidance and consent" of a body such as the UN.

The spirit behind these principles is clear, but in detail they are widely seen as challenging to put into practice.

One well understood shortcoming of the principles is that they largely presume that a candidate signal can be rather quickly verified to be both real and technological in origin, before its existence is widely known. While this could be true of a radio or laser signal, as we have seen, many technosignatures will be

[3] https://iaaspace.org/wp-content/uploads/iaa/Scientific Activity/setideclaration.pdf. These are sometimes loosely referred to as "the Post-Detection Protocols" but that term better refers to a still-not-established formal set or sets of steps researchers are bound to follow.

ambiguous and may take decades of study before a consensus builds that they are held to be from technological life.

Another is that the principles are somewhat contradictory: the state that practitioners are "free" to discuss their work to the public; that media inquires "should be" answered promptly and honestly; but also that there "should be" no disclosures until a signal is verified.

This needle may be impossible to thread. In an age of social media, yellow journalism, and rampant dis- and misinformation, especially on matters of alien life, it can be extremely challenging to successfully communicate the scientific rigor of searches being done and the nature of uncertainty in candidate signals. As the blc1 incident showed (Sections 12.5.5 and 20.7.1), candidate signals are likely to be learned about by the press whether the discoverers like it or not, and so researchers will likely not have the luxury of waiting for verification. The definition of media is now no longer a few print publications and some television, and so includes everything from social media accounts to *The New York Times*.

The principles continue to be actively discussed, and the IAA Permanent Committee on SETI is in the process of updating them.

The *Declaration of Principles* has signatories but no legal force: it is purely a voluntary declaration by SETI researchers as a commitment of their scientific integrity and transparency.

20.4 The Billingham Workshops

In 1991, as the High Resolution Microwave Survey was getting underway at NASA (Section 3.3.4), John Billingham initiated a series of Workshops on Cultural Aspects of SETI, held in Santa Cruz, California to examine post-detection policies and concerns in a multidisciplinary way. This effort probably remains the best funded and most ambitious effort to perform a thorough post-detection analysis.

The workshops focused on what the likely reactions would be to a successful SETI detection, how NASA could better inform the public about SETI, and what work needed to be done to answer these questions better in the future. The cancellation of HRMS shortly after the workshops concluded delayed the publication of their final report, which eventually happened under the thanks to the SETI Institute, which had picked up the mantle of HRMS in Project Phoenix. That report (Billingham 1999) described the process:

> The members of the working group were scholars with expertise in many disciplines—history, comparative religion, psychology, sociology, political science, law, education, anthropology, popular media studies, journalism—as well as SETI scientists…
>
> The members of the working group found this to be a more complex and challenging task than seemed apparent at first sight. They completed their work with an expanded list of topics yet to be examined, new questions that invite research, and new possibilities not yet explored.

The workshop report recommended NASA should:

- Examine, critically, analogous episodes of contact in human history as "approximate guides for insights about SETI," focusing on the transmission of ideas rather than physical encounters.
- Educate the public and press about SETI in the context of understanding our place in the Universe using a variety of popular media, including via games and fiction.
- Take care to have a plan and clearly communicate both regular null results and respond to false alarms and hoaxes.
- Take care to understand the language cultures across Earth have to describe the alien and strangers, and choose its own language carefully.
- Research popular perceptions of aliens, science and technology.
- Consider the lessons learned from how the public responded to Apollo and Viking.
- Do outreach and education to policymakers and institutions about SETI and the positive outcomes of success.
- Work with experts in the social sciences to address these issues.
- Remain humble and agnostic about the nature of ETI and its desire to contact or communicate with us.
- Create post-detection protocols both internally and among national and inter-governmental bodies.

In short, the report represented an optimistic and ambitious outline of post-detection efforts across many fronts; the cancellation of HRMS led to these efforts being significantly stalled and thwarted. Efforts since then have unfortunately been scattered and largely disorganized.

20.5 Impact of Detection on Humanity

20.5.1 A Profound Moment?

There is a widespread expectation that the success of SETI will have a profound effect on humanity, indeed the Morrison et al. (1977) report states of SETI's success (or failure!) "its consequences will be extraordinary." Certainly, the event will mark an epoch in human discovery and be front page news.

However, there has been little scholarly work on the topic of how profound most people will find the news, and (as we discussed in Section 18.1) the actual reactions of humans will likely be both highly varied and largely dependent on the contact scenario. Indeed, large fractions of humanity already believe, to various degrees of certainty, that aliens regularly visit the Earth, that they know someone who has made contact, that SETI or NASA has already succeeded in finding ETIs, or that the military or government is hiding physical evidence of ETIs. News of the discovery of a distant, unintelligible radio signal may strike many as barely worth their time.

It is possible, even likely, that the announcement of contact will come with very little information about the nature of the ETIs responsible for the technosignature. An interested public will naturally want to fit the discovery into their preconceived expectations for contact (Section 18.3), which will serve as a starting point for efforts to communicate the discovery.

20.5.2 Panic

In 1938, on the night before Halloween, the young American thespian Orson Welles was heard across America on CBS radio announcing the invasion of Earth by militant Martians. He had adapted the H. G. Wells novel *The War of the Worlds* (Section 18.3.3) for radio as a real-time news bulletin.

Many viewers, primed to believe in Martian life by Percival Lowell and others (Section 3.1.3) and by 40 years of science fiction, were genuinely frightened, with the *New York Times* declaring on its front page the next day "Radio Listeners in Panic, Taking War Drama as Fact." For a long time, it was widely accepted that the broadcast had produced widespread panic among the public, who were already nervous about the prospect of global war, naive to the power of the relatively new medium of radio plays, and taken in by Welles's stagecraft and mesmerizing voice.

Subsequent scholarship has significantly questioned the degree of panic the broadcast actually created (see Dick 2018, Chapter 1), but the incident is so well known it has set an expectation that the actual discovery of alien life would create similar panics. Dick (2018) writes:

> Panics related to anniversary rebroadcasts of "War of the Worlds" still occasionally occur, as do other mass panics involving erroneous reports of environmental contaminants, nuclear accidents, and other cultural concerns…[O]nly a small portion of the population needs to act on erroneous information over a short period to create large-scale disruptions to society. *Sociological study of the causes and effects of collective behavior are thus critically important to the study of the potential societal impact of discovering extraterrestrial life.*

Fascinatingly, in Wells's original book *The War of the Worlds*, the Martians are at first not met with panic at all. Indeed the protagonist finds it hard to persuade his countrymen that the Martians should be feared—even after hostilities begin:

> "Haven't you heard of the men from Mars?" said I; "the creatures from Mars?"
> "Quite enough," said the woman over the gate. "Thenks"; and all three of them laughed.

To modern readers, this seems like a strange and unlikely reaction to news of alien invasion, but we must remember that the characters in *The War of the Worlds* have not read *The War of the Worlds*, or been steeped in the countless stories of alien invasion it inspired. That novella takes place in a world before popular culture had trained us that alien invasion was a real and dangerous possibility to be feared. Indeed, Wells says as much in the opening chapter:

> No one would have believed in the last years of the nineteenth century that this world was being watched keenly and closely by intelligences greater than man's and yet as mortal as his own…
> And early in the twentieth century came the great disillusionment.

It is a testament to his storytelling that while Wells meant that last line literally only in the context of his fictional world, his Martians nonetheless taught those of us in the real world to fear aliens, as if his Martians had *actually* invaded. This is ironic because Wells's actual intention was at least partly, if not entirely, to use Martians as a metaphor to highlight the evils of colonialism (Section 18.4).

20.5.3 Government and Military Interference

A trope of science fiction is a hasty and overeager terrestrial military responding to contact with aliens with a show of force. In stories with warlike aliens this force is often, at first at least, inadequate (*Independence Day*, *Mars Attacks*, or *Aliens*); in stories with peaceful aliens militaries are often an antagonist, interfering with peaceful relations (*Close Encounters of the Third Kind*, *E.T.: The Extraterrestrial*, and *Contact*).

While such a response would seem to be most relevant only for first contact in the form of Close Encounters, Wisian & Traphagan (2020) pointed out the danger of militaries *mispercieving* even other forms of contact, such as the reception of a distant radio signal, as an opportunity for advanced technology development via information from a technologically "superior" species. In this concern, militaries might react even to a tentative detection of an alien signal by attempting to establish an information monopoly by using force and espionage to secure their own radio telescope facilities and destroy those of rivals.

Such an information monopoly is probably impossible (radio waves fall everywhere on the Earth, and radio receivers are ubiquitous), and the idea that contact with ETI would be immediately valuable to militaries rests on a very narrow and arguably extremely unlikely set of circumstances, but this nonetheless highlights the need for the education of policymakers, militaries, and governments, so they understand the methods and possibilities of SETI (as argued by Wright et al. 2023, but see Traphagan & Wisian 2023).

20.5.4 Impact on Religion

Religious questions of the existence of extraterrestrial life on religion run deep, and there is a widespread expectation that the discovery of extraterrestrial life, especially intelligent life, would have profound and potentially negative effects on the world's religions. One has only to look to the variety of religious cults that strongly feature UFOs in their belief systems to see how serious such ramifications could be. For instance, members of the Heaven's Gate cult committed mass suicide in 1997 in anticipation of being reincarnated on a nearby UFO they believed was traveling with Comet Hale-Bopp, which had recently made an apparition. Strong religious reactions to contact appear in many popular depictions of contact, for instance playing a major plot point in the book and film *Contact* (Box 18.1).

With respect to major world religions, we saw in Section 3.1.1 a glimpse at some of the debate in Christian and Islamic theology regarding the implications of literal alien life forms on other planets (as opposed to *divine* celestial beings like angels), but only recently with our understanding of other worlds and the possibility of finding life on them has the topic become rather concrete.

In the context of Christianity, a primary point of theological friction with the Plurality of Worlds doctrine regards the issue of salvation: Christ's salvation applies to humans but not other creatures on Earth, so in which category do we put intelligent aliens? Are they akin to humans far from Jerusalem during Christ's time who have not yet heard the Good Word, and so are thus targets for conversion and redemption? Or, not being descended from Adam and Eve, might they not have any need of salvation? Or might they have had their own Eden, their own Fall, and their own "alien Christ" that died for their sins? Or might the paradox only find resolution in the conclusion that such life must not exist?

To denominations that are committed to particular answers to these puzzles, especially if that answer is "they do not exist," the discovery of alien life might come as a theological shock and threaten their belief system. In this case, one might imagine an institutional hostility to the news akin to the Catholic Church's reaction to Galileo's observations supporting Copernicanism.[4] To more open-minded denominations, being able to ask an extraterrestrial species these questions might be a thrilling opportunity for a deeper understanding of humans' relationship with God and creation.

Or so one might imagine, but there has actually been very little research on the topic. For one thing, one must distinguish between the effect of such a discovery on theologians and institutional doctrines and the actual belief systems of most people, who might have at best passing familiarity with these corners of the theology underpinning their faith, and whose reactions might have little to do with their professed religious faith.

For instance many people are often surprised to hear that the Vatican astronomer, Brother Guy Consolmagno, sits on the science advisory board of the SETI Institute and is quite open-minded on the topic (just as many are surprised to hear the Vatican has an official astronomer and an observatory!). On the topic of the Vatican's position on alien life, Consolmagno is quite clear:

> There are no "official Vatican positions" on such topics. Nor is it a new idea; you can find people talking about this over the last thousand years…I have no idea how the Church would respond. It has been the

[4] Somewhat like the Bruno affair (Section 3.1.2), the Galileo affair is often caricatured as nothing more than the Catholic Church's stubborn persecution of a brave truth-teller for contradicting dogma. While that is certainly part of what happened, the actual charges against Galileo and nature of his persecution are more nuanced, and involved a large amount of politics, including Galileo's delight in disparaging his politically powerful critics in the church. Galileo appears himself to have been a good and faithful Catholic.

> subject of any number of science fiction stories; most of them not very good, I'm afraid.[5]

In 2009 American theologian Ted Peters conducted the *Peters ETI Religious Crisis Survey*[6] to assess the degree to which people of different belief systems felt *others* would react to the discovery of intelligent alien life. Peters (2011) indeed found a widespread expectation that such a discovery would be quite disruptive, and that this belief was especially strong among nonbelievers (see Figure 20.2)

The top part of this figure shows agreement with the statement "Even though my religious (or-nonreligious) viewpoint would remain unaffected, contact with extraterrestrials would so undercut traditional beliefs that the world's religions would face a crisis." Roughly 1/3 of religious people and 2/3 of nonbelievers that responded to the survey agreed with this statement, implying that nonbelievers are twice as likely to be concerned with this prospect as religious people, perhaps reflecting mistrust or misunderstanding of widespread religious belief on the topic.

The bottom part of the figure shows the self-reported degree of acceptance of the idea: over 80% of respondents in all belief systems disagreed with the statement that "official confirmation of the discovery of a civilization of intelligent beings living on another planet would so undercut my beliefs that my beliefs would face a crisis." Peters (2011) concludes "what passes for conventional wisdom may be the product of what nonreligious people say about religious people."

This study was small and warrants more thorough followup, but strongly suggests that while the discovery of alien life may be profound in many ways and indeed quite disruptive, widespread crises of faith by adherents of major religions should not be a major cause for concern.

20.6 The Rio Scale

20.6.1 The Scale and Its Origins

In 2000, Jill Tarter (Box 1.2) and Hungarian astronomer Iván Almár proposed a scale for communicating the importance of a claim of the detection of a technosignature at the 51st International Astronautical Congress, held in Rio de Janeiro, Brazil.[7] A modified version of their scale was adopted by the IAA SETI Committee in 2002.

Known as the "Rio Scale," it is based on the Torino scale, which is used by those who search for potentially hazardous asteroids as a way to communicate the hazard presented by a particular asteroid using the sort of risk matrix we encountered in Figure 19.7 in Section 19.4.4. The *hazard* here is a combination of the *likelihood* that a given asteroid will strike the Earth times the *severity*[8] of such an impact, which depends on the size of the asteroid (see Figure 20.3).

[5] From Cosnolmango's April 2013 blog post at https://brotherguy.livejournal.com/74394.html, retrieved 2024 Feb 18.

[6] https://counterbalance.org/etsurv/fullr-frame.html.

[7] Their paper was reprinted 11 years later as Almár & Tarter (2011).

[8] Risk or hazard is often phrased as "likelihood times impact" but "impact" here would then be a bit ambiguous (or punny), so for clarity we'll use "severity."

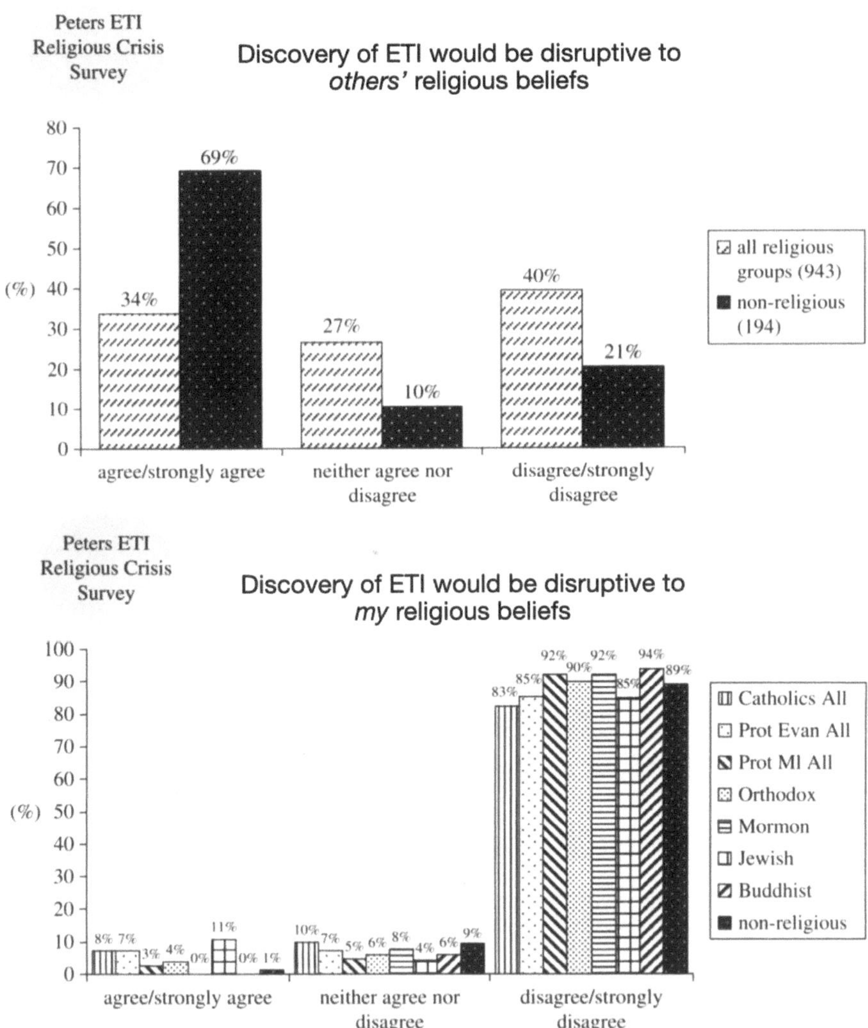

Figure 20.2. *Top:* Degree of agreement with the statement that the discovery of ETI would disrupt the world's religions, showing widespread agreement with this expectation, especially among nonbelievers. *Bottom:* Level of agreement with the statement that the *respondents' own* beliefs would face a crisis, showing that most would be unperturbed. This study suggests that concerns about the impact of contact on peoples' belief systems are both widespread and largely unwarranted. Reproduced with permission from Peters (2011), with annotation by the author. © 2011 The Royal Society.

Very small asteroids are always 0 because even if they are certain to strike they are too small to matter. Any impactor larger than 20 m is large enough to warrant a high score if its probability of impact is higher than 99%, with score increasing with severity. Below this probability, the hazard is scored as a combination of these two factors.

Newly discovered asteroids occasionally warrant a 1 on the scale, until their orbits are sufficiently well known that we can rule out a collision with Earth. As of this

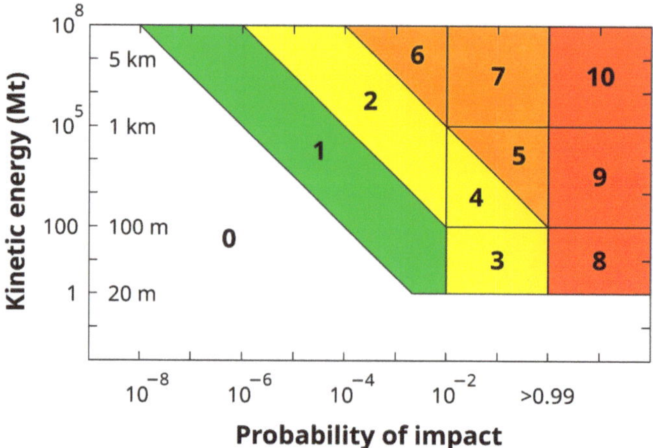

Figure 20.3. The Torino Scale, showing how the eleven-point scale maps in hazard space. The author obtained this image from the Wikimedia website where it is credited to Looxix and made available by Mikhail Ryazanov under a CC-BY-SA-3.0-migrated license.

writing, only thrice have asteroids ever risen to a higher level before subsequent observations reduced their scores to zero: Apophis scored as a 4 for four days in 2004; 2004 VD_{17} scored a 2 for a few months in 2006; and 2024 YR_4 spent almost a month at a score of 3.

A great value in the Torino scale is that it provides a clean and simple way to communicate risk to the public. The Rio scale seeks to do the same, not for risk but for the amount of attention one should give to a claim of the detection of alien life.

The essence of the Rio scale is to assess a *credibility* δ that the claim is correct or true, by a subjective impact on society, measured as a *consequence* score Q, which is a function of the type of signal found and the distance of the technology it implies. The product of these numbers is then the Rio Scale score (see Figure 20.4).

The Q score is the sum of three scores. The first is based on the "class of phenomenon":

6: Earth-specific message, or an ET artifact, capable of contact, or a physical encounter.

5: Omnidirectional message with decipherable information, or a functioning ET artifact or space probe.

4: Earth-specific beacon to draw our attention, or an ET artifact with a message to mankind.

3: Omnidirectional beacon designed to draw attention, or an ET artifact with a message of a general character.

2: Leakage radiation, without possible interpretation, or an ET artifact the purpose of which is understandable.

1: Traces of astroengineering, or any indication of technological activity by an extant or extinct civilization at any distance, or an ET artifact, the purpose of which is unknown.

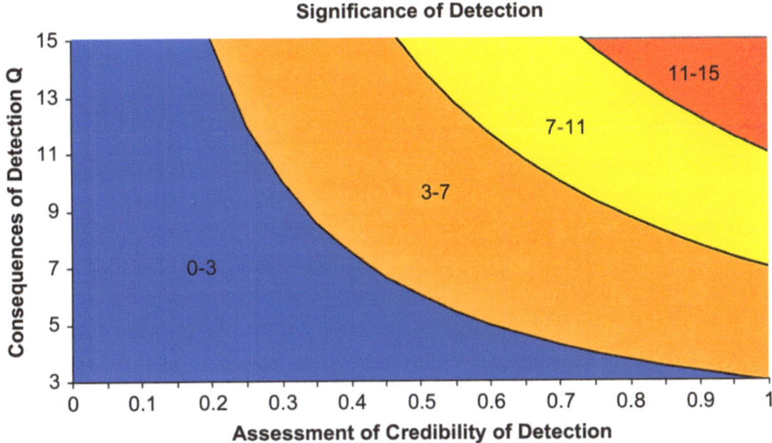

Figure 20.4. The Rio Scale, here shown in its original incarnation running from 0 to 15. It has since been rescaled to run from 0 to 10. Figure 3 from Almár & Tarter (2011), reprinted with permission from Elsevier.

The second is the "type of discovery":

5: SETI/SETA observation; steady phenomenon verifiable by repeated observation or investigation.
4: Non-SETI/SETA observation; steady phenomenon verifiable by repeated observation or investigation.
3: SETI/SETA observation; transient phenomenon that has been verified but never repeated.
2: Non-SETI/SETA observation; transient phenomenon that is reliable but never repeated.
1: From archival data; *a posteriori* discovery without possibility of verification.

The third is the apparent distance of the technology:

4: Within the solar system.
3: Within a distance which allows communication (at lightspeed) within a human lifetime.
2: Within the Galaxy.
1: Extragalactic.

These numbers are then summed and divided[9] by 1.5, so run from 0 to 10, and then multiplied by the credibility score δ, computed as:

1.0: Absolutely reliable, without any doubt.
0.7: Very probable, with verification already carried out.
0.5: Possible, but should be verified before taken seriously.

[9] The original scale, as shown in Figure 20.4, was not divided by 1.5, so ran from 1 to 15. Almár (2001) introduced alterations to the original 2000 proposal based on feedback, including this rescaling, because a 1–10 scale is more familiar to the public.

0.2: Very uncertain, but worthy of verification efforts.
0.0: Obviously fake or fraudulent.

20.6.2 Shortcomings of the Rio Scale, and Proposed Improvements

Although the Rio Scale ostensibly covers any kind of discovery, mentioning SETA (solar system SETI) and "traces of astroengineering," the scale is not immediately applicable to many potential kinds of SETI success.

For instance, consider a non-repeated radio signal detected with a single-dish telescope, like the Wow! Signal but with high confidence that it was real and from beyond the solar system. In this case, the actual origin of the signal may be unclear—the beam is wide and may contain many stars. In addition, there may be no way to establish whether it is omnidirectional or targeted, and if it is targeted whether Earth is the intended recipient.

Or, consider an inexplicably dimming star consistent with megastructures blocking its light. It might be an Arnold beacon (Section 15.7.1) conveying information to the Universe, or it might not be intended to attract attention at all.

Another critique is that the scale is linear with probability, so even a 10% chance of the most impactful possible discovery merits only a 1. This is in contrast to the Torino scale, where probability is mapped to the numbers logarithmically: even a 1.1% chance that a 2 km wide asteroid will strike Earth merits a 7, and only once probabilities drop below one in a million do the largest asteroids fall as far as 1 on the scale.

Several suggestions have been made to improve the scale, including "Rio 2.0" by Forgan et al. (2019). This scale is logarithmic in probability and has added nuance that allows for finer granularity for "everyday" SETI detections and candidates of the sort that come up frequently but have a very low probability of panning out. This new scale has not gained much traction or yet been adopted by the IAA.

20.7 False Alarms in the Media

Popular perceptions of SETI are often shaped by its appearances in the media, in documentaries and news articles. This often happens in the case of false alarms, when the media catches wind of a claim of a SETI candidate or the suggestion that some anomaly might be due to alien activity.

The most famous (probable?) false alarm is the Wow! Signal (Box 3.4), although this is not a good case study because the media and public became aware of it long after the fact, and it was presented as a mystery that might never be solved, not an imminent harbinger of contact.

20.7.1 blc-1

A better example is probably the blc-1 incident (Section 12.8). In 2020 the Breakthrough Listen (Section 3.5.1) team was analyzing data collected commensally at the Parkes telescope in Australia of the nearest star to Earth, Proxima, and quickly established that a weak signal in the data warranted careful consideration and followup.

It is worth reflecting here on the IAA Principles (Section 20.3), which state at this point the team was to:

> …make all efforts to verify the detection, using the resources available to the discoverer and with the collaboration of other investigators…
>
> …there should be no premature disclosures pending verification. Inquiries from the media and news organizations should be responded to promptly and honestly.

As the team was developing its analysis and followup plans, and despite their best efforts to deal with the matter without the attention of the press, *The Guardian* got wind of the activity and published a scoop entitled "Scientists looking for aliens investigate radio beam 'from nearby star': Tantalising 'signal' appears to have come from Proxima Centauri, the closest star to the Sun."

The article did not overhype the situation, stating:

> The latest "signal" is likely to have a mundane explanation too, but the direction of the narrow beam, around 980MHz, and an apparent shift in its frequency said to be consistent with the movement of a planet have added to the tantalising nature of the finding. Scientists are now preparing a paper on the beam, named BLC1, for Breakthrough Listen, the project to search for evidence of life in space, the Guardian understands.
>
> The beam that appears to have come from the direction of Proxima Centauri, a red dwarf star 4.2 lt-yr from Earth, has not been spotted since the initial observation, according to an individual in the astronomy community who requested anonymity because the work is ongoing. "It is the first serious candidate since the 'Wow! signal', they said.

Nonetheless, media scrutiny was intense, and the situation was widely reported, including in National Geographic and The Atlantic.

Eventually, the team published their analysis demonstrating that the signal was terrestrial radio interference, leading to another round of media interest including in *The New York Times* about how the team had established that this was a false alarm.

All in all, the media attention, at least from the larger and more responsible press, was careful and reserved, and the team managed it well. Nonetheless, the situation was a huge distraction and slowed down the analysis.

20.7.2 Boyajian's Star

In 2014, American astronomer Tabby Boyajian was studying data taken by the *Kepler* observatory (Section 12.4.1) that had been brought to her attention by members of the public participating in the *PlanetHunters* project on the *Zooniverse* citizen science platform. The *Kepler* science team had written algorithms to search the observatories data stream of 100,000 stars for signs of transiting planets, but they were aware that many other signals of astrophysical interest would be present in the data but overlooked without human eyeballs. The large volume of the data thus warranted a crowdsourced effort.

The star, originally given the designation KIC 8462852 by the *Kepler* team, showed large and irregular light curve variations, suggestive of large amounts of circumstellar material (see Figure 20.5). Unlike planetary transits, which have typical depths of <1%, many of the dimmings (or "dips") of this star were quite large, on one case over 20%. This could only be the case if the intervening material blocking the star's light were of similar size to the star, and mostly opaque.

Many of the dips also showed highly asymmetric profiles, suggesting that the intervening material was irregularly shaped, and not circular. Over the four years of the *Kepler* observations, many dips occur with varying shapes, depths, and durations, the most dramatic of which lasted over a week, suggesting either extremely large or slowly moving material.

There is a class of stars known as "dippers" that can show similar light curves, but stars of this class are universally young stars with irregular disks of circumstellar material (Section 2.13.1). When seen edge-on, these disks can block large portions of

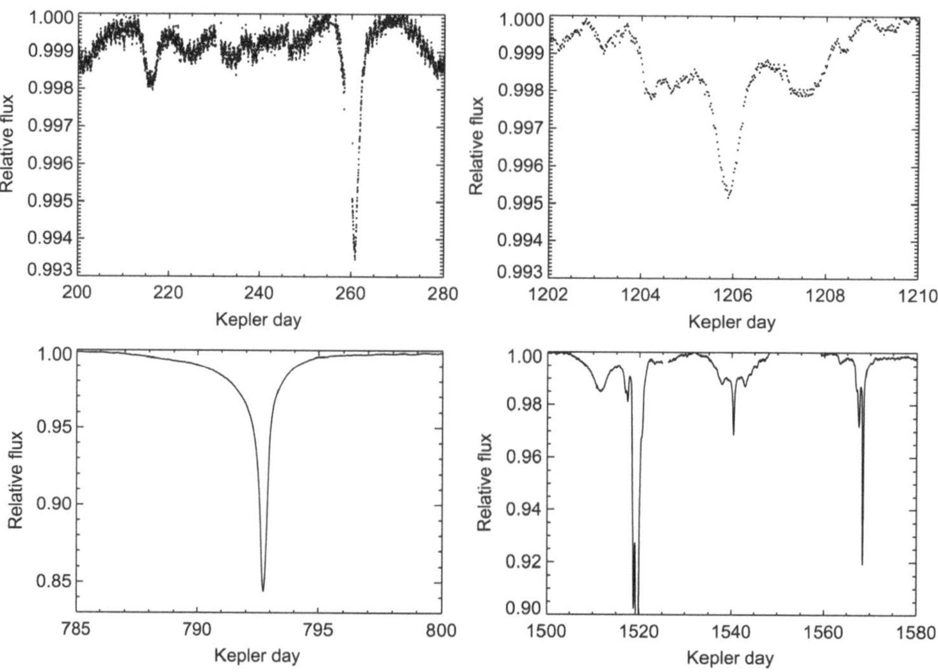

Figure 20.5. Figure 2 from Wright et al. (2016), after Figure 1 from Boyajian et al. (2016), showing some of the light variations from Boyajian's star. Note the differences in scale in the axes among the four panels. If these "dips" are caused by circumstellar material, that the material must have very large linear dimensions (on the order of a solar radius, so much larger than a planet, because the deepest dips are a significant fraction of the total flux), have non-circular aspects (because the dips are asymmetric), be largely opaque (again because the dips are a significant fraction of the total flux), and on irregular or very long period orbits (because no dips repeat regularly.)

the star's surface from our vantage in irregular ways that can produce very strange light curves. KIC 8462852, however, was very clearly not a young star and showed no other evidence of a disk, such as signs of accretion or an infrared excess associated with large amounts of circumstellar material—Boyajian showed quite conclusively that the star was an otherwise ordinary, middle-aged star.

Boyajian announced the discovery of this strange object in a preprint in 2015 (Boyajian et al. 2016), and hypothesized that the cause was "the passage of a family of exocomet or planetesimal fragments, all of which are associated with a single previous break-up event, possibly caused by tidal disruption or thermal processing." That is, they invoked a previously unobserved phenomenon, giant swarms of giant comets or icy planetessimals, whose cometary tails were blocking the light of the star as they orbited.

In the meantime, Boyajian had been collaborating with American astronomer Jason Wright[10] and the Breakthrough Listen team (Section 3.5.1) to explore the alternative hypothesis that this was an example of a star orbited by a swarm of megastructures of the sort we discussed in Section 15.7.2. Ross Andersen, a reporter for *The Atlantic*, heard about these efforts, and in 2015 October published a compelling piece online entitled "The Most Mysterious Star in Our Galaxy"[11] that included a pull quote from Wright: "It looked like the kind of thing you might expect an alien civilization to build."

This article caught widespread attention, and was mentioned on *The Late Show with Steven Colbert*,[12] a "Weekend Update" segment on *Saturday Night Live*, many news programs, and in many followup articles in the scientific and tabloid presses. At the time, there was no published paper discussing the megastructure hypothesis, and so Wright and his colleagues rushed to publish a preprint that included a section describing and defending the hypothesis (eventually published as Wright et al. 2016).

Boyajian's group had internally referred to the star not by its unmemorable KIC designation, but as "the WTF[13] star," a term they popularized and justified in the subtitle to Boyajian et al. (2016): "Where's the flux?." Internally, Wright's group's moniker was "Tabby's Star," a name that was picked up by the media. Wright et al. (2016) later gave the star the more formal name "Boyajian's Star," in keeping with the astronomical tradition of occasionally naming stars with the surname of the astronomer that first made the star famous.[14] This last designation is now commonly used in the literature,[15] although "Tabby's Star" is still commonly used in popular and less formal contexts.[16]

[10] Your author.

[11] https://www.theatlantic.com/science/archive/2015/10/the-most-interesting-star-in-our-galaxy/410023/ retrieved 2025 March 18.

[12] https://www.youtube.com/watch?v=3UDNOOJ3sr0.

[13] A common abbreviation and euphemism for the idiomatic expression of confusion "What the fuck?"

[14] As of this writing, it is the only star so named after a woman.

[15] And recognized by the SIMBAD Astronomical Database, the *de facto* standard in astronomy.

[16] Boyajian herself goes by Tabetha in formal contexts such as paper author lists, and Tabby elsewhere.

Scientific and media attention on the star was intense for months. Eventually, Boyajian was able to turn the popular interest into scientific productivity via a Kickstarter crowdfunding campaign to purchase time on the Las Cumbres Observatory network of telescopes to monitor the star for future dipping episodes. These efforts paid off in May 2017 when the telescopes were able to observe several dips from the ground at multiple wavelengths (Boyajian et al. 2018). Bodman et al. (2018) showed that these dips were much deeper at blue wavelengths, indicating that the intervening material was reddening the background light (Section 2.4.2), and so was likely ordinary astrophysical dust (Section 2.9), and not opaque megastructures.

20.7.3 'Oumuamua

In 2017 October, the PAN-STARRS1 telescope system detected an asteroid or comet receding from Earth unusually rapidly for its distance, and the object was given the temporary designation A2017U1. A team led by Karen Meech of the University of Hawai'i studying its orbit established that its velocity was higher than the escape velocity of the Sun (Section 6.2.3), demonstrating that it was an interstellar object passing through the solar system, the first macroscopic such object ever discovered.[17]

The episode immediately reminded many astronomers of the opening chapters of the Arthur C. Clarke novel *Rendezvous with Rama*, in which an interstellar object is discovered and given the name *Rama*, and turns out to be a large alien spacecraft passing through the solar system as part of a longer journey. Although many suggested that the discovery team name this object “Rama” as a nod to the book, the team ultimately settled on 'Oumuamua “which in Hawaiian reflects the way that this object is like a scout or messenger sent from the distant past to reach out to us” (Meech et al. 2017).

The eventual discovery of interstellar comets and asteroids had been anticipated for some time, and so this first member of the new class was met with significant scientific and media attention. In some ways, such as its colors and orbital trajectory, it conformed with prior expectations, but in others it was highly unusual compared to most solar system comets or asteroids. In particular, it was very small (perhaps 100 m across); its highly variable brightness meant that it appeared to be tumbling, and highly flattened or elongated (perhaps 10–20 m in its shortest dimension); and it showed strong non-gravitational accelerations (Sections 6.2.3 and 14.5.2). This acceleration was especially puzzling because of non-detections of any coma or tail consistent with cometary outgassing that could explain the it.

This led Bialy & Loeb (2018) to work out some of the physics behind the survival of thin objects in interstellar space, and suggest that 'Oumuamua might be a discarded bit of a light sail used for interstellar propulsion being accelerated via solar

[17] Its velocity and trajectory were inconsistent with it having been the product of a gravitational slingshot within the solar system, as can happen when asteroids have a close encounter with a planet (Schneider 2017; Wright 2017).

radiation pressure. Their initial publication was somewhat sober and even understated, mentioning 'Oumuamua only in the abstract and final section, however Israeli-American astrophysicist Avi Loeb (the second author of the paper) was much less equivocal in his media discussions of the topic, in public lectures on the topic, and eventually in a popular science book “Extraterrestrial: The First Sign of Intelligent Life Beyond Earth” (Loeb 2021).

Much of the media coverage of Loeb’s claims veered into the sensational, and Loeb himself appeared to welcome this, for instance headlining the 2023 “Contact in the Desert” UFO conference. Many planetary scientists and astronomers expressed strong dissatisfaction on social media and elsewhere in what they felt was unwarranted and irresponsible hype by Loeb about the probability that 'Oumuamua was an alien spacecraft, even after studies showed that a natural explanation for its behavior fit the data far better than an artificial one (Section 14.3.2).

Loeb further alienated some of the astronomical and SETI communities in a 2021 webinar[18] in which he had an argumentative encounter with Jill Tarter (Box 1.2) about support for SETI in general and his public stance on other SETI researchers’ work.[19] Loeb was able to turn the attention from this and other claims about interstellar objects into philanthropic funding to search for debris from interstellar spacecraft on the ocean floor and to conduct studies of UAPs as part of his Galileo Project (Section 18.7.7).[20]

20.7.4 Some Lessons of False Alarms

A primary lesson of these incidents is that it is important for practitioners in the field to be media savvy and not just understand how to manage the press, but be aware that SETI has such a strong popular appeal that there will be media interest in almost any aspect of it. It might be one of those rare serious scientific topics where there are more articles about it in the popular press than in the refereed journals!

Practitioners should thus be prepared to share a description of their work in a popular register any time they have a candidate signal, or even any time they have an interesting paper on a preprint server. It is also important to be prepared for premature media attention to an interesting SETI candidate. The kindling of broad public interest in the topic is so combustible that virtually any SETI paper or study, no matter how small its spark, can cause a media firestorm.

Another is that it is important to have some sort of professional publication describing a candidate, study, or hypothesis—until the Wright et al. (2016) preprint appeared, many astronomers contacted by the press for comment on Boyajian’s Star as a SETI candidate had nothing to go on beyond what they were hearing in the press, which hindered reporters’ ability to report the story thoroughly.

---

[18] https://youtu.be/bh3DnraBCqw.

[19] Loeb later apologized for his behavior.

[20] The Galileo Project’s tagline on its website is “daring to look through new telescopes.” Loeb’s book makes many comparisons between his work and that of Galileo.

Interestingly, it was not the work of Boyajian et al. (2018) and Bodman et al. (2018) showing the dips were caused by dust that quelled the popular interest in Boyajian's Star, but the much earlier non-detections of radio emission by Harp et al. (2016) using the Allen Telescope Array. The implication is that the public and media largely see radio searches for technosignatures as dispositive, meaning that our radio telescopes would detect any technology that exists at any star they point at. This is consistent with the perception that a great deal of searching has been done, that we have largely ruled out alien life in the Universe, and that therefore the Mystery of the Great Silence is difficult to resolve (Section 6.5.2).

Another lesson is that it can be very useful to communicate directly with colleagues and the public via social media and with reporters and editors at reputable media outlets. The media attention might have been much more sensational had the original *Atlantic* article on Boyajian's Star and work in the *Guardian* and *New York Times* on blc-1 not been so responsibly and accurately reported by experienced science reporters who worked closely with the teams on the stories.

Wright maintained a blog describing progress on studies of Boyajian's Star,[21] and he and his colleague, astronomer and science journalist Kimberly Cartier, wrote a print article in *Scientific American* explaining the best understanding of the star (Cartier & Wright 2017). Boyajian maintained public updates written for her team's Kickstarter backers that explained the followup science being done. This all allowed responsible reporters, professional astronomers, and the interested lay public to cut through the less responsible reporting by yellow journalists.

Finally, yellow journalism can be difficult or impossible to responsibly manage. Your author has stopped giving interviews to certain British tabloids after they twice either edited or paraphrased his quotes to imply or state the opposite of his clear meaning in order to generate hyperbolic headlines. Astronomers' amplification of sensational claims about alien life can certainly attract significant popular attention, but can also generate professional resentment, alienate one's colleagues, and feed the perception that work in SETI is undertaken primarily cynically, to attract attention and funding.

20.8 Who Speaks for Earth?

> There are other forms of intelligence on Earth, doctor. Only human arrogance would assume the message must be meant for Man.
>
> — Spock, *Star Trek IV: The Voyage Home* (in which the message was not, in fact, meant for humans at all, but for humpback whales.)

[21] https://sites.psu.edu/astrowright/tabbys-star-posts/.

20.8.1 What Should We Say?

In Section 19.3.3 we discussed how Drake and others involved in the construction and transmission of the Arecibo message felt that any more effective transmissions were best left to an ostensibly globally representative body such as the United Nations. In Section 19.2.4 we saw how Sagan found the process of working through the United Nations to collect messages for the Golden Records challenging. One might imagine that if the topic were a response to a signal that was likely to be received and considered by real aliens—instead of a mostly symbolic gesture like the Golden Records—that the process would be even more challenging.

Indeed, the topic of responding raises many of the same concerns as for METI that we discussed in Sections 19.4.1 and 19.5. Our response might tell an alien species something that they did not already know (for instance that a technological species exists on Earth, or something about our attitude toward the signal we detected). As with a *de novo* METI signal, might we worry that we might say "the wrong thing" and trigger a hostile response; that any contact with an alien species with a much older technological or cultural tradition than ours might be inherently disruptive and dangerous; and so on.

The problem of a proper response depends so strongly on the kind of signal or artifact we find that it is probably fruitless to hope to achieve global agreement on that response in advance, especially when so many of the representative bodies that should be involved have little reason to be engaged before a signal is detected. But as with the rest of post-detection work, it is proper and important to establish a set of principles that can guide the process in advance.

20.8.2 Does the Question Even Make Sense?

It is probably worthwhile to critically examine the very premise of the question "Who Speaks For Earth?" Science fiction certainly primes us to expect that Earth will be expected or need to have a single, representative voice, and that the aliens will have the same.

And this makes sense: interactions among human cultures and organizations at all levels from families to nations benefit greatly from the communal delegation of representation and authority to a single organization or person. Much of international diplomacy is premised on such delegation, and it is reasonable that interstellar diplomacy might proceed the same way.

But we should not assume aliens will expect, appreciate, or even understand such delegation. After all, if they have been monitoring our broadcasts they will understand that there are many contemporaneous human societies with different values. It might strike them as strange that one voice could represent them all. Even more than that: Earth hosts many intelligent species—will aliens expect or understand that humans speak for them, as well? Why would they?

So while we should certainly prepare for how humanity as a whole can generate a representative message from a single voice, we should also consider that proper interactions with alien species might involve a multitude of voices that do not necessarily agree or coordinate, potentially on both sides of the call (see Figure 20.6).

Figure 20.6. We might expect alien life to have very different forms of cognition from ours. This comic highlights how humanity's collective accomplishments might be seen—and dismissed—by aliens as an example of emergent intelligence without emergent consciousness, much like we might perceive the collective actions of an ant colony. From Saturday Morning Breakfast Cereal comics, courtesy of Zach Weinersmith.

20.9 Further Reading

Astrobiology, Discovery, and Social Impact: Dick (2018).
SETI and Post-Detection: Towards a New Research Roadmap: Denning et al. (2019).
Preparing for the Discovery of Life Beyond Earth: Denning et al. (2019).
"Meeting extraterrestrials: Scenarios of first contact from the perspective of exosociology": Anton et al. (2024).
Ten categories of post-detection decisions: Michaud (2003).
"First Do Nothing: A Passive Protocol for First Contact": Smith & Traphagan (2020).
"SETI and Democracy": Hatfield & Trueblood (2021).

Exercises

1. Screen the film *Contact*. Pay special attention to elements of the film that:
   (a) Represent or allude to real historical events.
   (b) Represent real people.
   (c) Depict the practice of SETI or astronomy generally.
   (d) Refer to the theory of SETI or astronomy generally.
   (e) Make reference to astronomy.
   (f) Follow common science fiction tropes.
2. Write up or otherwise present your observations in a format dictated by your instructor, making sure to highlight concepts covered in this class. Also consider:
   (a) In what ways does the film sacrifice veracity or otherwise make concessions to the necessities of storytelling and filmmaking in a big-budget popular film?
   (b) Investigate the film's use of footage of Bill Clinton. What is the provenance of this footage, and in what ways is that relevant to this course?

References

Almár, I. 2001, Review Meeting on SETI, 52nd International Astronautical Congress Paper #IAA-00-IAA9, *IAC*

Almár, I., & Tarter, J. 2011, AcAau, 68, 358

Anton, A., Elliott, J., & Schetsche, M. 2024, AcAau, 215, 308

Bialy, S., & Loeb, A. 2018, ApJL, 868, L1

Billingham, J. 1999, Social Implications of the Detection of an Extraterrestrial Civilization: A Report of the Workshops on the Cultural Aspects of SETI Held in Three Sessions, October 1991, May 1992, and September 1992 at Santa Cruz, California (Mountain View, CA: SETI Press)

Bodman, E., Wright, J., Boyajian, T., & Ellis, T. 2018, arXiv:1806.08842

Boyajian, T. S., LaCourse, D. M., Rappaport, S. A., et al. 2016, MNRAS, 457, 3988

Boyajian, T. S., Alonso, R., Ammerman, A., et al. 2018, ApJL, 853, L8

Cartier, K., & Wright, J. T. 2017, SciAm, 316, 36

Ćirković, M. M. 2016, AsBio, 16, 737
Cocconi, G., & Morrison, P. 1959, Natur, 184, 844
Denning, K., Berea, A., Capova, K. A., et al. 2019, 70th Int. Astronautical Congress (IAC) IAA Symp. on the Search for Extraterrestrial Intelligence (SETI)—The Next Steps, IAC–19 (Paris: IAF) A4,2,4,x52210
Dick, S. J. 2018, Astrobiology, Discovery, and Societal Impact, Cambridge Astrobiology (Cambridge: Cambridge Univ. Press)
Forgan, D., Wright, J., Tarter, J., et al. 2019, IJAsB, 18, 336
Harp, G. R., Richards, J., Shostak, S., et al. 2016, ApJ, 825, 155
Hatfield, P., & Trueblood, L. 2021, AcAau, 180, 596
Hynek, J. A. 1972, The UFO Experience: A Scientific Inquiry (Washington: H. Regnery)
Loeb, A. 2021, Extraterrestrial: The First Sign of Intelligent Life Beyond Earth (Boston, MA: Houghton Mifflin Harcourt)
Meech, K. J., Weryk, R., Micheli, M., et al. 2017, Natur, 552, 378
Michaud, M. A. G. 2003, SpPol, 19, 131
Morrison, P., Billingham, J., & Wolfe, J. 1977, The Search for Extraterrestrial Intelligence, SETI, Vol. 419 (Washington, DC: NASA)
Peters, T. 2011, RSPTA, 369, 644
Race, M. S. 2008, AcAau, 62, 71
Schneider, J. 2017, RNAAS, 1, 18
Smith, K. C., & Traphagan, J. W. 2020, SpPol, 54, 101389
Traphagan, J. W., & Wisian, K. W. 2023, SpPol, 63, 101536
Wisian, K. W., & Traphagan, J. W. 2020, SpPol, 52, 101377
Wright, J. T., Cartier, K. M. S., Zhao, M., Jontof-Hutter, D., & Ford, E. B. 2016, ApJ, 816, 17
Wright, J. T. 2017, RNAAS, 1, 38
Wright, J. T., Haramia, C., & Swiney, G. 2023, SpPol, 63, 101517

Afterword

A Personal Perspective

In this afterword I will take the author's prerogative to, perhaps a bit presumptuously, offer a bit of personal perspective on SETI, and advice and recommendations to those entering the field.

One of the most common questions I am asked is what I think the chances of "success" for SETI are, or the odds that I, specifically, will discover alien life. My most common and truthful answer is that I don't know. This is an area where we have almost nothing to go on beyond our own existence and the laws of nature, and as we have explored in Chapters 7–10, this is perhaps enough to build a hunch on, but not for any degree of certainty or consensus.

My implication is that SETI is worth pursuing regardless of one's personal belief in alien life. This is because SETI not an all-or-nothing gamble or a quest to prove we are not alone, and practitioners who advance the field technologically and algorithmically, place new upper limits on alien life, and discover new astrophysical phenomena along the way are not "failing," they are succeeding and doing good science.

Similarly, I am often asked what method of SETI—what particular technosignature—I think is the most promising, or the one I am excited about the most. My most common and truthful answer is to point to whatever upcoming search promises to cover the most parameter space. When some new window opens in wavelength or areal coverage or temporal or spectral resolution, we can search decades of the cosmic haystack quickly (Section 11.3). I also like to say that if I had to bet, success will come from some approach we haven't tried yet—perhaps one developed by a student after reading this textbook.

It is with this mentality that I have laid out a broad and perhaps idiosyncratic vision of SETI in the chapters above, attempting not to provide a comprehensive survey of the scientific literature on the topic of alien technological life, but to map out the landscape of where the field has been and can go.

doi:10.1088/2514-3433/adf5aech21

So while I will decline to recommend which *kinds* of technosignatures one should search for—I have my personal opinions and priors, but they may be as misguided as anyone else's—I will give general recommendations on features that any SETI project should follow.

Read the Literature

Because there is no formal SETI curriculum, it is difficult to find and appreciate the substantial body of work that has been done on any given topic within SETI. A good place to start is the SETI bibliography described in Section 1.3.

My own experience is that when I have what feels like an important new idea or insight, I can almost invariably find an old paper on it with zero or single digit citations, often in the *Journal of the British Interplanetary Society* or in some piece of gray literature. Indeed, many of my supposedly novel ideas have turned out to have been developed and published independently multiple times. For the field to grow, authors need to extend prior work, instead of (inadvertently) finding new ways to repeat old arguments.

This is not to say that the body of SETI literature is characterized by thousands of mostly unread papers of high quality: the corpus is highly heterogeneous, and many of the papers in it are obscure for good reason (Section 1.3.2). But even many of the lesser papers contain valuable nuggets, and if SETI is to be a true discipline, its practitioners need to build on the foundation of what has come before, and this begins with the elbow-grease work of literature searches and citation trees, and continues with proper acknowledgment of the priority of ideas.

Such work will help authors appreciate the variety and nuance of the work to date, and prevent one from attacking strawmen. Indeed, I have also found that many common challenges to ideas or assumptions commonly ascribed to "traditional SETI" are in fact as old as the field itself, and were often shared (or at least articulated) by the pioneers of the field.

Choose Theory Projects Carefully

To a theoretical astrophysicist or astrobiologist looking to work on SETI projects, I recommend those that "stay close to the data,"[1] for instance by using hypotheses about alien behavior to generate new search strategies, determine the detectability and likelihood of success of various proposed technosignatures, or by building frameworks for the interpretation of candidates and null results.

I think it is fair to write that it is *especially* important in a data-starved field like SETI, which has very little to go on, that its theory work stay grounded in observations. In the absence of data, it can be easy to develop a large corpus of work based on assumptions that may one day turn out to be false. While such work can still be worthwhile—there may be spinoff applications, or that work might be part of a broader theory landscape that ultimately lands on the correct way

[1] When I learned this admonition while being trained as an observer in graduate school, it was attributed to my academic "grand-adviser" George Herbig.

forward—the return on investment is low, which is an important consideration in a field as funding-starved as SETI.

I would also caution against too much more purely theoretical work on the Fermi Paradox (Chapter 6), Drake Equation (Chapter 8), and related concepts (Chapter 10). Much work has been done to "solve," "dissolve," or "resolve" the Fermi Paradox, and to repair, properly calculate, or extend the Drake Equation. Such works have been important for the justification of the field and informing how SETI fits into the broader cultural and scientific landscape, but when it comes to moving us forward in SETI I fear we have approached a point of diminishing returns that will persist until a detection is made.

After all, given how little we understand about the detectability of alien technology, it is not clear that the Fermi Paradox is a problem that even *needs* a solution, and the Drake Equation is too inherently heuristic and uncertain to give better direction to the field than it already has.

In my opinion, work in this direction is best focused on the problem of detectability, and on calculating what limits we can put on alien technology given the limited searching we have already done or could do (Section 11.3).

Think About the Nine Axes

The Nine Axes of Merit (Sheikh 2020; Section 11.4) provide a framework for thinking about good SETI strategies, and were not intended to be objective or to produce an unambiguous ranking. The weights one assigns the axes when designing a figure of merit with them will depend strongly on one's priors on the kind and abundance of technosignatures and life in the Universe. Nonetheless, I think any SETI project should pay special attention to three of them in particular.

Ancillary Benefits

The first is the ancillary benefits axis (Section 11.4). The primary purpose of SETI is, of course, to make a positive detection or, barring that, to place an upper limit on a particular kind of technosignature. Secondarily, some searches produce significant ancillary benefits either inevitably or by design. For instance, the Breakthrough Listen backend at Green Bank Observatory provides an outstanding platform for fast time domain radio astronomy of fast radio bursts and pulsars, and substantially contributes to those fields as a result (e.g., Michilli et al. 2018; Zhang et al. 2018; Gajjar et al. 2018; Price et al. 2019). Similarly, the PANOSETI project promises new insights into fast optical transients (e.g., Wright et al. 2018; Maire et al. 2019), and the Ĝ project identified many extreme, anomalous objects of presumably natural origin (Griffith et al. 2015).

This is consistent with Freeman Dyson's First Law of SETI Investigations: "Every search for alien civilizations should be planned to give interesting results even when no aliens are discovered" (Section 11.4). There are at least four good reasons for designing a SETI program to follow this advice.

The first two are for the good of the field. The first is the political difficulty of needing to use null results to justify ever increasing budgets for ever more sensitive searches. Indeed, Dyson's primary concern was that all-or-nothing experiments did not have sustainable funding models:

> It should be a fundamental principle in the planning of space operations, that every mission searching for evidence of life should have other exciting scientific objectives, so that the mission is worth flying whether or not it finds evidence of life. We should never repeat the mistake that was made with the Viking missions, whose advertised purpose was to give a definitive answer to the question whether life exists on Mars. After the Viking missions failed to find evidence of life, the further exploration of Mars was set back for 20 years. (Dyson 2003)

The second is that if SETI is to exploit its many synergies with astrophysics and take its proper home under the astrobiology umbrella, it needs to avoid being cloistered, meaning its researchers should be valued participants in those broader disciplines, known for more than the upper limits they produce on technosignatures.

The other two reasons are for the SETI researchers personally. SETI is an underfunded field, and training new researchers in it should give them skills transferable to other parts of astronomy and beyond where better job prospects lie.

And finally, as Drake (1965) advised those that would follow him:

> Our experience with Project Ozma showed that the constant acquisition of nothing but negative results can be discouraging. A scientist must have some flow of positive results, or his interest flags. Thus, any project aimed at the detection of intelligent extraterrestrial life should simultaneously conduct more conventional research. Perhaps time should be divided about equally between conventional research and the intelligent signal search. From our experience, this is the arrangement most likely to produce the quickest success.

This point about ancillary benefits can be made too strongly, however. SETI is an important field that should stand on its own merits, and not only to the degree that it supports other kinds of astronomy. Indeed, many forms of SETI will primarily produce little to no data amenable to significant astrophysical interpretation—most radio SETI data probably falls into this category. While it would be of course beneficial if those data had broader use, such searches have enough merit in other dimensions that they remain quite worthwhile.

Contrivance and Duration

I also recommend counting the number of assumptions one is making on the "contrivance/inevitability" axis. It can be easy to build a tower of assumptions about alien behavior that lead to a clean, unambiguous technosignature, and convince oneself that a search for that technosignature will likely succeed or else yield a strong null result. But tall towers teeter, and while it may be worthwhile exploring what is at the top, the utility of such a climb is often, in my experience, significantly less than the builder of the tower imagines. A good strategy is to run one's assumptions by a SETI skeptic before building a tower too tall.

Finally, think about the duration axis in terms of L in the Drake Equation. Consider the number densities and rates of a given technosignature required for us to have any reasonable chance of observing short-lived phenomena with low duty cycles (for instance, cataclysmic events), and consider whether a search for such phenomena can ever produce useful or interesting upper limits.

Engage Experts

SETI is closely related to other parts of astronomy. Especially when dealing with potential false positives—from natural confounders to terrestrial sources of interference—one works best as part of a team that includes people who know the instruments, data, and targets of the search best.

This engagement also helps to grow the field. I have found that many astronomers and astrobiologists are happy to help with a SETI project, but do not know how they can contribute. Respecting their expertise is a great way to bring them in and improve one's science.

This is especially true when wandering outside of one's domain, for instance for an astrophysicist considering the nature of intelligence (Section 7.5.3), the mathematical formulae for patterns of a species' expansion (Section 6.2), the nuances of interspecies communication (Section 19.5.2), the ethics of transmissions (Chapters 19 and 20), or the effect on humanity of a clear SETI detection (Section 20.5). These are well studied problems in the humanities and the life and social sciences, and while experts in those fields may not always have answers a physical scientist can immediately use for their purposes, they can help SETI practitioners bound such problems and prevent avoidable blunders.

Plan for Success, and for Null Results

For observational programs, I strongly recommend having a fully formed plan that goes beyond simply searching for a given technosignature. This means that one should plan for both success, and for a null result (I will not write "failure," since a null result can be an important result *per se* and not a failure at all.)

Planning for success does not just mean planning for what to do after one is convinced that one has definitive signs of extraterrestrial technology (although one needs to have a good plan for that as well, beyond simply "pop the champagne and call the press office"! Chapter 20). It also means planning what to do between the detection of a stunning, interesting, intriguing, confusing, or even merely curious signal, and the eventual conclusion that alien life has been found.

Searches should have detection thresholds tuned to generate a manageable number of candidate signals, and a robust follow-up program should be ready to pursue them. One should be prepared to document each signal and why it is or is not worthy of moving to the next stage of validation, and a pre-defined threshold for what constitutes sufficient evidence for a claim of a detection.

It is possible, and perhaps likely, that the first evidence of extraterrestrial technology will not be from an unambiguous "Wow! Signal," but from a hard-to-understand "Huh? Signal." In the latter case, it may take years or decades after the first public announcement of the signal before a consensus develops that alien technology has been detected. Managing such announcements will take care, and practitioners should be prepared to solicit and heed the advice of experts in risk communication, sociology or anthropology, and science communication to help manage the announcement. After all, the announcement of evidence for the existence of *technological* life may have significantly more amplified or different impact on the public than the announcements of potential biosignatures in the AH84001 meteorite or the clouds of Venus (both of which provide case studies that should be considered carefully.)

Planning for a null result is even more important, since that will be the more common outcome. Ideally, one would have calculated the significance of a null result *before one begins searching* a given data set as part of the justification for looking in the first place. Upper limits can be difficult to calculate rigorously (more difficult, often, than interpreting detections) but without such calculations a search has very little value to future researchers. Injection-recovery methods are a particularly useful tool in this regard.

Model-free anomaly searches (Section 11.2.7), especially those that employ machine learning, may have a particularly difficult time describing what, exactly, it is that they did not find, and thus parameterize exactly which technosignatures they have shown do not exist. But that problem is worth tackling.

Use This Plan to Help Frame the Search

Planning for success and null results is not just important for the progress of the field, but for the perception of the field, and therefore its future as a discipline.

If SETI is seen merely as the practice of uncritically asking "is it aliens?" about every anomaly, then it will be difficult to engage experts in useful data sets and leverage their expertise to advance the field, lest they become party to "crying wolf" at every poorly understood phenomenon. If the SETI literature is primarily a long string of null results and false alarms, then it will be seen as a risky endeavor, equivalent to buying a lottery ticket: something one might as well try, but not if it costs too much or will result in embarrassing headlines in the British tabloids (Section 20.7).

If, on the other hand, SETI develops a reputation for the rigorous calculation of upper limits and measurably advancing what we know is *not* out there, then its practitioners may be accepted as collaborators by even SETI skeptics (who, one would hope, would welcome such upper limits as confirming their priors). The goal should be to emulate the dark matter particle search community, which with each new instrument moves their upper limit thresholds lower and lower (see Wright & Oman-Reagan 2018, Appendix A therein, and Section 4.4.4). While they may never be able to rule out a particle of *any* cross section or mass, they can steadily shrink the remaining parameter space in which it can hide, provide clear markers of progress, and specifications for what the next useful experiment would look like. This can also help solve the funding paradox where it is the *lack* of results that must justify ever increasing efforts to press on.

Or consider a planetary scientist with expertise in high resolution imagery or subsurface radar, who may look skeptically on a request to collaborate on a search for alien artifacts on the Moon or Mars. They might reasonably worry that they will be asked to co-author another "Face on Mars" paper when a strange rock formation inevitably triggers a machine learning algorithm or community scientist's interest (Section 14.1.3).

If, however, they understand that the goal is to rigorously and quantitatively put upper limits on the *lack* of any such artifacts—while at the same time discovering new and interesting features that may or may not be such artifacts—then they are much more likely to appreciate the synergies of such collaboration, and to welcome abstracts and papers on the subject in their journals and at their conferences.

Stay Broad-Minded

We all come to the problem of SETI with very different priors for how SETI can succeed, and so will invariably encounter practitioners pursing what we feel are very unlikely or misguided paths to success. It helps to remember that the feeling may be mutual.

Avi Loeb and Clément Vidal have argued strongly that one should have very flat priors when considering natural versus artificial explanations for astrophysical phenomena. This at first glance would seem to be in violation of Carl Sagan's maxim that "extraordinary claims require extraordinary evidence." But that maxim is essentially a paraphrasing of Bayes's Theorem: the prior that any particular technosignature will be found is presumably very small, and so requires a large amount of evidence to overcome. Before applying the maxim to outré claims, we must be careful to distinguish between the claim that our prior for a particular claim is too small, and the claim that the posterior probability after considering a piece of evidence is near one.

Put another way, all great discoveries were, at some point, thought to be extremely unlikely, and the generation of unlikely hypotheses is thus an important, indeed essential, part of scientific progress. After all, part of the practice of science is being willing to update our priors regularly. Extraordinary and surprising objects or arguments can thus serve as case studies, and as opportunities for us to revisit those priors, even if those particular objects turn out in the end not to be examples of the phenomenon in question. For instance, my prior on the possibility of a prior indigenous technological species on Earth was once very small because I presumed such a species would have left traces that would be obvious to us today. After considering the matter and encountering the work of Schmidt & Frank (2019) (Section 9.1.5), I realize that this is not the case, and that I have very little evidentiary basis for finding the hypothesis unlikely. My prior on this possibility is still low, but it is no longer very close to zero (Wright 2018).

So while the "broad priors" stance can fairly be described as an "aliens of the gaps" philosophy, such an approach is a perfectly reasonable way to go about generating hypotheses in SETI, and we should welcome open-minded and creative papers that challenge our priors in the scientific literature *provided that* they are are accompanied by sound scientific reasoning and, optimally, specific suggestions for how to test the hypothesis. It is also just generally good practice to respect the range of "risk tolerance" among practitioners in the field, and ensure that some fraction of our collective efforts are "high-risk, high-reward" (Loeb 2010; Vickers 2020).

Stay Skeptical, but Not Cynical

Not all SETI researchers believe they will have a good chance of success in their lifetimes, but such a belief surely animates much of the field. It can therefore be challenging to maintain a scientist's proper, healthy skepticism about one's own work, especially when coming across a particularly intriguing signal.

I suspect everyone who engages in the practice long enough will come across what looks to be a Wow! Signal (Box 3.4) and, at least briefly, dream of the success that will follow. The proper response to such a discovery is a stance of extreme skepticism: if one is not one's own harshest critic, one may end up embarrassing oneself, and losing credibility for the field. It is at these moments that Sagan's maxim should have its strongest force.

But one should also not let the product of such false alarms be a cynicism that leads one to focus entirely on upper limits and dismiss all candidate signals as just so much noise before they are thoroughly examined. There is a wonder that brought many of us into the field that must be nurtured and protected against the discouragement of years or decades of null results that Drake (1965) warned about. One should cherish each false alarm and "Huh? signal" as an opportunity for hope and curiosity to flourish, "till human voices wake us, and we drown."[2]

References

Drake, F. D. 1965, Current Aspects of Exobiology, ed. G. Mamikunian, & M. H. Briggs (Oxford: Pergamon) 323

Dyson, F. J. 2003, IJAsB, 2, 103

Eliot, T. S. 1915, Poetry: A Magazine of Verse, ed. H. Monroe (Chicago, IL: Seymour, Daughaday and Company) 130

Gajjar, V., Siemion, A. P. V., Price, D. C., et al. 2018, ApJ, 863, 2

Griffith, R. L., Wright, J. T., Maldonado, J., et al. 2015, ApJS, 217, 25

Loeb, A. 2010, arXiv:1008.1586

Maire, J., Wright, S. A., Barrett, C. T., et al. 2019, AJ, 158, 203

Michilli, D., Seymour, A., Hessels, J. W. T., et al. 2018, Natur, 553, 182

Price, D. C., Foster, G., Geyer, M., et al. 2019, MNRAS, 486, 3636

Schmidt, G. A., & Frank, A. 2019, IJAsB, 18, 142

Sheikh, S. Z. 2020, IJAsB, 19, 237

[2] Eliot (1915).

Vickers, P. 2020, IJAsB, 19, 482
Wright, J. T. 2018, IJAsB, 17, 96
Wright, J. T., & Oman-Reagan, M. P. 2018, IJAsB, 17, 177
Wright, S. A., Horowitz, P., Maire, J., et al. 2018, Proc. SPIE, 10702, 107025I
Zhang, Y. G., Gajjar, V., Foster, G., et al. 2018, ApJ, 866, 149

Acknowledgement

It has been a busy journey from knowing almost nothing about the field to writing a textbook about it. It all started with an abandoned idea from my PhD adviser to look for Dyson spheres in 2MASS data, revived 10 years later in a brief conversation with Steinn Sigurðsson about a "fun" proposal we could submit to the Templeton Foundation to pull it off. Since then Steinn has been my accomplice on a lot of outré projects in SETI.

I thank Dawn Gelino for soliciting this textbook at AAS e-books, and her unwavering support along its long road to completion; and for her death-defying work to make the Houston workshop a success: the resulting report in many ways formed the template for this textbook. Thanks to Macy Huston for their help in the conception of this textbook and early work on it. I thank Pinchen Fan and Cayla Dedrick for their work as TAs in the undergraduate SETI course, ASTRO 476, and especially their work writing, editing, and typesetting the exercises and solutions for this book.

Most of my growth has been through my students, and much of the work in this book comes from what I've learned from them. I thank Sofia Sheikh for being the first student to go "all in" on SETI and showing me we could unapologetically have a graduate program that produced new PhDs in the field. Sofia paved the way for the next wave of graduate students, including Kimberly Cartier, Macy Huston, Nick Tusay, Pinchen Fan, and Winter Parts. This book largely exists so their example can be replicated everywhere.

I thank the students of my graduate and undergraduate SETI courses, especially the first few whose final projects have moved the field forward, including William Bowman, Caleb Cañas, Cayla Dedrick, Laura Duffy, Juniper Foote, Christian Gilbertson, Andrew Hyde, Shubham Kanodia, Stephen Kerby, Julia Lafond, Emily Lubar, Christian Gilbertson, Mariah MacDonald, Ella Mullikin, Michael Palumbo, Winter Parts, Phoebe Sandhaus, Hillary Smith, and Evan Sneed.

I thank the graduate and undergraduate researchers who have made the SETI bibliography possible and tried out new ideas, including Alan Reyes, Julia Lafond, Aidan Groenendaal, Lennon Nichol, and Ana Cristina Scigliani.

I am grateful to the many scholars outside the physical sciences that have helped challenge my assumptions about SETI and understand better its origins, future, and

doi:10.1088/2514-3433/adf5aech22

place in our world, including Doug Vakoch, Lisa Messeri, Denise Hertzing, Jacob Foster, Erica Cartmill, Greg Eghigian, and Gabriel Swiney. I am particularly indebted to Kathryn Denning, Rebecca Charbonneau, Michael Oman-Reagan, and Chelsea Haramia for their perspectives, scholarship, and engagement, and for helping me with many specific aspects of this book beyond my expertise.

Thanks to the Characterizing Atmospheric TechnosignatureS group's many members, past and present, for moving the field forward with hard science (while still entertaining discussions where no idea is too wild to consider) especially our fearless leader Adam Frank, and Manasvi Lingam, Thomas Beatty, and Ravi Kopparapu.

I'm grateful to many scholars that have joined the field and helped it grow, especially the Order of the Octopus for building the community of scholars that is the future of the field.

Special thanks to my co-authors who have pushed me and taught me, including David Kipping, Vishal Gajjar, Wael Farrah, Joe Lazio, Andrew Howard, Erik Zackrisson, Matías Suazo, Jonathan Carroll-Nellenback, Duncan Forgan, Eric Korpela, Iván Almár, Roger Griffith, Matt Povich, and Brendan Mullan.

Also invaluable in writing this has been the support and encouragement of the pioneering scientists and scholars in the field, especially Jill Tarter for her leadership, and her, Dan Werthimer, and Paul Horowitz for sharing their wisdom. I would thank Freeman Dyson and Frank Drake for their kinds words and encouragement in our brief but formative encounters.

I have to acknowledge the tolerance and support of the many scientists I've dragged along on my SETI adventures from whom I've learned so much, including Eric Mamajek, Darryl Seligman, and Jenn Macalady. Special thanks to Tabby Boyajian for being a game partner on the wild ride of trying to figure out "that" star.

I am grateful for the ongoing support of Penn State and its Eberly College of Science for funding the PSETI Center, giving me the space and freedom to move into SETI and produce this book. There are not many departments, colleges, or universities that would appreciate an assistant professor expected to produce grant money moving into a marginalized and largely unfunded field just before their tenure review, but mine have been nothing but supportive, which has been everything.

Also foundational has been the scientific, financial, and moral support of the Breakthrough Listen Initiative and the scientists at the Berkeley SETI Research Center, most especially Andrew Siemion and Steve Croft, who supported my sabbatical at Berkeley and with whom I always have a desk I can use in Campbell Hall.

I must give more than the boilerplate, perfunctory acknowledgment to the Astrophysics Data System (funded by NASA under Cooperative Agreement 80NSSC21M00561.) ADS is invaluable to my scholarship and the field of astronomy generally, and this book would have been only a shadow of what it is without ADS. I'm especially grateful to the staff who gave us an API key so we could beta test shared libraries for what would become the SETI bibliography, and for adding and editing abstracts for so many SETI papers (from the foundational to the trivial). Thanks to the Penn State Libraries for tracking down and scanning so many obscure articles, sometimes from scraps of references to long-forgotten journals.

I thank Erica Fox Brindley for help with translations and background in Chapter 3. Quote Investigator was invaluable ensuring that so many of the boxed quotes were correctly attributed. And where would any of us be without Wikipedia and its many tireless Wikipedians; I got too many ideas to count from their work, most notably the connection between the Water Hole and Schelling points.

And thank you to the many readers who offered feedback on the various manuscripts and papers not mentioned yet, including Frank Hirth, Shelley Wright, Naoki Seto, Michael Hippke, Jim Davenport, Andy Nilipour, and Dan Mills. I'm honored to have been the recipient of copious ink from Steven Dick's red pen in the history chapter—no one knows the history of the field better.

Thanks to my many editors at IOP Publishing who showed infinite patience as this manuscript grew far beyond its allotted length and deadlines; to Angelique Aitken and my family for their support; to Shut Up and Write! and its State College chapter members for helping me find the accountability and time to drag this tome across the finish line; and to Webster's Bookstore and Café for hosting us and providing a great work environment.

Much of the work I did to write this book ended up becoming papers published in AAS Journals before being edited, paraphrased, or sometimes simply copy-pasted into the final manuscript. In all of these instances I am the lead author of the originally published version of the text. Specifically, much of the Afterword and some of Chapters 8 and 11 is adapted or reprinted from Wright (2021) with permission from Elsevier; Chapter 9 from Wright et al. (2022); Sections 11.3 and 12.7.1 from Wright et al. (2018); and Chapter 15 from Wright (2020) (published in the Serbian Astronomical Journal) and Wright et al. (2014).

The manuscript for this book was prepared in Overleaf using the book LaTeX class. I thank the authors of the over 35 packages I used for sharing their work. There is no output from generative AI in this book.

References

Wright, J. T., Griffith, R. L., Sigurdsson, S., Povich, M. S., & Mullan, B. 2014, ApJ, 792, 27

Wright, J. T. 2020, SerAJ, 200, 1

Wright, J. T. 2021, AcAau, 188, 203

Wright, J. T., Kanodia, S., Lubar, E., et al. 2018, AJ, 156, 156

Wright, J. T., Haqq-Misra, J., Frank, A., et al. 2022, ApL, 927, L30

Epilogue

Cape Cod Light

Michael Elkins Hattersley.[1]

Dawn: earth is grey.
Overhead, a seagull catches the Sun,
Flames bronze, and dives behind scrub pines.
It will glide low over swamp rushes,
Bank across the dune, and settle at the water's edge.
Crabs wait, pursed muscles and clams,
To whom the golden seagull feels like death,
A conduit from sea to soil.
It becomes something new with each motion,
Plunging out of the sky.

It is a day to rise and go out to the Cape Cod Light
To watch the water work its will on the land.
We were all led down as children to the beach,
And bound, by a mutual gesture, to the sea.
Our awe: the miracle of light,
Plant light, sand light, bark bush and fire light,
Light on the several blues, greens, and whites of the sea.

Events intervened.
Grey dune shacks crumble into the sand
And more are built, the same. The stubby sandgrass
Creeps from green through yellow into August.
We can descend the dunes only by jumping,

[1] Used with permission.

doi:10.1088/2514-3433/adf5aech23

Printing accidental angels in the sand.
By the shore, souls claimed by ocean
Reach up through the waves to take the hands
Of the living, not at all like dead things,
More fluid than bones picked white.
We brush them aside, proceed, do their will.
We inherit what our parents didn't know.
It grows cool at sunset in August.
The sea blows hot and cold. The stars
That glint in the corners of your eyes
Evaporate looked at straight. Still,
You can chart a true course from them.

When storms spark
And the sky shakes itself like an angry old head,
When fog rolls in over the implacable light,
We calculate our position, and chart an according way.
In time, the full night sky will be lacerated with stars.
The sea will despoil itself on the white shore.
The enormous sandbar will hook around ahead
Into a harbor, where the fishing boats rise and fall.

The cries of the dead are stirring in the surf.
Anger keeps them here, each others' audience,
Wearing, from time to time, the bodies of the living.
Waves crack and slither on the shore,
Black, white and black. Dunes hunch,
Dark shoulders of earth in the night.
One by one, and graves slide into the sea.
The fat orange moon spills across the water
And the dead are assumed, the unbroken line of them
Moving solemnly as kings to a miracle. They remind us
How we are falling into the future, falling.

www.ingramcontent.com/pod-product-compliance
Lightning Source LLC
Chambersburg PA
CBHW041128230226
39985CB00004B/15
9780750347945